Selected Titles in This Series

16 **Victor P. Snaith, Editor,** Algebraic K-Theory, 1997

15 **Stephen P. Braham, Jack D. Gegenberg, and Robert J. McKellar, Editors,** Sixth Canadian conference on general relativity and relativistic astrophysics, 1997

14 **Mourad E. H. Ismail, David R. Masson, and Mizan Rahman, Editors,** Special functions, q-series and related topics, 1997

13 **Peter A. Fillmore and James A. Mingo, Editors,** Operator algebras and their applications, 1997

12 **Dan-Virgil Voiculescu, Editor,** Free probability theory, 1997

11 **Colleen D. Cutler and Daniel T. Kaplan, Editors,** Nonlinear dynamics and time series: Building a bridge between the natural and statistical sciences, 1997

10 **Jerrold E. Marsden, George W. Patrick, and William F. Shadwick, Editors,** Integration algorithms and classical mechanics, 1996

9 **W. H. Kliemann, W. F. Langford, and N. S. Namachchivaya, Editors,** Nonlinear dynamics and stochastic mechanics, 1996

8 **Larry M. Bates and David L. Rod, Editors,** Conservative systems and quantum chaos, 1996

7 **William F. Shadwick, Perinkulam Sambamurthy Krishnaprasad, and Tudor Stefan Ratiu, Editors,** Mechanics day, 1996

6 **Anna T. Lawniczak and Raymond Kapral, Editors,** Pattern formation and lattice gas automata, 1996

5 **John Chadam, Martin Golubitsky, William Langford, and Brian Wetton, Editors,** Pattern formation: Symmetry methods and applications, 1996

4 **William F. Langford and Wayne Nagata, Editors,** Normal forms and homoclinic chaos, 1995

3 **Anthony Bloch, Editor,** Hamiltonian and gradient flows, algorithms and control, 1994

2 **K. A. Morris, Editor,** Control of flexible structures, 1993

1 **Michael J. Enos, Editor,** Dynamics and control of mechanical systems: The falling cat and related problems, 1993

FIELDS INSTITUTE
COMMUNICATIONS

THE FIELDS INSTITUTE FOR RESEARCH IN MATHEMATICAL SCIENCES

Algebraic K-Theory

Victor P. Snaith
Editor

American Mathematical Society
Providence, Rhode Island

The Fields Institute for Research in Mathematical Sciences

The Fields Institute is named in honour of the Canadian mathematician John Charles Fields (1863–1932). Fields was a visionary who received many honours for his scientific work, including election to the Royal Society of Canada in 1909 and to the Royal Society of London in 1913. Among other accomplishments in the service of the international mathematics community, Fields was responsible for establishing the world's most prestigious prize for mathematics research—the Fields Medal.

The Fields Institute for Research in Mathematical Sciences is supported by grants from the Ontario Ministry of Education and Training and the Natural Sciences and Engineering Research Council of Canada. The Institute is sponsored by McMaster University, the University of Toronto, the University of Waterloo, and York University and has affiliated universities in Ontario and across Canada.

1991 *Mathematics Subject Classification.* Primary 19–06.

Library of Congress Cataloging-in-Publication Data

Algebraic K-Theory / Victor P. Snaith, editor.
p. cm. — (Fields Institute Communications, ISSN 1069-5265 ; v. 16)
Papers from the Second Great Lakes Conference on Algebraic K-Theory, held Mar. 1996 at the Fields Institute for Research in Mathematical Sciences in memory of Robert Wayne Thomason.
Includes bibliographical references.
ISBN 0-8218-0818-4 (alk. paper)
1. K-theory—Congresses. 2. Geometry, Algebraic—Congresses. 3. Topology—Congresses. I. Snaith, V. P. (Victor Percy), 1944– . II. Thomason, Robert Wayne, 1952–1995. III. Great Lakes Conference on Algebraic K-Theory (2nd : 1996 : Fields Institute for Research in Mathematical Sciences) IV. Series.
QA612.33.A384 1997
512′.55—dc21 97-20243
CIP

∞ The paper used in this book is acid-free and falls within the guidelines
established to ensure permanence and durability.
This publication was prepared by the Fields Institute.
Visit the AMS homepage at URL: `http://www.ams.org/`

10 9 8 7 6 5 4 3 2 1   02 01 00 99 98 97

Contents

Preface

The Second Great Lakes Conference on Algebraic K-Theory was hosted by The Fields Institute for Research in Mathematical Sciences in March 1996. It proved to be a mathematically very exciting occasion with many interesting new results being unveiled. The success of the meeting together with a desire to commemorate Bob Thomason, one of K-theory's most influential figures, led to this Proceedings volume.

As editor, I would like to express my thanks and those of the organizers of the conference to the staff of The Fields Institute who made this meeting so rewarding and enjoyable. In particular, I wish to extend special thanks to the President and Scientific Director, John Chadam, for inviting the Great Lakes K-Theory Conference to The Fields Institute, and to the Publications Assistant, Erna Unrau, for making the editing of this volume so easy for me.

Victor Snaith

Robert Wayne THOMASON
1952-1995

In the mid 1970's, when Bob Thomason was a star graduate student at Princeton University, algebraic K-theory was in a very tentative state. It had begun as Whitehead-ian algebra-to-capture-geometry, showing lots of low-dimensional promise. Then, with the work of Quillen, algebraic K-theory had suddenly expanded into a grand cosmological entity with its own, extremely fascinating, ontology and epistomology. It was clear that the subject constituted a major mathematical nexus but the evidence was in the form of hard-won hints about how Quillen's K-theory might unify such diverse phenomena as homotopy types of diffeomorphism groups of manifolds, special values of L-functions, polylogarithms, intersection theory of algebraic varieties, Weil cohomology theories and so on. As a result, no one was sure whether K-theory was the domain of topologists, algebraists or some other mathematical species. In many places this hardly mattered, the subject could still easily be shrugged off as a new fad with fine ambitions, lots of conjectured promise, but with no "big theorem" to mark some post-Quillen progress.

Bob Thomason was the one who was to provide such a "big theorem".

It must have been clear all the way up the educational ladder that Bob was exceptionally clever. For example, at eighteen he had won first prize in the North America wide competition entitled "Great Books of the Western World". Consequently, I was unaware that Bob was still a graduate student when I first met him around 1976. André Weil is reputed to have advised that one should learn from "the masters, not the students" – I can only conclude that he did not have Bob Thomason in mind!

My friendship with Bob got off to a rather shaky start. In 1975 topologists' contributions to algebraic K-theory often took the form of results about infinite loopspaces. Ib Madsen, Jørgen Tornehave and I had recently resolved all of Peter May's conjectures relating to the infinite loopspaces of geometric topology. Frank Adams, as an editor, suggested sending the paper to the Annals of Mathematics in Princeton. Midwestern homotopy theory was evidently not so popular in Princeton and the typescript gathered a lot of cobwebs, but an editor gave a copy to Bob to read. Eventually Adams took the paper away in disgust, publishing it a few weeks later in the Proceedings of the Cambridge Philosophical Society. Anyway, the point is that I was feeling anything but cheerful towards Bob at our first meeting, which took place at a conference in Evanston. However, within minutes of meeting Bob it was clear to anyone that here was someone who had no patience for intra Ivy League politics. Here was someone intent on doing mathematics for its own sake!

During the next few years, mostly through listening and not understanding, I learned a lot from Bob about the prevailing vision for K-theory. His intensity notwithstanding, Bob Thomason was always very approachable - focussed on research goals, he invariably had time patiently to explain things to lesser mortals such as myself. Also, he had a knack for putting across the *conceptual* picture. In the late 1970's he, Bill Dwyer, Eric Friedlander and I got involved in studying mod p K-theory which has been inflicted with Bott periodicity. When A is a commutative ring there is a map of graded rings, $\pi_*^S(BA_+) \longrightarrow K_*(A)$, from the stable homotopy of the classifying space of the units of A (with a disjoint point attached) to the algebraic K-theory of A. This map comes from Quillen's group-completion theorem and exists also for homotopy with coefficients. In order to say something about this map I had managed to calculate the stable homotopy when inflicted with Bott periodicity and had therefore inflicted algebraic K-theory with similar periodicities so as to be able to detect the asymptotic residues of this map. I had obtained a few rudimentary results about new elements in K-theory, but Bob immediately saw that my Bott periodicities were really connected with the periodicity in étale cohomology. For some time he had been trying to establish Quillen's prediction that K-theory was related to étale cohomology, after some low dimensional exceptions. Bob already possessed a number of very original constructions in which one did homological algebra with spectra and stable homotopy theory instead of groups and modules. With these techniques he had been trying to prove rather too much about the predicted connection. However, with the aid of our four-author paper, Bob soon had the required theorem - that Bott periodic algebraic K-theory mod p^n of any reasonable scheme had an Atiyah-Hirzebruch spectral sequence starting with étale cohomology. This was exactly a result of the much-sought-after type. It showed that, in high dimensions, the predicted connection between K-theory and cohomology existed.

This connection, together with its concomitant insights, served as the basis for many of Bob best theorems. Now he had a machine through which to funnel results from topological K-theory into algebraic K-theory, a sheet-anchor giving much of the stability of purpose to his later research. His subsequent string of important results forms the backbone of modern K-theory, particularly in its applications to arithmetic and algebraic geometry. His main theorem was a tour de force. It is very hard going and a proper study takes a long time, lots of work during which one passes through phases which Bob wryly described as the progression from "thrill seeker" to "reckless cheat" to "honest man". Personally, I doubt that I ever made it past the "reckless cheat" phase. However, many other authors have studied the ideas and methods. I am very grateful to Steve Mitchell who, in this volume, has written an "essay which will smooth the way for others" through Thomason's Theorem.

I always felt a certain sadness, and here I am not referring to his untimely death, about Bob's career. Before coming to North America it had never occurred to me that one must not only prove the "right" theorem but one must prove it in the "right" place. The fictions, fads and factions of mathematical society are quite clear to me now, but in those days I never understood why Bob's achievements went so comparatively unrewarded. To his K-theoretic friends, therefore, it came as no great surprise when, in October 1989, Bob moved to a position in the C.N.R.S. at the Unversité de Paris 7. With that move, Bob was mathematically where he felt most at home.

Sadly, at the height of his mathematical powers, Bob died, tragically. He would have been forty-three years old on November 5, 1995. A longtime diabetes sufferer, just before his birthday he went into diabetic shock and died in his Paris apartment. He will be missed immensely but, I like to think, mathematically his name will survive as long as does the subject algebraic K-theory, which he loved so much. And that will be a very long time!

Victor Snaith
February 1997

References

Dwyer, W., Friedlander, E., Snaith, V. and Thomason R. W. [1982], *Algebraic K-theory eventually surjects onto topological K-theory*, Invent. Math. **66**, 481–491. MR 84i:14015

Gillet, H. and Thomason R. W. [1984], *The K-theory of strict hensel local rings and a theorem of Suslin*, J. Pure Appl. Alg. **34**, 241–254. MR 86e:18014

Latch, D., Thomason, R. W. and Wilson S. [1979], *Simplicial sets from categories*, Math. Zeit. **164**, 195–214. MR 80e:55012

May, J. P. and Thomason, R. W. [1978], *The uniqueness of infinite loop space machines*, Topology **17**, 205–224. MR 80g:55015

Publications of Bob Thomason

[1974] *A note on spaces with normal product with some compact space*, Proc. AMS **44**, 509–510. MR 51#4165

[1977] *Homotopy colimits in* **Cat**, *with applications to algebraic K-theory and loop space theory*, Ph.D. thesis, 124 pages, Princeton University, available from University Microfilms, Ann Arbor. MI 48104

[1979] *Homotopy colimits in the category of small categories*, Math. Proc. Cambridge Philos. Soc. **85**, 91–109. MR 80b:18015

[1979] *Uniqueness of delooping machines*, Duke Math. J. **46**, 217–252. MR 80e:55013

[1980] **Cat** *as a closed model category*, Cahiers Top. Geom. Diff. **21**, 305–324. MR 82b:18005

[1979] *First quadrant spectral sequences in algebraic K-theory*, in Lecture Notes in Math. **763**, Springer-Verlag, 332–355. MR 81c:18018

[1982] *First quadrant spectral sequences in algebraic K-theory via homotopy colimits*, Comm. Algebra **10**, 1589–1668. 83k:18006

[1980] *Beware the phony multiplication on Quillen's* $\mathcal{A}^{-1}\mathcal{A}$, Proc. AMS **80**, 569–573. MR 81k:18010

[1989] *Algebraic K-theory and étale cohomology*, Ann. Scient. Éc. Norm. Sup., 437–552. **18**,, 4e série erratum **22**, 675–677, errata 91j:14013, MR 87k:14016

[1982] *The Lichtenbaum-Quillen conjecture for* $K/\ell_*[\beta^{-1}]$, in Current Trends in Algebraic Topology CMS Conf. Proc. vol. 2, Part 1, 117–139. MR 84f:18024

[1983] *Riemann-Roch for algebraic versus topological K-theory*, J. Pure Appl. Alg **27**, 87–109. MR 85c:14013

[1986] *Bott stability in algebraic K-theory*, AMS Contemp. Math., **55**, 389–406. MR 87m:18022

[1983] *The homotopy limit problem*, AMS Contemp. Math., **19**, 407–419. MR 84j:18012

[1984] *Absolute cohomological purity*, Bull. Soc. Math. France **112**, 397–406. MR 87e:14018

[1987] *Algebraic K-theory of group scheme actions*, Annals of Math. Study **113**, Princeton U. Press, 539–563. MR 89c:18016

[1986] *Lefschetz-Riemann-Roch theorem and coherent trace formula*, Invent. Math. **85**, 515–543. MR 87j:14028

[1986] *Comparison of equivariant algebraic and topological K-theory*, Duke Math. J. **53**, 795–825. MR 88h:18011

[1987] *Equivariant resolution, linearization and Hilbert's fourteenth problem over arbitrary base schemes*, Adv. Math. **65**, 16–34. MR 88g:14045

[1988] *Equivariant algebraic vs. topological K-homology Atiyah-Segal-style*, Duke Math. J. **56**, 589–636. MR 89f:14015

[1987] *The finite stable homotopy type of some topoi*, J. Pure Appl. Alg. **47**, 89–104. MR 88k:14010

[1989] *A finiteness condition equivalent to the Tate conjecture over* $\mathbb{F}_q$, in AMS Contemp. Math. **83**, 385–392. MR 90b:14029

[1989] *Survey of algebraic vs. étale topological K-theory*, in AMS Contemp. Math. **83**, 393–443. MR 90h:14029

[1991] *The local to global principle in algebraic K-theory*, in Proc. International Congress of Mathematicians, Kyoto 1990 Springer-Verlag **1**, 381–394. MR 93e:19010

[1992] *Une formule de Lefschetz en K-théorie équivariante algébrique*, Duke Math. J. **68**, 447–462. MR 93m:19007

[1992] *Le principe de scindage et l'inexistence de K-théorie de Milnor globale*, Topology **31**, 571–588. MR 93j:19005

[1993] *Les K-groupes d'un schéma éclaté et une formule d'intersection excédentaire*, Invent. Math. **112**, 195–215. MR 93k:19005

[1993] *Les K-groupes d'un fibré projectif*, in Algebraic K-theory and algebraic topology, NATO ASI Series C, vol.407, Kluwer, 243–248.

[1995] *The classification of triangulated subcategories*, Compositio Math., preprint, to appear.

[1995] *Symmetric monoidal categories model all connective spectra*, Theory Appl. Categories **1**, 78–118. (electronic journal http://www.tac.mta.ca/tac/)

Thomason, R. W. and Trobaugh, T. [1990] *Higher algebraic K-theory of schemes and of derived categories*, in The Grothendieck Festschrift III, Progress in Math. **88**, Birkhäuser, 247–435. MR 92f:19001

Thomason, R. W. and Trobaugh, T. [1988] *Le théorème de localisation en K-théorie algébrique*, C.R. Acad. Sci. (Paris) **307**, 829–831. MR 89m:18015

Thomason, R. W. and Wilson, S. [1980] *Hopf rings in the bar spectral sequence*, Quart. J. Math Oxford **31**, 507–511. MR 82f:55030

Ph.D. Theses Supervised by Bob Thomason

Harada, M. [1987] *A proof of the Riemann-Roch theorem*, Ph.D. thesis, Johns Hopkins University, Baltimore.

Yao, D. [1990] *Higher algebraic K-theory of admissible abelian categories and localization theorems*, Ph.D. thesis, Johns Hopkins University, Baltimore.

The Mathematics of Bob Thomason

Weibel, C. A. [1996] *Robert W. Thomason (1952-1995)*, A. M. Soc. Notices **43**, 860-862.

Weibel, C. A. [1997] *The Mathematical Enterprises of Robert W. Thomason*, Bull. A. M. Soc. **34(1)**, 1-13.

Fields Institute Communications
Volume **16**, 1997

Quaternionic Exercises in K-Theory Galois Module Structure

Ted Chinburg
Department of Mathematics
University of Pennsylvania
Philadelphia, PA USA 19066
ted@math.upenn.edu

Manfred Kolster
Department of Mathematics
McMaster University
Hamilton, Ontario, Canada L8S 4K1
kolster@mcmaster.ca

George Pappas
Department of Mathematics
Princeton University
Princeton, NJ USA 08544
gpappas@math.princeton.edu

Victor P. Snaith
Department of Mathematics
McMaster University
Hamilton, Ontario, Canada L8S 4K1
snaith@mcmaster.ca

Abstract. In [5] invariants of the Galois module structure of algebraic K-groups of rings of algebraic integers were introduced. These invariants live in the class-group of the group ring of the Galois groups and are conjectured to be related to Artin root numbers and special values of L-functions. Quaternion extensions are the test case for these conjectures. A formula is given, in terms of L-functions, for an infinite family of quaternion extensions.

1 Introduction

Let G be a finite group and denote by $\mathcal{CL}(\mathbf{Z}[G])$ the class-group of finitely generated $\mathbf{Z}[G]$-modules of finite projective dimension. Suppose that

$$X \longrightarrow A \longrightarrow B \longrightarrow Y$$

1991 *Mathematics Subject Classification.* Primary 11R70; Secondary 19F27.
TC and GP research partially supported by an NSF grant.
MK and VPS research partially supported by an NSERC grant.

is a 2-extension of finitely generated $\mathbf{Z}[G]$-modules representing an element of $Ext^2_{\mathbf{Z}[G]}(Y,X)$ inducing, via cup-product, isomorphisms between the Tate cohomology groups $H^i(G;Y)$ and $H^{i+2}(G;X)$ for all i. Under these circumstances a representative 2-extension may be chosen for which A and B are finitely generated $\mathbf{Z}[G]$-modules of finite projective dimension. The Euler characteristic of such a 2-extension is defined to be the element

$$[A] - [B] \in \mathcal{CL}(\mathbf{Z}[G]).$$

In the case when G is equal to $G(L/K)$, the Galois group of an extension of global fields, L/K, interesting examples of such Euler characteristics are afforded by the invariants, $\Omega(L/K,n)$ $(n = 1,2,3)$, constructed in [**4**]. Further examples, denoted by $\Omega_1(L/K,3)$, were constructed in ([**13**], Chapter 7) from 2-extensions in which Y and X are the algebraic K-groups of the ring of S-integers of L in dimensions 2 and 3, respectively. Here S is any finite $G(L/K)$-invariant set of finite primes containing those over primes of K which ramify in L/K, the construction being independent of the choice of S. In [**5**] this construction is generalised using K-groups in dimensions $2s$ and $2s+1$ to yield a family of Euler characteristics, denoted by $\Omega_s(L/K,3)$ for $s \geq 1$, to complete a pattern in which the original Chinburg invariant rightfully fits as $\Omega_0(L/K,3)$. On the other hand, associated to this situation, there is the Cassou-Noguès-Fröhlich class

$$W_{L/K} \in \mathcal{CL}(\mathbf{Z}[G(L/K)])$$

which is defined in terms of the root numbers of the L-functions of irreducible symplectic representations of $G(L/K)$ ([**2**], [**4**], [**13**], [**14**]).

Surprisingly, in the case of function fields in characteristic p, one can show that $W_{L/K} = \Omega_s(L/K,3)$ [**5**]. In the totally real number field case one can show that a similar relation holds in the classgroup of a maximal order of the rational group-ring of $G(L/K)$ ([**5**], [**13**], Chapter 7).

The purpose of this paper is to investigate the relationship between $W_{L/K}$ and $\Omega_1(L/K,3)$ in the case of an infinite family of totally real extensions of the rationals whose Galois groups are isomorphic to Q_8, the quaternion group of order eight. These extensions, denoted by $N/\mathbf{Q}$, will be constructed in Section 2. There will be one such extension for each prime, $p \equiv 3(\mathit{modulo}\ 8)$, which may be characterised as the unique totally real extension with $G(N/\mathbf{Q}) \cong Q_8$ in which only one odd prime (namely p) ramifies. The set S may be taken to be the set of primes of N over 2 and p.

The examples were chosen because they exhibit a number of simultaneous simplifications.

Firstly, being totally real, the invariant $\Omega_1(N/\mathbf{Q},3)$ may be defined by the method of [**13**] as the Euler characteristic of a canonical 2-extension of the form

$$K_3^{ind}(\mathcal{O}_{N,S}) \longrightarrow A \longrightarrow B \longrightarrow K_2(\mathcal{O}_{N,S})'$$

constructed in [**10**]. Here $K_3^{ind}(\mathcal{O}_{N,S}) \cong K_3^{ind}(N)$ denotes the indecomposable K_3-group and $K_2(\mathcal{O}_{N,S})'$ is equal to the kernel of the natural map, surjective in our examples, from $K_2(\mathcal{O}_{N,S})$ to the direct sum of the torsion in the K_2's of each of the real places of N. The groups A and B in this extension may be taken to be finite and cohomologically trivial. In fact, if $(\mathbf{Q}/\mathbf{Z})(2)$ is the Galois module given by the second Tate twist of the roots of unity there is a canonical isomorphism of the form [**10**]

$$H^1(N;(\mathbf{Q}/\mathbf{Z})(2)) \xrightarrow{\cong} K_2(N).$$

Let Ω_N denote the absolute Galois group of N. There is a short exact sequence of $\Omega_{\mathbf{Q}}$-modules

$$0 \longrightarrow (\mathbf{Q}/\mathbf{Z})(2) \longrightarrow Ind_{G_\infty}^{\Omega_{\mathbf{Q}}}((\mathbf{Q}/\mathbf{Z})(2)) \longrightarrow (\mathbf{Q}/\mathbf{Z})(2)_+ \longrightarrow 0$$

in which the first map is the diagonal embedding and G_∞ is the subgroup of order two generated by complex conjugation, c. The associated long exact cohomology sequence gives rise to a 2-extension of $\mathbf{Z}[G(N/\mathbf{Q})]$-modules of the form

$$(\mathbf{Q}/\mathbf{Z})(2)^{\Omega_N} \longrightarrow (Ind_{G_\infty}^{\Omega_{\mathbf{Q}}}((\mathbf{Q}/\mathbf{Z})(2)))^{\Omega_N}$$

$$\longrightarrow ((\mathbf{Q}/\mathbf{Z})(2)_+)^{\Omega_N} \longrightarrow H^1(N;(\mathbf{Q}/\mathbf{Z})(2))'$$

in which the last group denotes the kernel of the map from $H^1(N;(\mathbf{Q}/\mathbf{Z})(2)) \cong K_2(N)$ to the direct sum of the K_2's of each of the real places of N. There is a canonical isomorphism of the form [**10**]

$$(\mathbf{Q}/\mathbf{Z})(2)^{\Omega_N} \xrightarrow{\cong} K_3^{ind}(N)$$

and pulling back this extension by means of the canonical injection of $K_2(\mathcal{O}_{N,S})'$ into $K_2(N)$ yields the 2-extension whose Euler characteristic is equal to $\Omega_1(N/\mathbf{Q},3)$.

Secondly, there is an isomorphism of the form ([**14**], Chapter 5, Section 2)

$$\mathcal{CL}(\mathbf{Z}[Q_8]) \cong (\mathbf{Z}/4)^* \cong \{\pm 1\}.$$

The unique two dimensional irreducible complex representation, ν, of $G(N/\mathbf{Q}) \cong Q_8$ is symplectic and the root number class is given by the formula

$$W_{N/\mathbf{Q}} = W_{\mathbf{Q}}(\nu) \in \{\pm 1\} \cong \mathcal{CL}(\mathbf{Z}[Q_8])$$

where $W_{\mathbf{Q}}(\nu)$ is the Artin root number of ν define by the functional equation of the Artin L-function [**12**].

In all the quaternion examples, $N/\mathbf{Q}$, of 2.8, the Artin root number, $W_{\mathbf{Q}}(\nu)$, is trivial.

Our main result is the following computation of $\Omega_1(N/\mathbf{Q},3)$.

Theorem 1.1 *Let Σ denote the set of rational primes $\{2,p\}$ with $p \equiv 3$ (modulo 8) and let $N/\mathbf{Q}$ be the associated totally real Q_8-extension introduced above. Then*

$$\Omega_1(N/\mathbf{Q},3) = -\tilde{L}_{\mathbf{Q},\Sigma}(-1,\nu)(-1)^{(\frac{1}{4})log_2(z)} \in \mathcal{CL}(\mathbf{Z}[Q_8]) \cong (\mathbf{Z}/4)^*$$

where $z = \tilde{L}_{\mathbf{Q},\Sigma}(-1, Ind_{<C>}^{Q_8}(1))$. Here $L_{\mathbf{Q},\Sigma}(s,\rho)$ denotes the Artin L-function of a representation of $\Omega_{\mathbf{Q}}$ with the Euler factors at 2 and p removed and $C = G(N/\mathbf{Q}(\sqrt{2},\sqrt{p}))$, the centre of Q_8. Also, when the L-value is a non-zero rational number, $\tilde{L}_{\mathbf{Q},\Sigma}(s,\rho)$ denotes its odd-primary part. The element $(-1)^{(\frac{1}{4})log_2(z)}$ is equal to $\chi_+(z)$ where $\chi_+ : (\mathbf{Z}/8)^* \longrightarrow (\mathbf{Z}/4)^*$ is the surjective homomorphism with kernel equal to $\{\pm 1\}$ which appears in the proof of 4.5.

In general the rational numbers, $L_{\mathbf{Q},\Sigma}(-1,\nu)$, are unknown. However, when $p = 3$, $W_{\mathbf{Q}}(\nu) = 1$ and the computer calculations of [**8**] show that $L_{\mathbf{Q},\Sigma}(-1,\nu) = 2^2 \cdot 71$ while $L_{\mathbf{Q},\Sigma}(-1, Ind_{<C>}^{Q_8}(1)) = 1$. When $p = 11$ and $p = 19$ the numerology is similar and the facts may be rephrased in terms of the root number class.

Corollary 1.2 *When $p = 3$, $p = 11$ or $p = 19$ in Theorem 1.1*

$$\Omega_1(N/\mathbf{Q},3) = W_{N/\mathbf{Q}} \in \mathcal{CL}(\mathbf{Z}[Q_8]) \cong (\mathbf{Z}/4)^*,$$

both elements being trivial.

Theorem 1.1 and Corollary 1.2 will be proved in Section 5. The intermediate steps of the calculation are arranged in the following manner. In Section 2 we construct the Q_8 extensions, $N/\mathbf{Q}$, keeping track of the 2-primary part of $K_2(\mathcal{O}_N)$ and record the details of a Galois extension, $\mathbf{Q} \subset N \subset L$ which we shall require in Section 5 for the evaluation of the Euler characteristic of the 2-primary part of the 2-extension defining $\Omega_1(N/\mathbf{Q}, 3)$. In Section 3 we shall evaluate the Artin root numbers, $W_{\mathbf{Q}}(\nu)$. In Section 4 we develop a yoga which will enable us to evaluate the Euler characteristic of the 2-primary part of the 2-extension defining $\Omega_1(N/\mathbf{Q}, 3)$. In Section 5 we examine the cohomological description of $K_2(N)$ and describe as a representative 1-cocycle the basic element needed for the 2-primary yoga of Section 4. This involves the Galois extension, $L/\mathbf{Q}$, introduced in Section 2. The odd primary part of the Euler characteristic is evaluated using the solution, in [**17**], of the odd primary part of the Birch-Tate conjecture for totally real fields.

Part of this research was done while Victor Snaith was enjoying the hospitality of the University of Augsburg and the University of Edmonton. He is very grateful for this hospitality and, in particular, for many helpful conversations with Juergen Ritter and Al Weiss during those visits.

2 The Q_8-Examples

Let F be an algebraic number field with ring of integers $\mathcal{O}_F$. Let S denote the set of primes in F above 2, $\mathcal{O}_{F,S}$ the ring of S-integers with unit group, $\mathcal{O}^*_{F,S}$ and S-class group, $\mathcal{CL}(\mathcal{O}_{F,S})$. The canonical map to the K_2-groups of the real places

$$K_2(\mathcal{O}_F) \longrightarrow \oplus_{v\ real} \mathrm{Tors}\ K_2(F_v) \cong \oplus_{v\ real} \mathbf{Z}/2$$

is onto and we shall denote its kernel by $K_2(\mathcal{O}_F)'$. The 2-rank, rk_2, of $K_2(\mathcal{O}_F)$ may be calculated in general from the formula

$$rk_2(K_2(\mathcal{O}_F)) = r_1 + |S| - 1 + rk_2(\mathcal{CL}(\mathcal{O}_{F,S}))$$

where r_1 is the number of real places of F. Therefore we have the following result.

Proposition 2.1 *In the situation described above*

$$rk_2(K_2(\mathcal{O}_F)') = 0$$

if and only if $|S| = 1$ *and* $|\mathcal{CL}(\mathcal{O}_{F,S})|$ *is odd.*

Now suppose that F is totally real. If $rk_2(K_2(\mathcal{O}_F)') = 0$ then the group, $\mathcal{O}^*_{F,S}$, must contain units with independent signs at the real places. Moreover, by the S-unit theorem, there is precisely one class in $\mathcal{O}^*_{F,S} \otimes \mathbf{Z}/2$ represented by a totally positive unit. Let $\alpha_F \in \mathcal{O}^*_{F,S}$ be a representative for that class.

The following "going up" result is proved in [**1**].

Theorem 2.2 *Let F be a totally real number field such that* $rk_2(K_2(\mathcal{O}_F)') = 0$. *If N/F is a quadratic extension then* $rk_2(K_2(\mathcal{O}_N)') = 0$ *if and only if either*

(i) no odd prime ramifies in N/F

or

(ii) precisely one odd prime ramifies in N/F and α_F is not a norm from N.

We also have the following result.

Proposition 2.3 *If F is totally real and* $rk_2(K_2(\mathcal{O}_F)') = 0$ *then the Birch-Tate conjecture holds for F,*

$$\zeta_F(-1) = \frac{|K_2(\mathcal{O}_F)|}{|K_3^{ind}(F)|}.$$

Proof This follows from [**17**] for the odd-primary part and from [**11**] for the 2-primary part, since the 2-primary part of $K_2(\mathcal{O}_F)$ is elementary abelian.

Proposition 2.4 *Let p be an odd prime such that $p \equiv \pm 3$ (modulo 8). Let F denote the biquadratic extension of the rationals given by $F = \mathbf{Q}(\sqrt{2}, \sqrt{p})$. Then*

$$rk_2(K_2(\mathcal{O}_F)') = 0.$$

Proof Since $rk_2(K_2(\mathbf{Z}[\sqrt{2}])') = 0$ we may apply 2.2 with N/F replaced by the extension $\mathbf{Q}(\sqrt{2}, \sqrt{p})/\mathbf{Q}(\sqrt{2})$.

Since p is inert in $\mathbf{Q}(\sqrt{2})$ there is only one odd, ramified prime in $F/\mathbf{Q}(\sqrt{2})$. Moreover, $2 + \sqrt{2}$ represents the totally positive units modulo squares. Therefore, in condition (ii) of 2.2, $\alpha_{\mathbf{Q}(\sqrt{2})} = 2 + \sqrt{2}$.

If $\xi_n = exp(2\pi i/n)$ observe that $\alpha_{\mathbf{Q}(\sqrt{2})} = (\xi_{16} + \overline{\xi}_{16})^2$. Let $\mathbf{Q}^+(\xi_{16})$ denote the maximal real subfield of $\mathbf{Q}(\xi_{16})$. Now $\alpha_{\mathbf{Q}(\sqrt{2})}$ is a norm from $\mathbf{Q}(\sqrt{2}, \sqrt{p})$ if and only if $p \in \mathbf{Q}(\sqrt{2})$ is a norm from $\mathbf{Q}^+(\xi_{16})$. However, $p \equiv \pm 3$ (*modulo* 8) implies that $p^2 \equiv 9$ (*modulo* 16) so that the inertia degree of p in $\mathbf{Q}(\xi_{16})$ is four. Since p is inert in $\mathbf{Q}(\sqrt{2})$, the inertia degree of p in $\mathbf{Q}^+(\xi_{16})$ is also four. Hence neither p nor p^2 can be a norm in the extension $\mathbf{Q}^+(\xi_{16})/\mathbf{Q}$.

2.5 We shall now construct a specific quadratic extension, N, of the biquadratic field, $F = \mathbf{Q}(\sqrt{2}, \sqrt{p})$ such that $N/\mathbf{Q}$ is a totally real quaternion extension, unramified outside 2 and p. The details are given in 2.7. For the moment, however, we need only know that $N = F(\beta)$ where $\beta^2 = \alpha = (\sqrt{p})\epsilon$ and $(u + v\sqrt{p})^{-1}(2 + \sqrt{2}) = \epsilon \in \mathcal{O}_{F,S}^*$ is chosen so that α is totally positive. For such a field we have the following result.

Proposition 2.6 *In 2.5, $rk_2(K_2(\mathcal{O}_N)') = 0$.*

Proof Once again there is only one odd prime which ramifies in N/F. The totally positive units modulo squares are still represented by $\alpha_F = 2 + \sqrt{2}$ but this lies in the norms of N/F if and only if $\alpha = (\sqrt{p})\epsilon$ lies in the norms of $F^+(\xi_{16})/F$. This, in turn, implies that the norm $N_{F/\mathbf{Q}(\sqrt{2})}(\alpha)$ is a norm in the extension $\mathbf{Q}^+(\xi_{16})/\mathbf{Q}(\sqrt{2})$, which is equivalent to p being a norm in the extension $\mathbf{Q}^+(\xi_{16})/\mathbf{Q}(\sqrt{2})$, which is not true.

2.7 *The construction of N* Since $p \equiv 3$ (modulo 8), -2 is a norm from $\mathbf{Q}(\sqrt{p})$ and we may find u, v so that $-2 = u^2 - v^2 p$. Set $z = (\sqrt{2})/(u + v\sqrt{p})$. Then $z^2 = -1 + pv^2 - uv\sqrt{p}$ so that z is integral. This observation uses the fact that the class number of $\mathbf{Q}(\sqrt{p})$ is odd, which follows from Proposition 2.1 and Theorem 2.2. In fact, z is a unit. Set

$$\alpha = (\sqrt{p})z(1 + \sqrt{2}) = ((\sqrt{p})/(u + v\sqrt{p})) \cdot (\sqrt{2}(1 + \sqrt{2})).$$

Since $(\sqrt{p})/(u + v\sqrt{p})$ and $\alpha_F = \sqrt{2}(1 + \sqrt{2})$ are totally positive the same is true for α.

If $\beta^2 = \alpha$ then $N = F(\beta)$ is a totally real number field, which is unramified outside 2 and p. If x and y generate the Galois group, $G(\mathbf{Q}(\sqrt{2}, \sqrt{p})/\mathbf{Q})$, we have

$$\alpha y(\alpha) = ((\sqrt{p})/(u+v\sqrt{p}))\cdot(\sqrt{2}(1+\sqrt{2}))((-\sqrt{p})/(u-v\sqrt{p}))\cdot(\sqrt{2}(1+\sqrt{2})),$$

$$\alpha x(\alpha) = ((\sqrt{p})/(u+v\sqrt{p}))\cdot(\sqrt{2}(1+\sqrt{2}))((\sqrt{p})/(u+v\sqrt{p}))\cdot(-\sqrt{2}(1-\sqrt{2})),$$

$$\alpha xy(\alpha) = ((\sqrt{p})/(u+v\sqrt{p}))\cdot(\sqrt{2}(1+\sqrt{2}))((-\sqrt{p})/(u-v\sqrt{p}))\cdot(-\sqrt{2}(1-\sqrt{2}))$$

so that, modulo squares,

$$\alpha y(\alpha) = p \in \mathbf{Q}(\sqrt{2})^*/(\mathbf{Q}(\sqrt{2})^*)^2,$$

$$\alpha x(\alpha) = 2p \in \mathbf{Q}(\sqrt{p})^*/(\mathbf{Q}(\sqrt{p})^*)^2,$$

$$\alpha xy(\alpha) = 2 \in \mathbf{Q}(\sqrt{2p})^*/(\mathbf{Q}(\sqrt{2p})^*)^2.$$

These conditions constitute a well-known criterion (see [**9**], for example) for $N/\mathbf{Q}$ to be a quaternion extension.

In passing, we note that an integral basis for $N/\mathbf{Q}$ is given by

$$1, \sqrt{p}, z, z\sqrt{p}, \beta, \beta\sqrt{p}, \beta z, \beta z\sqrt{p}.$$

Therefore we have established the following result.

Theorem 2.8 *If $u, v \in \mathbf{Q}$ satisfy $-2 = u^2 - pv^2$ and*

$$\beta^2 = \sqrt{p}(1+\sqrt{2})(\sqrt{2}/(u+v\sqrt{p}))$$

then $N = \mathbf{Q}(\sqrt{2}, \sqrt{p}, \beta)$ is a totally real Galois extension of the rationals with Galois group isomorphic to

$$Q_8 = \{x, y \mid x^2 = y^2,\ xyx = y, x^4 = 1\}$$

acting in the following manner

$$x(\beta) = \beta/(\sqrt{2}+1), \quad y(\beta) = \beta((u+v\sqrt{p})/\sqrt{2}).$$

2.9 The action of x reverses and fixes the sign of $\sqrt{2}$ and $\sqrt{p}$, respectively, and y does the opposite. Let $\xi_n = exp(2\pi i/n)$ where $i = \sqrt{-1}$, as usual. The extension, $N(i)/\mathbf{Q}$, is Galois with $G(N(i)/\mathbf{Q}) \cong Q_8 \times \mathbf{Z}/2 < c >$ where c denotes complex conjugation and x, y extend to $N(i)$ by fixing i. The class group of $N(i)$ has odd order. This follows from ([**6**], Section 16.4) or may be proved directly from the cohomology sequences derived from the short exact sequence involving divisors and principal divisors which defines the class-group of $N(i)$. By ([**6**] Section 16.4), the prime ideal (β) ramified in $N(i)/N$. Therefore there exists $z_1 \in \mathcal{O}_{N(i)}$ generating the unique ideal of $N(i)$ above β and so $z_1 c(z_1) = \beta u_1$ for some $u_1 \in \mathcal{O}_N^*$. Since $x \in Q_8$ interchanges $1 \pm \sqrt{2}$ it must also interchange the ideals of $N(i)$ generated by z_1 and $c(z_1)$. Similarly, $y \in Q_8$ fixes each of these ideals. Therefore the primary decompositions of $(x+1)(z_1/c(z_1))$ and $(y-1)(z_1/c(z_1))$ are both trivial and being units of norm one at all infinite places implies that they are roots of unity in $N(i)$. When $p = 3$ the roots of unity in $N(i)$ are of order twenty-four and otherwise are of order eight. In all cases there exist integers, $a(x)$ and $a(y)$, such that

$$x(z_1/c(z_1)) = \xi_{24}^{a(x)}(z_1/c(z_1))^{-1}, \quad y(z_1/c(z_1)) = \xi_{24}^{a(y)} z_1/c(z_1).$$

If ω satisfies $\omega^8 = z_1/c(z_1)$ let $L = N(\xi_{64}, \omega)$ and $M = N(\xi_{64})$. $L/\mathbf{Q}$ is a Galois extension whose group is generated by elements X, Y, c and $\sigma \in G(L/M)$ where c

is complex conjugation and the action is given by the following formulae in which the integers, $\alpha(X)$ and $\alpha(Y)$, are multiples of 3 when $p > 3$:

$$\begin{aligned}
X(\omega) &= \xi_{192}^{\alpha(X)}\omega^{-1}, \\
Y(\omega) &= \xi_{192}^{\alpha(Y)}\omega, \\
X(\xi_{64}) &= \xi_{64}^{13}, \\
Y(\xi_{64}) &= \xi_{64}, \\
X(\beta) &= \beta/(\sqrt{2}+1), \\
Y(\beta) &= \beta((u+v\sqrt{p})/\sqrt{2}), \\
\sigma(\omega) &= \xi_8\omega.
\end{aligned}$$

The following fact concerning $G(L/\mathbf{Q})$ are straightforward to prove.

Proposition 2.10

(i) *If* $p = 3$ *then* $\alpha(X) + \alpha(Y) \equiv 0$ *(modulo 3)*,
(ii) $X^{-1}YXY^{-3} = \sigma^{-\alpha(Y)}$,
(iii) $X^{16} = \sigma^{4\alpha(X)}$,
(iv) $\alpha(Y) \equiv 0$ *(modulo 2)*,
(v) $Y^4 = \sigma^{\alpha(Y)/2}$,
(vi) $X\sigma X^{-1} = \sigma^3$,
(vii) $Y\sigma Y^{-1} = \sigma$.

We shall also need the following results, which are equally easily proved.

Proposition 2.11 *Let*

$$f : G(L/N(i)) \longrightarrow \mathbf{Z}/8$$

be the homomorphism given by the formula $\xi_8^{f(g)} = g(\omega)/\omega$. *Then*

(i) $f(\sigma) \equiv 1$ *(modulo 8)*,
(ii) $\alpha(Y) + 2\alpha(X) \equiv 0$ *(modulo 4)*,
(iii) $f(X^2Y^2) \equiv (9\alpha(Y) + 6\alpha(X))/4$ *(modulo 8)*,
(iv) $f(XYX^{-1}Y) \equiv -3\alpha(Y)/2$ *(modulo 8)*,
(v) $f(XY^{-3}X^{-1}Y) \equiv -3\alpha(Y)$ *(modulo 8)*.

Proposition 2.12 *For* f *as in 2.11 and* $h \in G(L/\mathbf{Q})$, *define*

$$h(f) : G(L/N(i)) \longrightarrow \mathbf{Z}/8$$

by the the formula $h(f)(g) = f(h^{-1}gh)$.

(i) $(X(f) - 3f)(\sigma) \equiv 0$ *(modulo 8)*,
(ii) $X(f)(X^2Y^2) \equiv f(XY^2X) \equiv 3(2\alpha(X) + \alpha(Y))/4$ *(modulo 8)*,
(iii) $(X(f) - 3f)(X^2Y^2) \equiv -3(\alpha(X) + 2\alpha(Y))$ *(modulo 8)*,
(iv) $(Y(f) - f)(\sigma) \equiv 0$ *(modulo 8)*,
(v) $Y(f)(X^2Y^2) \equiv f(Y^{-1}X^2Y^3) \equiv -3(\alpha(Y) - 2\alpha(X))/4$ *(modulo 8)*,
(vi) $(Y(f) - f)(X^2Y^2) \equiv -3\alpha(Y)$ *(modulo 8)*.

3 The Artin Root Numbers

In Theorem 2.8, we constructed a family of totally real Galois extensions of the form $N/\mathbf{Q}$ with $G(N/\mathbf{Q}) \cong Q_8$, the quaternion group of order eight. In this section we shall prove the following result.

Theorem 3.1 *In Theorem 2.8, let ν be the two-dimensional irreducible complex representation of $G(N/\mathbf{Q})$. Then the Artin root number of ν satisfies*

$$W_{\mathbf{Q}}(\nu) = 1.$$

Let p be a rational prime, $p \equiv 3$ (modulo 8), set $K = \mathbf{Q}(\sqrt{2})$ and let $N/\mathbf{Q}$ be the totally real quaternion extension of Theorem 2.8, containing $F = \mathbf{Q}(\sqrt{2}, \sqrt{p})$ and ramified only over 2 and p. Write $G = G(N/\mathbf{Q}) \cong Q_8$ and $\Gamma = G(N/K) \cong \mathbf{Z}/4$.

Since $p \equiv 3$ (modulo 8), p is inert in K. Let $\mathcal{P} = p\mathcal{O}_K$. Let $\mathcal{Q} = \sqrt{2}\mathcal{O}_K$ be the unique prime over 2 in K and define $\mathcal{F} = \mathcal{Q}^5\mathcal{P} = 4\mathcal{Q}\mathcal{P}$. If $\mathcal{A}$ is an integral ideal of K, let $\mathcal{CL}_{\mathcal{A}}(K)$ (resp. $\mathcal{CL}^+_{\mathcal{A}}(K)$) be the wide (resp. narrow) ideal class group of K of conductor $\mathcal{A}$. Since K has class number one, we have

$$\mathcal{CL}_{\mathcal{F}}(K) = \frac{(\mathcal{O}_K/\mathcal{Q}^5)^* \times (\mathcal{O}_K/\mathcal{P})^*}{Image(\mathcal{O}_K^*)}$$

and

$$\mathcal{CL}^+_{\mathcal{F}}(K) = \frac{(\mathcal{O}_K/\mathcal{Q}^5)^* \times (\mathcal{O}_K/\mathcal{P})^* \times \{\pm 1\}_{\infty_1} \times \{\pm 1\}_{\infty_2}}{Image^+(\mathcal{O}_K^*)}$$

where $\mathcal{O}_K^*$ is the group of units of K, $Image(\mathcal{O}_K^*)$ and $Image^+(\mathcal{O}_K^*)$ denote the images under the diagonal map of $\mathcal{O}_K^*$ in the numerators of the right-hand expressions. The infinite places of K are denoted by ∞_1 and ∞_2.

If A is a finite abelian group, let $A[2]$ denote the maximal quotient of A of order a power of two. Thus $A[2]$ is isomorphic to the two-Sylow subgroup of A.

Lemma 3.2 *The units, $\mathcal{O}_K^*$, are generated by -1 and $\epsilon = 1 + \sqrt{2}$. There is a unique isomorphism*

$$\Psi : \mathbf{Z}/4 \times \mathbf{Z}/2 \times \mathbf{Z}/2 \longrightarrow (\mathcal{O}_K/\mathcal{Q}^5)^*$$

defined by $\Psi(b_1, b_2, b_3) = \epsilon^{b_1} \cdot (-1)^{b_2} \cdot 5^{b_3}$. The group $(\mathcal{O}_K/\mathcal{P})^*[2]$ is cyclic of order 8 and generated by the image of ϵ. The group $\mathcal{CL}_{\mathcal{F}}(K)[2]$ is isomorphic to $\mathbf{Z}/4 \times \mathbf{Z}/2$, while $\mathcal{CL}^+_{\mathcal{F}}(K)[2]$ is isomorphic to $\mathbf{Z}/4 \times (\mathbf{Z}/2)^3$. Let i be a fixed fourth root of unity. There is a unique quartic character $\chi_{\mathcal{P}}$ of $(\mathcal{O}_K/\mathcal{P})^*$ such that $\chi_{\mathcal{P}}(\epsilon) = i$. There is a unique quartic character $\chi_{\mathcal{Q}}$ of $(\mathcal{O}_K/\mathcal{Q}^5)^*$ such that $\chi_{\mathcal{Q}}(\epsilon) = i^{-1}$, $\chi_{\mathcal{Q}}(-1) = 1$ and $\chi_{\mathcal{Q}}(5) = -1$. The character, $\chi_{\mathcal{Q}} \times \chi_{\mathcal{P}}$ of $(\mathcal{O}_K/\mathcal{Q}^5)^* \times (\mathcal{O}_K/\mathcal{P})^*$, is trivial on $Image(\mathcal{O}_K^*)$ and induces a quartic character, χ of $\mathcal{CL}_{\mathcal{F}}(K)$, which corresponds by class field theory to a quartic character of $\Gamma = Gal(L/K)$.

In particular, the field N is uniquely determined by p, and the conductor of N over K is $\mathcal{F}$.

Define T to be the real subfield of the field $\mathbf{Q}(\xi_{16})$ of 16-th roots of unity. Then $T = K\left(\sqrt{\epsilon\sqrt{2}}\right)$ is a quadratic extension of K conductor $\mathcal{Q}^5$ which, by class field theory, corresponds to a character λ of order two of $\mathcal{CL}_{\mathcal{F}}(K)$. Define $\chi' = \chi \cdot \lambda$, $\xi = \chi^2$ and let χ_0 be the trivial character of $\mathcal{CL}_{\mathcal{F}}(K)$. In the following table we will list the characters of the group $\mathcal{CL}_{\mathcal{F}}(K)$, their orders, the conductors of the associated primitive characters and the abelian extensions of K corresponding to these characters. The field T_1 is a totally real dihedral extension of $\mathbf{Q}$ containing

$F = \mathbf{Q}(\sqrt{2}, \sqrt{p})$ and ramified only over 2 and p. The field T_2 is the unique real cyclic quartic extension of $\mathbf{Q}$ of conductor $16p$ over $\mathbf{Q}$.

Character	*order*	*primitive conductor*	*fixed field*
χ_0	1	$\mathcal{O}_K$	$K = \mathbf{Q}(\sqrt{2})$
χ	4	$\mathcal{F} = \mathcal{Q}^5\mathcal{P}$	$N =$ *quaternion*
$\xi = \chi^2 = \chi'^2$	2	$\mathcal{Q}^2\mathcal{P}$	$F = \mathbf{Q}(\sqrt{2}, \sqrt{p})$
χ^{-1}	4	$\mathcal{F} = \mathcal{Q}^5\mathcal{P}$	$N =$ *quaternion*
λ	2	$\mathcal{Q}^5$	$T = \mathbf{Q}^+(\xi_{16})$
$\chi' = \chi\lambda$	4	$\mathcal{Q}^4\mathcal{P} = 4\mathcal{P}$	$T_1 =$ *dihedral*
χ'^{-1}	4	$\mathcal{Q}^4\mathcal{P}$	T_1
$\xi\lambda$	2	$\mathcal{F}$	T_2 (*quartic over* $\mathbf{Q}$)

The group of characters of $\mathcal{CL}^+_{\mathcal{F}}(K)$ is generated by the characters of $\mathcal{CL}_{\mathcal{F}}(K)$ together with the quadratic characters μ_1 and μ_2 associated to the quadratic extensions $K(\sqrt{-1})$ and $K\left(\sqrt{\sqrt{2}}\right)$ of K. The character μ_1 is totally negative, while μ_2 and $\mu_3 = \mu_1 \cdot \mu_2$ each ramify at exactly one infinite place, where μ_3 is associated to the quadratic extension $K\left(\sqrt{\sqrt{2}}\right)$ of K.

Proof It is easily checked that $\epsilon = 1 + \sqrt{2}$ is a fundamental unit for K. Each element of $(\mathcal{O}_K/\mathcal{Q}^5)^*$ is represented by a unique sum $1 + \sum_{i=1}^4 a_i(\sqrt{2})^i$ in which $a_i \in \{0, 1\}$. In $(\mathcal{O}_K/\mathcal{Q}^5)^*$ one has $\epsilon^4 = (1 + \sqrt{2})^4 = 1 + 12\sqrt{2} + 16 \equiv 1$ and $(-1)^2 \equiv 5^2 \equiv 1$. Thus the map Ψ in the statement of Lemma 3.2 is well-defined. Suppose $\Psi(b_1, b_2, b_3) = (1 + \sqrt{2})^{b_1} \cdot (-1)^{b_2} \cdot 5^{b_3} \equiv 1$ for some $(b_1, b_2, b_3) \in \mathbf{Z}/4 \times \mathbf{Z}/2 \times \mathbf{Z}/2$. Viewing this congruence modulo $2\mathcal{O}_K$ shows $b_1 \equiv 0$ (modulo 2). Since $(1 + \sqrt{2})^2 = 1 + 2\sqrt{2} + 2$ and $-1 = 1 - 2 = 1 + 2 + 4$, we can then view the congruence modulo $4\mathcal{O}_K$ to conclude $b_1 = 0 \in \mathbf{Z}/4$ and $b_2 = 0 \in \mathbf{Z}/2$. Now the congruence is $(1 + 4)^{b_3} \equiv 1$ so that $b_3 = 0 \in \mathbf{Z}/2$. Thus Ψ is injective and by counting orders we conclude that Ψ is an isomorphism. It follows that there is a unique character $\chi_{\mathcal{Q}}$ of order four of $(\mathcal{O}_K/\mathcal{Q}^5)^*$ such that $\chi_{\mathcal{Q}}(\epsilon) = i^{-1}$, $\chi_{\mathcal{Q}}(\epsilon)(-1) = 1$ and $\chi_{\mathcal{Q}}(5) = -1$.

We have also shown that the subgroup of $\mathcal{O}_K^*$ having trivial image in $(\mathcal{O}_K/\mathcal{Q}^5)^*$ is generated by ϵ^4.

Since $p \equiv 3$ (modulo 8), the order of $\mathcal{O}_K/\mathcal{P}$ is $p^2 \equiv 9$ (modulo 16). Thus $(\mathcal{O}_K/\mathcal{P})^*[2]$ is cyclic of order 8, while $(\mathbf{Z}/p)^*[2]$ has order 2. We have to show that ϵ generates $(\mathcal{O}_K/\mathcal{P})^*[2]$. Since $\mathcal{CL}_{\mathcal{P}}(K) = (\mathcal{O}_K/\mathcal{P})^*/Image(\mathcal{O}_K^*)$, this is the same as showing that $H = \mathcal{CL}_{\mathcal{P}}(K)[2]$ is trivial. On the contrary, suppose that H is non-trivial. Then H is a non-zero module of 2-primary order for the cyclic group $G(K/\mathbf{Q}) \cong \mathbf{Z}/2$. Hence the covariants of H with respect to $G(K/\mathbf{Q})$ are non-trivial. We conclude that there is a non-trivial quotient, H', of H of order two on which $G(K/\mathbf{Q})$ acts trivially. Let F' be the quadratic extension of K corresponding by class field theory to H'. Then F' is Galois over $\mathbf{Q}$ and $G(K/\mathbf{Q})$ acts trivially by conjugation on the group $G(F'/K) = H'$. Since $G(K/\mathbf{Q})$ is cyclic, this forces F' to be an abelian extension of degree four over $\mathbf{Q}$. Since F' contains K and F'/K can ramify only over $\mathcal{P}$, $F'/\mathbf{Q}$ can ramify only over 2 and p and the conductor of F' over $\mathbf{Q}$ divides $8p$. Since F' is totally real, it corresponds by class field theory to a quotient of order 4 of the wide class group $\mathcal{CL}_{8p\mathbf{Z}}(\mathbf{Z}) = (\mathbf{Z}/8p)^*/\{\pm 1\}$. Since $p \equiv 3$ (modulo 8), the two-Sylow subgroup of the latter group has order 4. Therefore F' must be the maximal real abelian 2-extension of $\mathbf{Q}$ of conductor dividing $8p$, which is $F = \mathbf{Q}(\sqrt{2}, \sqrt{p})$. However, F/K ramifies over $\mathcal{Q} = \sqrt{2}\mathcal{O}_K$, contradicting the fact that F'/K can ramify only over $\mathcal{P}$. Thus ϵ generates $(\mathcal{O}_K/\mathcal{P})^*[2]$, as required.

This discussion showed that $(\mathcal{O}_K/\mathcal{Q}^5)^*/Image(\mathcal{O}_K^*)$ has order two and ϵ^4 generates the subgroup of $\mathcal{O}_K^*$ having trivial image in $(\mathcal{O}_K/\mathcal{Q}^5)^*$. Thus there is a short exact sequence of the form

$$(\mathcal{O}_K/\mathcal{P})^*[2]/ < \epsilon^4 > \longrightarrow \mathcal{CL}_{\mathcal{F}}(K)[2] \longrightarrow (\mathcal{O}_K/\mathcal{Q}^5)^*/Image(\mathcal{O}_K^*) \cong \mathbf{Z}/2.$$

The group on the left has order four since $(\mathcal{O}_K/\mathcal{P})^*[2] \cong \mathbf{Z}/8$, generated by ϵ. Since $(\mathbf{Z}/p)^*[2] \cong \mathbf{Z}/2$, each element of this group is a 4th power in $(\mathcal{O}_K/\mathcal{P})^*[2]$ and therefore $\chi_{\mathcal{P}}$ is trivial on $(\mathbf{Z}/p)^*[2]$. In particular, $\chi_{\mathcal{P}} = 1$.

We have now arranged that the character, $\chi_{\mathcal{Q}} \times \chi_{\mathcal{P}}$, is trivial on both -1 and ϵ. Hence this character descends to a character, χ, of $\mathcal{CL}_{\mathcal{F}}(K) \cong \mathbf{Z}/4 \times \mathbf{Z}/2$. Also χ has order four, since it is non-trivial on the subgroup of squares in $\mathcal{CL}_{\mathcal{F}}(K)$.

We now have to check that χ is a quaternion character. For this it suffices to show that χ is trivial on elements of $Norm_{K/\mathbf{Q}}((\mathcal{O}_K/\mathcal{Q}^5)^* \times (\mathcal{O}_K/\mathcal{P})^*)$ but not on some residue class represented by a rational integer m, because the transfer on the first homology group distinguishes the quaternion and dihedral groups of order 8 (c.f. [**15**], pp. 275-76). We know that $\chi_{\mathcal{P}}$ is trivial on $(\mathbf{Z}/p)^* = Norm_{K/\mathbf{Q}}(\mathcal{O}_K/\mathcal{P})^*$, since this group is contained in the group of fourth powers of $(\mathcal{O}_K/\mathcal{P})^*$. The group, $Norm_{K/\mathbf{Q}}(\mathcal{O}_K/\mathcal{Q}^5)^*$, is contained in $\{\pm 1\} \in \mathbf{Z}/8^*$ since $K = \mathbf{Q}(\sqrt{2})$ and $\mathcal{Q}^5$ divides $8\mathcal{O}_K = \mathcal{Q}^6$. Thus, since $\chi_{\mathcal{Q}}(-1) = 1$ by construction, we see that χ is trivial on $Norm_{K/\mathbf{Q}}((\mathcal{O}_K/\mathcal{Q}^5)^* \times (\mathcal{O}_K/\mathcal{P})^*)$. By construction we have $\chi_{\mathcal{Q}}(5) = -1$ and $\chi_{\mathcal{P}}(5) = 1$, since $5 \in (\mathbf{Z}/p)^*$. Hence $\chi(5) = -1$ and therefore χ is a quaternion character, as required.

The quartic extension, N'/K, which corresponds to χ by classfield theory is thus a quaternion extension of $\mathbf{Q}$ which contains K and ramifies only over 2 and p. Since $\mathcal{CL}_{\mathcal{F}}(K) \cong \mathbf{Z}/4 \times \mathbf{Z}/2$, there is only one square of a character of order four of $\mathcal{CL}_{\mathcal{F}}(K)$. Thus χ^2 must be the quadratic character associated to the extension F/K, since the conductor of $F = \mathbf{Q}(\sqrt{2}, \sqrt{p})$ over K is $2p\mathcal{O}_K = \mathcal{Q}^2\mathcal{P}$. Thus N' is a quaternion extension of $\mathbf{Q}$ containing F. By [**7**], N' is a quadratic twist of the extension N constructed in Theorem 2.8. Suppose N' is the twist of N by $\mathbf{Q}(\sqrt{m})$ for some square-free integer, m. Since N and N' are both totally real and ramified only over 2 and p, $\mathbf{Q}(\sqrt{m})$ must be a real quadratic field unramified outside 2 and p. The only m for which this is true are $m = 2, p, 2p$. All of these $\mathbf{Q}(\sqrt{m})$ are subfields of $F \subset N \cap N'$. Therefore $N = N'$. By construction of χ, the conductor of N/K is equal to $\mathcal{F}$.

The characters χ and χ^{-1} correspond by class field theory to the two quartic characters of $G(N/K)$. The imprimitive quadratic character, $\xi = \chi^2$, corresponds to the quotient $G(F/K)$ of $G(N/K)$ and thus has primitive conductor $\mathcal{Q}^2\mathcal{P}$. Suppose, in the above construction of χ, we replaced $\chi_{\mathcal{Q}}$ by the unique character, $\chi'_{\mathcal{Q}}$, of $(\mathcal{O}_K/(\mathcal{Q}^5)^*$ such that $\chi'_{\mathcal{Q}}(\epsilon) = i^{-1}$, $\chi'_{\mathcal{Q}}(-1) = 1$ and $\chi'_{\mathcal{Q}}(5) = 1$. Then $\chi'_{\mathcal{Q}} \times \chi_{\mathcal{P}}$ descends to an imprimitive quartic character, χ', of $\mathcal{CL}_{\mathcal{F}}(K)$. The associated primitive character has conductor $\mathcal{Q}^4\mathcal{P} = 4p\mathcal{O}_K$. Since χ' is trivial on $(\mathbf{Z}/8)^* \times (\mathbf{Z}/p)^*$, one finds that χ' corresponds to the dihedral extension, $T_1/\mathbf{Q}$, containing F.

Define $\lambda = \chi/\chi'$. The $\mathcal{P}$-components of χ and χ' are equal so that λ has trivial $\mathcal{P}$-component. One sees that, since $\chi_{\mathcal{Q}}(5) = -1 = -\chi'_{\mathcal{Q}}$, the conductor of the primitive character, λ^{prim}, associated to λ is $\mathcal{Q}^5 = 4\mathcal{Q}$. By our calculations of the images of ϵ and -1 in $(\mathcal{O}_K/\mathcal{Q}^5)^*$, the order of $\mathcal{CL}_{\mathcal{Q}^5}(K) = (\mathcal{O}_K/\mathcal{Q}^5)^*/Image(\mathcal{O}_K^*)$ is 2. Hence λ^{prim} is the unique non-trivial character of conductor $\mathcal{Q}^5$. Since

$Norm_{K/\mathbf{Q}}(1+\mathcal{Q}^5) \subset 1+16\mathbf{Z}$, we see that the extension $\mathbf{Q}^+(\xi_{16})/K$ corresponds to a quadratic character of $\mathcal{CL}_{\mathcal{Q}^5}(K)$, so this character must be λ^{prim}.

Each quadratic extension T'/K which is unramified outside 2 and ∞ must have the form $T' = K(\sqrt{\alpha})$ for some α in the group $(\mathcal{O}_K[1/2])^*$ of 2-units of K. Here $(\mathcal{O}_K[1/2])^*$ is generated by $\epsilon = 1+\sqrt{2}$, -1 and $\sqrt{2}$ and the totally positive elements of $(\mathcal{O}_K[1/2])^*$ are generated by $\epsilon \cdot \sqrt{2}$ and $((\mathcal{O}_K[1/2])^*)^2$. Since $\mathbf{Q}^+(\xi_{16})/K$ is a totally real quadratic extension, we must have $\mathbf{Q}^+(\xi_{16}) = K\left(\sqrt{\epsilon \cdot \sqrt{2}}\right)$.

It is not hard to check that the group of characters $\{1, \mu_1, \mu_2, \mu_3\}$ defined at the end of Lemma 3.2 have conductors dividing $\mathcal{F}$ and all possible signs at infinity. Thus these characters and those of $CL_{\mathcal{F}}(K)$ generate the character group of $\mathcal{CL}^+_{\mathcal{F}}(K)$ and Lemma 3.2 is proved.

Proof of Theorem 3.1 The irreducible two-dimensional representation ν of $G = G(N/\mathbf{Q})$ equals $Ind_\Gamma^G(\chi)$ and we must show that the associated Artin root numbers satisfy

$$W_{\mathbf{Q}}(\nu) = W_K(\chi) = 1.$$

For each place, v of K, let χ_v be the local component at v of the idèle class character of K corresponding to χ. Then χ_v is unramified if v is not equal to one of the places $v_{\mathcal{Q}}$ or $v_{\mathcal{P}}$ determined by $\mathcal{Q}$ or $\mathcal{P}$, respectively. If v is a finite place let $\mathcal{O}_{K,v}$ denote the ring of integers of K_v and let $r_v : \mathcal{O}^*_{K,v} \longrightarrow (\mathcal{O}_{K,v_{\mathcal{Q}}}/f(\chi_{v_{\mathcal{Q}}}))^*$ be the homomorphism given by reduction modulo the conductor, $f(\chi_{v_{\mathcal{Q}}})$ of $\chi_{v_{\mathcal{Q}}}$. For $\mathcal{R} = \mathcal{Q}, \mathcal{P}$ we may view $\chi_{\mathcal{R}}$ as the character on $(\mathcal{O}_{K,v_{\mathcal{R}}}/f(\chi_{v_{\mathcal{R}}}))^*$ given by

$$\chi_{v_{\mathcal{R}}}(\alpha_{\mathcal{R}}) = \chi_{\mathcal{R}}(r_{v_{\mathcal{R}}}(\alpha_{\mathcal{R}}^{-1})) \tag{1}$$

for all $\alpha_{\mathcal{R}} \in \mathcal{O}^*_{K,v_{\mathcal{R}}}$.

In order to determine $\chi_{\mathcal{Q}}$ and $\chi_{\mathcal{P}}$ completely it is sufficient to find their values on uniformising parameters. Since $\prod_v \chi_v(\sqrt{2}) = \prod_v \chi_v(p) = 1$ we find that

$$\chi_{v_{\mathcal{Q}}}(\sqrt{2}) = \chi_{v_{\mathcal{P}}}(\sqrt{2})^{-1}, \ \chi_{v_{\mathcal{P}}}(p) = \chi_{v_{\mathcal{Q}}}(p)^{-1}. \tag{2}$$

Since $p \equiv 3$ (modulo 8), the class of 2 in $(\mathbf{Z}/p)^*[2] \cong \mathbf{Z}/2$ is non-trivial so that $\sqrt{2}$ has order 4 in $(\mathcal{O}_K/\mathcal{P})^*[2]$. Now $(\mathcal{O}_K/\mathcal{P})^*[2]$ is cyclic of order 8 and $\chi_{\mathcal{P}}$ is a quartic character of this group so that (2) implies that $\chi_{v_{\mathcal{P}}}(\sqrt{2})^{-1} = -1$. Since $\chi_{\mathcal{Q}}(-1) = 1 = -\chi_{\mathcal{Q}}(5)$, we have $\chi_{\mathcal{Q}}(p) = \chi_{\mathcal{Q}}(3) = -1$. Thus for the uniformisers $\sqrt{2}$ and p at $v_{\mathcal{Q}}$ and $v_{\mathcal{P}}$, respectively, we find from (1) and (2) that

$$\chi_{v_{\mathcal{Q}}}(\sqrt{2}) = -1 \quad \text{and} \quad \chi_{v_{\mathcal{P}}}(p) = -1. \tag{3}$$

Note that $\chi_{v_{\mathcal{Q}}}(-1) = 1$, so that the choice of $\sqrt{2}$ does not matter in this equality. We fix one choice in what follows.

We have local conductors

$$f(\chi_{v_{\mathcal{Q}}}) = \sqrt{2}^5 O_{k,v_{\mathcal{Q}}} \quad \text{and} \quad f(\chi_{v_{\mathcal{P}}}) = pO_{k,v_{\mathcal{P}}}.$$

Let χ' be the dihedral character defined in Lemma 3.2. By construction, $\chi'_{\mathcal{P}} = \chi_{\mathcal{P}}$ as characters of $(\mathcal{O}_K/\mathcal{P})^*$. We have $\chi'_{\mathcal{Q}} = \chi_{\mathcal{Q}}\lambda_{\mathcal{Q}}$ as characters of $(\mathcal{O}_K/\mathcal{Q}^5)^*$, where $\lambda_{\mathcal{Q}}$ is the quadratic character $\lambda_2 \circ Norm_{K/\mathbf{Q}}$ when λ_2 is a fixed choice of quartic character of $(\mathbf{Z}/16)^*/\{\pm 1\}$. Let χ'_v be the local component at v of the idèle class character of K corresponding to the primitive character of conductor $\mathcal{Q}^4\mathcal{P}$ which induces χ'. By construction,

$$\chi'_{v_{\mathcal{Q}}}(\alpha_{\mathcal{Q}}) = \chi'_{\mathcal{Q}}(r_{v_{\mathcal{Q}}}(\alpha_{\mathcal{Q}})^{-1}) = \chi_{\mathcal{Q}}(r_{v_{\mathcal{Q}}}(\alpha_{\mathcal{Q}})^{-1}) \cdot \lambda_{\mathcal{Q}}(r_{v_{\mathcal{Q}}}(\alpha_{\mathcal{Q}})^{-1}) \tag{4}$$

for $\alpha_{\mathcal{Q}} \in \mathcal{O}^*_{K,v_{\mathcal{Q}}}$ and

$$\chi'_{v_{\mathcal{P}}}(\alpha_{\mathcal{P}}) = \chi'_{\mathcal{P}}(r_{v_{\mathcal{P}}}(\alpha_{\mathcal{P}})^{-1}) = \chi_{\mathcal{P}}(r_{v_{\mathcal{P}}}(\alpha_{\mathcal{P}})^{-1}) \tag{5}$$

for $\alpha_{\mathcal{P}} \in \mathcal{O}^*_{K,v_{\mathcal{P}}}$. As in the case of χ, we find

$$\chi'_{v_{\mathcal{Q}}}(\sqrt{2}) = \chi'_{v_{\mathcal{P}}}(\sqrt{2})^{-1} = \chi_{v_{\mathcal{P}}}(\sqrt{2})^{-1} = -1. \tag{6}$$

Since $p^2 \equiv 9$ (modulo 16), we find that $\lambda_{\mathcal{Q}}(p) = \lambda_2(p^2) = -1$. Thus (4), (2) and (3) give

$$\chi'_{v_{\mathcal{P}}}(p) = \chi'_{v_{\mathcal{Q}}}(p)^{-1} = -\chi_{v_{\mathcal{Q}}}(p)^{-1} = 1. \tag{7}$$

We have local conductors

$$f(\chi'_{v_{\mathcal{Q}}}) = \sqrt{2}^4 \mathcal{O}_{K,v_{\mathcal{Q}}} \quad \text{and} \quad f(\chi'_{v_{\mathcal{P}}}) = p\mathcal{O}_{K,v_{\mathcal{P}}}.$$

By the properties of Artin root numbers (c.f. [**12**], [**15**], pp. 254-55) one has

$$W_{\mathbf{Q}}(V) = W_K(\chi) = \prod_v W(\chi_v) \quad \text{and} \quad W_K(\chi') = \prod_v W(\chi'_v) \tag{8}$$

where the product is over all of the places v of K and $W(\chi_v)$, $W(\chi'_v)$ are the local root numbers of χ_v and χ'_v at v. Since χ' is quartic and has fixed field a dihedral extension T_1 of degree eight of $\mathbf{Q}$, the induced representation $Ind_{G(T_1/K)}^{G(T_1/\mathbf{Q})}(\chi')$ is the two-dimensional irreducible representation V' of $G(T_1/\mathbf{Q})$. Since V' is orthogonal, the Fröhlich-Queyrut Theorem (c.f. [**15**], pp. 276-80) gives

$$W_K(\chi') = W_{\mathbf{Q}}(V') = 1. \tag{9}$$

Combining (8) and (9), we see that

$$W_{\mathbf{Q}}(V) = W_K(\chi) = W_K(\chi)/W_K(\chi') = \prod_v W(\chi_v)/W(\chi'_v). \tag{10}$$

If v is an infinite place of K, then v is real and χ_v, χ'_v are unramified so that $W(\chi_v) = W(\chi'_v) = 1$ by ([**7**], p. 33). Suppose now that v is a finite place of K different from $v_{\mathcal{Q}}$ and $v_{\mathcal{P}}$. Then v is unramified over $\mathbf{Q}$ and χ_v and χ'_v are unramified at v. Hence ([**7**], p. 34-35, Equations 5.5 and 5.8) shows that $W(\chi_v) = W(\chi'_v) = 1$.

Suppose now that $v = v_{\mathcal{P}}$. Then χ_v and χ'_v each have conductor $f = \mathcal{P} = p\mathcal{O}_{K,v}$, and the relative difference of K_v is $\mathcal{D}_{K_v} = \mathcal{O}_{K,v}$. Following ([**7**], p.34) we define $c = p$, so $c\mathcal{O}_{K,v} = f\mathcal{D}_{K,v}$. Let $\overline{\chi}_v$ be the complex conjugate of χ_v. By ([**7**], Equation 5.8, p.35 and Equation 5.5, p. 34),

$$W(\chi_v) = \tau(\overline{\chi}_v)/p^{\frac{1}{2}} \quad \text{and} \quad W(\chi'_v) = \tau(\overline{\chi}'_v)/p^{\frac{1}{2}} \tag{11}$$

where the Gauss sums on the right are defined as follows. Let ψ_{K_v} be the standard additive character of K_v (c.f. [**7**], p. 34, [**12**], [**15**], p. 359). Then

$$\tau(\overline{\chi}_v) = \sum_u \overline{\chi}_v\left(\frac{u}{c}\right)\psi_{k_v}\left(\frac{u}{c}\right) \quad \text{and} \quad \tau(\overline{\chi}'_v) = \sum_u \overline{\chi}'_v\left(\frac{u}{c}\right)\psi_{k_v}\left(\frac{u}{c}\right) \tag{12}$$

where the sums are taken over a set of representatives u for $\mathcal{O}^*_{K,v}$ modulo $1 + f\mathcal{O}_{K,v} = 1 + p\mathcal{O}_{K,v}$. From (1), (3), (5) and (7) we have $\chi_v\left(\frac{u}{c}\right) = \chi_v(u)/\chi_v(p) = -\chi'_v(u)/\chi'_v(p) = -\chi'_v(\frac{u}{c})$ for all units $u \in \mathcal{O}^*_{K,v}$. Thus (11) and (12) show that $W(\chi_v) = -W(\chi'_v)$. Formula (10) now simplifies to

$$W_{\mathbf{Q}}(V) = -W(\chi_{v_{\mathcal{Q}}})/W(\chi'_{v_{\mathcal{Q}}}). \tag{13}$$

For the remainder of the proof we fix $v = v_{\mathcal{Q}}$. The standard additive character, ψ_2, of $\mathbf{Q}_2$ is defined by $\psi_2(\mathbf{Z}_2) = 1$ and $\psi_2(2^{-r}) = e^{2\pi i/2^r}$ for $r \geq 0$. Define the additive character $\psi_{K_v} = \psi_2 \circ Tr_{K_v/\mathbf{Q}_2}$. The different, $\mathcal{D}_{K_v}$, equals $(\sqrt{2})^3\mathcal{O}_{K,v}$. The conductor, f, of χ_v is $(\sqrt{2})^5\mathcal{O}_{K,v}$ so, if we set $c = (\sqrt{2})^8 = 2^4$, then $c\mathcal{O}_{K,v} = \mathcal{D}_{K_v}f$. By ([7], p. 34-35),

$$W(\chi_v) = 2^{-5/2}\tau(\overline{\chi}_v) = 2^{5/2}\sum_u \chi_v\left(\frac{u}{c}\right)\psi_{k_v}\left(\frac{u}{c}\right) \tag{14}$$

where the sum is over a set of representatives u for $\mathcal{O}^*_{K,v}$ modulo $1+f = 1+(\sqrt{2})^5\mathcal{O}_{K,v}$. By (1) and (2)

$$\overline{\chi}_v(u/c) = \overline{\chi}_v(u)\overline{\chi}_v(c)^{-1} = \overline{\chi}_v(u)\chi_v(c) = \chi_{\mathcal{Q}}(r_v(u))\chi_v(c) \tag{15}$$

where $r_v : \mathcal{O}^*_{K,v} \longrightarrow (\mathcal{O}_{K,v}/f)^* = (\mathcal{O}_K/\mathcal{Q}^5)^*$ is the reduction map. By our construction of $\chi_{\mathcal{Q}}$, $\chi_v(u) = \chi_v(-u) = -\chi_v(5u) = -\chi_v(-5u)$ for all local units u. By Lemma 3.2, $S = \{1, \epsilon, \epsilon^2, \epsilon^3\}$ is a set of representatives for $(\mathcal{O}^*_{K,v}/(1+f))/\{\pm 1, \pm 5\}$, where $\epsilon = 1+\sqrt{2}$. Thus we can write (14) as

$$W(\chi_v) = 2^{-5/2}\chi_v(c)\sum_{s\in S}\overline{\chi}_v(s)\Bigg(\psi_{K_v}\Big(\frac{s}{c} + \psi_{K_v}\Big(\frac{-s}{c}\Big) - \psi_{K_v}\Big(\frac{5s}{c}\Big) - \psi_{K_v}\Big(\frac{-5s}{c}\Big)\Big)\Bigg). \tag{16}$$

For $u \in \mathcal{O}^*_{K,v} = (\mathbf{Z}_2[\sqrt{2}])^*$, $Tr_{K_v/\mathbf{Z}_2}(\frac{u}{c}) = Tr_{K_v/\mathbf{Z}_2}(\frac{u}{16})$ is in $8^{-1}\mathbf{Z}_2$, so $\psi_{K_v}(\frac{u}{c}) = \psi_2(Tr_{K_v/\mathbf{Z}_2}(\frac{u}{16}))$ is a (possibly not primitive) eighth root of unity. Fix $\zeta = exp(2\pi i/8)$. We have the following table:

Table A: Traces and character values for χ and $v = v_{\mathcal{Q}}$.

$s \in S$	$\overline{\chi}_v(s) = \chi_{\mathcal{Q}}(r_v(s))$	$t = \psi_{K_v}(\frac{s}{c})$	$t + t^{-1} - t^5 - t^{-5}$
1	1	ζ	$2\sqrt{2}$
$\epsilon = 1+\sqrt{2}$	i^{-1}	ζ	$2\sqrt{2}$
$\epsilon^2 = 1+2+2\sqrt{2}$	-1	ζ^3	$-2\sqrt{2}$
$\epsilon^3 = 7+5\sqrt{2}$	i	ζ^{-1}	$2\sqrt{2}$

We have $\chi_v(c) = \chi_v(2^4) = 1$. Thus, using Table A to evaluate (16) we obtain

$$W(\chi_{v_{\mathcal{Q}}}) = 2^{-5/2}\cdot 1\cdot(1\cdot(2\sqrt{2}) + i^{-1}\cdot(2\sqrt{2}) + (-1)\cdot(-2\sqrt{2}) + i\cdot(2\sqrt{2})) = 1. \tag{17}$$

We now must evaluate $W(\chi'_v)$. The conductor of χ'_v is $f' = (\sqrt{2})^4\mathcal{O}_{K,v}$. Hence we choose $c' = (\sqrt{2})^7 = 8\sqrt{2}$ in order to have $c'\mathcal{O}_{K,v} = f'\cdot\mathcal{D}_{K_v}$. The formula for $W(\chi'_v)$ is

$$W(\chi'_v) = 2^{-4/2}\tau(\overline{\chi}'_v) = 2^{-2}\sum_u \overline{\chi}'_v\left(\frac{u}{c'}\right)\psi_{K_v}\left(\frac{u}{c'}\right) \tag{18}$$

where the sum is over a set of representatives u for $\mathcal{O}^*_{K,v}$ modulo $1+f' = 1+(\sqrt{2})^4\mathcal{O}_{K,v}$. One has $\overline{\chi'}_v(u/c') = \overline{\chi'}_v(u)\overline{\chi'}_v(c')^{-1} = \overline{\chi'}_v(u)\chi'_v(c')$. By our construction of χ'_v, $\chi'_v(u) = \chi'_v(-u)$ for all local units u. By Lemma 3.2, $S = \{1, \epsilon, \epsilon^2, \epsilon^3\}$ is a set of representatives for $(\mathcal{O}^*_{K,v}/(1+f'))/\{\pm 1\}$. Thus we can rewrite (18) as

$$W(\chi'_v) = 2^{-2}\chi'_v(c')\sum_{s\in S}\chi'_v(s)\left(\psi_{K_v}\left(\frac{s}{c'}\right) + \psi_{K_v}\left(\frac{-s}{c'}\right)\right). \tag{19}$$

For $u \in \mathcal{O}^*_{K,v} = (\mathbf{Z}_2[\sqrt{2}])^*$, $Tr_{K_v/\mathbf{Q}_2}(\frac{u}{c'}) = Tr_{K_v/\mathbf{Q}_2}(\frac{u}{8\sqrt{2}})$ is in $2^{-2}\mathbf{Z}_2$ so that $\psi_{K_v}(\frac{u}{c'})$ is a fourth root of unity. Fix $\zeta' = exp(2\pi i/4)$. By (4), we have the following table:

Table B: Traces and character values for χ' and $v = v_Q$.

$s \in S$	$\overline{\chi}'_v(s) = \chi'_Q(r_v(s))$	$t = \psi_{K_v}(\frac{s}{c'})$	$t + t^{-1}$
1	1	1	2
$\epsilon = 1 + \sqrt{2}$	i^{-1}	ζ'	0
$\epsilon^2 = 1 + 2 + 2\sqrt{2}$	-1	$\zeta'^2 = -1$	-2
$\epsilon^3 = 7 + 5\sqrt{2}$	i	ζ'	0

We have $\chi'_v(c') = \chi'_v((\sqrt{2})^7) = (-1)^7 = -1$ by (6). Therefore, on using Table B to evaluate (19) we obtain

$$W(\chi'_{v_Q}) = 2^{-2} \cdot (-1) \cdot (1 \cdot 2 + i^{-1} \cdot 0 + (-1) \cdot -2 + i \cdot 0) = -1. \tag{20}$$

Substituting (17) and (20) into (13) yields

$$W_K(\chi) = -W(\chi_{v_Q})/W(\chi'_{v_Q}) = 1,$$

which completes the proof of Theorem 3.1.

4 The 2-Primary Euler Characterstic

We have the quaternion group of order eight,

$$Q_8 = \{x, y \mid x^2 = y^2,\ y^4 = 1,\ xyx^{-1} = y^{-1}\}.$$

If $a \in \mathbf{Z}$ is an integer such that $a^2 \equiv 1$ (*modulo* 2^n) write $(\mathbf{Z}/2^n)_a$ for the $\mathbf{Z}[Q_8]$-module given by the cyclic group of order 2^n on which x acts by multiplication by a and y acts trivially.

If $N/\mathbf{Q}$ is one of the quaternion extensions of Theorem 2.8 with x and y as in 2.9 then

$$K_3^{ind}(N) \cong (\mathbf{Q}/\mathbf{Z})(2)^{\Omega_N} \cong \mathbf{Z}/16$$

and the module structure is that of $(\mathbf{Z}/16)_9$, since x acts by multiplication by 13^2 on $(\mathbf{Q}/\mathbf{Z})(2)$, and $K_3^{ind}(N) \cong K_3^{ind}(\mathbf{Q}(\sqrt{2}))$ which is fixed by y. If S consists of the primes P_2 and P_p of N over 2 and p respectively, the K-theory localisation sequence yields a short exact sequence of the form

$$0 \longrightarrow K_2(\mathcal{O}_N) \longrightarrow K_2(\mathcal{O}_{N,S}) \longrightarrow (\mathcal{O}_N/P_2)^* \oplus (\mathcal{O}_N/P_p)^* \longrightarrow 0.$$

Since the residue fields at P_2 and P_p are of order 2 and p^2 respectively and $p \equiv 3$ (*modulo* 8) the 2-primary torsion of $K_2(\mathcal{O}_{N,S})$ is isomorphic to $(\oplus_1^8 \mathbf{Z}/2) \oplus \mathbf{Z}/8$, by Proposition 2.6. The 2-primary torsion in $K_2(\mathcal{O}_{N,S})'$ is cyclic of order eight and the module structure is that of $(\mathbf{Z}/8)_3$ since the image of $1 + \sqrt{2}$ is of order eight in $(\mathcal{O}_N/P_p)^*$.

Therefore the 2-primary part of the 2-extension defining $\Omega_1(N/\mathbf{Q}, 3)$ has the form

$$0 \longrightarrow (\mathbf{Z}/16)_9 \longrightarrow A' \longrightarrow B' \longrightarrow (\mathbf{Z}/8)_3 \longrightarrow 0$$

where A' and B' are finite, cohomologically trivial $\mathbf{Z}[Q_8]$-modules.

In this section we will develop a yoga for evaluating the Euler characteristic

$$[A'] - [B'] \in \mathcal{CL}(\mathbf{Z}[Q_8]) \cong (\mathbf{Z}/4)^*.$$

Define free $\mathbf{Z}[Q_8]$-modules, $\tilde{D}_0$ and $\tilde{D}_1$, by

$$\tilde{D}_0 = \mathbf{Z}[Q_8] < a_0 >, \quad \tilde{D}_1 = \mathbf{Z}[Q_8] < b_0, b'_0, b''_0 > .$$

We have the start of a free resolution of the form

$$\ldots \longrightarrow \tilde{D}_2 \longrightarrow \tilde{D}_1 \stackrel{\tilde{d}}{\longrightarrow} \tilde{D}_0 \stackrel{\nu}{\longrightarrow} (\mathbf{Z}/8)_3 \longrightarrow 0$$

where $\nu(1) \equiv 1$ (*modulo* 8), $\tilde{d}(b_0) = (x-3)a_0$, $\tilde{d}(b_0') = (y-1)a_0$ and $\tilde{d}(b_0'') = 8a_0$. The class of the 2-primary 2-extension is represented by a 2-cocycle, $j : \tilde{D}_2 \longrightarrow (\mathbf{Z}/16)_9$, which factorises to give a $\mathbf{Z}[Q_8]$-module homomorphism, F, well-defined up to the addition of maps which extend over $\tilde{D}_1$ and fitting into a commutative diagram of the form

$$\begin{array}{ccccccc}
Ker(\tilde{d}) & \longrightarrow & \tilde{D}_1 & \stackrel{\tilde{d}}{\longrightarrow} & \tilde{D}_0 & \stackrel{\nu}{\longrightarrow} & (\mathbf{Z}/8)_3 \\
\downarrow F & & \downarrow & & \downarrow & & \downarrow 1 \\
(\mathbf{Z}/16)_9 & \longrightarrow & A' & \longrightarrow & B' & \longrightarrow & (\mathbf{Z}/8)_3
\end{array}$$

Lemma 4.1 *Any choice of* $\mathbf{Z}[Q_8]$*-module homomorphism,* $F : Ker(\tilde{d}) \longrightarrow (\mathbf{Z}/16)_9$*, is surjective and* $P = Ker(F)$ *is a finitely generated, projective* $\mathbf{Z}[Q_8]$*-module such that*

$$[A'] - [B'] = [P] \in \mathcal{CL}(\mathbf{Z}[Q_8]) \cong (\mathbf{Z}/4)^*.$$

Proof The result is easily proved (c.f. [**3**]) once we have shown that F is surjective. To see this let

$$\ldots \stackrel{d_2}{\longrightarrow} D_2 \stackrel{d_1}{\longrightarrow} D_1 \stackrel{d_0}{\longrightarrow} D_0 \longrightarrow \mathbf{Z} \longrightarrow 0$$

be a free $\mathbf{Z}[Q_8]$-module resolution of the trivial module, $\mathbf{Z}$. Let $z \in H^4(Q_8; (\mathbf{Z}/8)_3) \cong \mathbf{Z}/2$ be a generator. The cup-product with the class of the 2-primary 2-extension

$$H^4(Q_8; (\mathbf{Z}/8)_3) \longrightarrow H^6(Q_8; (\mathbf{Z}/16)_9)$$

is an isomorphism. We may represent z by a cocycle of the form $D_4 \longrightarrow Ker(d_2) \stackrel{\phi}{\longrightarrow} (\mathbf{Z}/8)_3$. Starting with ϕ we may construct a chain map of the free resolution

$$\ldots \stackrel{d_6}{\longrightarrow} D_6 \stackrel{d_5}{\longrightarrow} D_5 \stackrel{d_4}{\longrightarrow} D_4 \longrightarrow Ker(d_2) \longrightarrow 0$$

into the resolution of $(\mathbf{Z}/8)_3$ which results in a homomorphism, $\delta(z) : D_6 \longrightarrow \tilde{D}_2 \longrightarrow Ker(\tilde{d})$. The image of z under the cup-product isomorphism is represented by the cocycle $F\delta(z) : D_6 \longrightarrow (\mathbf{Z}/16)_9$. Therefore f must be surjective if every cocycle representing the generator of $H^6(Q_8; (\mathbf{Z}/16)_9)$ is surjective.

From ([**15**], p. 19) the standard free $\mathbf{Z}[Q_8]$-resolution of $\mathbf{Z}$ has

$$D_7 = \mathbf{Z}[Q_8] < e >, \ D_6 = \mathbf{Z}[Q_8] < c, c' >, \ D_5 = \mathbf{Z}[Q_8] < b, b' >$$

with $d_5(c) = (1+x)b - (1+y)b'$, $d_5(c') = (xy+1)b + (x-1)b'$ and $d_6(e) = (x-1)c - (xy-1)c'$. From these formulae one finds easily that the generator of $H^6(Q_8; (\mathbf{Z}/16)_9)$ is represented by any homomorphism from D_6 to $(\mathbf{Z}/16)_9$ of the form $(c, c') \mapsto (2a+1, 2b+1)$, all of which are surjective.

4.2 In $\tilde{D}_1 = \mathbf{Z}[Q_8] < b_0, b_0'.b_0'' >$ we may define some elements by the following formulae:

$$
\begin{aligned}
A &= (x-3)b_0'' - 8b_0,\\
B &= (y-1)b_0'' - 8b_0',\\
C &= (1+y+y^2+y^3)b_0',\\
X_1 &= (1-y^3)b_0 + (3y^3 - xy^3 - xy - xy^2)b_0',\\
X_2 &= (y-y^3)b_0 - (3y+3y^2+xy+xy^2)b_0',\\
X_3 &= (y^2-y^3)b_0 - (3y^2+xy)b_0',\\
X_4 &= (x-xy^3)b_0 + (3xy^3+y^2)b_0',\\
X_5 &= (xy-xy^3)b_0 + (y+y^2-3xy-3xy^2)b_0',\\
X_6 &= (xy^2-xy^3)b_0 - (y^3+3xy^2)b_0',\\
X_7 &= (3y^3+xy^3)b_0 - (8y^3-3xy-3xy^2)b_0' + b_0''.
\end{aligned}
$$

The following results are straightforward to verify.

Lemma 4.3 *In terms of the elements of 4.2 a* $\mathbf{Z}$*-basis for* $Ker(\tilde{d})$ *is given by*

$$\{y^iA, 0 \le i \le 3; y^jB, 0 \le j \le 2; x^kC, 0 \le k \le 1; X_l, 1 \le l \le 7\}.$$

Proposition 4.4 *The following relations hold in* $Ker(\tilde{d})$*:*

(i) $yX_3 = -X_1 - xC$,
(ii) $y^2X_3 = X_1 - X_2 - 3C$,
(iii) $(y^3+1)X_3 = X_2$,
(iv) $xX_3 = X_6$,
(v) $xyX_3 = -X_4$,
(vi) $xy^2X_3 = X_4 - X_5 - 3xC$,
(vii) $xy^3X_3 = X_5 - X_6 - C$,
(viii) $xX_1 = X_4 - C$,
(ix) $xX_2 = X_5 - C$,
(x) $(x-3)X_7 = A + 8X_1 + X_2 + 3C$,
(xi) $(y-1)X_7 = B + 3X_1 + X_6 + 3xC$.

We shall now proof our first evaluation of the Euler characteristic.

Theorem 4.5 *The only 2-extensions of the form*

$$0 \longrightarrow (\mathbf{Z}/16)_9 \longrightarrow A' \longrightarrow B' \longrightarrow (\mathbf{Z}/8)_3 \longrightarrow 0$$

having an Euler characteristic defined in the class group,

$$\mathcal{CL}(\mathbf{Z}[Q_8]) \cong (\mathbf{Z}/4)^*$$

are represented by surjective 2-cocycles, $F : Ker(\tilde{d}) \longrightarrow (\mathbf{Z}/16)_9$ of the following form in which $s, z \in \mathbf{Z}$.

Z	$F(Z)$ (*modulo* 16)
A	$-9(2s+1)$
B	$-6(2s+1)$
C	$(-3-4z)(2s+1)$
X_1	$(2+4z)(2s+1)$
X_2	$2(2s+1)$
X_3	$9(2s+1)$
X_4	$-(2s+1)$
X_5	$(-1-4z)(2s+1)$
X_6	$(2s+1)$
X_7	$-2z(2s+1)$

The Euler characteristic corresponding to such a cocycle is non-trivial if and only if $z \equiv 0, 1$ (modulo 4).

Proof Using Proposition 4.4 and the fact that F must be surjective, it is straightforward to show that F has the form described in the table. Of course, this does not guarantee that all values of s and z even give rise to a $\mathbf{Z}[Q_8]$-module homomorphism. With hindsight we shall see that they all do, but we do not need this fact since we are only concerned with the (non-empty) set of F's representing 2-extensions which have Euler characteristics in the class-group.

A $\mathbf{Z}$-basis for $\mathrm{Ker}(\tilde{d})$ is given by $y^i(A+9X_6)$ $(0 \leq i \leq 3)$, $y^j(B+6X_6)$ $(0 \leq j \leq 2)$, $X_7 - f(X_7)X_6$, $C+(3+4z)X_6$, $xC+(11+4z)X_6$, $X_1-(2+4z)X_6$, X_2-2X_6, X_3-9X_6, X_4+X_6, $X_5+(1+4z)X_6$ and X_6. All but the last element are in $P = ker(F : Ker(\tilde{d}) \longrightarrow (\mathbf{Z}/16)_9)$ so that a $\mathbf{Z}$-basis for P is given by $y^i(A+9X_6)$ $(0 \leq i \leq 3)$, $y^j(B+6X_6)$ $(0 \leq j \leq 2)$, $X_7 - f(X_7)X_6$, $C+(3+4z)X_6$, $xC+(11+4z)X_6$, $X_1-(2+4z)X_6$, X_2-2X_6, X_3-9X_6, X_4+X_6, $X_5+(1+4z)X_6$ and $16X_6$.

Now consider the surjective map, $\pi : P \longrightarrow \mathbf{Z}[Q_8] < b_0'' >$, given by projecting each element to its b_0''-coordinate. The kernel of this map, $Q = ker(\pi)$, has a $\mathbf{Z}$-basis given by $C+(3+4z)X_6$, $xC+(11+4z)X_6$, $X_1-(2+4z)X_6$, X_2-2X_6, X_3-9X_6, X_4+X_6, $X_5+(1+4z)X_6$ and $16X_6$. If P is projective then Q is projective of rank one.

Let $Q_{\pm} = \{z \in Q | x^2(z) = \pm z\}$.

We are going to manufacture a projective submodule, S, of Q such that Q/S is a finite group of odd order and for which we can evaluate the class of S in the classgroup.

If Q is projective then $Q_+ = (1+x^2)Q$ is a free $\mathbf{Z}[Q_8^{ab}]$-module and a basis element may be found by the following process. Let $\sigma = (1+x+x^2+x^3)(1+y)$, $\lambda = (1+x+x^2+x^3)(1-y)$, $\mu = (1+y+y^2+y^3)(1-x)$ and $\rho = (1+xy+(xy)^2+(xy)^3)(1-y)$ in $\mathbf{Z}[Q_8]$. A basis element for Q_+ will be given by any $(1+x^2)z$ such that σz, λz, μz and ρz are generators for σQ, λQ, μQ and ρQ, respectively. To evaluate σz as z varies through the generators we merely set $x = 1 = y$ in the formulae. Accordingly we obtain the following table.

Z	n $(\sigma Z = n\sigma 8b_0')$
$C + (3+4z)X_6$	$-1-2z$
$xC + (11+4z)X_6$	$-5-2z$
$X_1 - (2+4z)X_6$	$1+2z$
$X_2 - 2X_6$	0
$X_3 - 9X_6$	4
$X_4 + X_6$	0
$X_5 + (1+4z)X_6$	$-1-2z$
$16X_6$	-8

To evaluate λz we set $x = 1$, $y = -1$ in the formulae.

Z	n $(\lambda Z = n\lambda(2b_0 - 2b_0'))$
$C + (3+4z)X_6$	$3+4z$
$xC + (11+4z)X_6$	$11+4z$
$X_1 - (2+4z)X_6$	$-1-4z$
$X_2 - 2X_6$	-2
$X_3 - 9X_6$	-8
$X_4 + X_6$	2
$X_5 + (1+4z)X_6$	$1+4z$
$16X_6$	16

Similarly, to evaluate μz we set $x = -1$, $y = 1$ in the formulae.

Z	n $(\mu Z = n\mu 2b_0')$
$C + (3+4z)X_6$	$5+4z$
$xC + (11+4z)X_6$	$9+4z$
$X_1 - (2+4z)X_6$	$1-4z$
$X_2 - 2X_6$	-4
$X_3 - 9X_6$	-10
$X_4 + X_6$	0
$X_5 + (1+4z)X_6$	$5+4z$
$16X_6$	16

Finally, setting $x = -1 = y$, we obtain the following table for ρz.

Z	n $(\rho Z = n\rho(4b_0' - 2b_0))$
$C + (3+4z)X_6$	$3+4z$
$xC + (11+4z)X_6$	$11+4z$
$X_1 - (2+4z)X_6$	$-3-4z$
$X_2 - 2X_6$	-2
$X_3 - 9X_6$	-10
$X_4 + X_6$	2
$X_5 + (1+4z)X_6$	$1+4z$
$16X_6$	16

From these tables we see that $(1+x^2)(C+(3+4z)X_6)$ generates a free $\mathbf{Z}[Q_8^{ab}]$-submodule of Q_+ with quotient isomorphic to

$$R = \mathbf{Z}/(1+2z) \times \mathbf{Z}/(3+4z) \times \mathbf{Z}/(5+4z) \times \mathbf{Z}/(3+4z).$$

This $\mathbf{Z}[Q_8]$-module is cohomologically trivial and, since Q_+ is Q/Q_-, we may manufacture an exact sequence of the form

$$0 \longrightarrow S \longrightarrow Q \longrightarrow R \longrightarrow 0$$

in which S is projective and $S_+ = \mathbf{Z}[Q_8^{ab}] < (1+x^2)(C+(3+4z)X_6) >$ and $S_- = Q_-$.

If $\chi_+ : (\mathbf{Z}/8)^* \longrightarrow \{\pm 1\}$ has kernel equal to $\{\pm 1\}$ then the class of R in $\mathcal{CL}(\mathbf{Z}[Q_8]) \cong \{\pm 1\}$ is equal to $\chi_+(|R|) = \chi_+((1+2z)(5+4z))$, by the formula of ([**14**] Theorem 5.2.23; [**3**], Section 4.3.6, p. 44; [**13**], Lemma 6.3.7, [**16**], p. 88) which is restated for the reader's convenience in 5.4. Hence $\chi_+(|R|) \in (\mathbf{Z}/4)^*$ depends only on z (modulo 4) and is non-trivial precisely when $z \equiv 0, 1$ (modulo 4).

According to ([**14**], Chapter 5, Section 2) the class of S is determined by comparing a basis element of S_+ with one for S_- in the following manner. If we write $U_i = (1-x^2)X_i$ then $S_- = Q_- = (1-x^2)Q$ and is generated over the integral quaternions, $\mathbf{H_Z}$, by $(3+4z)U_6$, $(11+8z)U_6$ (since $(1-x^2)C = 0$), $U_1 - (2+4z)U_6$, $U_2 - 2U_6$, $U_3 - 9U_6$, $U_4 + U_6$, $U_5 + (1+4z)U_6$ and $16U_6$. However, by 4.4(v) and (vii), $U_5 = U_4 + U_6$ and 4.4 easily implies that U_6 is a basis element for S_- over $\mathbf{H_Z}$. To compare bases we take the basis of S_+ and lift it from $S/S_- \cong S_+$ to $C + (3+4z)X_6 \in S$. Next we multiply this lifting by $(1-x^2)$ to give

$$(1-x^2)(C+(3+4z)X_6) \in S_-.$$

This is an integral quaternion times the basis element, U_6. In fact, it is $(3+4z)U_6$. The class of S in $\mathcal{CL}(\mathbf{Z}[Q_8]) \cong (\mathbf{Z}/4)^*$ is equal to the reduced norm of the quaternionic multiplier – in this case $(3+4z)^2$ – which is trivial in $(\mathbf{Z}/4)^*$. This completes the proof since the exact sequence defining S implies the relation $[Q] = [R] + [S] \in \mathcal{CL}(\mathbf{Z}[Q_8])$.

The following result gives a yoga whereby to extract the Euler characteristic directly from the 2-extension sequence in Theorem 4.5.

Corollary 4.6 *Suppose that we have a 2-extension of $\mathbf{Z}[Q_8]$-modules of the form*

$$(\mathbf{Z}/16)_9 \longrightarrow A' \longrightarrow B' \longrightarrow (\mathbf{Z}/8)_3$$

which possesses an Euler characteristic, as in Theorem 4.5.

Choose any element, $\gamma \in B'$, which maps to a generator of $(\mathbf{Z}/8)_3$. Hence there exist elements $b, b' \in A'$ mapping to $(x-3)\gamma$ and $(y-1)\gamma$ respectively. Form

$$x_5 = (xy - xy^3)b + (y + y^2 - 3xy - 3xy^2)b' \in A'$$

and

$$x_6 = (xy^2 - xy^3)b - (y^3 + 3xy^2)b' \in A'.$$

In fact x_5, x_6 lie in $(\mathbf{Z}/16)_9$ and are units (modulo 16). Furthermore, there exists an integer, z, such that

$$x_5 = x_6(-1-4z) \in (\mathbf{Z}/16)_9.$$

The value of z (modulo 4) is independent of the choices and depends only on the equivalence class of the 2-extension. Euler characteristic, $[A'] - [B']$ of Theorem 4.5, is non-trivial if and only if $z \equiv 0, 1$ (modulo 4).

Proof It is easy to see that x_5 and x_6 lie in $(\mathbf{Z}/16)_9$. From the discussion in Section 3, of the homomorphism, $F : Ker(\tilde{d}) \longrightarrow (\mathbf{Z}/16)_9$, we see that we may perform the yoga in the resolution for $(\mathbf{Z}/8)_3$ to obtain $X_5, X_6 \in Ker(\tilde{d})$ and set $x_i = F(X_i)$ for $i = 5, 6$. These choices are units (modulo 16) by Theorem 4.5, from which it is easily seen that the same is true for all choices. Similarly, for these elements $x_5 \equiv x_6(-1 - 4z)$ (modulo 16). Therefore the result follows from the straightforward verification that all choices yield the same ratio, $x_5/x_6 \in (\mathbf{Z}/16)^*$.

5 Evaluation of $\Omega_1(N/\mathbf{Q}, 3)$

Let $p \equiv 3$ (modulo 8) be a prime and let $N/\mathbf{Q}$ be the associated Q_8-extension constructed in Section 2. Let z and ω be as in 2.9. Therefore ω satisfies $\omega^8 = z/c(z)$ and we have a Galois extension, $L = N(\xi_{64}, \omega)$ over $\mathbf{Q}$, with $G(L/\mathbf{Q}) =< X, Y, \sigma, c >$. As explained in Section 3, the 2-Sylow subgroup of $K_2(\mathcal{O}_{N,S})'$ is cyclic of order eight. It is a subgroup of $K_2(N)'$ which is in turn a subgroup of

$$H^1(N; (\mathbf{Q}/\mathbf{Z})(2)) \cong K_2(N).$$

In order to apply the 2-primary yoga of Corollary 4.6 we need to describe an element of order eight in $H^1(N; (\mathbf{Q}/\mathbf{Z})(2))$ which corresponds to a generator of the 2-Sylow subgroup of $K_2(\mathcal{O}_{N,S})'$. Observe that the absolute Galois group, Ω_N, is generated by $\Omega_{N(i)}$ and c, so that it acts trivially on $\mu_8 \otimes \mu_8 = \mu_8(2)$ inside $(\mathbf{Q}/\mathbf{Z})(2)$. Therefore the homomorphism of 2.11,

$$f : G(L/N(i)) \longrightarrow \mathbf{Z}/8$$

given by the formula $\xi_8^{f(g)} = g(\omega)/\omega$, extends to $G(L/N) \cong G(L/N(i)) \times \mathbf{Z}/2 < c >$ by setting $f(c) = 0$ and defines a cohomology class

$$[f] \in H^1(N; \mu_8(2)).$$

Proposition 5.1 *The image of $[f]$ in $H^1(N; (\mathbf{Q}/\mathbf{Z})(2))$ corresponds under the isomorphism with $K_2(N)$ to a generator of the 2-Sylow subgroup of $K_2(\mathcal{O}_{N,S})'$.*

Proof The Steinberg symbol, $\{z, \xi_8\} \in K_2(N(i))$ has order eight because it's tame symbol (the image under the boundary map in the localisation sequence) is zero except in multiplicative group of the residue field corresponding to z, $(\mathcal{O}_{N(i)}/(z))^*$. This residue field has order p^2 where the tame symbol is equal to ξ_8. The transfer, $Tr_{N(i)/N}(\{z, \xi_8\}) \in K_2(N)$ also has order eight since its image in $K_2(N(i))$ is $(1 + c)(\{z, \xi_8\})$ whose tame symbol is concentrated in $(\mathcal{O}_{N(i)}/(z))^* \oplus (\mathcal{O}_{N(i)}/(z))^*$ and is given there by $(\xi_8, c(\xi_8))$. On the other hand, since transfer commutes with the map to the sum of the K_2's of the infinite places and since K_2 of the complex numbers is trivial, the element $Tr_{N(i)/N}(\{z, \xi_8\})$ lies in $K_2(N)'$. Therefore this element is a generator of the 2-Sylow subgroup of $K_2(\mathcal{O}_{N,S})'$.

Now let us consider the cohomological side.

Let F be a field of characteristic different from 2 and let F^s denote a separable closure of F. Following Tate, consider the exact sequence

$$0 \longrightarrow \mu_{2^n} \longrightarrow (F^s)^* \xrightarrow{2^n} (F^s)^* \longrightarrow 0$$

and the resulting coboundary homomorphism

$$d_F : H^0(F; (F^s)^*) \cong F^* \longrightarrow H^1(F; \mu_{2^n}).$$

This coboundary is defined by sending $y \in F^*$ to the 1-cocycle, $d_F : \Omega_F = G(F^s/F) \longrightarrow \mu_{2^n}$ given by $d_F(y)(g) = g(w)/w$ where $w \in (F^s)^*$ satisfies $w^{2^n} = y$. Therefore, if $n = 3$ and $F = N(i)$, we have $d_F(y)(g) = g(w)/w$ where $w^8 = y$.

Tate constructed a map of the form

$$H^1(F; \mathbf{Q}_2/\mathbf{Z}_2(2)) \longrightarrow K_2(F)$$

which is an isomorphism onto the 2-primary torsion subgroup. This map is characterised in the following manner. There is a map

$$h : K_2(F) \longrightarrow H^2(F; \mathbf{Z}_2(2))$$

defined on Steinberg symbols by $h(\{a,b\}) = d_F(a) \cup d_F(b)$. The other map is characterised by the fact that the composition

$$H^1(F; (\mathbf{Q}_2/\mathbf{Z}_2)(2)) \longrightarrow K_2(F) \xrightarrow{h} H^2(F; \mathbf{Z}_2(2))$$

is equal to the coboundary, δ, associated to the exact sequence

$$0 \longrightarrow \mathbf{Z}_2(2) \longrightarrow \mathbf{Q}_2(2) \longrightarrow (\mathbf{Q}_2/\mathbf{Z}_2)(2) \longrightarrow 0.$$

Here $\mathbf{Z}_2(2) = \mathbf{Z}_2 \otimes \mathbf{Z}_2$ and $\mathbf{Z}_2$ is the Galois module given by the inverse limit of the groups of two-primary roots of unity in F^s.

From the following commutative diagram, with $F = N(i)$,

$$\begin{array}{ccc}
H^1(F; \mathbf{Z}_2(1)) \otimes H^0(F; \mu_8) & \longrightarrow & H^1(F; (\mathbf{Q}_2/\mathbf{Z}_2)(2)) \\
\downarrow 1 \otimes \delta & & \downarrow \delta \\
H^1(F; \mathbf{Z}_2(1)) \otimes H^1(F; \mathbf{Z}_2(1)) & \longrightarrow & H^2(F; \mathbf{Z}_2(2))
\end{array}$$

we see that the Steinberg symbol, $\{z, \xi_8\}$, corresponds in $H^1(N(i); \mathbf{Q}_2/\mathbf{Z}_2(2))$ to the cohomology class represented by the 1-cocycle

$$t : \Omega_{N(i)} \longrightarrow \mu_8(2)$$

given by $t(g) = (g(w)/w) \otimes \xi_8$ where $w^8 = z$. The cohomological transfer and the K_2 transfer homomorphisms coincide under Tate's isomorphism. Therefore the cohomology class we want is given by

$$Tr_{N(i)/N}([t]) \in H^1(N; \mu_8(2)).$$

The transfer in group cohomology for a subgroup of index two is described explicitly in ([**15**], Chapter 4) from which one finds that a 1-cocycle representing $Tr_{N(i)/N}([t])$ is given by the following formula, for $g \in \Omega_{N(\sqrt{-1})}$,

$$\begin{aligned}
Tr_{N(i)/N}([t])(g) &= c(c^{-1}gc(w)/w) \otimes c(\xi_8) \text{“}+\text{''}\ (g(w)/w) \otimes \xi_8 \\
&= Tr_{N(i)/N}([t])(gc) \\
&= c(c^{-1}gc(w)/w) \otimes c(\xi_8) \text{“}+\text{''}\ (gc^2(w)/w) \otimes \xi_8
\end{aligned}$$

where “$+$'' refers to “addition” in $\mu_8 \otimes \mu_8$. However, in $\mu_8 \otimes \mu_8$,

$$c(c^{-1}gc(w)/w) \otimes c(\xi_8) = (gc(w)/c(w)) \otimes \xi_8^{-1} = (c(w)/gc(w)) \otimes \xi_8$$

and

$$(c(w)/gc(w)) \otimes \xi_8 \text{“}+\text{''}\ (g(w)/w) \otimes \xi_8 =$$

$$((g(w/c(w)))/(w/c(w))) \otimes \xi_8 = (g(\omega)/\omega) \otimes \xi_8$$

where $\omega^8 = z/c(z)$, as required.

5.2 We have to apply the 2-primary yoga of 4.6 to the cohomology sequence

$$(\mathbf{Q}_2/\mathbf{Z}_2)(2)^{\Omega_N} \longrightarrow Ind_{<c>}^{\Omega_{\mathbf{Q}}}((\mathbf{Q}_2/\mathbf{Z}_2)(2))^{\Omega_N} \longrightarrow$$

$$(\mathbf{Q}_2/\mathbf{Z}_2)(2)_+^{\Omega_N} \longrightarrow H^1(N;(\mathbf{Q}_2/\mathbf{Z}_2)(2)) \longrightarrow \ldots .$$

Recall that, by 2.9, we have a Galois extension, $L/\mathbf{Q}$, containing $N/\mathbf{Q}$ with group $G(L/\mathbf{Q}) \cong< X, Y, \sigma, c > \cong < X, Y, \sigma > \times < c >$. Define $E = L(\xi_{128})$ then $E/\mathbf{Q}$ is also Galois with $G(E/\mathbf{Q}) \cong G(E/\mathbf{Q}(i)) \times < c >$ and $G(E/\mathbf{Q}(i)) \cong < X, Y, \sigma >$ where the action of X, Y, σ is given by the same formulae as for $L/\mathbf{Q}$. In particular, Y and σ fix ξ_{128} and $X(\xi_{128}) = \xi_{128}^{13}$. We shall perform the yoga in the cohomology sequence obtained by replacing $\Omega_{\mathbf{Q}}$, Ω_N and $(\mathbf{Q}_2/\mathbf{Z}_2)(2)$ by $G(E/\mathbf{Q})$, $G(E/N)$ and $\mu_{128}(2)$, respectively. This is permissible since the generator of order eight in $K_2(\mathcal{O}_{N,S})'$ originates in $H^1(N;\mu_8(2))$ and is represented by the homomorphism

$$G(L/N) \cong G(L/N(i)) \times < c > \longrightarrow G(L/N(i)) \stackrel{f}{\longrightarrow} \mu_8 \otimes \mu_8 = \mu_8(2).$$

It will be convenient to write $\mu_{128}(2)$ additively as $\mathbf{Z}/128$ with X, Y, σ and c acting by multipilcation by $169, 1, 1$ and 1, respectively. From the short exact sequence

$$0 \longrightarrow \mathbf{Z}/128 \longrightarrow Ind_{<c>}^{G(E/\mathbf{Q})}(\mathbf{Z}/128) \longrightarrow (\mathbf{Z}/128)_+ \longrightarrow 0$$

we form the cohomology sequence

$$(\mathbf{Z}/128)^{G(E/N)} \stackrel{\phi}{\longrightarrow} Ind_{<c>}^{G(E/\mathbf{Q})}(\mathbf{Z}/128)^{G(E/N)} \longrightarrow$$

$$(\mathbf{Z}/128)_+^{G(E/N)} \stackrel{\delta}{\longrightarrow} H^1(G(E/N);\mathbf{Z}/128) \longrightarrow \ldots .$$

Here $\phi(a) = \sum_{h \in G(E/\mathbf{Q}(i))} h \otimes h^{-1}(a)$ for $a \in \mathbf{Z}/128$.

The bar resolution ([**15**], Chapter 1) begins,

$$B_1G(E/N) = \mathbf{Z}[G(E/N)] < [g] \mid g \in G(E/N) > \stackrel{d}{\longrightarrow} \mathbf{Z}[G(E/N)] \longrightarrow \mathbf{Z} \longrightarrow 0.$$

The cocycle, f, corresponds to $f \in Hom_{\mathbf{Z}[G(E/N)]}(B_1G(E/N), \mathbf{Z}/128)$ given by $f([g]) = f(g)$. Suppose that $\gamma \in Ind_{<c>}^{G(E/\mathbf{Q})}(\mathbf{Z}/128)$ maps to $[\gamma] \in (\mathbf{Z}/128)_+^{G(E/N)}$. Then, by definition, $\delta([\gamma]) = [f]$ if $d^*(\gamma) = \phi_*(f)$ where

$$d^* : Ind_{<c>}^{G(E/\mathbf{Q})}(\mathbf{Z}/128) \longrightarrow Hom_{\mathbf{Z}[G(E/N)]}(B_1G(E/N), Ind_{<c>}^{G(E/\mathbf{Q})}(\mathbf{Z}/128))$$

is given by $d_*(\gamma)[g] = g(\gamma) - \gamma$ and $\phi_*(f) = \phi(f(g))$.

Define such a γ by the formula

$$\gamma = \sum_{v \in G(E/N(i))} \sum_{\epsilon=0,1} \sum_{0 \le i \le 3} vX^{\epsilon}Y^i \otimes f(v^{-1}).$$

Then, if $g \in G(E/N(i))$,

$$\begin{aligned} g(\gamma) - \gamma &= \sum_{v \in G(E/N(i))} \sum_{\epsilon=0,1} \sum_{0 \le i \le 3} gvX^{\epsilon}Y^{i} \otimes f(v^{-1}) - vX^{\epsilon}Y^{i} \otimes f(v^{-1}) \\ &= \sum_{v \in G(E/N(i))} \sum_{\epsilon=0,1} \sum_{0 \le i \le 3} vX^{\epsilon}Y^{i} \otimes f(v^{-1}g) - vX^{\epsilon}Y^{i} \otimes f(v^{-1}) \\ &= \sum_{v \in G(E/N(i))} \sum_{\epsilon=0,1} \sum_{0 \le i \le 3} gvX^{\epsilon}Y^{i} \otimes (f(v^{-1}) + f(g) - f(v^{-1}) \\ &= \sum_{v \in G(E/N(i))} \sum_{\epsilon=0,1} \sum_{0 \le i \le 3} gvX^{\epsilon}Y^{i} \otimes (gvX^{\epsilon}Y^{i})^{-1} f(g) \\ &= \phi(f(g)) \end{aligned}$$

since v, Y and X act trivially on $\mathbf{Z}/8 = \mu_8 \otimes \mu_8$. Similarly $gc(\gamma) - \gamma = \phi(f(g)) = \phi(f(gc))$. Hence $\delta([\gamma]) = [f]$.

If $g \in G(E/\mathbf{Q})$ the action on the 1-cocycle, f, is given by $g(f)(v) = g(f(g^{-1}vg))$ for $v \in G(E/N)$. From 2.12(iii) and (vi) one may show that $(X-3)f$ and $(Y-1)f$ are coboundaries. That is, there exist $\beta_x, \beta_y \in \mathbf{Z}/128$, whose identities will be determined when needed, such that

$$X(f)(v) - 3f(v) = v(\beta_x) - \beta_x$$

and

$$Y(f)(v) - f(v) = v(\beta_y) - \beta_y$$

for all $v \in G(E/N(i))$. Therefore, in $Ind_{<c>}^{G(E/\mathbf{Q})}(\mathbf{Z}/128)$,

$$\begin{aligned} X(\gamma) &= \sum_{v,\epsilon,i} XvX^{\epsilon}Y^{i} \otimes f(v^{-1}) \\ &= \sum_{v,\epsilon,i} vX^{\epsilon+1}Y^{i} \otimes f(X^{-1}v^{-1}X) \\ &= \sum_{v,\epsilon,i} vX^{\epsilon+1}Y^{i} \otimes (3f(v^{-1}) + v^{-1}(\beta_x) - \beta_x). \end{aligned}$$

Hence

$$
\begin{aligned}
(X-3)\gamma = & \sum_{v,\epsilon,i} vX^{\epsilon+1}Y^i \otimes (v^{-1}(\beta_x)-\beta_x) + 3\sum_{v,\epsilon,i} v(X^{\epsilon+1}Y^i - X^{\epsilon}Y^i)\otimes f(v^{-1}) \\
= & \sum_{v,\epsilon,i} vX^{\epsilon+1}Y^i \otimes (169^{\epsilon+1}(vX^{\epsilon+1}Y^i)^{-1}\beta_x - \beta_x) \\
& +3\sum_{v} v(X^2-1)(1+Y+Y^2+Y^3)\otimes f(v^{-1}) \\
= & \sum_{v,\epsilon,i} vX^{\epsilon+1}Y^i \otimes ((169^{\epsilon+1}-1)(vX^{\epsilon+1}Y^i)^{-1}\beta_x + (vX^{\epsilon+1}Y^i)^{-1}\beta_x - \beta_x) \\
& +3\sum_{v}(vX^2Y^2 - v)\otimes f(v^{-1}) + 3\sum_{v}(vX^2Y^2Y - vY)\otimes f(v^{-1}) \\
& +3\sum_{v}(vX^2Y^{-2}Y^2 - vY^2)\otimes f(v^{-1}) \\
& +3\sum_{v}(vX^2Y^{-2}Y^3 - vY^3)\otimes f(v^{-1}) \\
= & \ \phi(\beta_x) + \sum_{v,\epsilon,i} vX^{\epsilon+1}Y^i \otimes ((169^{\epsilon+1}-1)(vX^{\epsilon+1}Y^i)^{-1}\beta_x - \beta_x) \\
& +3\sum_{v} v(1+Y)\otimes f(X^2Y^2) + 3\sum_{v} v(Y^2+Y^3)\otimes f(X^2Y^{-2}),
\end{aligned}
$$

since v acts trivially on $8(\mathbf{Z}/128) = \mathbf{Z}/16$. There, since $vX^{\epsilon+1}Y^i$ multiplies by $169^{\epsilon+1}$, $(X-3)[\gamma]$ is the image of the $G(E/N)$-invariant element

$$
\begin{aligned}
b = & -\sum_{v,\epsilon,i} vX^{\epsilon+1}Y^i \otimes 169^{-\epsilon-1}\beta_x \\
& +3\sum_{v} v(1+Y)\otimes f(X^2Y^2) + 3\sum_{v} v(Y^2+Y^3)\otimes f(X^2Y^{-2}).
\end{aligned}
$$

Similarly

$$
\begin{aligned}
Y(\gamma) = & \sum_{v,\epsilon,i} YvX^{\epsilon}Y^i \otimes f(v^{-1}) \\
= & \sum_{v,\epsilon,i} vYX^{\epsilon}Y^i \otimes f(Y^{-1}v^{-1}Y) \\
= & \sum_{v,\epsilon,i} vYX^{\epsilon}Y^i \otimes (f(v^{-1}) + v^{-1}(\beta_y) - \beta_y)
\end{aligned}
$$

and

$$(Y-1)\gamma = \sum_{v,\epsilon,i} YvX^{\epsilon}Y^{i} \otimes (v^{-1}(\beta_y) - \beta_y)$$

$$+\sum_{v,\epsilon,i} v(YX^{\epsilon}Y^{i} - X^{\epsilon}Y^{i}) \otimes f(v^{-1})$$

$$= \sum_{v,\epsilon,i} vYX^{\epsilon}Y^{i} \otimes (169^{\epsilon}(vYX^{\epsilon+1}Y^{i})^{-1}\beta_y - \beta_y)$$

$$+\sum_{v} v((Y^4-1) + (YX-X)(1+Y+Y^2+Y^3)) \otimes f(v^{-1})$$

$$= \phi(\beta_y) - \sum_{v,\epsilon,i} vYX^{\epsilon}Y^{i} \otimes 169^{-\epsilon}\beta_y$$

$$+\sum_{v} v \otimes f(Y^4) + \sum_{v} v(XY^2 + XY + X) \otimes f(YXYX^{-1})$$

$$+\sum_{v} vXY^3 \otimes f(YXY^{-3}X^{-1}).$$

Set

$$b' = -\sum_{v,\epsilon,i} vYX^{\epsilon}Y^{i} \otimes 169^{-\epsilon}\beta_y$$

$$+\sum_{v} v \otimes f(Y^4) + \sum_{v} v(XY^2 + XY + X) \otimes f(YXYX^{-1})$$

$$+\sum_{v} vXY^3 \otimes f(YXY^{-3}X^{-1}).$$

Next, bearing in mind that $a \in \mathbf{Z}/8$ corresponds to $16a \in \mathbf{Z}/128$, it is straightforward to verify the following formulae, using Propositions 2.11 and 2.12,

$$\begin{aligned}
f(Y^4) &\equiv 8\alpha(Y),\\
f(X^2Y^2) &\equiv 36\alpha(Y) + 24\alpha(X),\\
f(X^2Y^{-2}) &\equiv 28\alpha(Y) + 24\alpha(X),\\
f(YXY^{-3}X^{-1}) &\equiv -48\alpha(Y),\\
f(YXYX^{-1}) &\equiv -24\alpha(Y).
\end{aligned}$$

Also, from 2.12(vi), $\beta_y \equiv -3\alpha(Y)$, since X^2Y^2 acts like 169^2 and $169^2 - 1 \equiv 16$ (modulo 128). In order to evaluate

$$x_5 = (xy - xy^3)b + (y + y^2 - 3xy - 3xy^2)b' \in \mathbf{Z}/16$$

and

$$x_6 = (xy^2 - xy^3)b - (y^3 + 3xy^2)b' \in \mathbf{Z}/16$$

we first observe that, for $a \in \mathbf{Z}/16$,

$$\phi(a) = \sum_{v,i} vY^i \otimes a + \sum_{v,i} vXY^i \otimes 9a.$$

Therefore, in order to evaluate the inverse images under ϕ of x_5 and x_6 it suffices to compute the multiples of $\sum_v v \otimes 8$ which occur. Furthermore, the presence of the factor $\sum_v v$ means that we may manipulate the X's and Y's as if they were in the group-ring of Q_8. We find easily that the term in x_5 is $-(8/169)\alpha(Y)/2$ and in x_6 is $(8/169)\alpha(Y)/2$. Since these are 2-adic units we see that $\alpha(Y)/2$ is an odd integer and

$$x_5 = (-1)x_6 \in \mathbf{Z}/16.$$

This corresponds to $z \equiv 0$ (modulo 4) in Corollary 4.6 and we have established the following result.

Proposition 5.3 *Let $N/\mathbf{Q}$ be one of the quaternion extensions of Theorem 2.8 and let*

$$0 \longrightarrow (\mathbf{Z}/16)_9 \longrightarrow A' \longrightarrow B' \longrightarrow (\mathbf{Z}/8)_3 \longrightarrow 0$$

be a 2-extension representing the 2-primary part of the 2-extension defining $\Omega_1(N/\mathbf{Q}, 3)$. Then

$$[A'] - [B'] = -1 \in \{\pm 1\} \cong \mathcal{CL}(\mathbf{Z}[G(N/\mathbf{Q})]).$$

In order to evaluate the Euler characteristic of the odd primary part of the 2-extension defining $\Omega_1(N/\mathbf{Q}, 3)$ we shall use the following result.

Proposition 5.4 *([**14**], Theorem 5.2.23, [**3**], Section 4.3.6, p. 44; [**13**] Lemma 6.3.7, [**16**], p. 88)*

Let X be a finite group of odd order with a Q_8-action then

$$[X] \in \mathcal{CL}(\mathbf{Z}[Q_8]) \cong (\mathbf{Z}/4)^*$$

is given by

$$[X] = \chi_+(|X_+|)(\sqrt{|X_-|}) = (\sqrt{|X_-|})(-1)^{(1/4)log_2(|X_+|)} \in (\mathbf{Z}/4)^*$$

where $X_\pm = \{z \in X | x^2(z) = \pm z\}$ and $0 \leq (\sqrt{|X_-|})$ is the positive square root of the cardinality of X_-. Alternatively, if $z \in \mathbf{Z}$, $(-1)^{(1/4)log_2(|X_+|)} \in (\mathbf{Z}/4)^*$ is equal to $\chi_+(z)$ where $\chi_+ : (\mathbf{Z}/8)^* \longrightarrow (\mathbf{Z}/4)^*$ is the surjective homomorphism with kernel equal to $\{\pm 1\}$.

Proof of Theorem 1.1 The class, $\Omega_1(N/\mathbf{Q}, 3)$, is defined as the Euler characteristic of a 2-extension of the form

$$0 \longrightarrow K_3^{ind}(\mathcal{O}_{N,S}) \longrightarrow A \longrightarrow B \longrightarrow K_2(\mathcal{O}_{N,S})' \longrightarrow 0$$

where S is the set of primes of N over $\{2, p\}$ and A, B are finite groups. The Euler characteristic is the sum of that of the 2-primary part – non-trivial by Proposition 5.3 – and that of the odd primary part, which we shall denote by

$$0 \longrightarrow k_3^{ind}(\mathcal{O}_{N,S}) \longrightarrow A'' \longrightarrow B'' \longrightarrow k_2(\mathcal{O}_{N,S})' \longrightarrow 0.$$

Since all these groups are finite and of odd order we have

$$[A''] - [B''] = [k_3^{ind}(\mathcal{O}_{N,S})] - [k_2(\mathcal{O}_{N,S})'] \in \mathcal{CL}(\mathbf{Z}[Q_8]) \cong (\mathbf{Z}/4)^*.$$

Now we shall apply the analysis of ([**13**], Section 7.2) to calculate

$$[A''] - [B''] \in \mathcal{CL}(\mathbf{Z}[Q_8]) \cong (\mathbf{Z}/4)^*.$$

Let $\chi_1, \chi_2 : Q_8 \longrightarrow \{\pm 1\}$ be given by $\chi_1(x) = -1 = \chi_2(y)$, $\chi_1(y) = 1 = \chi_2(x)$. Then, if X is a finite Q_8-group of odd order and if X^λ denotes the λ-eigenspace of X

$$|X_+| = |X^1 \oplus X^{\chi_1} \oplus X^{\chi_2} \oplus X^{\chi_1\chi_2}| = |X^1||X^{\chi_1}||X^{\chi_2}||X^{\chi_1\chi_2}|$$

and the same is true for the formal difference of two such X's. Therefore, using Wile's proof of the odd primary part of the Birch-Tate conjecture [**17**] and arguing as in ([**13**], Section 7.2),

$$|\{k_2(\mathcal{O}_{N,S})' - k_3^{ind}(\mathcal{O}_{N,S})\}_+| = \frac{|k_2(\mathcal{O}_{N,S})'_+|}{|k_3^{ind}(\mathcal{O}_{N,S})_+|}$$

$$= \prod_{\lambda=1,\chi_1,\chi_2,\chi_1\chi_2} \frac{|(k_2(\mathcal{O}_{L,S})')^\lambda|}{|k_3^{ind}(\mathcal{O}_{L,S})^\lambda|}$$

$$= \pm \tilde{L}_{\mathbf{Q},\Sigma}(-1, Ind_{<x^2>}^{Q_8}(1)),$$

where $\Sigma = \{2, p\}$ and $\tilde{L}$ is as in Theorem 1.1. The ± 1-sign is irrelevant for our purposes, since $\chi_+(-1) = 1$. Therefore we find that

$$\chi_+(|\{k_2(\mathcal{O}_{N,S})' - k_3^{ind}(\mathcal{O}_{N,S})\}_+|) = \chi_+(\tilde{L}_{\mathbf{Q},\Sigma}(-1, Ind_{<x^2>}^{Q_8}(1))).$$

Next we must decide the *positive* square root of $|\{k_2(\mathcal{O}_{N,S})' - k_3^{ind}(\mathcal{O}_{N,S})\}_-|$. By the argument of ([**13**], Section 7.2), if χ is the non-trivial character on the centre of Q_8, then $X_- = X^\chi$ and

$$|\{k_2(\mathcal{O}_{N,S})' - k_3^{ind}(\mathcal{O}_{N,S})\}_-| = (\prod_{q\in\Sigma}(-1)^{dim(V_q)} det(F_q|V_q))\tilde{L}_{\mathbf{Q},\Sigma}(-1, V)$$

where $V = Ind_{<x^2>}^{Q_8}(\chi) = 2\nu$ and ν is the unique 2-dimensional irreducible representation of Q_8. Since this is a positive number, the sign must be positive and

$$|\{k_2(\mathcal{O}_{N,S})' - k_3^{ind}(\mathcal{O}_{N,S})\}_-| = \tilde{L}_{\mathbf{Q},\Sigma}(-1, \nu)^2,$$

whose positive square root is $W_{\mathbf{Q}}(\nu)\tilde{L}_{\mathbf{Q},\Sigma}(-1,\nu)$. Here $W_{\mathbf{Q}} : R(G(N/\mathbf{Q})) \longrightarrow \{\pm 1\}$ is the Artin root number.

Combining these facts with Theorem 3.1 and Propositions 5.3-5.4 completes the proof of Theorem 1.1.

Proof of Corollary 1.2 Consider the case of Theorem 1.1 when $p = 3$.

The extension, $N/\mathbf{Q}$, is totally ramified at 2 with decomposition and inertia groups equal to Q_8 and a trivial lifted Frobenius. At 3 the decomposition group is Q_8, the inertia group is $< y >$ and the lifted Frobenius is $x \in Q_8$. Hence, if $V = Ind_{<x^2>}^{Q_8}(1)$,

$$L_{\mathbf{Q},\Sigma}(-1, V) = L_{\mathbf{Q}}(-1, V) det(1 - 2(I|V^{Q_8})) det(1 - 3(x|V^{<y>}))$$

$$= \zeta_{\mathbf{Q}(\sqrt{2},\sqrt{3})}(-1)(-1)(1-9)$$

$$= 8\zeta_{\mathbf{Q}(\sqrt{2},\sqrt{3})}(-1)$$

and $\tilde{L}_{\mathbf{Q},\Sigma}(-1, V)$ is equal to the odd primary part of $\zeta_{\mathbf{Q}(\sqrt{2},\sqrt{3})}(-1)$. By the Birch-Tate conjecture, the absolute value of this is equal to the odd primary part of the ratio of the orders of K_2 and and K_3^{ind} of the integers in $\mathbf{Q}(\sqrt{2}, \sqrt{3})$. The odd torsion in these groups is each cyclic of order three. Hence

$$\Omega_1(N/\mathbf{Q}, 3) = -W_{\mathbf{Q}}(\nu)\tilde{L}_{\mathbf{Q},\Sigma}(-1, \nu) \in \mathcal{CL}(\mathbf{Z}[Q_8]) \cong (\mathbf{Z}/4)^*.$$

Similarly, since ν is irreducible with non-trivial central character,

$$L_{\mathbf{Q},\Sigma}(-1, \nu) = L_{\mathbf{Q}}(-1, \nu)det(1 - 2(I|\nu^{Q_8}))det(1 - 3(x|\nu^{<y>})) = L_{\mathbf{Q}}(-1, \nu)$$

and also, by [**8**],

$$2^4 \cdot 71^2 = \zeta_N(-1) = L_{\mathbf{Q}}(-1, V)L_{\mathbf{Q}}(-1, \nu)^2$$

so that $\tilde{L}_{\mathbf{Q}}(-1, \nu) = \pm 71$ and the sign is equal to $W_{\mathbf{Q}}(\nu) = 1$, by Theorem 3.1, so that $\tilde{L}_{\mathbf{Q},\Sigma}(-1, \nu) \equiv -1$ (modulo 4), as required.

When $p = 11$ [**8**] tells us that

$$L_{\mathbf{Q}}(-1, V) = \zeta_{\mathbf{Q}(\sqrt{2}, \sqrt{11})}(-1) = 3^{-1} \cdot 7 \cdot 23$$

and

$$L_{\mathbf{Q}}(-1, V)L_{\mathbf{Q}}(-1, \nu)^2 = \zeta_N(-1) = 2^4 \cdot 3 \cdot 7 \cdot 11^2 \cdot 23 \cdot 109^2.$$

In the formula of Theorem 1.1 one finds that the $log_2(L_{\mathbf{Q},\Sigma}(-1, V)$-term is non-trivial while the $L_{\mathbf{Q},\Sigma}(-1, \nu)$-term is trivial and again $\Omega_1(N/\mathbf{Q}, 3)$ and $W_{\mathbf{Q}}(\nu)$ are both trivial, as required.

When $p = 19$ [**8**] tells us that

$$L_{\mathbf{Q}}(-1, V) = \zeta_{\mathbf{Q}(\sqrt{2}, \sqrt{11})}(-1) = 3^{-1} \cdot 19 \cdot 41$$

and

$$L_{\mathbf{Q}}(-1, V)L_{\mathbf{Q}}(-1, \nu)^2 = \zeta_N(-1) = 2^4 \cdot 3^3 \cdot 19 \cdot 41 \cdot 1993^2.$$

Again these values lead to the conclusion that $\Omega_1(N/\mathbf{Q}, 3)$ and $W_{\mathbf{Q}}(\nu)$ are both trivial.

References

[1] Berger, R. [1990], *Quadratic extensions of number fields with elementary abelian 2-Prim $K_2(\mathcal{O}_F)$ of smallest rank*, J. Number Theory **34 (3)**, 284-292.

[2] Cassou-Nogués, Ph., Chinburg, T., Fröhlich, A. and Taylor, M. J. [1991], *L-functions and Galois modules*, London Math. Soc. 1989 Durham Symposium, L-functions and Arithmetic, Cambridge University Press , 75-139.

[3] Chinburg, T. [1989], *The Analytic Theory of Multiplicative Galois Structure*, Mem. A. M. Soc. **395**.

[4] Chinburg, T. [1985], *Exact sequences and Galois module structure*, Annals Math. **121**, 351-376.

[5] Chinburg, T., Kolster, M., Pappas, G. and Snaith, V. P. [1995], *Galois structure of K-groups of rings of integers*, C. R. Acad. Sci. Paris, t. **320**, Série I, 1435-1440.

[6] Conner, P. E. and Hurrelbrink, J. [1988], *Class Number Parity*, World Scientific Series in Mathematics **8**.

[7] Fröhlich, A. [1972], *Artin root numbers and normal integral bases for quaternion fields*, Inventiones Math. **17**, 143-66.

[8] Hooper, J. and Kolster, M. [1995], *Some values at $s = -1$ of zeta functions of quaternion fields*, McMaster University preprint.

[9] Jensen, C. U. and Yui, N. [1987], *Quaternion extensions*, Conf. Alg. Geom. and Comm. Alg. in honour of M. Nagata, 155-182.

[10] Kahn, B. [1993], *Descente Galoisienne et K_2 des corps de nombres*, K-theory **7 (1)**, 55-100.

[11] Kolster, M. [1989], *A relation between the 2-primary parts of the Main Conjecture and the Birch-Tate Conjecture*, Can. Math. Bull. **32 (12)** 248-251.

[12] Martinet, J. [1977], *Character theory and Artin L-functions*, London Math. Soc. 1975 Durham Symposium, Algebraic Number Fields, Academic Press, 1-87.

[13] Snaith, V. P. [1994], *Galois Module Structure*; The Fields Institute Monograph Series Volume 2, A.M.S., Providence, R. I.

[14] Snaith, V. P. [1994], *Explicit Brauer Induction (with applications to algebra and number theory)*; Cambridge Studies in Advanced Math. **42**, Cambridge University Press.

[15] Snaith, V. P. [1989], *Topological Methods in Galois Representation Theory*, C. M. Soc. Monographs, Wiley.

[16] Taylor, M. J. [1984], *Class Groups of Group Rings*, L.M.Soc. Lecture Notes Series **91**, Cambridge University Press.

[17] Wiles, A. [1990], *The Iwasawa conjecture for totally real fields*, Annals of Math. **131**, 493-540.

Fields Institute Communications
Volume **16**, 1997

The Bloch-Ogus–Gabber Theorem

Jean-Louis Colliot-Thélène
Université Paris-Sud
CNRS, URA D0752
Mathématiques, Bâtiment 425
91405 Orsay Cedex France
colliot@math.u-psud.fr

Raymond T. Hoobler
City College of New York
New York, NY 10031 USA
rthcc@cunyvm.cuny.edu

Bruno Kahn
Institut de Mathématiques de Jussieu
Equipe Théories Géométriques
Université Paris 7, Case 7012
75251 Paris Cedex 05 France
kahn@math.jussieu.fr

To the memory of Robert W. Thomason

Abstract. The purpose of this paper is twofold. Firstly, we give an exposition of Gabber's proof of the Bloch-Ogus theorem for étale cohomology of a smooth variety X over a field k, with torsion coefficients defined over k. Secondly, we discuss the axioms used in that proof (étale excision and cohomology of the projective line), and apply them to other cohomology theories. We give several extensions of the theorem: the torsion hypothesis is irrelevant, the basic exact sequences for semi-local rings of X are universally exact in the sense of Grayson, and they remain exact after multiplying X by an arbitrary k-variety.

Introduction

Stemming directly from Quillen's proof of the Gersten conjecture in the geometric case [**43**], the Bloch-Ogus theorem is a fundamental result of modern algebraic geometry. Its simplest consequence is that, for a local ring A of a smooth algebraic variety with function field K, $h^*(A)$ injects into $h^*(K)$ for all "cohomology theories" h^* satisfying a list of natural axioms (étale cohomology with coefficients in roots of unity is such a theory).

1991 *Mathematics Subject Classification.* Primary 14F20; Secondary 19E08, 19E20, 14-02, 19-02.

The Bloch-Ogus theorem, briefly described, is as follows. Given a smooth algebraic variety X and a cohomology theory h^* as above, filtration by codimension of support yields *Cousin complexes* which form the E_1-terms of the *coniveau spectral sequence*, converging to $h^*(X)$ (compare [**24**]). Restriction of the Cousin complexes to the open subsets of X defines complexes of flasque Zariski sheaves. The Bloch-Ogus theorem says that these complexes of sheaves are *acyclic*, except in degree 0 where their cohomology is the Zariski sheaf $\mathcal{H}^*$ associated to $U \mapsto h^*(U)$. This identifies the E_2-term of the coniveau spectral sequence to $H^*(X_{Zar}, \mathcal{H}^*)$. By a standard argument, the proof reduces to proving an "effacement theorem" for the version of h^* with supports.

Bloch and Ogus' proof of the effacement theorem [**2**] uses a geometric presentation lemma devised by Quillen for his proof of the analogous Gersten conjecture for algebraic K-theory (a strengthening of Noether's normalization theorem), see [**43**, Lemma 7.5.12]. This step reduces the proof to a "split" case. However, Quillen's simple K-theoretic argument in this split case has no counterpart in the cohomology setting, and is replaced in [**2**] by a much longer argument, involving a delicate diagram chase (*op. cit.*, Section 5).

At the beginning of the eighties, Gabber provided a different proof of the effacement theorem for étale cohomology, which has now appeared in print as [**14**] (he considers there a more general situation of vanishing cycles). Gabber uses another, more powerful variant of the Noether normalization theorem, which reduces one to a situation where the ambiant space is the affine line over some base and the relevant closed subset is finite over the base. In this special case, Gabber's argument to prove effacement is different from that of Bloch-Ogus. It essentially uses the section at infinity coming from an embedding of the affine line into the projective line, as well as a computation of the cohomology of the latter. A geometric presentation lemma very close to Gabber's was independently found at about the same time by Ojanguren, who used it to prove Gersten-like properties of quadratic forms [**40**]. Unlike Gabber, Ojanguren does not use the projective line, but rather the affine line and homotopy invariance proved for the Witt ring by Karoubi; see remark after Lemma 4.1.3. This kind of idea has been reemployed to go further in that direction, notably for non-abelian cohomology [**8**], [**7**]. See also the articles of Nashier [**37**] and Dutta [**11**].

Gabber's proof makes it clear that the Bloch-Ogus theorem holds for considerably more general cohomology theories than those considered in [**2**], notably in cases where purity or homotopy invariance does not hold. As an example, his method was then used by Gros and Suwa to prove Gersten's conjecture for logarithmic Hodge-Witt sheaves [**22**].

Somewhat later, Grayson, looking at Quillen's proof again, observed that the argument gives more: the K-theoretic Zariski sheaves of complexes analogous to the sheafified Cousin complexes have stalks which are not only exact, but even "universally exact" [**19**]. This means that they remain exact after applying any finitely presented additive (not necessarily exact) functor to them. Gabber also states that the sheaves of Cousin complexes for étale cohomology have universally exact stalks [**14**, 1.6], but he does not elaborate. Finally, by unrelated methods, Gillet, using Suslin's rigidity theorem for the algebraic K-theory with coefficients of henselian discrete valuation rings, obtained a proof of Gersten's conjecture for the K-theory with coefficients of an arbitrary discrete valuation ring [**16**].

The scope of this paper is twofold. Firstly, we give a simple and detailed exposition of the proof of the Bloch-Ogus theorem for étale cohomology, following

Gabber's method: this is done in Part I. In contrast with [**2**], étale homology is not used. Secondly, we give in Part II an axiomatic treatment of Gabber's argument. We show that it applies to any "cohomology theory with supports" which satisfies two simple axioms: étale excision and an axiom dubbed "key lemma". The latter follows either from homotopy invariance or from a good behaviour of the cohomology of $\mathbf{P}^1$.

In Section 1, we construct the Cousin complexes and the coniveau spectral sequence. In Section 2, we formulate the effacement theorem for torsion coefficients defined over the base field, and derive the Bloch-Ogus theorem from it. In Section 3, we prove a geometric presentation theorem which is a little stronger than Gabber's. In Section 4, we prove Gabber's effacement theorem in a special case, and deduce it in general thanks to the geometric presentation theorem.

In Section 5, we set up the axioms described above. In Section 6, we formulate a "universally exact" version of the Bloch-Ogus theorem for cohomology theories with supports which are defined by a "substratum" (see subsections 5.1 and 5.2). In Section 7, we give several examples to which our framework applies. Classical examples include étale cohomology with finite coefficients, Betti cohomology, de Rham cohomology, algebraic K- and G-theory. Other examples include Hodge, de Rham-Witt, Hodge-Witt and logarithmic Hodge-Witt cohomology, Rost's cycle cohomology and Voevodsky's motivic cohomology. In Section 8, we give a few corollaries which partially motivated this paper (see Table of Contents): the one in subsection 8.1 was announced in [**7**], and those in subsections 8.2 and 8.4 were used in [**30**].

There are 3 appendices. In Appendix A, we extend the effacement theorem to arbitrary (not necessarily torsion) complexes of sheaves coming from the small étale site of k. This includes a self-contained proof of Gabber's computation of the étale cohomology of the projective line [**13**], which was not given in Section 4. In Appendix B, we prove a refined version of the Bloch-Ogus theorem over a semi-local Dedekind domain, Gillet style. Appendix C is technical: it exposes a homological theory of unbounded complexes of objects of an abelian category, which allows for a smoother exposition in Sections 5 and 6.

As the expert reader will already be aware of after having read this introduction, we do not claim much originality in many results and proofs given here. The main purpose of writing this text was for us to make those easily accessible to the general public. We strongly encourage the reader to also have a look at V. Voevodsky's preprint [**52**], whose methods and results are different yet closely related.

We thank Ofer Gabber for a number of very useful comments, and Fabien Morel for helpful discussions on the topic of Appendix C.

Contents

Part I Etale Cohomology

1 The Coniveau Spectral Sequence

Let X be a scheme and A a sheaf of abelian groups on the small étale site of X. In this section, we recall the construction of the coniveau spectral sequence over X with coefficients in A in a leisurely way.

1.1 An exact couple. First consider a chain of closed subsets of X

$$\vec{Z} : \emptyset \subset Z_d \subset Z_{d-1} \subset \cdots \subset Z_0 = X.$$

Take the convention that $Z_i = \emptyset$ for $i > d$ and $Z_i = X$ for $i < 0$. For a pair $(Z_{p+1} \subset Z_p)$, we have a long exact sequence of cohomology with supports:

$$\cdots \to H^{p+q}_{Z_{p+1}}(X, A) \xrightarrow{i^{p+1,q-1}} H^{p+q}_{Z_p}(X, A)$$
$$\xrightarrow{j^{p,q}} H^{p+q}_{Z_p - Z_{p+1}}(X - Z_{p+1}, A) \xrightarrow{k^{p,q}} H^{p+q+1}_{Z_{p+1}}(X, A) \to \cdots .$$

We construct an exact couple $C_{\vec{Z}}(D, E, i, j, k)$ [**27**, Chapter VIII, Section 4], [**35**, Chapter 2, Section 2.3] by setting

$$D^{p,q} = H^{p+q}_{Z_p}(X, A) \qquad \text{and} \qquad E^{p,q} = H^{p+q}_{Z_p - Z_{p+1}}(X - Z_{p+1}, A):$$

$$\begin{array}{ccc} D^{p+1,q-1} & \xrightarrow{i^{p+1,q-1}} & D^{p,q} \\ {\scriptstyle k^{p,q}} \nwarrow & {\scriptstyle (0,+1)} & \swarrow {\scriptstyle j^{p,q}} \\ & E^{p,q} & \end{array}$$

This exact couple yields a spectral sequence of cohomological type, converging to $D^{0,n} = H^n(X, A)$ with respect to the filtration

$$F^p = \operatorname{Im}[H^n_{Z_p}(X, A) \to H^n(X, A)] = \operatorname{Ker}[H^n(X, A) \to H^n(X - Z_p, A)].$$

The E_1-terms of this spectral sequence are $E_1^{p,q} = E^{p,q}$, and the differential $d_1^{p,q} : E_1^{p,q} \to E_1^{p+1,q}$ is the composite

$$H^{p+q}_{Z_p - Z_{p+1}}(X - Z_{p+1}, A) \xrightarrow{k} H^{p+q+1}_{Z_{p+1}}(X, A) \xrightarrow{j} H^{p+q+1}_{Z_{p+1} - Z_{p+2}}(X - Z_{p+2}, A).$$

1.2 Passing to the limit. We now assume that X is equidimensional and noetherian of dimension d and that for all p, $\operatorname{codim}_X Z_p \geq p$. Order the set of $(d+1)$-tuples $\vec{Z}$ by $\vec{Z} \leq \vec{Z'}$ if $Z_p \subseteq Z'_p$ for all p. The construction of the exact couple $C_{\vec{Z}}$ is covariant with respect to this ordering. Passing to the limit now defines a new exact couple $\underset{\to}{C}$ with

$$\underset{\to}{D}^{p,q} = \varinjlim_{\vec{Z}} H^{p+q}_{Z_p}(X, A) := H^{p+q}_{X^{(p)}}(X, A), \text{ and}$$

$$\underset{\to}{E}^{p,q} = \varinjlim_{\vec{Z}} H^{p+q}_{Z_p - Z_{p+1}}(X - Z_{p+1}, A)$$

where $X^{(p)}$ denotes the set of points of codimension p in X. The following lemma describes the second direct limit more concretely:

Lemma 1.2.1 *a) If $T_1, \dots, T_r$ are pairwise disjoint closed subsets of X, then*

$$\bigoplus H^*_{T_i}(X, A) \xrightarrow{\sim} H^*_{\bigcup T_i}(X, A).$$

b) We have

$$\underset{\to}{E}^{p,q} \simeq \coprod_{x \in X^{(p)}} H^{p+q}_x(X, A) \tag{1.1}$$

where, for $x \in X^{(p)}$, $H^{p+q}_x(X, A) := \varinjlim_{U \ni x} H^{p+q}_{\bar{x} \cap U}(U, A)$.

Proof a) By induction on r we may assume $r = 2$. We have a commutative diagram

$$\begin{array}{ccccc}
 & & H^*_{T_2}(X, A) & & \\
 & & \downarrow & \searrow & \\
H^*_{T_1}(X, A) & \longrightarrow & H^*_{T_1 \cup T_2}(X, A) & \longrightarrow & H^*_{T_2}(X - T_1, A) \\
 & \searrow & \downarrow & & \\
 & & H^*_{T_1}(X - T_2, A) & &
\end{array}$$

in which the row and column are exact and the two diagonal maps are isomorphisms by excision. The claim follows.

b) Note that, if the irreducible components of codimension p of Z_p are $Y_1, \dots, Y_r$, then $Z_p \setminus Z_{p+1} = \coprod (Y_i \setminus Z_{p+1})$ (disjoint union) as soon as Z_{p+1} contains the $Y_i \cap Y_j$ and the higher codimensional components of Z_p. The isomorphism now follows from a). □

The spectral sequence associated to the exact couple $\underset{\rightarrow}{C}$ still converges to $H^*(X,A)$. It is called the *coniveau spectral sequence* (compare [**2**, Remark 3.10], [**15**, p. 239]):

$$E_1^{p,q} = \coprod_{x\in X^{(p)}} H_x^{p+q}(X,A) \Rightarrow H^{p+q}(X,A). \tag{1.2}$$

The associated filtration

$$N^pH^n(X,A) = \operatorname{Im}(H^n_{X^{(p)}}(X,A) \to H^n(X,A))$$

is called the *coniveau filtration* or *filtration by codimension of support.* Its E_1-terms yield *Cousin complexes*:

$$\begin{array}{l} 0 \to \coprod_{x\in X^{(0)}} H_x^q(X,A) \xrightarrow{d_1^{0,q}} \coprod_{x\in X^{(1)}} H_x^{1+q}(X,A) \xrightarrow{d_1^{1,q}} \cdots \\ \qquad \cdots \xrightarrow{d_1^{p-1,q}} \coprod_{x\in X^{(p)}} H_x^{p+q}(X,A) \xrightarrow{d_1^{p,q}} \cdots . \end{array} \tag{1.3}$$

In the next section, we shall need:

Lemma 1.2.2 *For all n, p, the presheaf*

$$U \mapsto \coprod_{x\in U^{(p)}} H_x^n(U,A)$$

is a sheaf for the Zariski topology of X. This sheaf is flasque and can be identified with

$$\coprod_{x\in X^{(p)}} i_{x*}H_x^n(X,A)$$

where i_x is the immersion $x \hookrightarrow X$ and the abelian group $H_x^n(X,A)$ is considered as a (constant) sheaf on x for the Zariski topology.

Proof For $x \in X^{(p)}$, define a presheaf F_x on the category of Zariski open subsets of X by

$$F_x(U) = \begin{cases} H_x^n(U,A) & \text{if } U \ni x \\ 0 & \text{if } U \not\ni x. \end{cases}$$

By definition of $H_x^n(X,A)$ (see Lemma 1.2.1 b)), $F_x(U) = F_x(X)$ if $U \ni x$, hence F_x is the sheaf $i_{x*}H_x^n(X,A)$, which is obviously flasque. □

Suppose now that X is a smooth, irreducible variety over a field k, A is a locally constant, constructible sheaf and the stalks of A are m-torsion, with m prime to the characteristic of k. We shall use cohomological purity to transform these complexes into ones which involve only étale cohomology without supports. For $i \in \mathbf{Z}$, we write

$$A(i) = A \otimes \mu_m^{\otimes i}$$

where μ_m is the sheaf of m-th roots of unity. Let Z be a smooth irreducible closed subvariety of X of codimension p. Cohomological purity ([**36**, ch. VI, Sections 5 and 6], [**SGA4 1/2**, p. 63, Theorem V.3.4]) then gives canonical isomorphisms:

$$H_Z^n(X,A) \xleftarrow{\sim} H^{n-2p}(Z,A(-p)).$$

Noting that, for an arbitrary irreducible closed subvariety Z of X, the intersection $Z \cap U$ is smooth for small enough open subsets U, hence defines a smooth pair $Z \cap U \subset U$, this yields isomorphisms

$$H_x^{p+q}(X, A) \simeq H^{q-p}(k(x), A(-p))^1$$

for $x \in X^{(p)}$ (where $k(x)$ is the residue field of x), so that the complex (1.3) takes the following, perhaps more familiar form (compare [**2**, Proposition 3.9]):

$$\begin{aligned} 0 \to H^q(k(X), A) \to \coprod_{x \in X^{(1)}} H^{q-1}(k(x), A(-1)) \to \cdots \\ \cdots \to \coprod_{x \in X^{(p)}} H^{q-p}(k(x), A(-p)) \to \cdots . \end{aligned} \tag{1.4}$$

Here, $k(X)$ is the function field of X. Note that $H^{q-p}(k(x), A(-p))$ is simply Galois cohomology of the residue field of x. So the E_1-terms of the coniveau spectral sequence have taken an especially simple form. Note also that

$$E_1^{p,q} = 0 \text{ for } p > q.$$

2 The Effacement Theorem and the Bloch-Ogus Theorem

2.1 Effaceable sheaves. In this paper, we are interested in a special property of the sheaf A:

Definition 2.1.1 Let X be a variety over k. Let $t_1, \ldots, t_r \in X$ be a finite number of points contained in some affine open subset of X. An étale sheaf A over X is *effaceable at* $t_1, \ldots, t_r$ if the following condition is satisfied:

Given $p \geq 0$, for any small enough affine open neighbourhood W of $t_1, \ldots, t_r$ and any closed subset $Z \subseteq W$ of codimension $\geq p+1$, there exists a smaller open neighbourhood of $t_1, \ldots, t_r$, $U \subseteq W$, and a closed subset $Z' \subseteq U$ containing $Z \cap U$ such that

(1) $\operatorname{codim}_U(Z') \geq p$;
(2) the composite $H_Z^n(W, A) \to H_{Z \cap U}^n(U, A) \to H_{Z' \cap U}^n(U, A)$ is 0 for all $n \geq 0$.

The sheaf A is *effaceable* if it is effaceable at $t_1, \ldots, t_r$ for all $t_1, \ldots, t_r$ as above.

This condition looks very technical, but it has far-reaching consequences:

Proposition 2.1.2 *Let $R = \mathcal{O}_{X,(t_1,\ldots,t_r)}$ be the semi-local ring of X at $(t_1, \ldots, t_r)$ and $Y = \operatorname{Spec} R$. Suppose the sheaf A is effaceable at $t_1, \ldots, t_r$. Then, in the exact couple defining the coniveau spectral sequence for (Y, A), the map $i^{p,q}$ is identically 0 for all $p > 0$. In particular,*

$$E_2^{p,q} = \begin{cases} H^q(Y, A) & \text{if } p = 0 \\ 0 & \text{if } p > 0. \end{cases}$$

[1] Strictly speaking, this argument is only valid when the ground field k is perfect. Otherwise a closed point of X whose residue field is inseparable over k will produce a counterexample to the statement just before this equation. However, if k is imperfect, the isomorphism will hold after passing to its perfect closure. Since étale cohomology is invariant under purely inseparable extensions [**36**, Chapter II, p. 77, Remark 3.17], the isomorphism holds in general. Compare [**2**, Remark 4.7].

The Cousin complex (1.3) *yields an exact sequence:*

$$0 \to H^q(Y,A) \xrightarrow{e} \coprod_{x\in Y^{(0)}} H^q_x(Y,A) \xrightarrow{d_1^{0,q}} \coprod_{x\in Y^{(1)}} H^{q+1}_x(Y,A) \xrightarrow{d_1^{1,q}} \cdots .$$

Proof Consider the diagram

$$\begin{array}{ccccc}
H^n_Z(W,A) & \longrightarrow & H^n_{Z\cap U}(U,A) & \longrightarrow & H^n_{Z'\cap U}(U,A) \\
\downarrow & & \downarrow & & \downarrow \\
H^n_{W^{(p+1)}}(W,A) & \longrightarrow & H^n_{U^{(p+1)}}(U,A) & \longrightarrow & H^n_{U^{(p)}}(U,A) \\
 & & \downarrow & & \downarrow \\
 & & H^n_{Y^{(p+1)}}(Y,A) & \longrightarrow & H^n_{Y^{(p)}}(Y,A).
\end{array}$$

The composition of arrows in the first row is identically 0 for any n. Therefore, the compositions $H^n_Z(W,A) \to H^n_{Y^{(p+1)}}(Y,A) \to H^n_{Y^{(p)}}(Y,A)$ are 0. Passing to the limit over Z, this gives that the compositions $H^n_{W^{(p+1)}}(W,A) \to H^n_{Y^{(p+1)}}(Y,A) \to H^n_{Y^{(p)}}(Y,A)$ are 0. Passing to the limit over W, we get that the map

$$H^n_{Y^{(p+1)}}(Y,A) \xrightarrow{i^{p+1,n-p-1}} H^n_{Y^{(p)}}(Y,A)$$

is itself 0. □

Corollary 2.1.3 (The Bloch-Ogus theorem) *Let A be effaceable on X. Then, the E_2-term of the coniveau spectral sequence for (X,A) is*

$$E_2^{p,q} = H^p_{\mathrm{Zar}}(X, \mathcal{H}^q(A))$$

where $\mathcal{H}^q(A)$ is the Zariski sheaf associated to the presheaf $U \mapsto H^q(U,A)$.

Proof Consider the complex of flasque Zariski sheaves associated to the Cousin complexes (1.3) (compare Lemma 1.2.2):

$$0 \to \coprod_{x\in X^{(0)}} i_{x*}H^q_x(X,A) \to \coprod_{x\in X^{(1)}} i_{x*}H^{1+q}_x(X,A) \to \cdots \to \coprod_{x\in X^{(p)}} i_{x*}H^{p+q}_x(X,A) \to \cdots . \tag{2.1}$$

Proposition 2.1.2 implies that (2.1) is a resolution of $\mathcal{H}^q(A)$, with global Sections (1.3). The conclusion follows. □

2.2 The effacement theorem. The main result of [**14**] is:

Theorem 2.2.1 *(Gabber) For X smooth over k, any torsion sheaf (on the small étale site of X) of the form p^*A_0 is effaceable, where $p : X \to \operatorname{Spec} k$ is the structural morphism.*

Specializing to twisted roots of unity and using sequence (1.4), we get a more familiar case:

Corollary 2.2.2 *(Bloch-Ogus, [**2**]) Let X be smooth, irreducible over k, R and Y as in Proposition 2.1.2, and let m be an integer prime to the characteristic of k.*

Then, for all $i \in \mathbf{Z}$ and $q \geq 0$, we have an exact sequence:

$$0 \to H^q(Y, \mu_m^{\otimes i}) \to H^q(k(Y), \mu_m^{\otimes i}) \to \coprod_{x \in Y^{(1)}} H^{q-1}(k(x), \mu_m^{\otimes(i-1)}) \to \cdots .$$

Remarks 2.2.3

(1) In Appendix A we shall remove the hypothesis that A_0 is torsion in Theorem 2.2.1.

(2) Effaceable sheaves have the following trivial stability properties:
- A sheaf A is effaceable at $t_1, \ldots, t_r$ if and only if $A_{|W}$ is effaceable for a small enough open neighbourhood W of $t_1, \ldots, t_r$ (effaceability is a local condition).
- Let A, B be two étale sheaves over X. Then $A \oplus B$ is effaceable if and only if A and B are.

(3) Over a smooth k-variety X, it is not true that all étale sheaves are effaceable. As an example, take $k = \mathbf{R}$ and for X the affine line $\mathbf{A}^1_{\mathbf{R}}$. Let $f : X' \to X$ be the two-fold covering given by the equation $x^2 + y^2 = 0$, where x is the parameter of $\mathbf{A}^1_{\mathbf{R}}$, and let $A = f_*\mathbf{Z}/2$. Let Y be the localization of X at 0 and $Y' = X' \times_X Y$. Since f is finite, there is an isomorphism

$$H^*(Y, A) \xrightarrow{\sim} H^*(Y', \mathbf{Z}/2).$$

On the other hand, the structural morphism $Y' \to \operatorname{Spec} \mathbf{R}$ is split by the inclusion of the closed point of Y', hence $H^*(Y', \mathbf{Z}/2)$ contains $H^*(\mathbf{R}, \mathbf{Z}/2) \neq 0$ as a direct summand. However, the two generic points η'_1, η'_2 of Y' are isomorphic to $\operatorname{Spec} \mathbf{C}(x)$, hence the Kummer theory class of -1 in $H^1(Y', \mathbf{Z}/2)$ goes to 0 in both $H^1(\eta'_1, \mathbf{Z}/2)$ and $H^1(\eta'_2, \mathbf{Z}/2)$. Correspondingly, if η denotes the generic point of Y, the map

$$H^1(Y, A) \to H^1(\eta, A)$$

is not injective. One can give a similar example with $X = \mathbf{A}^2_{\mathbf{R}}$ and X' defined by the equation $x^2 + y^2 + z^2 = 0$ (from H^2 onwards), etc. See also [**9**, p. 173].

(4) One may however produce effaceable sheaves which are more general than those of Theorem 2.2.1:

Proposition 2.2.4 *Let $\tilde{X} \xrightarrow{f} X$ be a finite map between schemes of pure dimension d with $\tilde{X}$ smooth, and let B be an étale sheaf over $\tilde{X}$. If B is effaceable at $f^{-1}(\{t_1, \ldots, t_r\})$, then f_*B is effaceable at $t_1, \ldots, t_r$.*

Proof Let $T = \{t_1, \ldots, t_r\}$, Z be as in Definition 2.1.1, $\tilde{T} = f^{-1}(T)$ and $\tilde{Z} = f^{-1}(Z)$. Apply the effacement theorem to $(\tilde{X}, \tilde{T}, \tilde{Z}, B)$ and get a pair $(\tilde{U}, \tilde{Z}')$ such that $\tilde{T} \subset \tilde{U}$ and the composition

$$H^n_{\tilde{Z}}(\tilde{X}, B) \to H^n_{\tilde{Z} \cap \tilde{U}}(\tilde{U}, B) \to H^n_{\tilde{Z}' \cap \tilde{U}}(\tilde{U}, B)$$

is 0 for all $n \geq 0$. Let

$$U = X - f(\tilde{X} - \tilde{U})$$
$$Z' = f(\tilde{Z}')$$

so that $T \subset U$, $Z \subset Z'$, $\operatorname{codim}_X Z' \geq p$ and $f^{-1}(U) \subseteq \tilde{U}$, $\tilde{Z}' \subseteq f^{-1}(Z')$. We then get a commutative diagram

$$
\begin{array}{ccccccc}
H^n_Z(X, f_*B) & \longrightarrow & H^n_{Z\cap U}(U, f_*B) & & \longrightarrow & & H^n_{Z'\cap U}(U, f_*B) \\
 & & & & & & \downarrow \wr \\
\wr \downarrow & & & & & & H^n_{f^{-1}(Z')\cap f^{-1}(U)}(f^{-1}(U), B) \\
 & & & & & & \uparrow \\
H^N_{\tilde{Z}}(\tilde{X}, B) & \longrightarrow & H^n_{\tilde{Z}\cap\tilde{U}}(\tilde{U}, B) & \xrightarrow{0} & H^n_{\tilde{Z}'\cap\tilde{U}}(\tilde{U}, B) & \longrightarrow & H^n_{f^{-1}(Z')\cap\tilde{U}}(\tilde{U}, B)
\end{array}
$$

where the left vertical map and top right vertical map are isomorphisms by Shapiro's lemma for étale cohomology (exactness of f_* for a finite morphism). Proposition 2.2.4 follows. □

Note that in the proof of Proposition 2.2.4, it is not necessary to assume X smooth. So it provides non-smooth cases in which the effacement theorem holds.

Corollary 2.2.5 (Shapiro's lemma for Zariski cohomology) *Let $X' \xrightarrow{f} X$ be a finite flat map between two smooth varieties over k, and let $\mathcal{F}$ be a sheaf of abelian groups over X' for the Zariski topology. Suppose that $\mathcal{F}$ is of the form $\mathcal{H}^q(A)$ for some effaceable étale sheaf A over X'. Then the natural homomorphism*

$$H^p_{\text{Zar}}(X, f_*\mathcal{F}) \to H^p_{\text{Zar}}(X', \mathcal{F})$$

is an isomorphism for any $p \geq 0$.

Proof Shapiro's lemma for étale cohomology yields isomorphisms of cohomology groups with supports, for $Z \subset X$ a closed subset and $Z' = f^{-1}(Z)$:

$$H^p_Z(X, f_*A) \xrightarrow{\sim} H^p_{Z'}(X', A).$$

Localizing, we get isomorphisms of the E_1-terms of the coniveau spectral sequences for A (over X') and f_*A (over X):

$$\coprod_{x\in X^{(p)}} H^{p+q}_x(X, f_*A) \xrightarrow{\sim} \coprod_{x\in X'^{(p)}} H^{p+q}_x(X', A).$$

This isomorphism of Cousin complexes induces an isomorphism of their homology groups:

$$H^p_{\text{Zar}}(X, \mathcal{H}^q(f_*A)) \xrightarrow{\sim} H^p_{\text{Zar}}(X', \mathcal{H}^q(A))$$

by the Bloch-Ogus theorem over semi-local rings (note that f_*A is effaceable by Proposition 2.2.4). Finally, there is an isomorphism of Zariski sheaves

$$\mathcal{H}^q(f_*A) \xrightarrow{\sim} f_*\mathcal{H}^q(A)$$

which is merely Shapiro's lemma for étale cohomology applied at the local rings of X. □

Remarks 2.2.6

(1) An alternative argument would be to show that $R^q f_*\mathcal{F} = 0$ for $q > 0$. This is what we shall do in subsection 8.4, removing the flatness hypothesis.

(2) "Shapiro's lemma for Zariski cohomology" is false for arbitrary Zariski sheaves! For example, take $X = \mathbf{A}^1_k$ and X' some 2-fold covering of X split at 0 (e.g. $X' = \mathbf{A}^1_k$, $f(x) = x^2 - 1$ if $\operatorname{char} k \neq 2$). Let Y be the localization of X at 0 and $Y' = f^{-1}(Y)$. Let $\mathcal{F} = j_! A$ (extension by 0), where $j : \eta \to X'$ is the inclusion of the generic point and A is a constant sheaf. Then we have:

$$H^i(Y, f_* \mathcal{F}) = 0 \text{ for } i > 0 \quad (\text{since } Y \text{ is local});$$
$$H^1(Y', \mathcal{F}) = A.$$

The latter is easily seen by a Čech cohomology computation.

The proof of Theorem 2.2.1 is given in Sections 3 and 4. We shall in fact prove something slightly stronger (and simpler to state):

Theorem 2.2.7 (Effacement Theorem) *Let X be a smooth, affine variety over k, $t_1, \dots, t_r \in X$ a finite number of points, $p \geq 0$ an integer and Z a closed subvariety of codimension $\geq p+1$. Let A be a sheaf of torsion abelian groups over the (small) étale site of X. Assume that $A = p^* A_0$, where $p : X \to \operatorname{Spec} k$ is the structural morphism and A_0 is a $Gal(k_s/k)$-module. If k is infinite, then there exist an open subset U of X containing all t_i and a closed subvariety $Z' \subseteq X$ containing Z such that*

(1) $\operatorname{codim}_X(Z') \geq p$;

(2) *the map* $H^n_{Z \cap U}(U, A) \to H^n_{Z' \cap U}(U, A)$ *is 0 for all* $n \geq 0$.

If k is finite, then there exists (U, Z') as above such that (at least) the composite

$$H^n_Z(X, A) \to H^n_{Z \cap U}(U, A) \to H^n_{Z' \cap U}(U, A) \tag{2.2}$$

is 0 for all $n \geq 0$.

Remark 2.2.8 We would like to point out that (contrary to the definition of effaceability) the statement in Theorem 2.2.7 is not local: the proof by no means implies that the map of Theorem 2.2.7 (2) remains 0 when U is replaced by a smaller open set. Therefore, if in Theorem 2.2.7 one replaces X by Y as in Proposition 2.1.2, it is not at all clear that the conclusion still holds. In other words, given a closed subset $Z \subset Y$ of codimension $\geq p+1$, although the maps $H^n_Z(Y, A) \to H^n_{Y^{(p)}}(Y, A)$ are all 0 by the proof of Proposition 2.1.2, it is not clear whether one can find a single $Z' \subset Y$ as in Theorem 2.2.7 such that the maps $H^n_Z(Y, A) \to H^n_{Z'}(Y, A)$ are 0. This shows the subtlety of the situation and probably why Gersten's conjecture is so difficult for general regular local rings of dimension ≥ 2.

3 Some Geometry

3.1 The geometric presentation theorem. The key to the proof of Theorem 2.2.7 is a geometric presentation theorem which follows from lemmas of Gabber [**14**] supplemented with some remarks of Gros and Suwa [**22**]. This section is devoted to a detailed proof of this theorem. For simplicity, we write $\mathbf{A}^n$ for $\mathbf{A}^n_k$ and $S \times S'$ rather than $S \times_k S'$ for the product of two k-schemes S and S'.

Theorem 3.1.1 (Geometric Presentation Theorem) *Let X be a smooth, affine, irreducible variety of dimension d over an infinite field k; let $t_1, \dots, t_r \in X$ be a finite set of points and Z a closed subvariety of codimension > 0. Then there exists a map $\varphi = (\psi, v) : X \to \mathbf{A}^{d-1} \times \mathbf{A}^1$, an open set $V \subset \mathbf{A}^{d-1}$, and an open set $U \subset \psi^{-1}(V)$ containing $t_1, \dots, t_r$ such that*

(1) $Z \cap U = Z \cap \psi^{-1}(V)$;

(2) *$\psi|_Z$ is finite;*
(3) *$\varphi|_U$ is étale and defines a closed immersion $Z \cap U \hookrightarrow \mathbf{A}^1_V$;*
(4) *$\varphi(t_i) \notin \varphi(Z)$ if $t_i \notin Z$ $(1 \le i \le r)$;*
(5) *$\varphi^{-1}(\varphi(Z \cap U)) \cap U = Z \cap U$.*

(If no t_i lies on Z, it is quite possible that $Z \cap U$ is empty in Theorem 3.1.1. See remark after the proof of Lemma 3.5.1.)

Corollary 3.1.2 *With notation as in Theorem 3.1.1, $\psi|_{Z\cap U} : Z \cap U \to V$ is a finite morphism and one has a cartesian square:*

$$\begin{array}{ccc} Z \cap U & \hookrightarrow & U \\ \wr\downarrow & & \downarrow \varphi|_U \\ \varphi(Z \cap U) & \hookrightarrow & \mathbf{A}^1_V \end{array}$$

where the horizontal arrows are closed immersions, the left vertical one is an isomorphism and the right vertical one is étale. (One could say that φ defines an étale neighbourhood of $Z \cap U \subset \mathbf{A}^1_V$, that is, it induces an analytic isomorphism along $Z \cap U$.)

This theorem and its corollary can be summarized by the diagram below:

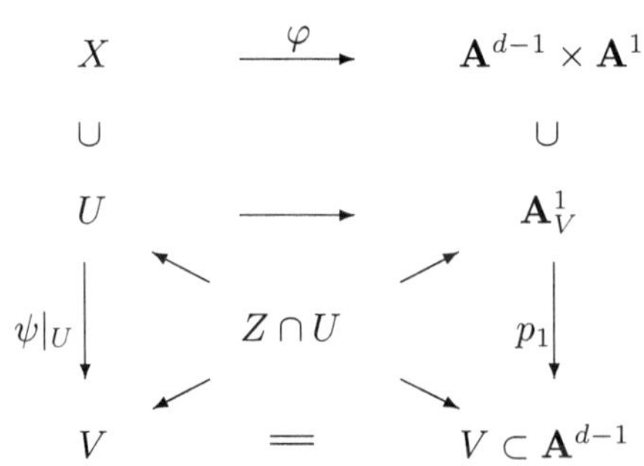

We shall see that the pair (Z', U) of Theorem 2.2.7 may be taken as $(\psi^{-1}(\psi(Z)), U)$ where ψ and U are as in Theorem 3.1.1.

The proof of Theorem 3.1.1 is essentially a (quite involved) series of exercises of commutative algebra and elementary algebraic geometry; it elaborates on those of [**14**, Section 3] and [**22**, Section 2]. We briefly outline it:

(1) Reduction to the case where the t_i are closed points and Z a principal divisor.
(2) Securing ψ.
(3) Securing φ.
(4) Constructing V.
(5) Constructing U.

3.2 Reduction to closed points.

Lemma 3.2.1 *a) With notation as in Theorem 3.1.1, there exist closed points $s_1, \dots, s_r \in X$ such that*

(1) *$s_i \in \overline{\{t_i\}}$ for all i;*

(2) $s_i \notin Z$ *if* $t_i \notin Z$.

b) Let the t_i *be closed. Then there is a non-zero* $f \in \Gamma(X, \mathcal{O}_X)$ *such that* $Z \subset V(f)$ *and* $t_i \notin V(f)$ *if* $t_i \notin Z$.

Proof a) follows from [**34**, p. 34, Theorem 5.5] and b) follows from [**6**, Chapter II, Section 1, Proposition 2]. □

Reduction. Thanks to Lemma 3.2.1, we may assume that the t_i's are closed points in Theorem 3.1.1 (note that since $\psi|_Z$ is finite, $\varphi|_Z$ is also finite, hence $\varphi(Z)$ is closed; moreover, any U containing the s_i's also contains the t_i's) and also that Z is a principal divisor. Therefore:

Hypothesis. *From now on, we assume that*

- *the* t_i*'s are closed in* X*;*
- Z *is a principal divisor in* X.

Let $A = \Gamma(X, \mathcal{O}_X)$, so that $X = \operatorname{Spec} A$. Write A as a quotient of $k[X_1, \dots, X_N]$ (for N large enough), hence X as a closed subvariety of $\mathbf{A}^N$. Any $u \in k[X_1, \dots, X_N]$ defines a morphism $\mathbf{A}^N \to \mathbf{A}^1$. This is the case in particular for $u \in E$, where E is the sub-vector space of $k[X_1, \dots, X_N]$ spanned by $X_1, \dots, X_N$. Hence any r-tuple $\varphi = (u_1, ..., u_r) \in E^r$ defines a morphism $\mathbf{A}^N \to \mathbf{A}^r$. Composing it with the closed immersion $X \hookrightarrow \mathbf{A}^N$, we get a morphism $X \to \mathbf{A}^r$ that we still denote by φ.

Let $\mathcal{E}$ be the affine variety associated to E (so that $E = \mathcal{E}(k)$). We may view $\mathcal{E}$ as the dual space of $\mathbf{A}^N$. We shall in fact prove the following more precise statement:

Theorem 3.2.2 *Let* $X, Z, t_1, \dots, t_r$ *be as in Theorem 3.1.1. We assume that the* t_i *are closed and that* Z *is a principal divisor. Then there exists a non-empty Zariski open subset* Ω *of* $\mathcal{E}^{d-1} \times \mathcal{E}$*, such that for any* $\varphi = (\psi, v) \in \Omega(k)$*:*

(1) $\psi|_Z : Z \to \mathbf{A}^{d-1}$ *is finite;*

(2) φ *is étale at all* t_i $(1 \leq i \leq r)$ *and at all points of the finite set* $S = (\bigcup_{1 \leq i \leq r} \psi^{-1}(\psi(t_i))) \cap Z$*;*

(3) $\varphi|_S : S \to \varphi(S)$ *is radicial;*

(4) *For* $1 \leq i \leq r$, $t_i \notin Z \Rightarrow \varphi(t_i) \notin \varphi(Z)$.

Moreover, for any $\varphi \in E^d$ *satisfying 1)–4), there exists a pair* (U, V) *such that* (φ, U, V) *satisfies the remaining conditions of the Geometric Presentation Theorem.*

(Note that $\Omega(k) \neq \emptyset$ since k is infinite.)

Condition 3) means that $\varphi|_S$ separates the points in S and that for $P \in S$, the residue field extension $k(P)/k(\varphi(P))$ is purely inseparable. This theorem slightly improves on the lemmas of Gabber and Gros-Suwa: it says that the set of φ satisfying 1)–4) contains a Zariski open set, while the former only say that this set is non-empty.

Remark 3.2.3 It can be shown that Theorem 3.2.2 holds even without assuming the t_i's to be closed and Z to be a principal divisor. We don't need this refinement here.

3.3 Securing ψ.

Lemma 3.3.1 *(compare* [**18**, Proposition 1.1]*) There exists a non-empty open set* $\Omega_1 \subseteq \mathcal{E}^{d-1}$ *such that, for* $\psi \in \Omega_1(k)$, $\psi|_Z : Z \to \mathbf{A}^{d-1}$ *is finite.*

Proof Let $B = k[Z]$ be the affine algebra of Z, and x_i the image of X_i in B. For $\psi = (u_1, \dots, u_{d-1}) \in E^{d-1}$, $\psi|_Z$ is finite if and only if, for all i, x_i is integral over $k[\overline{u}_1, \dots, \overline{u}_{d-1}]$, where $\overline{u}_j$ is the image of u_j in B.

Let $K = k(\mathcal{E}^{d-1})$ be the function field of $\mathcal{E}^{d-1}$ and η the generic point of $\mathcal{E}^{d-1}$. We view η as a rational point of $\mathcal{E}^{d-1}$ over K, so that $\eta = (\eta_1, \dots, \eta_{d-1})$ with $\eta_i \in \mathcal{E}(K)$. For simplicity, we still write η_j for the image of η_j in $K \otimes_k B$. Since $\dim Z \leq d-1$, there is for all i a non-zero algebraic relation in $K \otimes_k B$:

$$f_i(\eta_1, \dots, \eta_{d-1}, x_i) = 0$$

with $f_i \in K[T_1, \dots, T_d]$. We claim that f_i can be chosen so that it gives an integral dependence relation on x_i. To see this, we argue as in [**34**, p. 262, proof of Lemma 2]: let $n = \deg f_i$ and $f_i^{(n)}$ be the homogeneous part of degree n of f_i. Since k is infinite, we can find $(t_1, \dots, t_{d-1}) \in k^{d-1}$ such that $f_i^{(n)}(t_1, \dots, t_{d-1}, 1) \neq 0$ in K. Letting $g_i = f_i(T_1 + t_1 T_d, \dots, T_{d-1} + t_{d-1} T_d, T_d)$, the coefficient of T_d in g_i is $f_i^{(n)}(t_1, \dots, t_{d-1}, 1)$ and we get

$$g_i(\eta_1', \dots, \eta_{d-1}', x_i) = 0$$

with $\eta_j' = \eta_j - t_j x_i$. Substituting back η_j instead of η_j' gives the desired integral dependence relation. Therefore there exists a polynomial $g_i' \in k[\mathcal{E}^{d-1}][T_1, \dots, T_d]$ such that

- $g_i'(\eta_1, \dots, \eta_{d-1}, x_i) = 0 \in k[\mathcal{E}^{d-1}] \otimes_k B$;
- the coefficient a_i of $T_d^{\deg g_i'}$ in g_i' is $\neq 0$.

Then, we may take

$$\Omega_1 = \left\{(u_1, \dots, u_{d-1}) \in \mathcal{E}^{d-1} \mid a_i(u_1, \dots, u_{d-1}) \neq 0 \text{ for } 1 \leq i \leq N\right\}. \qquad \square$$

3.4 Securing φ.

Lemma 3.4.1 *Assume k is algebraically closed. With notation as in Theorem 3.1.1, there exists a non-empty open set $\Omega_2 \subseteq \mathcal{E}^d$ such that, for $\varphi = (\psi, \upsilon) \in \Omega_2(k)$, φ is étale at $t_i, 1 \leq i \leq p$ and all points of S.*

Proof Recall that the Jacobian criterion says that $\varphi = (u_1, ..., u_d)$ is étale at $x \in X$ if and only if $(du_1 \wedge ... \wedge du_d) |_x \neq 0$ in $\Omega^d_{X/k} \otimes k(x)$. Throughout this proof, for a $\varphi \in \mathcal{E}^d$, we write $\varphi = (\psi, \upsilon)$ with $\psi = (u_1, \dots, u_{d-1}) \in \mathcal{E}^{d-1}$ and $\upsilon = u_d \in \mathcal{E}$.

For $i = 1, \dots, p$, define

$$T^i = \left\{(\varphi, y) \in \mathcal{E}^d \times Z \mid \psi(y) = \psi(t_i) \text{ and } (du_1 \wedge ... \wedge du_d) |_y = 0\right\},$$

i.e., for any k-algebra R,

$$T^i(R) = \left\{(\varphi, y) \in (E \otimes_k R)^d \times Z(R) \mid \psi(y) = \psi(t_i) \in R^{d-1} \text{ and } (du_1 \wedge ... \wedge du_d)|_y = 0\right\},$$

and also

$$T'^i = \left\{\varphi \in \mathcal{E}^d \mid (du_1 \wedge ... \wedge du_d)|_{t_i} = 0\right\}.$$

It is clear that T^i is a closed subset of $\mathcal{E}^d \times Z$. By Chevalley's theorem, its projection on $\mathcal{E}^d$:

$$F^i = \left\{\varphi \in \mathcal{E}^d \mid \text{ for some } y \in Z, \psi(y) = \psi(t_i) \text{ and } (du_1 \wedge ... \wedge du_d) |_y = 0\right\}$$

is constructible. On the other hand, T'^i is closed in $\mathcal{E}^d$. Let $\Omega_2^i = \mathcal{E}^d - (\overline{F}^i \cup T'^i)$. By definition, $\varphi \in \Omega_2^i$ implies that φ is étale at t_i and all points of $\psi^{-1}(\psi(t_i)) \cap Z$.

We shall show that $\dim T^i < \dim \mathcal{E}^d$, hence that $\dim \overline{F}^i < \dim \mathcal{E}^d$, and also that $\dim T'^i < \dim \mathcal{E}^d$, hence that Ω_2^i is non-empty. For $y \in X$, let

$$F_y^i = \left\{\varphi = (u_1, ..., u_d) \in \mathcal{E}^d \mid \psi(y) = \psi(t_i) \text{ and } (du_1 \wedge ... \wedge du_d) |_y = 0\right\} \subseteq \mathcal{E}^d \times_k k(y).$$

For $y = t_i$, this is just T'^i. For $y \in Z$ this is the fibre at y of the projection $T^i \to Z$. To prove that $\dim T^i < \dim \mathcal{E}^d$, it is enough to show that for all $y \in Z$, $\operatorname{codim}_{\mathcal{E}^d_{k(y)}} F^i_y > \dim \overline{\{y\}}$ (It would be enough to let y run through the projections of the generic points of T^i.)

Let y be such a point. First assume $y \neq t_i$. Then the linear space $\mathcal{H} = \{u \in \mathcal{E}_{k(y)} \mid u(y) = u(t_i)\}$ is of codimension one in $\mathcal{E}_{k(y)}$ and so $\mathcal{H}^{d-1}$ is of codimension $d-1$ in $\mathcal{E}^{d-1}_{k(y)}$. Since the differentials du, $u \in \mathcal{E}$, generate $\Omega^1_{X/k}$ at each point of X, the subspace spanned by $du, u \in \mathcal{H}$, is of codimension ≤ 1 in $\Omega^1_{X/k} \otimes k(y)$. In particular we can find $u_i \in \mathcal{H}(k(y)) \subseteq k(y)[X_1, \ldots, X_N]$ $(i = 1, \ldots, d-1)$ such that $du_1, \ldots, du_{d-1} \in \Omega^1_{X/k} \otimes k(y)$ are independent at y. Complete this system by a $u_d \in \mathcal{E}(k(y))$ such that $du_1, \ldots, du_d$ are linearly independent at y. Then $\varphi = (u_1, \ldots, u_d) \in \mathcal{E}(k(y))^d$ is such that:

- $\psi(y) = \psi(t_i)$;
- $(du_1 \wedge \ldots \wedge du_d) \mid_y \neq 0$.

Thus F^i_y is of codimension > 0 in $\mathcal{H}^{d-1} \times \mathcal{E}_{k(y)}$ and so is of codimension at least $d > \dim Z$ in $\mathcal{E}^d_{k(y)}$. As $\dim Z \geq \dim \overline{\{y\}}$, we are done.

The case $y = t_i$ is easier, since then $\dim \overline{\{y\}} = 0$: we need only use the fact that $\Omega^1_{X/k}$ is generated by $d\mathcal{E}$ at t_i. This also shows that $\dim T'^i < \dim \mathcal{E}^{d-1}$ for all i, which we wanted. Thus $\Omega^i_2 \neq \emptyset$.

Let $\Omega_2 = \bigcap_{i=1}^p \Omega^i_2$. For $\varphi \in \Omega_2(k)$, φ is étale at all t_i and all points of $\psi^{-1}(\psi(t_i)) \cap Z$ $(1 \leq i \leq p)$. This completes the proof of Lemma 3.4.1. $\square$

Lemma 3.4.2 *Assume k is algebraically closed. There exist non-empty open sets $\Omega_3, \Omega_4 \subseteq \mathcal{E}^d$ such that:*

(1) *For $\varphi = (\psi, v) \in \Omega_3(k)$, $\varphi|_S$ is injective.*
(2) *For $\varphi \in \Omega_4(k)$ and $1 \leq i \leq p$, $t_i \notin Z \Rightarrow \varphi(t_i) \notin \varphi(Z)$.*

Proof For $s, s' \in X(k)$, consider

$$M(s,s') = \left\{ (\psi, v, t, t') \in \mathcal{E}^{d-1} \times \mathcal{E} \times (Z \times Z \setminus \Delta_Z) \mid \left\{ \begin{array}{c} \psi(t) = \psi(s) \\ \psi(t') = \psi(s') \\ v(t) = v(t') \end{array} \right\} \right\},$$

where Δ_Z is the diagonal of $Z \times Z$. Let p be the projection of $M(s,s')$ on $\mathcal{E}^d = \mathcal{E}^{d-1} \times \mathcal{E}$. Let us fix $s, s' \in X(k)$ and for simplicity write M instead of $M(s,s')$. Let p be the projection of M on $\mathcal{E}^d = \mathcal{E}^{d-1} \times \mathcal{E}$; we are going to show that p is not dominant.

Let q be the composite of p with the projection $\mathcal{E}^{d-1} \times \mathcal{E} \to \mathcal{E}^{d-1}$, $\eta = \operatorname{Spec} K$ the generic point of $\mathcal{E}^{d-1}$ and M_η the generic fibre of q. Then p induces a map $M_\eta \to \mathcal{E}_K$ which is dominant if p is dominant.

Let $\overline{K}$ be an algebraic closure of K. This gives rise to the geometric point $\overline{\eta} : \operatorname{Spec} \overline{K} \to \operatorname{Spec} K \to \mathcal{E}^{d-1}$, and we have an induced morphism $M_{\overline{\eta}} \to \mathcal{E}_{\overline{K}}$ (where $M_{\overline{\eta}} = M_\eta \otimes_K \overline{K}$), which is still dominant if p is dominant.

Now the point $\overline{\eta}$ defines a $\overline{K}$-morphism $\psi_0 : Z_{\overline{K}} \to \mathbf{A}^{d-1}_{\overline{K}}$ which is finite as a consequence of 3.3.1. View $s \in X(k)$ as a point of $X(\overline{K})$. Let y_α $(\alpha = 1, \ldots, n)$ be the finitely many ($\overline{K}$-rational) points of $Z_{\overline{K}}$ such that $\psi_0(y_\alpha) = \psi_0(s)$. Let z_β $(\beta = 1, \ldots, m)$ be the finitely many ($\overline{K}$-rational) points of $Z_{\overline{K}}$ such that $\psi_0(z_\beta) =$

$\psi_0(s')$. We have:

$$\begin{aligned} M_{\overline{\eta}} &= \left\{ (v,t,t') \in (\mathcal{E} \times (Z \times Z \setminus \Delta_Z)) \times_k \overline{K} \mid \left\{ \begin{array}{c} \psi_0(t) = \psi_0(s) \\ \psi_0(t') = \psi_0(s') \\ v(t) = v(t') \end{array} \right\} \right. \\ &= \bigcup_{y_\alpha \neq z_\beta} V_{\alpha,\beta} \end{aligned}$$

where

$$V_{\alpha,\beta} = \{v \in \mathcal{E}_{\overline{K}} \mid v(y_\alpha) = v(z_\beta)\} \times \{(y_\alpha, z_\beta)\}.$$

Thus the projection $p(M_{\overline{\eta}}) \subset \mathcal{E}_{\overline{K}}$ decomposes as

$$p(M_{\overline{\eta}}) = \bigcup_{y_\alpha \neq z_\beta} \{v \in \mathcal{E}_{\overline{K}} \mid v(y_\alpha) = v(z_\beta)\}.$$

This is a finite union of proper linear subspaces of $\mathcal{E}_{\overline{K}}$. Hence the projection map $M_{\overline{\eta}} \to \mathcal{E}_{\overline{K}}$ is not dominant, so $p : M \to \mathcal{E}^{d-1} \times \mathcal{E}$ is not dominant and the Zariski closure of the constructible set $p(M)$ is a proper closed subset of $\mathcal{E}^d$.

Now let Ω_3 be the complement of $\overline{\bigcup_{i,j} p(M(t_i,t_j))}$ in $\mathcal{E}^d$: this is a proper open set that satisfies condition 1) of Lemma 3.4.2.

The proof of condition 2) is entirely similar, using the sets

$$N(s) = \{(\varphi, t) \in \mathcal{E}^d \times Z \mid \varphi(t) = \varphi(s)\} \quad (s \notin Z),$$

for $s = t_i$, $t_i \notin Z$. □

Lemma 3.4.3 *There exists a non-empty open subset Ω of $\mathcal{E}^d$ such that all $\varphi \in \Omega(k)$ satisfy conditions 1)–4) of Theorem 3.2.2.*

Proof Let $\overline{k}$ be an algebraic closure of k, $\Omega_1 \subseteq \mathcal{E}^{d-1}_{\overline{k}}$ as in Lemma 3.3.1 and $\Omega_2, \Omega_3, \Omega_4 \subseteq \mathcal{E}^d_{\overline{k}}$ as in Lemmas 3.4.1 and 3.4.2. Let $\overline{\Omega} = (\Omega_1 \times \mathcal{E}) \cap \Omega_2 \cap \Omega_3 \cap \Omega_4$. There exists a finite normal extension K/k such that $\overline{\Omega}$ is defined over K. Let Ω be the intersection of its conjugates under $\mathrm{Aut}(K/k)$. This is a non-empty open subset of $\mathcal{E}^d_K$, which is defined over the radicial closure L of k in K. But Ω is even defined over k, since we can raise the equations of its complement to an appropriate p^n-th power, where $p = \operatorname{char} k$. Then $\Omega_{\overline{k}} \subseteq \overline{\Omega}$.

Let $\varphi \in \Omega(k)$. By construction, φ satisfies conditions 1)–4) of Theorem 3.2.2 after extending scalars to $\overline{k}$. We conclude the proof of Lemma 3.4.3 by observing that each of these conditions descends to k. □

3.5 Constructing V.

Lemma 3.5.1 *Let $\varphi = (\psi, v) \in E^d$ satisfy conditions 1)–4) of Theorem 3.2.2. Then there exists $V \subseteq \mathbf{A}^{d-1}$ such that*

(1) *φ is étale at all points of $Z \cap \psi^{-1}(V)$;*
(2) *$\varphi|_{Z \cap \psi^{-1}(V)} \to \mathbf{A}^1_V$ is a closed immersion;*
(3) *$\psi(t_i) \in V$ for all i.*

Proof Let Z_φ be the intersection of Z with the (closed) locus where φ is not étale. By condition 2) of Theorem 3.2.2, $Z_\varphi \cap \psi^{-1}(\psi(t_i)) = \emptyset$ for all i. By condition 1), $\psi|_Z$ is finite and $\psi(Z_\varphi)$ is closed in $\mathbf{A}^{d-1}$. Its complement $V_1 \subseteq \mathbf{A}^{d-1}$ is such that φ is étale at all points of $Z \cap \psi^{-1}(V_1)$, and that $\psi(t_i) \in V_1$ for all i. Consider the commutative diagram

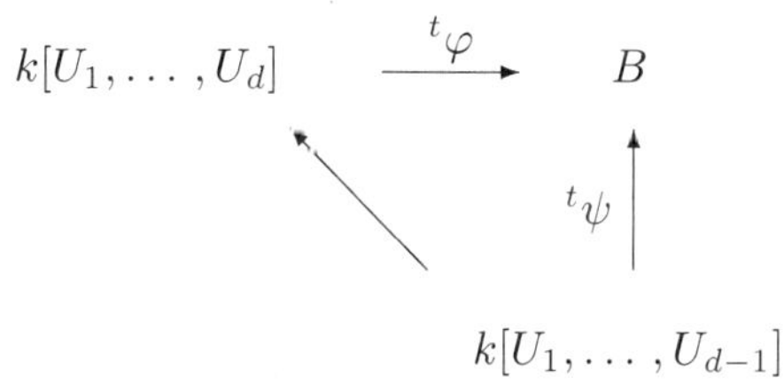

where $B = k[Z]$ as above, and $^t\varphi, {}^t\psi$ are the homomorphisms corresponding to φ and ψ on coordinate rings. Let $\mathfrak{p}_1, \dots, \mathfrak{p}_r$ be the maximal ideals of $k[U_1, \dots, U_{d-1}]$ corresponding to $\psi(t_1), \dots, \psi(t_r)$. Let $\mathfrak{a} = \bigcap \mathfrak{p}_i$.

Since φ is étale at all points of S and $\varphi|_S : S \to \varphi(S)$ is radicial,

$$k[U_1, \dots, U_d]/\mathfrak{a}k[U_1, \dots, U_d] \to B/\mathfrak{a}B$$

is an isomorphism. Since B is a finitely generated $k[U_1, \dots, U_{d-1}]$-module, by Nakayama's lemma there is an $f \in k[U_1, \dots, U_{d-1}] - (\mathfrak{p}_1 \cup \dots \cup \mathfrak{p}_r)$ such that $k[U_1, \dots, U_d][1/f] \to B[1/f]$ is surjective. Let $V_2 = \{f \neq 0\} \subseteq \mathbf{A}^{d-1}$: then V_2 contains $\psi(t_i)$ for all i and has the property that φ induces a closed immersion $Z \cap \psi^{-1}(V_2) \hookrightarrow \mathbf{A}^1_{V_2}$.

Let $V = V_1 \cap V_2$. Then φ is étale at all points of $Z \cap \psi^{-1}(V)$, induces a closed immersion $Z \cap \psi^{-1}(V) \hookrightarrow \mathbf{A}^1_V$ and contains $\psi(t_i)$ for all i. Hence V satisfies conditions 1), 2), 3) of Lemma 3.5.1. □

Remark 3.5.2 If $S = \emptyset$, $Z \cap \psi^{-1}(V)$ may well be empty.

3.6 Constructing U.

Lemma 3.6.1 *With notation as in Lemma 3.5.1, let*

$$\Phi = \varphi^{-1}\Big(\varphi\big(Z \cap \psi^{-1}(V)\big)\Big) - Z \cap \psi^{-1}(V).$$

Then:

(1) *Φ is closed in $\psi^{-1}(V)$;*
(2) *$U_1 = \psi^{-1}(V) - \Phi$ contains all the t_i, satisfies $Z \cap \psi^{-1}(V) = Z \cap U_1$ and $\varphi^{-1}\big(\varphi(Z \cap U_1)\big) \cap U_1 = Z \cap U_1$.*

Proof For simplicity, let $T = Z \cap \psi^{-1}(V)$. By Lemma 3.5.1, $\varphi|_{\varphi^{-1}(\varphi(T))} : \varphi^{-1}(\varphi((T)) \to \varphi(T)$ is étale and $T \to \varphi(T)$ is an isomorphism. Therefore T is open in $\varphi^{-1}(\varphi((T))$ and $\Phi = \varphi^{-1}(\varphi(T)) - T$ is closed in $\varphi^{-1}(\varphi(T))$. Since $\psi|_Z$ is finite, $\psi|_T$ is finite over V and $\varphi|_T$ is finite over $\mathbf{A}^1_V$. So $\varphi(T)$ is closed in $\mathbf{A}^1_V$, hence in $\mathbf{A}^1_{\psi(T)}$. It follows that $\varphi^{-1}(\varphi(T))$ is closed in $\varphi^{-1}(\mathbf{A}^1_{\psi(T)}) = \psi^{-1}(\psi(T))$. Thus Φ is closed in $\psi^{-1}(\psi(T))$. Still by finiteness of $\psi\,|_Z$, $\psi^{-1}(\psi(T))$ is closed in $\psi^{-1}(V)$. This proves 1).

If $t_i \in Z$, then $t_i \notin \Phi$, hence $t_i \in U_1$; if $t_i \notin Z$, $t_i \notin \Phi$ by condition 4) of Theorem 3.2.2. In both cases, U_1 is a Zariski neighborhood of t_i. It is obvious that $Z \cap U_1 = Z \cap \psi^{-1}(V)$; the last condition of Lemma 3.6.1 holds by construction. □

End of Proof of Theorem 3.2.2. Let $U_2 \subseteq X$ be the (open) locus where φ is étale. By Lemma 3.5.1, $t_1, \dots, t_r \in U_2$ and $Z \cap \psi^{-1}(V) \subseteq U_2$. It follows that $U = U_1 \cap U_2$, with U_1 as in Lemma 3.6.1, also satisfies condition 2) of this lemma; moreover, $\varphi|_U$ is étale. So the triple (φ, V, U) satisfies all conditions of Theorem 3.1.1. □

4 Proof of the Effacement Theorem

For the convenience of the reader, we state the effacement theorem once again:

Effacement Theorem. *Let X be a smooth, affine variety over a field k, $t_1, \dots, t_r \in X$ a finite number of points, $p \geq 0$ an integer and Z a closed subvariety of codimension $\geq p+1$. Let A be a sheaf of torsion abelian groups over the étale site of X. Assume that $A = p^* A_0$, where $p : X \to \operatorname{Spec} k$ is the structural morphism and A_0 is a $Gal(k_s/k)$-module. If k is infinite, then there exists an open subset U of X containing all t_i and a closed subvariety $Z' \subseteq X$ containing Z such that*

(1) $\operatorname{codim}_X(Z') \geq p$;

(2) *the map* $H^n_{Z\cap U}(U, A) \to H^n_{Z'\cap U}(U, A)$ *is* 0 *for all* $n \geq 0$.

If k is finite, then there exists (U, Z') as above such that (at least) the composite

$$H^n_Z(X, A) \to H^n_{Z\cap U}(U, A) \to H^n_{Z'\cap U}(U, A)$$

is 0 for all $n \geq 0$.

4.1 A key lemma.

Key lemma (compare [**13**, Lemma 2], [**22**, p. 621]) *Let V be a k-scheme and A be as above. Let $\pi : \mathbf{A}^1_V \to V$, $\tilde{\pi} : \mathbf{P}^1_V \to V$ be the natural projections, $j : \mathbf{A}^1_V \to \mathbf{P}^1_V$ the inclusion and $s_\infty : V \to \mathbf{P}^1_V$ the section at infinity of $\tilde{\pi}$. Finally, let $F \subset V$ be a closed subset of V. Assume that V and $V - F$ are quasi-compact and quasi-separated. Then the diagram*

$$\begin{array}{ccc} H^n_{\mathbf{A}^1_F}(\mathbf{A}^1_V, A) & & \\ & \nwarrow{\scriptstyle \pi^*} & \\ \uparrow{\scriptstyle j^*} & & H^n_F(V, A) \\ & \nearrow{\scriptstyle s^*_\infty} & \\ H^n_{\mathbf{P}^1_F}(\mathbf{P}^1_V, A) & & \end{array}$$

is commutative.

Proof We may clearly limit ourselves to the cases where

(1) A is torsion prime to the characteristic of k, or

(2) A is p-primary torsion, where $p > 0$ is the characteristic of k.

In case (1), we use the following computation of $H^n(\mathbf{P}^1_V, A)$. For $i \in \mathbf{Z}$, let $A(i) = \varinjlim_m \operatorname{Hom}(\mu_m^{\otimes -i}, A)$. Recall the étale first Chern class of $\mathcal{O}(1)$ mod m:

$$c^{(m)} = c_1(\mathcal{O}(1))_m \in H^2(\mathbf{P}^1_V, \mu_m^{\otimes i})$$

defined as the boundary of the canonical class $[\mathcal{O}(1)] \in \operatorname{Pic}(\mathbf{P}^1_V) = H^1(\mathbf{P}^1_V, \mathbb{G}_m)$ in the long exact sequence associated to the Kummer exact sequence

$$1 \to \mu_m \to \mathbb{G}_m \xrightarrow{m} \mathbb{G}_m \to 1.$$

As V is quasi-compact and quasi-separated, the cup-products $H^i(V, {}_mA) \xrightarrow{\tilde{\pi}^*} H^i(\mathbf{P}^1_V, {}_mA) \xrightarrow{\cdot c^{(m)}} H^{i+2}(\mathbf{P}^1_V, {}_mA(1))$ for various m fit together and give a map

$$H^i(V, A) \xrightarrow{c_1(\mathcal{O}(1))} H^{i+2}(\mathbf{P}^1_V, A(1)).$$

We then have:

Proposition 4.1.1 *(compare* [**SGA5**, exposé VII, Corollary 2.2.4]*) In case (1), the natural map*

$$H^n(V,A)\oplus H^{n-2}(V,A(-1)) \xrightarrow{(\tilde{\pi}^*,c_1(\mathcal{O}(1)))} H^n(\mathbf{P}^1_V,A)$$

is an isomorphism for all $n\geq 0$. □

In case (2), things are even simpler:

Proposition 4.1.2 *(compare* [**13**, Lemma 3]*) In case (2), the natural map*

$$H^n(V,A)\to H^n(\mathbf{P}^1_V,A)$$

is an isomorphism for all $n\geq 0$. □

Using the exact sequence for cohomology with supports and applying Propositions 4.1.1 and 4.1.2 to V and $V-F$, we get canonical decompositions:

$$H^n_{\mathbf{P}^1_F}(\mathbf{P}^1_V,A)\simeq\begin{cases} H^n_F(V,A)\oplus H^{n-2}_F(V,A(-1)) & \text{in case (1)}\\ H^n_F(V,A) & \text{in case (2).}\end{cases} \tag{4.1}$$

The key lemma now follows from

Lemma 4.1.3 *Suppose A is torsion prime to the characteristic of k. In the diagram of the key lemma, the restrictions of s_∞^* and j^* to the factor $H^{n-2}_F(V,A(-1))$ of* (4.1) *are* 0.

Indeed, the map $H^{n-2}_F(V,A(-1))\to H^n_{\mathbf{P}^1_F}(\mathbf{P}^1_V,A)$ is given by cup-product by the first Chern class of $\mathcal{O}(1)$. But $s_\infty^*\mathcal{O}(1)$ is trivial, and so is $j^*\mathcal{O}(1)$. □

Remark 4.1.4 If A is torsion invertible on V, the key lemma has a much simpler proof: in this case π^* is an isomorphism by homotopy invariance [**36**, Chapter VI, p. 240, Corollary 4.20] (for the definition of acyclicity, see [**36**, Chapter VI, p. 232, Section 4]). Applying this homotopy invariance to the projection $(\mathbf{P}^1-\{1\})_V\to V$, we see that s_0 and s_∞, as right inverses of this projection, induce the same map on cohomology. Now replace s_∞^* by s_0^* in the diagram of the key lemma and add a map on the top

$$\begin{array}{ccc}
H^n_F(V,A) & & \\
\wr\uparrow s_0^* & & \\
H^n_{\mathbf{A}^1_F}(\mathbf{A}^1_V,A) & & \\
 & \underset{\simeq}{\nwarrow}{\scriptstyle \pi^*} & \\
\uparrow j^* & & H^n_F(V,A) \\
 & \nearrow{\scriptstyle s_0^*} & \\
H^n_{\mathbf{P}^1_F}(\mathbf{P}^1_V,A) & &
\end{array}$$

to make its commutativity obvious.

So Gabber's argument going via the cohomology of $\mathbf{P}^1$ can be thought of as a substitute for homotopy invariance when the latter does not hold.

See [**7**, Section 5] for more on the homotopy invariant point of view, notably in non-abelian situations.

4.2 The proof.

Theorem 4.2.1 *Let V be a k-scheme, F a closed subset of V and F' a closed subset of $\mathbf{A}^1_F$ such that the projection $f : F' \to F$ is finite.*

$$\begin{array}{ccccc} F' & \hookrightarrow & \mathbf{A}^1_F & \overset{i'}{\hookrightarrow} & \mathbf{A}^1_V \\ & \overset{f}{\searrow} & \downarrow{\scriptstyle \pi_F} & & \downarrow{\scriptstyle \pi} \\ & & F & \overset{i}{\hookrightarrow} & V \end{array}$$

Assume that V and $V - F$ are quasi-compact and quasi-separated. Then, for any torsion étale sheaf A of abelian groups on the small étale site of V, the map

$$H^n_{F'}(\mathbf{A}^1_V, A) \to H^n_{\mathbf{A}^1_F}(\mathbf{A}^1_V, A)$$

is identically 0.

Caution. When we write $H^n_{F'}(\mathbf{A}^1_V, A)$ and so on, we really mean $H^n_{F'}(\mathbf{A}^1_V, \pi^* A)$ and so on, where π^* denotes the pull-back morphism from sheaves over the small étale site of V to that of $\mathbf{A}^1_V$ via π. Being this fastidious would quickly become cumbersome notationally; hence we allow ourselves to abbreviate $\pi^* A$ to A, and similarly for the other morphisms to V; this should cause no confusion. Another way to present things is to consider the morphism α from the big étale site to the small étale site of V. What we do is consider $\alpha^* A$ and restrict it to the small étale site of T for any V-scheme T (and then we sneakily change the notation $\alpha^* A$ back to A).

Proof Note that $s_\infty(V) \cap F' = \emptyset$. Therefore we can factor s_∞ into

$$s_\infty = k \circ s'$$

where k is the open immersion $\mathbf{P}^1_V - F' \hookrightarrow \mathbf{P}^1_V$, and insert the diagram above into the bigger commutative diagram:

$$\begin{array}{ccccc} H^n_{F'}(\mathbf{A}^1_V, A) & \overset{\alpha}{\longrightarrow} & H^n_{\mathbf{A}^1_F}(\mathbf{A}^1_V, A) & & \\ & & & \nwarrow{\scriptstyle \pi^*} & \\ \text{excision} \uparrow \simeq & & \uparrow{\scriptstyle j^*} & & H^n_F(V, A) \\ & & & \nearrow{\scriptstyle s^*_\infty} & & \nwarrow{\scriptstyle s'^*} \\ H^n_{F'}(\mathbf{P}^1_V, A) & \overset{\beta}{\longrightarrow} & H^n_{\mathbf{P}^1_F}(\mathbf{P}^1_V, A) & & \overset{k^*}{\longrightarrow} & H^n_{\mathbf{P}^1_F - F'}(\mathbf{P}^1_V - F', A) \end{array}$$

where the bottom row is part of an exact sequence for cohomology with supports. Since $k^* \circ \beta = 0$, it obviously follows that $\alpha = 0$. □

Proof of Theorem 2.2.7. We may assume X irreducible. Suppose first k infinite. Let ψ, φ, V, U be as in Theorem 3.1.1, and set $Z' = \psi^{-1}(\psi(Z))$. We apply Theorem 4.2.1 with $V = V$, $F = \psi(Z)$ and $F' = \varphi(Z \cap U)$.

In the commutative diagram

$$\begin{array}{ccc} H^n_{Z\cap U}(U,A) & \longrightarrow & H^n_{Z'\cap U}(U,A) \\ \varphi^* \uparrow \wr & & \varphi^* \uparrow \\ H^n_{F'}(\mathbf{A}^1_V, A) & \longrightarrow & H^n_{\mathbf{A}^1_F}(\mathbf{A}^1_V, A) \end{array}$$

the left vertical map is an isomorphism by Corollary 3.1.2 and étale excision ([**36**, Chapter III, p. 92, Proposition 1.27] and [**8**, Proposition 4.4]), and the bottom horizontal map is 0 by Theorem 4.2.1. So the top horizontal map is 0 as well.

Suppose now that k is finite. We reduce to the infinite case by the following standard argument. Let p, q be two distinct prime numbers and let K_1, K_2 denote respectively the $\mathbf{Z}_p$ and $\mathbf{Z}_q$-extensions of k. Let $(\psi_1, \varphi_1, V_1, U_1)$, $(\psi_2, \varphi_2, V_2, U_2)$ be as in Theorem 3.1.1 and Z'_1, Z'_2 be as above for (X_{K_1}, Z_{K_1}) and (X_{K_2}, Z_{K_2}) respectively. There are finite subextensions $k \subset k_1 \subset K_1$, $k \subset k_2 \subset K_2$ such that $(\psi_1, \varphi_1, V_1, U_1, Z'_1)$ and $(\psi_2, \varphi_2, V_2, U_2, Z'_2)$ are respectively defined over k_1 and k_2. Note that the effacement theorem holds respectively over k_1 and k_2 with these choices, by the above. Define

$$U = X - \left(p_1(X_{k_1} - U_1) \cup p_2(X_{k_2} - U_2)\right), \; Z' = p_1(Z'_1) \cup p_2(Z'_2)$$

where $p_1 : X_{k_1} \to X$, $p_2 : X_{k_2} \to X$ are the two projections. In other words, U and $X - Z'$ are those parts of $(U_1)_{\overline{k}} \cap (U_2)_{\overline{k}}$ and $X_{\overline{k}} \setminus \left((Z_1)_{\overline{k}} \cup (Z_2)_{\overline{k}}\right)$ which are rational over k. Note that U contains all t_i's. We have

$$U_{k_i} \subseteq U_i,\; Z_i \subseteq Z_{k_i}\; (i = 1, 2).$$

Considering the commutative diagrams $(i = 1, 2)$

$$\begin{array}{ccc} H^n_{(Z\cap U)_{k_i}}(U_{k_i}, A) & \longrightarrow & H^n_{(Z'\cap U)_{k_i}}(U_{k_i}, A) \\ \uparrow & & \uparrow \\ H^n_{Z\cap U_i}(U_i, A) & \xrightarrow{\;0\;} & H^n_{Z'\cap U_i}(U_i, A) \\ \uparrow & & \\ H^n_Z(X, A) & & \end{array}$$

shows that the composite $H^n_Z(X,A) \to H^n_{Z_{k_i}}(X_{k_i}, A) \to H^n_{(Z\cap U)_{k_i}}(U_{k_i}, A) \to H^n_{(Z'\cap U)_{k_i}}(U_{k_i}, A)$ is 0 for $i = 1, 2$. Equivalently, the composite $H^n_Z(X,A) \to H^n_{Z\cap U}(U,A) \to H^n_{Z'\cap U}(U,A) \to H^n_{(Z'\cap U)_{k_i}}(U_{k_i}, A)$ is 0 for $i = 1, 2$. But the composite

$$H^n_{Z'\cap U}(U,A) \to H^n_{(Z'\cap U)_{k_i}}(U_{k_i}, A) \to H^n_{Z'\cap U}(U,A)$$

equals multiplication by $[k_i : k]$, where the second map is transfer. Since these two degrees are coprime, this shows that the composite $H^n_Z(X,A) \to H^n_{Z\cap U}(U,A) \to H^n_{Z'\cap U}(U,A)$ is 0 indeed. □

Part II Other Cohomology Theories

5 Axiomatizing Gabber's Proof

5.1 Basic axioms. In [**2**], Bloch and Ogus prove their main theorem not only for étale cohomology with coefficients twisted roots of unity, but also for other "cohomology theories with supports". Counterexample 2.2.3 (3) and the method of proof in the present paper show that this point of view should be taken seriously. On the other hand, the Bloch-Ogus axioms are very complicated, and the present proof shows that many of them are unnecessary. In this section, we want to indulge in the exercise of finding a convenient and simpler set of axioms which is enough to make the proof of the effacement theorem work.

Let $\mathcal{S}_k$ be a full subcategory of the category of algebraic k-schemes, stable under étale extensions. In practice, $\mathcal{S}_k$ will be either the category Var/k of all separated algebraic k-schemes or the category Sm/k of smooth k-schemes. Let $\mathcal{P}_k$ the category of pairs (X, Z), where $X \in \mathcal{S}_k$ and Z is a closed subset of X. By definition, a morphism $(X', Z') \xrightarrow{f} (X, Z)$ of $\mathcal{P}_k$ is any k-morphism $f : X' \to X$ such that $f^{-1}(Z) \subseteq Z'$ (example: $X' = X$, f the identity, $Z \subseteq Z'$).

The most naïve is to ask for a "cohomology theory with supports"

$$(X, Z) \mapsto h^*_Z(X)$$

a contravariant functor from $\mathcal{P}_k$ to $\mathbf{Z}$-graded abelian groups, satisfying conditions abstracted from the proof of Theorem 2.2.7. It is more natural, however, and more powerful, as we shall see in section 6, to give such a theory a cohomological support

$$(X, Z) \mapsto C_Z(X)$$

where $C_Z(X)$ is, for example, a complex of abelian groups. It may be useful to allow $C_Z(X)$ to be a complex of objects in more general abelian categories $\mathcal{A}$ than abelian groups, for example if we want to have some ring action on the situation. So we give the following general definitions:

Definition 5.1.1 Let $\mathcal{A}$ be an abelian category.
a) A *cohomology theory with supports* is a contravariant functor $(X, Z) \mapsto h^*_Z(X)$ from $\mathcal{P}_k$ to $\mathcal{A}$ satisfying:

For any triple $Z \subseteq Y \subseteq X$, where Y, Z are closed in X, there is a long exact sequence ($q \in \mathbf{Z}$)

$$\cdots \to h^q_Z(X) \to h^q_Y(X) \to h^q_{Y-Z}(X - Z) \to h^{q+1}_Z(X) \to \cdots$$

which is natural in (X, Y, Z) in an obvious sense.
b) A *substratum* is a contravariant functor $X \mapsto C(X)$ from $\mathcal{S}_k$ to complexes of objects of $\mathcal{A}$.

In b), define $C_Z(X)$ as the homotopy fibre of $C(X) \to C(X - Z)$, i.e., $C[-1]$, where C is the mapping cone of this morphism. This definition is natural in (X, Z). For all triples (X, Y, Z) as in a), the sequence of complexes

$$0 \to C_Z(X) \to C_Y(X) \to C_{Y-Z}(X - Z) \to 0$$

is exact up to homotopy ([**20**, p. 47, Proposition 5.12], [**29**, p. 32, I.4.22]). Hence, defining

$$h^q_Z(X) = H^q(C_Z(X)) \tag{5.1}$$

yields a cohomology theory with supports in the sense of a).

Recall that a contravariant functor T to an additive category is *additive* if it commutes with finite coproducts, i.e., $T(X\coprod Y)\to T(X)\times T(Y)$ is an isomorphism for all X, Y.

We now introduce a first axiom for a cohomology theory with supports h^* (resp. a substratum C) in the sense of Definition 5.1.1.

COH1 (Etale excision). h^* is additive and for any diagram

$$\begin{array}{ccc} & & X' \\ & \nearrow & \Big\downarrow f \\ Z & \hookrightarrow & X \end{array}$$

where f is étale and $f^{-1}(Z)\xrightarrow{f} Z$ is an isomorphism, the induced map

$$h^q_Z(X)\xrightarrow{f^*} h^q_Z(X')$$

is an isomorphism for all q.

SUB1 (Etale Mayer-Vietoris). C is additive and for Z, X', X, f as in a), the commutative square

$$\begin{array}{ccc} C(X') & \xrightarrow{v} & C(X'-Z) \\ f\Big\uparrow & & f\Big\uparrow \\ C(X) & \xrightarrow{u} & C(X-Z) \end{array}$$

is homotopy cartesian.

Recall that a commutative square of complexes

$$\begin{array}{ccc} A & \longrightarrow & B \\ \Big\uparrow & & \Big\uparrow \\ C & \longrightarrow & D \end{array}$$

is *homotopy cartesian* if the natural map from the mapping cone of $[C\to A\oplus D]$ to B is a homotopy equivalence.

Lemma 5.1.2 *The square of axiom **SUB1** is homotopy cartesian if and only if the induced map*

$$C_Z(X)\xrightarrow{f} C_Z(X')$$

is a homotopy equivalence.

Proof This follows from the triangulated category version of the nine diagram. More precisely, consider the map of exact triangles

$$\begin{array}{ccccc}
C(X')\oplus C(X-Z) & \xrightarrow{diag(v,Id)} & C(X'-Z)\oplus C(X-Z) & \longrightarrow & C_Z(X')[1] \\
{\scriptstyle (f,-u)}\uparrow & & {\scriptstyle (f,-Id)}\uparrow & & {\scriptstyle f}\uparrow \\
C(X) & \xrightarrow{u} & C(X-Z) & \longrightarrow & C_Z(X)[1]
\end{array}$$

in the homotopy category $K(\mathcal{A})$. By [**29**, Proposition 5.6 of chapter XI], we can complete this diagram, up to isomorphism, into

$$\begin{array}{ccccc}
D & \longrightarrow & D' & \longrightarrow & D'' \\
\uparrow & & \uparrow & & \uparrow \\
C(X')\oplus C(X-Z) & \xrightarrow{diag(v,Id)} & C(X'-Z)\oplus C(X-Z) & \longrightarrow & C_Z(X')[1] \\
{\scriptstyle (f,-u)}\uparrow & & {\scriptstyle (f,-Id)}\uparrow & & {\scriptstyle f}\uparrow \\
C(X) & \xrightarrow{u} & C(X-Z) & \longrightarrow & C_Z(X)[1]
\end{array}$$

in which all rows and columns are exact triangles. It is clear that:

- the middle top vertical map induces an isomorphism $C(X'-Z) \xrightarrow{\sim} D'$.
- $D'' \simeq 0 \iff C_Z(X) \xrightarrow{\sim} C_Z(X')$.
- **SUB1** holds $\iff D \xrightarrow{\sim} D' \iff D'' \simeq 0$.

The claim follows. □

Remarks 5.1.3

(1) We say that a cohomology theory (resp. a substratum) satisfies *Zariski excision* (resp. *Zariski Mayer-Vietoris*) if axiom **COH1** (resp. **SUB1**) holds when we let f run through open immersions. The obvious analogue of Lemma 5.1.2 (ordinary excision) holds.

(2) Note that in **SUB1** one can replace the condition "C is additive" by "$C(\emptyset) = 0$" (take the case $X' = Z$ in the commutative square).

(3) Definition 5.1.1 allows us to set up a coniveau exact couple and spectral sequence as in Section 1. Zariski excision allows us to recognize the E_1-terms of the coniveau spectral sequence in the form of Equation (1.1), producing *Cousin complexes* in the sense of [**24**] by Zariski sheafification. In particular, if h^* satisfies Zariski excision, we get a convergent *coniveau spectral sequence*, for all $X \in \mathcal{S}_k$

$$E_1^{p,q} = \coprod_{x\in X^{(p)}} h_x^{p+q}(X) \Rightarrow h^{p+q}(X)$$

analogous to (1.2), where $h^n_x(X) = \varinjlim_{U \ni x} h^n_{\overline{x} \cap U}(U)$ (note that $E_1^{p,q} = 0$ for $p \notin [0, \dim X]$).

As we shall see in Section 7, if h^* is defined by a substratum, Zariski excision gives rise to an (a priori unrelated) *Brown-Gersten* spectral sequence as well.

Lemma 5.1.4 *For a substratum C and the associated cohomology theory with supports h^*:*
*a) Axiom **SUB1** implies axiom **COH1**.*
b) The converse is true if, for all X, $C(X)$ is fibrant in the sense of Definition C.1.1 b) of Appendix C.

Proof Part a) follows trivially from Lemma 5.1.2. Part b) also follows from this lemma, Lemma C.1.4 b) and Corollary C.2.7. □

Remarks 5.1.5

(1) Obviously, Lemma 5.1.4 holds when étale Mayer-Vietoris is replaced by Zariski Mayer-Vietoris.
(2) By Theorem C.3.1, if $\mathcal{A}$ verifies axiom AB5 and has a generator in the sense of [**23**, 1.5 and 1.6] and if moreover countable products are exact in $\mathcal{A}$ (for example, $\mathcal{A}$ satisfies AB4*), then there exists another substratum F and a natural transformation $C \xrightarrow{\psi} F$ such that, for all X,
   (a) $F(X)$ is fibrant;
   (b) ψ_X is a monomorphism and a quasi-isomorphism.
   This applies to the case where $\mathcal{A}$ is the category of left R-modules over some ring [**23**, Section 1].

Let us now introduce our second axiom. To do this, we need an assumption on $\mathcal{S}_k$.

Assumption 5.1.6

a) $\operatorname{Spec} k \in \mathcal{S}_k$.
b) If $X \in \mathcal{S}_k$, then $\mathbf{P}^1_X \in \mathcal{S}_k$.

Lemma 5.1.7 *If $\mathcal{S}_k$ satisfies Assumption 5.1.6, then*
a) $X \in \mathcal{S}_k \Rightarrow \mathbf{A}^1_X \in \mathcal{S}_k$.
b) For any $n \geq 1$, the open subsets of $\mathbf{A}^n_k$ are in $\mathcal{S}_k$. □

In axioms **COH2** and **SUB2**, we assume that $\mathcal{S}_k$ satisfies Assumption 5.1.6.

COH2 ("Key lemma" for cohomology) Let V be an open subset of $\mathbf{A}^n_k$ (for some n) and

$$\begin{array}{ccccc} \mathbf{A}^1_V & \xrightarrow{j} & \mathbf{P}^1_V & \xleftarrow{s_\infty} & V \\ & {\scriptstyle\pi}\searrow & \downarrow{\scriptstyle\tilde{\pi}} & \swarrow{\scriptstyle =} & \\ & & V & & \end{array}$$

be the diagram representing the inclusion of $\mathbf{A}^1_V$ and the section at infinity into $\mathbf{P}^1_V$. Let F be a closed subset of V. Then the diagram of the key lemma of subsection 4.1

$$\begin{array}{ccc} h^q_{\mathbf{A}^1_F}(\mathbf{A}^1_V) & & \\ & \nwarrow \pi^* & \\ j^* \uparrow & & h^q_F(V) \\ & \nearrow s^*_\infty & \\ h^q_{\mathbf{P}^1_F}(\mathbf{P}^1_V) & & \end{array}$$

is commutative.

SUB2 ("Key lemma" for substrata) Let V, F be as in axiom **COH2**. Then the diagram

$$\begin{array}{ccc} C_{\mathbf{A}^1_F}(\mathbf{A}^1_V) & & \\ & \nwarrow \pi^* & \\ j^* \uparrow & & C_F(V) \\ & \nearrow s^*_\infty & \\ C_{\mathbf{P}^1_F}(\mathbf{P}^1_V) & & \end{array}$$

is homotopy commutative.

We give a last definition.

Definition 5.1.8 Let $X \in \mathcal{S}_k$ be affine and let $t_1, \ldots, t_r \in X$ be a finite set of points. A cohomology theory with supports h^* (resp. a substratum C) over k is *strictly effaceable at* $(t_1, \ldots, t_r)$ if, given $p \geq 0$, for any open neighbourhood $W \subseteq X$ of $t_1, \ldots, t_r$ and for any closed subset $Z \subseteq W$ of codimension $\geq p+1$, there exist an open neighbourhood $U \subseteq W$ of $t_1, \ldots, t_r$ and a closed subset $Z' \subseteq W$ containing Z such that $\operatorname{codim}_W(Z') \geq p$ and the map $h^q_{Z \cap U}(U) \to h^q_{Z' \cap U}(U)$ is 0 for all $q \in \mathbf{Z}$ (resp. the map $C_{Z \cap U}(U) \to C_{Z' \cap U}(U)$ is nullhomotopic). It is *strictly effaceable* if this condition is satisfied for any $(X, t_1, \ldots, t_r)$ as above, with X smooth.

Example 5.1.9 Suppose k infinite and let A be a sheaf of torsion abelian groups over the small étale site of $\operatorname{Spec} k$. By Theorem 2.2.7, the cohomology theory with supports

$$(X, Z) \mapsto H^*_Z(X_{\text{ét}}, \alpha^* A)$$

is strictly effaceable, where α is the projection of the big étale site on the small étale site.

The following theorem is immediate from the arguments of Section 4.

Theorem 5.1.10 *Let k be infinite and S_k satisfy Assumption 5.1.6. A cohomology theory with supports h^* (resp. a substratum C) satisfying étale excision **COH1** (resp. étale Mayer-Vietoris **SUB1**) and the key lemma for cohomology **COH2** (resp. the key lemma for substrata **SUB2**) is strictly effaceable.*

The point here is that, in diagrams where there is an excision map as in subsection 4.2, one can use a homotopy inverse of this map to show that the map one wishes to be nullhomotopic is indeed nullhomotopic.

Corollary 5.1.11 *Let k be infinite and $\mathcal{S}_k$ verify Assumption 5.1.6, and let the cohomology theory with supports h^* satisfy axioms **COH1** and **COH2**. Then, for any smooth $X \in \mathcal{S}_k$, the Cousin complexes are flasque resolutions of the Zariski sheaves $\mathcal{H}^q$ associated to the presheaves $U \mapsto h^q(U)$, and the E_2-terms of the spectral sequence of Remark 5.1.3 (3) are*

$$E_2^{p,q} = H^p_{\mathrm{Zar}}(X, \mathcal{H}^q).$$

There is a need for something extra, like transfer maps, to deal with finite fields: see Theorem 6.2.5.

Remarks 5.1.12

(1) The axioms above are much more economical than those of Bloch and Ogus in [**2**, Section 1]. Definition 5.1.1 corresponds to axioms (1.1.1) and (1.1.2) of Bloch-Ogus. Axiom **COH1** corresponds to axiom (1.1.3). Axiom **COH2** has no counterpart in [**2**], but might be compared with [**2**, (1.5)]. On the other hand, we do not need to introduce any twists, nor a corresponding homology theory. This means that purity, let alone Poincaré duality, is irrelevant for strict effaceability.

(2) However, the "key lemma" we axiomatized in **COH2** and **SUB2** is unsatisfactory, because it is not obvious how to check it in practice. Moreover, **COH2** need not imply **SUB2** even for fibrant substrata. In subsections 5.3 and 5.4, we introduce stronger axioms that do not have this defect.

5.2 Spectra.

In order to include algebraic K-theory in the formalism of this section, it is necessary to consider substrata with values not only in complexes, but also in the category $\mathcal{E}$ of *spectra* in the sense of algebraic topology. We refer to [**4**, Section 2] for the definition of such a suitable category $\mathcal{E}$, provided with an appropriate closed model structure (fibrations, cofibrations, weak equivalences). Recall (e.g. [**50**, 5.32]) that the Dold-Kan correspondence gives rise to an embedding DK of the category of complexes of abelian groups into the category of spectra such that $\pi_n(DK(C^\cdot)) = H^{-n}(C^\cdot)$ for any complex $C^\cdot$ of abelian groups.

The preceding subsection "extends" to substrata with values in spectra by

- replacing "complexes of objects of $\mathcal{A}$" by "objects of $\mathcal{E}$" in Definition 5.1.1 b);
- defining $h^q_Z(X) = \pi_{-q}(C_Z(X))$ in (5.1);
- replacing the cases in Lemma 5.1.4 b) by "For all $X \in \mathcal{S}_k$, the spectrum $C(X)$ is fibrant and cofibrant."
- replacing "Lemma C.1.4 b)" in the proof of Lemma 5.1.4 b) by "the following fact: a weak equivalence between two fibrant and cofibrant spectra is a homotopy equivalence".
- replacing Remark 5.1.5 (2) by "For any substratum $C : \mathcal{S}_k \to \mathcal{E}$, there exist two substrata C', C'' and natural transformations $C \xrightarrow{\varphi} C' \xleftarrow{\varphi'} C''$ such that for all X,
  (a) $C'(X)$ is fibrant;
  (b) $C''(X)$ is fibrant and cofibrant;
  (c) φ_X, φ'_X are weak equivalences.

  This follows from the folklore result, for which we have no reference, that any map between objects of $\mathcal{E}$ can be factored *in a functorial way* into a cofibration followed by a trivial fibration, and also into a trivial cofibration followed by a fibration (small object argument)."

In the next sections, we shall allow substrata C to take their values in spectra, and comment on this only when necessary.

5.3 Homotopy invariance. In this subsection we discuss two new axioms **COH3/SUB3** and **COH4/SUB4** for cohomology theories/substrata. As we shall see in Section 7, the axiom **COH3/SUB3** below is satisfied by many theories. Axiom **COH4/SUB4** is auxiliary and merely serves to give a smooth proof that **COH3** $\Rightarrow$ **COH2** (resp. **SUB3** $\Rightarrow$ **SUB2**).

We assume that $\mathcal{S}_k$ satisfies Assumption 5.1.6.

COH3 (Homotopy invariance for cohomology). Suppose that V, π are as in axiom **COH2**. Then $h^q_F(V) \xrightarrow{\pi^*} h^q_{\mathbf{A}^1_F}(\mathbf{A}^1_V)$ is an isomorphism for all q.

COH4 (Rigidity for cohomology). Let $V, \tilde{\pi}, F$ be as in axiom **COH2**, and let s_0, s_∞ be the sections at 0 and ∞ of $\tilde{\pi}$. Then $s_0^*, s_\infty^* : h^q_{\mathbf{P}^1_F}(\mathbf{P}^1_V) \to h^q_F(V)$ coincide for all q.

SUB3 (Homotopy invariance for substrata). Let V, π be as in axiom **COH2**. Then $C(V) \xrightarrow{\pi^*} C(\mathbf{A}^1_V)$ is a homotopy equivalence.

SUB4 (Rigidity for substrata). Let $V, \tilde{\pi}, F$ be as in axiom **COH2**, and let s_0, s_∞ be the sections at 0 and ∞ of $\tilde{\pi}$. Then $s_0^*, s_\infty^* : C_{\mathbf{P}^1_F}(\mathbf{P}^1_V) \to C_F(V)$ are homotopic.

Lemma 5.3.1 *Let h^* be the cohomology theory with supports associated to the substratum C.*
a) Axiom **SUB3** *implies axiom* **COH3**.
b) If $C(X)$ is fibrant and cofibrant for all X, then axiom **COH3** *implies axiom* **SUB3**.

Proof Part a) : Axiom **SUB3** implies the same property for substrata with support, by the same argument as in the proof of Lemma 5.1.2. Part b): the proof is the same as for Lemma 5.1.4. □

Proposition 5.3.2 *a) Axiom* **COH3** *implies axiom* **COH2** *and axiom* **COH2** *implies axiom* **COH4**.
b) Axiom **SUB3** *implies axiom* **SUB2** *and axiom* **SUB2** *implies axiom* **SUB4**.

Proof Part a): compare Remark 4.1.4. Part b) is analogous but we give a detailed proof for the convenience of the reader. First we show that **SUB2** implies **SUB4**. Generally, for $x \in \mathbf{P}^1(k)$ (resp. $x \in \mathbf{A}^1(k)$), let us denote by s_x (resp. s'_x) the section of $\tilde{\pi}$ (resp. of π) determined by x. We can complete the diagram of **SUB2** as

$$\begin{array}{ccc} C_F(V) & & \\ {\scriptstyle s'^*_0}\uparrow & & \\ C_{\mathbf{A}^1_F}(\mathbf{A}^1_V) & & \\ & \nwarrow{\scriptstyle \pi^*} & \\ {\scriptstyle j^*}\uparrow & & C_F(V). \\ & \nearrow{\scriptstyle s^*_\infty} & \\ C_{\mathbf{P}^1_F}(\mathbf{P}^1_V) & & \end{array} \qquad (5.2)$$

It is then clear that the vertical composition is s_0^* while the right composition is s_∞^*.

We now show that **SUB3** implies **SUB4**. Axiom **SUB3** implies that ${s'}_x^*$ is a homotopy inverse of π^* for all $x \in \mathbf{A}^1(k)$. Consider now the inclusion $j_1 : \mathbf{A}_X^1 \hookrightarrow \mathbf{P}_X^1$ given by $t \mapsto t/(t-1)$. We have $j_1(0) = 0$ and $j_1(1) = \infty$, or in other words:

$$j_1 \circ s_0' = s_0, \qquad j_1 \circ s_1' = s_\infty.$$

Since ${s'}_0^*$ and ${s'}_1^*$ are homotopic, it follows that s_0^* and s_∞^* are homotopic.

Finally, we show that **SUB3** implies **SUB2**. Using **SUB3** $\Rightarrow$ **SUB4**, we may replace, up to homotopy, s_∞^* by s_0^* in diagram 5.3, which then becomes obviously commutative. But **SUB3** implies that ${s'}_0^*$ is a homotopy equivalence, hence the triangle of **SUB2** is homotopy commutative as desired. □

5.4 Cohomology of $\mathbf{P}^1$. In order to express our axiom on the cohomology of $\mathbf{P}^1$, we need to introduce more material. We still assume $\mathcal{S}_k$ to satisfy Assumption 5.1.6.

a) *Cohomology theories.* We suppose given a cohomology theory h^*, a cohomology theory e^* and, for any $(X, Z) \in \mathcal{P}_k$, a map

$$\operatorname{Pic} X \to \operatorname{Hom}(e_Z^*(X), h_Z^*(X))$$

which is natural in (X, Z) (we do not require this map to be additive). Taking $X = \mathbf{P}_V^1$, $Z = \mathbf{P}_F^1$, we get a homomorphism

$$e^*_{\mathbf{P}^1_F}(\mathbf{P}^1_V) \xrightarrow{[\mathcal{O}(1)]-[\mathcal{O}]} h^*_{\mathbf{P}^1_F}(\mathbf{P}^1_V)$$

hence, composing with $\tilde{\pi}^*$, a homomorphism

$$e_F^*(V) \xrightarrow{\alpha(V,F)} h^*_{\mathbf{P}^1_F}(\mathbf{P}^1_V)$$

natural in (V, F).

COH5 (cohomology of $\mathbf{P}^1$, cohomological version) Assume that $V, F, \tilde{\pi}$ are as in axiom **COH2**. Then the natural map

$$h_F^q(V) \oplus e_F^q(V) \xrightarrow{(\tilde{\pi}^*, \alpha(V,F))} h^q_{\mathbf{P}^1_F}(\mathbf{P}^1_V)$$

is an isomorphism for all q.

b) *Substrata.* We suppose given a substratum C, a substratum D and, for any $X \in \mathcal{S}_k$, a map

$$\operatorname{Pic} X \to \operatorname{Hom}_{\mathcal{E}}(D(X), C(X)) \qquad (5.3)$$

natural in X, where $\mathcal{E}$ is either the category of complexes of objects of our abelian category $\mathcal{A}$ or the category of spectra of subsection 5.2. Taking $X = \mathbf{P}^1_V$ we get a map (for spectra, in the stable homotopy category)

$$D(\mathbf{P}^1_V) \xrightarrow{[\mathcal{O}(1)]-[\mathcal{O}]} C(\mathbf{P}^1_V)$$

hence, composing with $\tilde{\pi}^*$, a map (for spectra, in the stable homotopy category)

$$D(V) \xrightarrow{\alpha_V} C(\mathbf{P}^1_V)$$

natural in V.

SUB5 (cohomology of $\mathbf{P}^1$, substratum version) Assume that $V, \tilde{\pi}$ are as in axiom **COH2**. Then the natural map (for spectra, in the stable homotopy category)

$$C(V) \oplus D(V) \xrightarrow{(\tilde{\pi}^*, \alpha_V)} C(\mathbf{P}^1_V)$$

is a homotopy equivalence.

(To be correct, we should use wedge $\vee$ rather than direct sum $\oplus$ in **SUB5** when C and D are given by spectra.)

Remarks 5.4.1

(1) The map (5.4) induces, by functoriality, a map on cones

$$\operatorname{Pic} X \to \operatorname{Hom}_{\mathcal{E}}(D_Z(X), C_Z(X))$$

for any $(X, Z) \in \mathcal{P}_k$. Hence we get a map (for spectra, in the stable homotopy category)

$$C_F(V) \oplus D_F(V) \xrightarrow{(\tilde{\pi}^*, \alpha_V)} C_{\mathbf{P}^1_F}(\mathbf{P}^1_V)$$

generalizing that of axiom **SUB5**, and the latter implies by the usual argument (cf. proof of Lemma 5.1.2) that this generalized map is a homotopy equivalence as well.

(2) Axiom **COH5** implies that the cohomology theory e^* is uniquely determined by h^* up to isomorphism. For example, e^* verifies Zariski (resp. étale) excision if h^* does. Similarly, axiom **SUB5** implies that D is uniquely determined by C up to homotopy. But the action of Pic is not determined by these axioms in an obvious way.

Lemma 5.4.2 *Let h^* be the cohomology theory with supports associated to the substratum C and e^* the cohomology theory with supports associated to the substratum D.*
a) Axiom **SUB5** *implies axiom* **COH5**.
b) If $C(X)$ is fibrant and cofibrant for all X, then axiom **COH5** *implies axiom* **SUB5**.

Proof a) follows from Remark 5.4.1 (1); b) is proven as in Lemma 5.1.4 b). □

Proposition 5.4.3 *a) Axiom* **COH5** *implies axiom* **COH2**.
b) Axiom **SUB5** *implies axiom* **SUB2**.

Proof For a), compare proof of Lemma 4.1.3. Part b) is similar but we give a detailed proof, as in subsection 5.3. By Remark 5.4.1, we are reduced to checking that in the diagram

$$\begin{array}{ccc}
 & C_{\mathbf{A}^1_F}(\mathbf{A}^1_V) & \\
 & {\scriptstyle j^*}\uparrow & \nwarrow{\scriptstyle \pi^*} \\
 & & C_F(V) \\
 & & \nearrow{\scriptstyle s_\infty^*} \\
C_F(V)\oplus D_F(V) & \xrightarrow{(\tilde\pi^*,\alpha_V)} \; C_{\mathbf{P}^1_F}(\mathbf{P}^1_V) &
\end{array}$$

the two paths from $C_F(V)\oplus D_F(V)$ to $C_{\mathbf{A}^1_F}(\mathbf{A}^1_V)$ are homotopic. It is enough to check this on both components $C_F(V)$ and $D_F(V)$. On $C_F(V)$ this is trivial (the two paths are actually equal). On $D_F(V)$, the two paths are nullhomotopic, because the pull-backs of $\mathcal{O}(1)$ by s_∞ and j are both trivial. □

5.5 Generating new theories out of old. The following remarks show how to construct some strictly effaceable cohomology theories and substrata. Here substrata take their values either in $C(\mathcal{A})$, where $\mathcal{A}$ is a suitable abelian category, or in the category of spectra $\mathcal{E}$ of subsection 5.2.

(1) Let $\mathcal{S}_k = Var/k$, let h^* (resp. C) be a cohomology theory with supports (resp. a substratum) over $\mathcal{S}_k$, and let $T\in Var/k$. Define a new cohomology theory with supports h^T (resp. substratum C^T) by

$$(h^T)^*_Z(X) = h^*_{Z\times_k T}(X\times_k T)$$

(resp.

$$C^T(X) = C(X\times_k T)).$$

Assume that h^* (resp. C) satisfies axiom **COH1** (resp. **SUB1**) (for all k-schemes). Then h^T (resp. C^T) also does. This is obvious.

Suppose now that h^* (resp. C) satisfies axiom **COHi** (resp. **SUBi**) for some $i\in\{2,3,5\}$, not only for open subsets of $\mathbf{A}^n_k$ but for any $V\in Var/k$. Then the same holds for h^T (resp. C^T). This is equally obvious.

(2) Let $C\xrightarrow{f} C'$ be a morphism of substrata, and let C'' be the homotopy fibre of f. Then, for $i=1,3$, if two among C, C', C'' verify axiom **SUBi**, so does the third. This is *not* clear (and probably wrong) for axioms **SUB2** and **SUB4**, or for "strictly effaceable". As for axiom **SUB5**, the following holds:
Let D, D' be the substrata attached respectively to C and C' in **COH5**. Assume given a morphism $f: D\to D'$ such that, for any $X\in\mathcal{S}_k$ and $\alpha\in\operatorname{Pic}X$, the diagram

$$\begin{array}{ccc}
D(X) & \xrightarrow{\alpha_*} & C(X) \\
{\scriptstyle f_X}\downarrow & & \downarrow{\scriptstyle f_X} \\
D'(X) & \xrightarrow{\alpha_*} & C'(X)
\end{array}$$

commutes. Let D'' be the homotopy fibre of $D \xrightarrow{f} D'$. From the assumption above, we get a natural transformation

$$\operatorname{Pic} X \to \operatorname{Hom}(D''(X), C''(X)).$$

Then, if two among the pairs $(C, D), (C', D'), (C'', D'')$ (together with the actions of Pic) verify axiom **COH5**, so does the third.

(3) Let $(h_\alpha)_{\alpha \in A}$ (resp. $(C_\alpha)_{\alpha \in A}$) be a filtered direct system of cohomology theories with supports (resp. substrata) and $h = \varinjlim h_\alpha$ (resp. $C = \varinjlim C_\alpha$). If all h_α (resp. all C_α) verify axiom **COHi** for some i (resp. **SUBi** for $i = 1, 3$), then so does h^* (resp. C). The same claim for **SUB2** and **SUB4** in the case of substrata is not clear. As for **COH5**, we must request that the D_α attached to the C_α form a direct system compatible with that of the C_α via the actions of Pic.

(4) Let C be a substratum, and suppose given a direct system of substrata

$$\cdots \to C^{(n)} \to C^{(n+1)} \to \cdots \qquad (n \geq 0)$$

with a homotopy equivalence $\varinjlim C^{(n)} \xrightarrow{\sim} C$. Suppose that $C^{(0)}$ and, for all n, the homotopy fibre of $C^{(n)} \to C^{(n+1)}$ satisfies axiom **SUBi** for $i = 1, 3$ or 5 (for **SUB5**, we request analogous conditions on the D's, as above). Then so does C. This follows by induction from Remarks 5.5 and 5.5.

6 Universal Exactness

In this section, we want to show how strict effaceability of a substratum (rather than a cohomology theory) implies not only exactness, but even *universal exactness* of the associated Cousin complexes. Recall that a complex $A^\cdot$ is *contractible* if there is a homotopy from the identity to 0 on $A^\cdot$.

6.1 Generalities. We take from [**19**] the definition of universal exactness, actually in slightly greater generality:

Definition 6.1.1 Let $\mathcal{A}$ be an abelian category. A complex $C^\cdot$ of objects of $\mathcal{A}$ is *universally exact* if the following condition is satisfied:

For any abelian category $\mathcal{B}$ and any additive functor $T : \mathcal{A} \to \mathcal{B}$ commuting with filtering direct limits, the complex $T(C^\cdot)$ is exact.

(In the case $\mathcal{A}$ is the category of left modules over a ring, one should compare this notion with Lazard's pure sequences [**32**, Chapter I, Section 2, esp. Theorem 2.3].)

Note that the exactness of $T(C)$ is automatic if T is exact, but we only require it to be *additive*. Here are some examples:

Examples 6.1.2

(1) A contractible complex is universally exact. Indeed, any additive functor will transform a homotopy into a homotopy.

(2) Let $\mathcal{A}$ be an abelian category, and let $\mathcal{B}$ be an abelian category satisfying AB5. For any filtering direct limit $C^\cdot$ of universally exact complexes, and for any exact functor $T : \mathcal{A} \to \mathcal{B}$ commuting with filtering direct limit, the complex $T(C^\cdot)$ is exact.

(3) Let $C^\cdot : \ 0 \to C^0 \to \cdots \to C^{n-1} \to C^n \to 0$ be a bounded exact complex, so that $B^i(C^\cdot) \xrightarrow{\sim} Z^i(C^\cdot)$ for all i. Suppose all the exact sequences $0 \to Z^i(C^\cdot) \to C^i \to B^{i+1}(C) \to 0$ are filtering direct limits of split exact

sequences. Then $C^{\cdot}$ is universally exact. This follows from the previous two examples.

Conversely:

Proposition 6.1.3 *Suppose $\mathcal{A}$ satisfies AB5 and any object of $\mathcal{A}$ is a filtering direct limit of finitely presented objects (e.g. $\mathcal{A}$ is the category of left modules over a ring). Then any bounded universally exact complex $C^{\cdot}$ of objects of $\mathcal{A}$ can be described as in Example 6.1.2 (3).*

Recall that an object X of $\mathcal{A}$ is of finite presentation if the functor $Y \mapsto \operatorname{Hom}(X, Y)$ commutes with direct limits.

Proof Applying Definition 6.1.1 with $T =$ identity, we see that $C^{\cdot}$ is exact. Let X be a finitely presented object of $\mathcal{A}$. Applying the functor

$$T(M) = \operatorname{Hom}(X, M)$$

to $C^{\cdot}$, we see in particular that $\operatorname{Hom}(X, C^{n-1}) \to \operatorname{Hom}(X, C^n)$ is surjective. It follows that, for any $f : X \to C^n$, the pull-back of the exact sequence

$$0 \to Z^{n-1}(C^{\cdot}) \to C^{n-1} \to C^n \to 0 \tag{D}$$

by f is split. By the assumption in Proposition 6.1.3, (D) is a filtering direct limit of split exact sequences, and in particular is universally exact. This now implies that the sequence

$$0 \to C^0 \to \cdots \to C^{n-2} \to B^{n-1}(C^{\cdot}) \to 0$$

is universally exact. We get the conclusion by induction on n. □

6.2 Universal exactness of Cousin complexes.

Theorem 6.2.1 *Let $\mathcal{S}_k$ satisfy Assumption 5.1.6. Let $X \in \mathcal{S}_k$ be an affine variety, $t_1, \ldots, t_r \in X$ a finite set of points and h^* a cohomology theory with supports on $\mathcal{P}_k$. Suppose that h^* is given by a substratum C which is strictly effaceable at $t_1, \ldots, t_r$. Then the Cousin complexes*

$$0 \to h^q(Y) \xrightarrow{e} \coprod_{x \in Y^{(0)}} h^q_x(Y) \xrightarrow{d_1^{0,q}} \coprod_{x \in Y^{(1)}} h^{q+1}_x(Y) \xrightarrow{d_1^{1,q}} \ldots$$

are universally exact, where $Y = \operatorname{Spec} \mathcal{O}_{X, t_1, \ldots, t_r}$.

The proof uses the following well-known lemma:

Lemma 6.2.2 *Let $\mathcal{T}$ be a triangulated category and $A \to B \to C \to A[1]$ an exact triangle in $\mathcal{T}$. Suppose that the map $C \to A[1]$ is 0. Then the map $B \to C$ has a section. ("Every epimorphism is split".)*

Proof Apply the functor $\operatorname{Hom}(C, ?)$ to the triangle and get an exact sequence

$$\operatorname{Hom}(C, B) \to \operatorname{Hom}(C, C) \xrightarrow{0} \operatorname{Hom}(C, A[1]).$$

which shows that $\operatorname{Hom}(C, B) \to \operatorname{Hom}(C, C)$ is surjective. Let $s : C \to B$ be an element that maps to Id_C. Then by definition, s is a section, as wanted. □

To prove Theorem 6.2.1, we go a little more carefully than in the proof of Proposition 2.1.2. We note that the Cousin complex of Theorem 6.2.1 is obtained by pasting together complexes

$$0 \to h^q_{Y^{(p)}}(Y) \to \coprod_{y \in Y^{(p)}} h^q_y(Y) \to h^{q+1}_{Y^{(p+1)}}(Y) \to 0 \tag{6.1}$$

which in turn are obtained as direct limits of the complexes

$$0 \to h^q_{Z' \cap W}(W) \to h^q_{(Z'-Z)\cap W}(W \setminus Z) \to h^{q+1}_{Z\cap W}(W) \to 0 \qquad (6.2)$$

coming from the long exact cohomology sequence of Definition 5.1.1. Here W varies among the open neighbourhoods of $(t_1, \ldots, t_r)$ and $Z \subseteq Z' \subseteq W$ vary among closed subsets of codimensions respectively $\geq p$ and $\geq p+1$.

Lemma 6.2.3 *Let $t_1, \ldots, t_r$ and Y be as in Theorem 6.2.1. Suppose the substratum C is strictly effaceable at $t_1, \ldots, t_r$. Then for any $p \geq 0$ and $q \in \mathbf{Z}$, the complex* (6.1) *is a direct limit of split exact sequences, where h^* is the cohomology theory associated to C. In particular, it is universally exact.*

Proof Let $W \subseteq X$ be an open neighbourhood of $(t_1, \ldots, t_r)$, $Z \subseteq W$ a closed subset of codimension $\geq p$ and take U, Z' as given by Definition 5.1.8. In the triangulated category $K(\mathcal{A})$ of complexes of objects of $\mathcal{A}$ up to homotopy ([**26**, Chapter I] and [**20**, Section 5]), or in the homotopy category of $\mathcal{E}$ if C is given by spectra [**4**], consider the triangle or fibre sequence

$$C_{Z'\cap U}(U) \to C_{(Z'-Z)\cap U}(U \setminus Z) \to C_{Z\cap U}(U)[1] \xrightarrow{0} C_{Z'\cap U}(U)[1].$$

Here $C_{Z'\cap U}(U)[1]$ means $\Sigma C_{Z'\cap U}(U)$ if C is given by spectra. Lemma 6.2.2 shows that the map $C_{(Z'-Z)\cap U}(U \setminus Z) \to C_{Z\cap U}(U)[1]$ has a homotopy section. Correspondingly, the sequence

$$0 \to h^q_{Z'\cap U}(U) \to h^q_{(Z'-Z)\cap U}(U \setminus Z) \to h^{q+1}_{Z\cap U}(U) \to 0$$

is split exact for all q. And such sequences are cofinal in the direct system of complexes (6.2). □

Corollary 6.2.4 *Suppose k is infinite and $\mathcal{S}_k$ verifies Assumption 5.1.6. Let h^* be a cohomology theory with supports on $\mathcal{P}_k$, with values in an abelian category $\mathcal{A}$ satisfying axiom AB5 and having a generator. If h^* satisfies axioms* **COH1** *(étale excision) and either* **COH3** *(homotopy invariance) or* **COH5** *(cohomology of $\mathbf{P}^1$), and can be defined by a substratum of complexes or spectra (the latter assuming $\mathcal{A} = \{$abelian groups$\}$), then the Cousin complexes of Theorem 6.2.1 are universally exact for X smooth.*

Proof If h^* can be defined by a substratum of complexes, it can be defined by a fibrant substratum C by Remark 5.1.5 (2). Similarly, for a substratum of spectra, it can be defined by a fibrant and cofibrant substratum by subsection 5.2. By Lemmas 5.1.4 b), 5.3.1 b) and 5.4.2 b), C satisfies axioms **SUB1** and either **SUB3** or **SUB5**, hence axiom **SUB2** by Propositions 5.3.2 and 5.4.3. By Theorem 5.1.10, it is strictly effacable. The corollary now follows from Theorem 6.2.1.

Note that Theorem 6.2.1 does not cover the case of finite fields. For this, we introduce another axiom, which was already used in Section 4:

COH6 For any finite field extension ℓ/k and any $(X, Z) \in \mathcal{P}_k$, there is given a map

$$\mathrm{Cor}_{\ell/k} : h^*_{Z_\ell}(X_\ell) \to h^*_Z(X)$$

such that $\mathrm{Cor}_{\ell/k} \circ \mathrm{Res}_{\ell/k} = [\ell : k]$, where $\mathrm{Res}_{\ell/k}$ corresponds to extension of scalars. This map is natural in $(X, Z) \in \mathcal{P}_k$.

Theorem 6.2.5 *Let k be a finite field and h^* a cohomology theory with supports on $\mathcal{P}_k$, with values in an abelian category satisfying axiom AB5 and having a generator. Suppose $\mathcal{S}_k$ verifies Assumption 5.1.6 and h^* satisfies axioms* **COH1**,

COH6 *and either* **COH3** *or* **COH5**, *and can be defined by a substratum of complexes or spectra. Then, for any connected smooth affine $X \in \mathcal{S}_k$ and any finite set of points $t_1, \cdots, t_p \in X$, the Cousin complexes*

$$0 \to h^q(Y) \xrightarrow{e} h^q_\eta(Y) \xrightarrow{d_1^{0,q}} \coprod_{x \in Y^{(1)}} h^{q+1}_x(Y) \xrightarrow{d_1^{1,q}} \cdots$$

are universally exact, where $Y = \operatorname{Spec} \mathcal{O}_{X,t_1,\dots,t_p}$.

Proof Extend C to $\mathcal{S}_K$ for infinite algebraic extensions K/k by setting $C(X) = \varinjlim C(X_0 \otimes_{k_0} \ell)$, where k_0 is a suitable finite subextension of K such that $X = X_0 \otimes_{k_0} K$ for some X_0, and ℓ runs through the finite subextensions of K/k_0. This extends h^* to a cohomology theory with supports on $\mathcal{P}_K$, admitting a substratum and satisfying axioms **COH1**, **COH6** and either **COH3** or **COH5**. By Corollary 6.2.4, the Cousin complexes of Theorem 6.2.1 are universally exact for K-varieties.

Let T be an additive functor with values in some abelian category satisfying AB5 and which commutes with filtering direct limits. We have to prove that the complex

$$0 \to T(h^q(Y)) \xrightarrow{e} T(h^q_\eta(Y)) \xrightarrow{d_1^{0,q}} \coprod_{x \in Y^{(1)}} T(h^{q+1}_x(Y)) \xrightarrow{d_1^{1,q}} \cdots \tag{6.3}$$

is acyclic, for X a smooth k-variety and Y as in Theorem 6.2.1. We use the same trick as in Section 4. Let p_1, p_2 be two different primes and K_1, K_2 the $\mathbf{Z}_{p_1}$ and $\mathbf{Z}_{p_2}$-extensions of k respectively. Let A be some homology group of the sequence (6.3). For an algebraic extension K/k, let $(6.3)_K$ denote (6.3) "pushed over K". By assumption on T, we have

$$A_K = \varinjlim A_\ell$$

where ℓ runs through finite subextensions of K/k. On the other hand, the transfer condition shows that, if $[K : k] = N < +\infty$, then

$$N \operatorname{Ker}(A \to A_K) = 0.$$

It follows that $\operatorname{Ker}(A \to A_{K_i})$ is p_i-primary torsion and therefore $\operatorname{Ker}(A \to A_{K_1} \oplus A_{K_2}) = 0$. Finally, since K_1 and K_2 are infinite, we have $A_{K_1} = A_{K_2} = 0$ as observed above. So $A = 0$, as was to be proven. □

7 Examples

7.1 Hypercohomology of sheaves. In this subsection as in subsection 7.5, the category $\mathcal{S}_k$ need not satisfy Assumption 5.1.6.

7.1.1 Let ν be a Grothendieck topology on $\mathcal{S}_k$. To a complex of sheaves of abelian groups $\mathcal{C}$ over ν one can associate a cohomology theory with supports h^*, given by the ν-hypercohomology of $\mathcal{C}$ with supports:

$$h^*_Z(X) = \mathbb{H}^*_Z(X_\nu, \mathcal{C}).$$

7.1.2 Let $f : \mathcal{C} \to \mathcal{C}'$ be a morphism. Then f induces a morphism f_* of associated cohomology theories. If f is a quasi-isomorphism, f_* is an isomorphism in the following two cases:

- $\mathcal{C}$ and $\mathcal{C}'$ are bounded below;
- for all $X \in \mathcal{S}_k$, the ν-cohomological dimension of X is finite.

Indeed, we have a morphism of hypercohomology spectral sequences:

$$\begin{array}{ccccc}
{}'E_2^{p,q} & = & H^p_Z(X_\nu, \mathcal{H}^q(\mathcal{C}')) & \Rightarrow & \mathbb{H}^{p+q}_Z(X_\nu, \mathcal{C}') \\
 & & \uparrow{\scriptstyle f_*} & & \uparrow{\scriptstyle f_*} \\
E_2^{p,q} & = & H^p_Z(X_\nu, \mathcal{H}^q(\mathcal{C})) & \Rightarrow & \mathbb{H}^{p+q}_Z(X_\nu, \mathcal{C})
\end{array}$$

which is an isomorphism on E_2-terms by assumption. Here, $\mathcal{H}^q(\mathcal{C})$ and $\mathcal{H}^q(\mathcal{C}')$ are the cohomology sheaves of $\mathcal{C}$ and $\mathcal{C}'$ respectively. In both cases, the two spectral sequences converge, hence the map on abuttments is an isomorphism.

7.1.3 A cohomology theory given by a complex of ν-sheaves $\mathcal{C}$ can always be defined by a substratum C of complexes of abelian groups (this is in fact the way ν-hypercohomology is defined!) There are several well-known constructions for C:

- Suppose $\mathcal{C}$ is bounded below. Choose a Cartan-Eilenberg injective resolution $\mathcal{I}$ of $\mathcal{C}$ and define $C(X) = Tot(\mathcal{I})(X)$, where $Tot(\mathcal{I})$ is the total complex associated to the bicomplex $\mathcal{I}$. Note that the terms of $C(X)$ are injective abelian groups and $C(X)$ is bounded below. Hence $C(X)$ is fibrant in the sense of Definition C.1.1 (compare Proposition C.1.2).

- Replace $\mathcal{C}$ by a fibrant complex of sheaves $\mathcal{F}$. By Theorem C.3.1, we can do this functorially in $\mathcal{C}$. Define now $C(X)$ as $\mathcal{F}(X)$; note that $C(X)$ is fibrant for all X as a complex of abelian groups. This construction does not require $\mathcal{C}$ to be bounded below.

- The Godement resolution. Suppose the topos associated to ν has enough points (this is the case for Zariski, Nisnevich, étale, complex topologies). To $\mathcal{C}$ one associates a new complex of sheaves

  $$T\mathcal{C} : U \mapsto \prod_{f\in\Pi} \prod_{f^*(U)} f^*\mathcal{C}$$

  where Π is the set of points of ν (compare [**50**, 1.31], especially for set-theoretic problems). The terms of this complex are flabby in the sense of [**36**, Example III.1.9 (c)]. Iterating T yields a cosimplicial complex of flabby sheaves $T^{\cdot}\mathcal{C}$, which in turn yields a bicomplex of flabby sheaves $T^*\mathcal{C}$ in the usual way. One defines $C(X) = Tot(T^*\mathcal{C})(X)$. This is essentially the object denoted by $\mathbb{H}^{\cdot}(X, \mathcal{C})$ in [**50**].

By the usual arguments, there is a commutative diagram (for $\mathcal{C}$ bounded below)

$$\begin{array}{ccc}
 & & Tot(\mathcal{I}) \\
 & \nearrow & \uparrow \\
\mathcal{C} & & \\
 & \searrow & \\
 & & Tot(T^*\mathcal{C})
\end{array}$$

in which the vertical map induces a quasi-isomorphism on global sections.

The last two constructions are natural in $\mathcal{C}$. All constructions have the following virtue: if $\mathcal{C}' \to \mathcal{C} \to \mathcal{C}''$ defines an exact triangle in the derived category of ν-sheaves, then so does $C'(X) \to C(X) \to C''(X)$, for all X, in the derived category of abelian groups, where C', C, C'' are the associated substrata.

7.1.4 Let C be associated to $\mathcal{C}$ as in 7.1.3, and suppose we sheafify it for the ν-topology. In the first construction C is already the complex of sheaves $Tot(\mathcal{I})$. In the second one, the stalk of $T^{\cdot}\mathcal{C}$ at a point x is homotopy equivalent to the constant cosimplicial complex of abelian groups defined by $\mathcal{C}_x$. In both cases, the resulting complex of sheaves is quasi-isomorphic to $\mathcal{C}$.

7.1.5 Instead of taking complexes of sheaves of abelian groups, one can take complexes of sheaves with values in an abelian category with enough injectives, or *sheaves of spectra* [**50**] in the line of 5.2. All the above holds in these contexts, mutatis mutandis. In the case of spectra, for 7.1.2 use the spectral sequence of [**50**, Proposition 1.36]. For the second construction in 7.1.3, use [**50**, Definition 1.33]. Nisnevich repeats these constructions in [**38**], because he uses a different notion of point of a topos for the Nisnevich topology.

7.1.6 Let ν' be another Grothendieck topology on $\mathcal{S}_k$ which is finer than ν. Then the identity functor of $\mathcal{S}_k$ defines a morphism of sites $\nu' \xrightarrow{\alpha} \nu$. If $\mathcal{C}$ is a complex of sheaves for the ν' topology, there is an isomorphism

$$\mathbb{H}^*_Z(X_{\nu'}, \mathcal{C}) \xrightarrow{\sim} \mathbb{H}^*_Z(X_\nu, R\alpha_*\mathcal{C})$$

where $R\alpha_*\mathcal{C}$ is the total direct image of $\mathcal{C}$ (in the derived category). In the case $\mathcal{C}$ is a sheaf of spectra, one should use the object $\mathbb{R}^{\cdot}\alpha\mathcal{C}$ of [**50**, def. 1.55] instead of $R\alpha_*\mathcal{C}$, cf [**50**, th. 1.56]. So we can view ν'-hypercohomology as ν-hypercohomology.

7.1.7 Suppose that ν is the big Zariski site of Spec k. Then ν-hypercohomology of $\mathcal{C}$ verifies Zariski excision. Similarly, suppose that ν is the big Nisnevich site Nis on Spec k. Then ν-hypercohomology verifies étale excision, i.e., axiom **COH1**. This is known when $\mathcal{C}$ is reduced to a single sheaf (for the Nisnevich case, cf. [**8**, Proposition 4.4], which applies to Nisnevich cohomology; recall that the proofs of [**36**, Proposition III.1.27] and [**38**, Theorem 1.27] have a gap). In general, the proof follows from a comparison of convergent hypercohomology spectral sequences, as in 7.1.2. The two spectral sequences converge without boundedness conditions on $\mathcal{C}$, because the Zariski or Nisnevich cohomological dimensions of k-schemes of finite type are finite [**38**]. See [**50**, Example 1.49] for Zariski excision in the case of a sheaf of spectra (in [**38**] Nisnevich does not give the corresponding statement for étale excision explicitly).

7.1.8 By 7.1.6 and 7.1.7, ν-hypercohomology satisfies Zariski (resp. étale) excision as soon as ν is finer than the Zariski (resp. the Nisnevich) topology.

7.2 Generating new theories out of old, continued.

Let $\mathcal{C}$ be as in 7.1.1, let A be a bounded below complex of abelian groups, viewed as a complex of constant Nisnevich sheaves, and let $\mathcal{C}' = \mathcal{C} \overset{L}{\otimes} A$ (in the derived category). Then, if the cohomology theory associated to $\mathcal{C}$ verifies axiom **COH3** or **COH5**, the same is true for $\mathcal{C}'$. In the case of axiom **COH5**, if $\mathcal{D}$ is a complex of sheaves associated to $\mathcal{C}$, we associate to $\mathcal{C}'$ the complex $\mathcal{D}' = \mathcal{D} \overset{L}{\otimes} A$ and take for the action of Pic the original action tensored by A. The claim can be justified in a few steps:

(a) $A = \mathbf{Z}[0]$. This is trivial.

(b) $A = \mathbf{Z}/n$. Follows from the previous case, item (2) of subsection 5.5, 7.1.3 and the exact sequence $0 \to \mathbf{Z} \to \mathbf{Z} \to \mathbf{Z}/n \to 0$.

(c) A consists of a single finitely generated abelian group placed in degree zero. Follows from the previous cases.
(d) A consists of a single abelian group placed in degree 0. Follows from the previous case and item (3) of subsection 5.5 by a passage to the limit.
(e) The general case. Follows from the previous case and item (4) of subsection 5.5.

In case $\mathcal{C}$ is a sheaf of spectra as in 7.1.5, one has the same by taking for $\mathcal{D}$ the sheaf of spectra $\mathcal{C} \wedge A$, where A is an arbitrary spectrum, viewed as a constant sheaf of spectra. The method is the same, reducing to the case where A is an Eilenberg-Mac Lane spectrum by dévissage from its Postnikov tower, cf [**50**, proof of Theorem 1.46].

7.3 Homotopy invariant examples.

In all examples of this subsection, $\mathcal{S}_k = Var/k$.

(1) *Etale cohomology with coefficients in a sheaf defined over k and torsion prime to the characteristic of k.* Etale excision follows from 7.1.8 and homotopy invariance from [**36**, Corollary VI.4.20]. More generally, by 7.1.8 and item (4) of subsection 5.5, one may take étale hypercohomology of a bounded below complex of sheaves whose cohomology is torsion prime to the characteristic of k.
(2) (over **C**:) *Classical hypercohomology with coefficients in a bounded below complex of abelian groups.* Here, étale excision again comes from 7.1.8 and the fact that, for X a **C**-variety, the topological space $X(\mathbf{C})$ essentially maps to the small étale site of X (cf. [**SGA4-III**, exposé XI, 4.0]). Homotopy invariance is known for **Z** as coefficients, and the general case follows from Remark 7.2.
(3) ($\operatorname{char} k = 0$:) *De Rham cohomology.* Recall that, for a k-variety X, $H^*_{dR}(X/k) = \mathbb{H}^*_{\mathrm{Zar}}(X, \Omega^{\cdot}_{X/k})$, where $\Omega^{\cdot}_{X/k}$ is the de Rham complex. To check étale excision, we note that, since the $\Omega^i_{X/k}$ are coherent sheaves, the maps $\mathbb{H}^*_{\mathrm{Zar}}(X, \Omega^{\cdot}_{X/k}) \to \mathbb{H}^*_{\mathrm{Nis}}(X, \Omega^{\cdot}_{X/k})$ are isomorphisms [**36**, Remark III.3.8], so we can apply 7.1.8. Actually, since $\operatorname{char} k = 0$, we even have purity [**25**]. Homotopy invariance is proven in [**25**, Remark, p. 54].
(4) *Motivic cohomology.* Let k admit resolution of singularities in the sense of [**12**, Definition 3.4] (for example, $\operatorname{char} k = 0$), and let $i \geq 0$. In [**49**, Section 2], motivic cohomology of weight i is defined by $H^*_Z(X, \mathbf{Z}(i)) = \mathbb{H}^*_Z(X_{\mathrm{cdh}}, \mathbf{Z}(i)_{\mathrm{cdh}})$ for $(X, Z) \in \mathcal{P}_k$, where cdh is the Grothendieck topology introduced in [**12**, Definition 3.2], $\mathbf{Z}(i)$ is a certain complex of presheaves with transfers with homotopy invariant cohomology presheaves in the sense of [**52**] and $\mathbf{Z}(i)_{\mathrm{cdh}}$ is its sheafification for the cdh topology. Therefore motivic cohomology is given by a substratum, is homotopy invariant and satisfies étale excision since the cdh topology is stronger than the Nisnevich topology. It also satisfies purity by [**49**, Proposition 2.4].
(5) *Cycle modules.* Let M_* be a cycle module in the sense of [**44**]. For $Z \in Var/k$ and $j \in \mathbf{Z}$, denote by $C_{\cdot}(Z, M_j)$ the (homological) Gersten complex associated to Z and ending with $\coprod_{x \in X_{(0)}} M_j(k(x))$. For X of pure dimension d, define

$$C(X) = C^{\cdot}(X, M_j) = C_{\cdot}(X, M_{j-d})$$

viewed as a cohomological complex. For $Z \subset X$ a closed subset, there is an obvious short exact sequence of complexes

$$0 \to C_\cdot(Z, M_{j+d}) \to C^\cdot(X, M_j) \to C^\cdot(X - Z, M_j) \to 0.$$

So $C_\cdot(Z, M_{j+d})$ is homotopy equivalent to the homotopy fibre $C_Z(X)$. In particular, the substratum C verifies étale excision (axiom **SUB 1**), and even purity. By a result of Rost [**44**, Proposition 8.6], it is also homotopy invariant. Therefore, the Gersten complexes on a smooth semi-local scheme are universally exact ([**44**, Theorem 6.1] for the exactness). Universal exactness actually follows directly from replacing the cycle module M_* by $T \circ M_*$, where T is a given additive functor which commutes with filtering direct limits.

This example includes as a special case Milnor's K-groups, Milnor's K-groups mod m, m-torsion in Milnor's K-groups ...

(6) *Algebraic G- (= K'-)theory.* This is the only case in this list of examples where the substratum is given by spectra, not complexes. Etale excision **COH1** is implied by the much stronger *localization theorem* of Quillen, akin to purity [**43**, Proposition 7.3.2]. Homotopy invariance **COH3** also follows from Quillen [**43**, Proposition 7.4.1]. By smashing the algebraic G-theory spectrum by the Moore spectrum $M(\mathbf{Z}/n)$, we get the case of algebraic G-theory with coefficients $\mathbf{Z}/n$ (compare subsection 7.2).

7.4 Non homotopy invariant examples. In all examples of this subsection, $\mathcal{S}_k = Var/k$, except in Examples (4) and (5) where $\mathcal{S}_k = Sm/k$.

(1) *Etale (hyper-)cohomology with bounded below coefficients coming from k.* As above, étale excision follows from 7.1.8. Axiom **COH5** is proven in Appendix A. More precisely, in subsection A.2, we define an étale sheaf $\mathbf{Q}/\mathbf{Z}(-1)$ (over the big étale site of $\operatorname{Spec}\mathbf{Z}$) and a map

$$\operatorname{Pic} X \to \Gamma(X, R\operatorname{Hom}_{\acute{e}t}(\mathbf{Q}/\mathbf{Z}(-1)[-3], \mathbf{Z})) \tag{7.1}$$

for any scheme X. Let now $\mathcal{C}_0$ be a bounded below complex of sheaves over the *small* étale site of $\operatorname{Spec} k$, and $\mathcal{C}$ its inverse image to the *big* étale site. Let h^* be the cohomology theory with supports defined by $\mathcal{C}$ and e^* the cohomology theory with supports defined by $\mathcal{C} \overset{L}{\otimes} \mathbf{Q}/\mathbf{Z}(-1)[-3]$. The map (7.1) induces a map $\operatorname{Pic} X \to \operatorname{Hom}(e, h)$, and we show in subsection A.3 that this map satisfies axiom **COH5**.

(2) *Hodge and de Rham cohomology in any characteristic.* Etale excision is seen as above. Axiom **COH5** is due to Illusie: if h^* is the cohomology theory associated to the Zariski sheaf $\Omega^i_{X/k}$, then axiom **COH5** holds for h^* with e^* associated to $\Omega^{i-1}_{X/k}[-1]$ (i.e., $e^j_Z(X) = H^{j-1}_Z(X_{\mathrm{Zar}}, \Omega^{i-1}_{X/k})$). Here the map $\operatorname{Pic} X \to \operatorname{Hom}(e, h)$ is given by cup-product with the first Chern class, defined through the map $H^1_{\mathrm{Zar}}(X, \mathcal{O}^*_X) \xrightarrow{d\log} H^1_{\mathrm{Zar}}(X, \Omega^1_{X/k})$. Using item (4) of subsection 5.5, one can then extend axiom **COH5** to the de Rham-Witt complex itself, or truncations of it. Compare [**21**, p. 22, Proof of (4.2.7)]. Note that X need not be smooth since, in [**SGA7**, exposé XI, Theorem 1.1], X is arbitrary. Note also that, even in characteristic 0, Hodge cohomology is not homotopy invariant. In characteristic p one can then "escalate the ladder" to get the same result for Deligne-Illusie's Hodge-Witt and de Rham-Witt cohomology ([**28**], compare [**21**]).

(3) (char $k = p$) *Logarithmic Hodge-Witt and de Rham-Witt cohomology* [**22**]. Etale excision is proven as above. Axiom **COH5** follows from [**21**, Theorem I.2.1.11] and is proven there in the spirit of item (2) of subsection 5.5, using the description of the logarithmic de Rham-Witt pro-complex as Frobenius fixed points of the de Rham-Witt pro-complex [**21**, I. (1.3.2)] .

(4) *Cohomology of a torus.* Let $\mathbb{T}$ be a k-torus. Consider the cohomology theory with supports h^* given by the sheaf associated to $\mathbb{T}$ on the big étale site of $\operatorname{Spec} k$. As above, h^* satisfies étale excision. Let $M = \operatorname{Hom}(\mathbb{G}_m, \mathbb{T})$ be the group of cocharacters of $\mathbb{T}$, also viewed as a big étale sheaf. Let e^* be the cohomology theory with supports given by

$$e^q_Z(X) = H^{q-1}_Z(X_{ét}, M).$$

Cup-product defines a map

$$\operatorname{Pic} X \to \operatorname{Hom}_X(e, h)$$

for all X. It can be shown that axiom **COH5** is verified for h, e and this natural transformation.

(5) *Etale weight-two motivic cohomology.* For an affine scheme X, let $\Gamma(X, 2)$ be the weight-two motivic complex introduced by Lichtenbaum in [**33**]. For X smooth, its Zariski sheafification $\Gamma(2)_{\mathrm{Zar}}$ is quasi-isomorphic to $\tau^{>0}\mathbf{Z}(2)_{\mathrm{Zar}}$, where $\mathbf{Z}(2)_{\mathrm{Zar}}$ is the Zariski sheafification of the complex of 7.3 (4). The main steps in the proof of this are [**1**, Theorem 7.2], [**48**] and [**53**, proof of Proposition 4.9 and subsection 4.3]. (Conjecturally $\Gamma(2)_{\mathrm{Zar}}$ and $\mathbf{Z}(2)_{\mathrm{Zar}}$ coincide). It is proven in [**31**] that the cohomology theory $(X, Z) \mapsto \mathbb{H}^*_Z(X_{ét}, \Gamma(2)_{ét})$ satisfies axiom **COH5**, where $\Gamma(2)_{ét}$ is the étale sheafification of $X \mapsto \Gamma(X, 2)$.

(6) *Algebraic K^B-theory.* Here K^B denotes the Bass extension of Quillen's algebraic K-theory, which coincides with the latter for regular Noetherian schemes, see [**51**, Section 6]. Axiom **COH1** is one of the main results of Thomason-Trobaugh: it applies generally to X, X' quasi-compact and quasi-separated such that $X - Z$ is quasi-compact as well [**51**, Theorems 7.1 and 7.4]. It would be wrong for ordinary K-theory in general. Axiom **COH5** follows from [**51**, Theorem 7.3]. Just as in Example 7.3 (6) , we get algebraic K^B-theory with finite coefficients by smashing with a Moore spectrum.

7.5 More on hypercohomology and excision. This subsection can be considered as a sequel to subsection 7.1.

7.5.1 Let ν be a Grothendieck topology on $\mathcal{S}_k$. Suppose that we now start with a substratum of complexes C. Sheafifying C for the ν-topology, we get a complex of ν-sheaves $\mathcal{C}$. Choosing an injective right Cartan-Eilenberg resolution $\mathcal{I}$ of $\mathcal{C}$, the augmentation $\mathcal{C} \to Tot(\mathcal{I})$ yields an augmentation

$$C(X) \to \Gamma(X, \mathcal{C}) \to \Gamma(X, Tot(\mathcal{I})).$$

Similarly, if ν has enough points, the Godement resolution construction of 7.1.3 gives a natural transformation

$$C(X) \to \mathbb{H}^{\cdot}(X_\nu, \mathcal{C}). \tag{7.2}$$

By analogy with 7.1.4, we may ask the question:

Question. When is (7.2) a quasi-isomorphism?

This problem has no simple solution in general; however we shall explain that it has one when ν is either the Zariski or the Nisnevich topology.

In both cases, an obvious necessary condition is that the cohomology theory h^* associated to C satisfies Zariski (resp. étale) excision, since $\mathbb{H}^{\cdot}(X_\nu, \mathcal{C})$ does by 7.1.7. The remarkable fact is that this condition is sufficient:

Theorem 7.5.1 *Suppose that* $\nu = \mathrm{Zar}$ *(resp.* $\nu = \mathrm{Nis}$*). Then* (7.2) *is a quasi-isomorphism if and only if the cohomology theory* h^* *associated to* C *satisfies Zariski (resp. étale) excision. In other terms, a cohomology theory with supports which admits a substratum made of complexes satisfies Zariski (resp. étale) excision if and only if it can be defined by Zariski (resp. Nisnevich) hypercohomology of a complex of sheaves.*

This theorem is either a consequence or an easy analogue of

Theorem 7.5.2 (Brown-Gersten–Thomason–Nisnevich) *Suppose that* $\nu = \mathrm{Zar}$ *(resp.* $\nu = \mathrm{Nis}$*) and let* C *be a substratum of spectra over* $\mathcal{S}_k$*. Then* (7.2) *is a quasi-isomorphism if and only if the cohomology theory* h^* *associated to* C *satisfies Zariski (resp. étale) excision. In other terms, a cohomology theory with supports which admits a substratum made of spectra satisfies Zariski (resp. étale) excision if and only if it can be defined by Zariski (resp. Nisnevich) hypercohomology of a sheaf of spectra.*

Proof See [**5**] and [**50**, 2.5] for the Zariski case, [**38**] for the Nisnevich case. □

By 7.1.4, a complex of sheaves $\mathcal{C}$ defining h^* can be chosen as the sheaf associated to the presheaf $U \mapsto C(U)$, where C is a substratum defining h^*.

Corollary 7.5.3 *Under the conditions of Theorem 7.5.2, there is for any* $X \in \mathcal{S}_k$ *a spectral sequence*

$$E_2^{p,q} = H^p(X_\nu, \mathcal{H}^q) \Rightarrow h^{p+q}(X),$$

where $\nu = \mathrm{Zar}$ *or* Nis *and* $\mathcal{H}^q$ *is the* ν *sheaf associated to the presheaf* $U \mapsto h^q(U)$*.*

Proof This is just the hypercohomology spectral sequence for the cohomology of $\mathbb{H}^{\cdot}(X_\nu, \mathcal{C})$. □

Example 7.5.4 Let μ be a Grothendieck topology on $\mathcal{S}_k$ which is finer than ν, and let $\alpha : \mu \to \nu$ be the corresponding morphism of sites. Let $\mathcal{D}$ be a complex of sheaves (or sheaf of spectra) for the μ-topology, and take $\mathcal{C} = R\alpha_* \mathcal{D}$ (or $\mathbb{R}^{\cdot}\alpha\mathcal{D}$). There is a canonical quasi-isomorphism (or weak equivalence)

$$\mathbb{H}^{\cdot}(X_\nu, \mathcal{C}) \simeq \mathbb{H}^{\cdot}(X_\mu, \mathcal{D})$$

and we recover the Leray spectral sequence for the morphism α.

Note that, just as the spectral sequence of Example 5.1.3 (3), this spectral sequence is defined for arbitrary, not necessarily smooth, $X \in \mathcal{S}_k$. The two spectral sequences have a priori nothing to do with each other. In other words, the comment in [**50**, last section of p. 467] misses the point. The Bloch-Ogus–Gabber theorem implies that, when X is smooth, they have isomorphic E_2-terms. Moreover, they actually coincide in this case for many theories (Deligne, unpublished, cf. [**2**, footnote p. 195], Gillet-Soulé, [**17**]). See also Paranjape [**41**].

7.5.2 Let h^* be the cohomology theory with supports associated to some complex of Nisnevich sheaves $\mathcal{C}$. As seen above, h^* satisfies axiom **COH1**. By 7.1.2, h^* only depends, up to isomorphism, on the class of $\mathcal{C}$ in the *derived category* $\mathcal{D}(\mathrm{Nis})$ of the category of Nisnevich sheaves. Indeed, the Nisnevich cohomological dimension of a scheme of finite Krull dimension is finite [**38**]. For the convenience of the reader, we reformulate axioms **COH3** and **COH5** purely in terms of $\mathcal{C}$ (viewed in $\mathcal{D}(\mathrm{Nis})$):

D3 Let $\pi : \mathbf{A}^1_k \to \operatorname{Spec} k$ be the structural map. Then $\mathcal{C} \xrightarrow{\sim} R\pi_*(\mathcal{C}_{|\mathbf{A}^1})$.

To formulate axiom **D5**, note that, if h^* satisfies étale excision and axiom **COH5**, the associated cohomology theory e^* satisfies étale excision as well by Remark 5.4.1 (2). If h^* is given by a substratum C, then e^* is given by the substratum $D(X) = \operatorname{Ker}(C(\mathbf{P}^1_X) \xrightarrow{s^*_\infty} C(X))$. By Theorem 7.5.2, both h^* and e^* are given by Nisnevich hypercohomology of complexes of sheaves $\mathcal{C}$ and $\mathcal{D}$.

D5 a) There exists an object $\mathcal{D} \in \mathcal{D}(\mathrm{Nis})$ and, for all $X \in \mathcal{S}_k$, a map

$$\operatorname{Pic} X \to \operatorname{Hom}_{\mathcal{D}(\mathrm{Nis}_{|X})}(\mathcal{D}_{|X}, \mathcal{C}_{|X})$$

natural in X.

b) For $X = \mathbf{P}^1_k$, the map of a) induces a morphism

$$\mathcal{D}_{|\mathbf{P}^1} \xrightarrow{[\mathcal{O}(1)]-[\mathcal{O}]} \mathcal{C}_{|\mathbf{P}^1}.$$

Consider the adjoint map

$$\mathcal{D} \xrightarrow{\alpha} R\tilde{\pi}_*\mathcal{C}_{|\mathbf{P}^1}.$$

Then the map

$$\mathcal{C} \oplus \mathcal{D} \xrightarrow{(\varepsilon,\alpha)} R\tilde{\pi}_*\mathcal{C}_{|\mathbf{P}^1}$$

is a (quasi-)isomorphism, where ε is the unit (adjunction) map.

If $\mathcal{C}$ is a sheaf of spectra, one should replace $R\pi_*$ and $R\tilde{\pi}_*$ by $\mathbb{R}^{\cdot}\pi$ and $\mathbb{R}^{\cdot}\tilde{\pi}$ in axioms **D3** and **D5**.

8 A Selection of Corollaries

8.1 Multiplying by a fixed variety. Let T be a (not necessarily smooth) k-variety. The following theorem gives concrete illustrations of item (1) in subsection 5.5.

Theorem 8.1.1 *Let Y be the spectrum of a semi-local ring of a smooth, connected k-variety, as in Proposition 2.1.2. Let n be prime to* $\operatorname{char} k$ *and $i \in \mathbf{Z}$. Then, with notation as in Proposition 2.1.2, there are universally exact sequences:*

$$0 \to H^q(Y \times_k T, \mu_n^{\otimes i}) \xrightarrow{e} H^q_{\eta \times_k T}(Y \times_k T, \mu_n^{\otimes i}) \xrightarrow{d_1^{0,q}} \coprod_{x \in Y^{(1)}} H^{q+1}_{x \times_k T}(Y \times_k T, \mu_n^{\otimes i}) \xrightarrow{d_1^{1,q}} \cdots \tag{8.1}$$

$$0 \to G_q(Y \times_k T) \xrightarrow{e} G_q(k(Y) \otimes_k T) \xrightarrow{d_1^{0,q}} \coprod_{x \in Y^{(1)}} G_{q-1}(k(x) \otimes_k T) \xrightarrow{d_1^{1,q}} \cdots \tag{8.2}$$

$$0 \to K_q^B(Y \times_k T) \xrightarrow{e} K_q^{B,\eta\times_k T}(Y \times_k T) \xrightarrow{d_1^{0,q}} \coprod_{x\in Y^{(1)}} K_{q-1}^{B,x\times_k T}(Y \times_k T) \xrightarrow{d_1^{1,q}} \cdots \tag{8.3}$$

where (8.1) *is étale cohomology. If, moreover, T is smooth, there are universally exact sequences:*

$$\begin{aligned} 0 \to H^q(Y \times_k T, \mu_n^{\otimes i}) \xrightarrow{e} H^q(k(Y) \otimes_k T, \mu_n^{\otimes i}) \xrightarrow{d_1^{0,q}} \\ \coprod_{x\in Y^{(1)}} H^{q-1}(k(x) \otimes_k T, \mu_n^{\otimes(i-1)}) \xrightarrow{d_1^{1,q}} \cdots \end{aligned} \tag{8.4}$$

$$0 \to K_q(Y \times_k T) \xrightarrow{e} K_q(k(Y) \otimes_k T) \xrightarrow{d_1^{0,q}} \coprod_{x\in Y^{(1)}} K_{q-1}(k(x) \otimes_k T) \xrightarrow{d_1^{1,q}} \cdots . \tag{8.5}$$

Proof After item (1) of subsection 5.5, the exactness of (8.1) and (8.3), as well as the same sequence as (8.3) with G instead of K, follows from Examples 7.3 (1), 7.3 (6) and 7.4 (6). Universal exactness follows from Section 6. We have (8.2) by purity of G-theory [**43**, Proposition 7.3.2]. When T is smooth we have purity for étale cohomology, hence (8.4), and the K-groups with support identify with G-groups with support, hence (8.3) yields (8.5). □

Remarks 8.1.2

(1) We could of course state (8.3) and (8.5) for K-theory with finite coefficients.
(2) The reader is invited to apply this principle to other examples (e.g. $\mathcal{K}$-cohomology, compare [**7**, Theorem 5.2.5]).
(3) In the étale case, the use of item (1) of subsection 5.5 can be replaced by the isomorphisms

$$\mathbb{H}^*_{ét}(Y \times_k T, \mu_n^{\otimes i}) \simeq \mathbb{H}^*_{ét}(Y, Rf_*((\mu_n^{\otimes i})_{|T}))$$

where $f : T \to \operatorname{Spec} k$ is the structural map, noting that the complex of sheaves $Rf_*((\mu_n^{\otimes i})_{|T})$ is defined over k. In Appendix B, we shall prove an analogue of Theorem 8.1.1, replacing the projection $Y \times_k T \to Y$ by a not necessarily constant map $X \xrightarrow{\pi} Y$, provided
- π is proper and smooth;
- $\dim Y = 1$.

One may ask whether the condition $\dim Y = 1$ is necessary. This issue is being investigated by Panin.

8.2 Galois action.

Proposition 8.2.1 *Let $\mathcal{S}_k$ satisfy Assumption 5.1.6. Let R be a ring and h^* a cohomology theory with supports on $\mathcal{S}_k$ with values in the category of R-modules. Let X, $t_1, \dots, t_r$, Y be as in Theorem 6.2.1. Suppose h^* is given by a substratum C which is strictly effaceable at $t_1, \dots, t_r$. Finally, let M be a left R-module. Then,* *for any q, s, the complex*

$$0 \to Tor_s^R(M, h^q(Y)) \xrightarrow{e} \coprod_{x \in Y^{(0)}} Tor_s^R(M, h_x^q(Y)) \xrightarrow{d_1^{0,q}} \coprod_{x \in Y^{(1)}} Tor_s^R(M, h_x^{q+1}(Y)) \xrightarrow{d_1^{1,q}} \cdots .$$

is exact.

Proof This is an immediate consequence of Theorem 6.2.1. □

Theorem 8.2.2 *Let ℓ/k be a finite Galois extension and $G = Gal(\ell/k)$. Assume $\mathcal{S}_k$ satisfies Assumption 5.1.6. Let h^* be a cohomology theory with supports satisfying axioms **COH1** and either **COH3** or **COH5**, plus **COH6** if k is finite. Suppose h^* is given by a substratum C. Let $X, t_1, \dots, t_r, Y$ be as in Theorem 6.2.1, with X smooth; denote by Y_ℓ the pull-back of Y over ℓ. Then, at least if C is given by complexes of abelian groups, the complex*

$$0 \to H_n(G, h^q(Y_\ell)) \to \coprod_{x \in Y_\ell^{(0)}} H_n(G, h_x^q(Y_\ell)) \to \coprod_{x \in Y_\ell^{(1)}} H_n(G, h_x^{q+1}(Y_\ell)) \to \cdots$$

is exact for all $q, n \geq 0$.

Proof Consider the new cohomology theory with support and substratum h_ℓ^*, C_ℓ given by

$$h_Z^q(X)_\ell = h_{Z_\ell}^q(X_\ell)$$
$$C(X)_\ell = C(X_\ell).$$

Then h_ℓ^* naturally takes its values in the category of $\mathbf{Z}[G]$-modules, and clearly satisfies the same set of axioms as h^*. If C is given by complexes of abelian groups, then C_ℓ takes its values in the category of complexes of $\mathbf{Z}[G]$-modules. The claim then follows from Corollary 6.2.4 and Theorem 6.2.5, applying the functor $H_n(G, ?)$ to a universally exact sequence just as for Proposition 8.2.1. □

If C is given by spectra, then C_ℓ takes its values in the category $\mathcal{E}^G$ of G-spectra. We can then get away similarly if functorial factorizations similar to those in $\mathcal{E}$ (see subsection 5.2) are available in $\mathcal{E}^G$, provided with a suitable closed model category structure. This is closely related to Thomason's unfinished approach to model structures on functor categories, as outlined in Weibel [**54**].

Theorem 8.2.2 applies in particular to étale cohomology. It also applies to algebraic K-theory provided one fixes the remark of the last paragraph. In the former case, specializing to coefficients twisted roots of unity and using purity, we get the following, which was needed in [**30**] (precisely for a finite base field!):

Corollary 8.2.3 *Let $X, t_1, \dots, t_r, Y$ be as in Theorem 8.2.2, with X irreducible; denote by Y_ℓ the pull-back of Y over ℓ. Then, for all $q, n \geq 0$, the complex*

$$0 \to H_n(G, H^q(Y_\ell, \mu_m^{\otimes i})) \to H_n(G, H^q(\ell(Y), \mu_m^{\otimes i})) \to \coprod_{x \in Y_\ell^{(1)}} H_n(G, H^{q-1}(\ell(x), \mu_m^{\otimes (i-1)})) \to \cdots$$

is exact. □

Remark 8.2.4 One might be tempted to extend this result to the case of any Galois étale covering (not only those coming from the base field) by using Proposition 2.2.4, but this fails. The point is that Proposition 2.2.4 will give homotopies, but not necessarily G-equivariant homotopies.

8.3 Zariski cohomology and Nisnevich cohomology.

Theorem 8.3.1 *(Nisnevich* [**39**, Theorem 0.12]*) a) Let h^* be a cohomology theory with supports satisfying axioms **COH1**, **COH2** and also axiom **COH6** if the base field k is finite. For $i \in \mathbf{Z}$, let $\mathcal{H}^i_{\mathrm{Zar}}$ (resp. $\mathcal{H}^i_{\mathrm{Nis}}$) be the sheaf associated to the presheaf $U \mapsto h^i(U)$ on the big Zariski (resp. Nisnevich) site of* $\operatorname{Spec} k$*. Then, for all smooth $X \in \mathcal{S}_k$ and $n \geq 0$, the natural map*

$$H^n_{\mathrm{Zar}}(X, \mathcal{H}^i_{\mathrm{Zar}}) \to H^n_{\mathrm{Nis}}(X, \mathcal{H}^i_{\mathrm{Nis}})$$

is bijective.

This theorem applies notably to algebraic K-theory, étale cohomology and all examples listed in subsections 7.3 and 7.4.

Proof We need a lemma:

Lemma 8.3.2 *Let $X \in \mathcal{S}_k$ and $\mathcal{M}$ be the category of étale morphisms $U \xrightarrow{f} X$. For $x \in X$ and $i \in \mathbf{Z}$, let h^i_x denote the presheaf on $\mathcal{M}$*

$$f \mapsto \coprod_{y \in f^{-1}(x)} h^i_y(U).$$

Then h^i_x is a sheaf for the Nisnevich topology on $\mathcal{M}$.

Proof For simplicity, let us write $h^i_x(U)$ instead of $h^i_x(f)$. It is enough to show that, if f covers X at x, i.e., if there exists $x' \in f^{-1}(x)$ such that $k(x) \xrightarrow{\sim} k(x')$, then the sequence

$$0 \to h^i_x(X) \xrightarrow{\varphi} \coprod_{y \in f^{-1}(x)} h^i_y(U) \xrightarrow{\psi} \coprod_{z \in (f \times_X f)^{-1}(x)} h^i_z(U \times_X U)$$

is exact. Write $f^{-1}(x) = \{x'\} \cup T$. From étale excision it is easy to deduce that the map

$$h^i_x(X) \to h^i_{x'}(U)$$

is bijective. This shows that φ is split injective. It is now enough to show that the quotient complex

$$0 \to 0 \to \coprod_{y \in T} h^i_y(U) \xrightarrow{\psi'} \coprod_{z \in (f \times_X f)^{-1}(x)} h^i_z(U \times_X U)$$

is exact, i.e., that ψ' is injective. But we can decompose the set $(f \times_X f)^{-1}(x)$ into

$$\{(x', x')\} \cup \{x'\} \times T \cup T \times \{x'\} \cup T'.$$

Here we note that, for any $y \in f^{-1}(x)$, the schemes $x' \times_X y$ and $y \times_X x'$ are spectra of fields, because $k(x') = k(x)$; we abbreviate these schemes by (x', y) and (y, x'). By étale excision again, the maps

$$h^i_y(U) \to h^i_{(x',y)}(U \times_X U)$$
$$h^i_y(U) \to h^i_{(y,x')}(U \times_X U)$$

given respectively by the first and the second projection are bijective. The injectivity of ψ' follows. □

Proof of Theorem 8.3.1. a) Let $\tilde{X}_{\mathrm{Zar}}$ be the restriction of the big Zariski site of X to the category of schemes étale over X and $\alpha : X_{\mathrm{Nis}} \to \tilde{X}_{\mathrm{Zar}}$ be the natural projection. It is obvious that $\alpha^*\mathcal{H}^i_{\mathrm{Zar}} = \mathcal{H}^i_{\mathrm{Nis}}$. Therefore, applying α^* to the resolution (analogous to) (2.1) of $\mathcal{H}^i_{\mathrm{Zar}}$ yields a resolution of $\mathcal{H}^i_{\mathrm{Nis}}$. For $x \in X$, one has clearly an isomorphism of functors

$$\alpha^* i_{x*}^{\mathrm{Zar}} \simeq i_{x*}^{\mathrm{Nis}} \alpha^*$$

hence $\alpha^*(2.1)$ can be identified to the complex of Nisnevich sheaves

$$0 \to \coprod_{x\in X^{(0)}} i_{x*}^{\mathrm{Nis}} h_x^q(X) \to \coprod_{x\in X^{(1)}} i_{x*}^{\mathrm{Nis}} h_x^{1+q}(X) \to \cdots \to \coprod_{x\in X^{(p)}} i_{x*}^{\mathrm{Nis}} h_x^{p+q}(X) \to \cdots. \tag{8.6}$$

It is clear that i_{x*}^{Nis} is an exact functor for all $x \in X$. Therefore, the n-th Nisnevich cohomology of the p-th term of this complex is

$$\coprod_{x\in X^{(p)}} H^n_{\mathrm{Nis}}(x, h_x^{p+q}(X)).$$

But the Nisnevich cohomological dimension of a field is 0, hence this group is 0 for $n > 0$. It follows that the terms of (8.6) are acyclic. Finally, Lemma 8.3.2 shows that the global sections of (8.6) are (the analogue of) (2.1). □

Corollary 8.3.3 *Under the assumptions of Theorem 8.3.1, the Zariski and Nisnevich Brown-Gersten spectral sequences of Corollary 7.5.3 coincide.*

Indeed, they are compatible and their E_2-terms coincide. □

8.4 Shapiro's lemma.

Theorem 8.4.1 *Let h^* be a strictly effaceable cohomology theory with supports and let $i \in \mathbf{Z}$. Let $\mathcal{H}^i$ denote the Zariski sheaf associated to the presheaf h^i. Let $f : Y \to X$ be a finite morphism, with Y smooth. Then $R^q f_*\mathcal{H}^i = 0$ for $q > 0$.*

Proof We can compute $R^q f_*\mathcal{H}^i$ by using the (flasque) Cousin resolution $Cous$ of $\mathcal{H}^i$ over Y. But the stalk of f_*Cous at a point $x \in X$ is none other than the "stalk" of $Cous$ at $f^{-1}(x)$, i.e, $\Gamma(\mathcal{O}_{Y,f^{-1}(x)}, Cous)$, which is exact. □

8.5 Birational invariance.

Theorem 8.5.1 *Let h^* be a cohomology theory with supports satisfying axioms **COH1**, **COH2** and also axiom **COH6** if the base field k is finite. Let $X \in \mathcal{S}_k$ be smooth and let $H^*(X, \mathcal{H}^i)$ denote either of the groups $H^*_{\mathrm{Zar}}(X, \mathcal{H}^i_{\mathrm{Zar}}), H^*_{\mathrm{Nis}}(X, \mathcal{H}^i_{\mathrm{Nis}})$ of Theorem 8.3.1 (they coincide by this theorem). Then, for all $i \in \mathbf{Z}$, $H^0(X, \mathcal{H}^i)$ is a birational invariant of smooth proper varieties $X \in \mathcal{S}_k$.*

Proof a) By Corollary 5.1.11, the functor $X \mapsto h^i(X)$ satisfies "codimension 1 purity" for regular local rings of a smooth variety in the sense of [**7**, Definition 2.1.4 (b)] (a cohomology class which is unramified at points of codimension 1 is unramified everywhere locally). The claim now follows from [**7**, Proposition 2.1.8]. □

8.6 Rational invariance.

Let $\mathcal{S}_k = Sm/k$, and let h^* be a cohomology theory with supports on $\mathcal{P}_k$. Assume h^* satisfies axioms **COH1** (étale excision) and **COH2** (key lemma), the latter for all $V \in Sm/k$. If k is finite, assume h^* also satisfies axiom **COH6**. We then have the following theorem:

Theorem 8.6.1 *Let X, Y be two smooth integral k-varieties, with respective function fields $k(X), k(Y)$, and let $p : X \to Y$ be a proper morphism. Assume that the generic fibre X_η of p is $k(Y)$-birational to d-dimensional projective space $\mathbf{P}^d_{k(Y)}$. Then, for any $i \in \mathbf{Z}$, the map*

$$H^0(Y, \mathcal{H}^i) \xrightarrow{p^*} H^0(X, \mathcal{H}^i)$$

is an isomorphism.

Proof For any smooth integral k-variety Z, with generic point η, we have by definition

$$h^i_\eta(Z) = \varinjlim_{U \subseteq Z} h^i(U)$$

where U runs through the nonempty open subsets of Z. We denote this group by $h^i(k(Z))$. Corollary 5.1.11 yields an exact sequence

$$0 \to H^0(Z, \mathcal{H}^i) \to h^i(k(Z)) \to \coprod_{x \in Z^{(1)}} h^{i+1}_x(Z). \tag{8.7}$$

We may replace $\mathbf{P}^d_{k(Y)}$ by the d-fold self-product $(\mathbf{P}^1_{k(Y)})^d$ in the assumption of Theorem 8.6.1. By hypothesis, there exists a birational map

$$(\mathbf{P}^1_Y)^d \cdots\cdots\to X$$

over Y. Since $(\mathbf{P}^1_Y)^d$ is regular and p is proper, this rational map extends to a Y-morphism

$$U \xrightarrow{f} X$$

where U is an open subset of $(\mathbf{P}^1_Y)^d$ containing all points of codimension 1 (valuative criterion of properness, cf. [**26**, Theorem II. 4.7]). The exact sequence (8.7) then shows that the restriction map

$$H^0((\mathbf{P}^1_Y)^d, \mathcal{H}^i) \to H^0(U, \mathcal{H}^i)$$

is an isomorphism.

Let ζ (resp. ξ) be the generic point of X (resp. $(\mathbf{P}^1_Y)^d$ and U). We have a commutative diagram

$$\begin{array}{ccccc}
h^i(k(X)) & \xrightarrow{f^*} & h^i(k(\mathbf{P}^1_Y)^d) & = & h^i(k(\mathbf{P}^1_Y)^d) \\
\| & & \| & & \| \\
h^i_\zeta(X) & \xrightarrow{f^*} & h^i_\xi(U) & = & h^i_\xi((\mathbf{P}^1_Y)^d) \\
\cup & & \cup & & \cup \\
H^0(X,\mathcal{H}^i) & \longrightarrow & H^0(U,\mathcal{H}^i) & = & H^0((\mathbf{P}^1_Y)^d,\mathcal{H}^i) \\
& {\scriptstyle p^*}\nwarrow & \uparrow & \nearrow & \\
& & H^0(Y,\mathcal{H}^i) & &
\end{array}$$

in which the vertical inclusions follow from (8.7). Since f is birational, f^* is an isomorphism. It is thus enough to prove Theorem 8.6.1 in the case $X = (\mathbf{P}^1_Y)^d$. By induction on d, we may assume $d = 1$.

We first deal with the special case $Y = \operatorname{Spec} k$. To begin with, the natural map

$$h^i(k) \to h^i(k(\mathbf{P}^1_k))$$

is injective. If k is infinite, this follows from the classical section argument, since any open subset of $\mathbf{P}^1_k$ contains a rational point. If k is finite, axiom **COH6** provides a variant of this argument, since any open subset of $\mathbf{P}^1_k$ contains two closed points of coprime degrees.

On the other hand, since $\mathbf{P}^1_k$ is of dimension one, we have an exact sequence

$$h^i(\mathbf{P}^1_k) \to h^i(k(\mathbf{P}^1_k)) \to \coprod_{x \in (\mathbf{P}^1_k)^{(1)}} h^{i+1}_x(\mathbf{P}^1_k)$$

hence, from (8.7):

$$H^0(\mathbf{P}^1_k, \mathcal{H}^i) = \operatorname{Im}(h^i(\mathbf{P}^1_k) \to h^i(k(\mathbf{P}^1_k))).$$

The map $h^i(\mathbf{P}^1_k) \to h^i(k(\mathbf{P}^1_k))$ obviously factors through $h^i(\mathbf{A}^1_k)$. By axiom **COH2**, it even factors through $h^i(k)$, hence Theorem 8.6.1 in this case.

In the general case, let $\eta = \operatorname{Spec} k(Y)$ denote the generic point of Y. Note that any smooth $k(Y)$-variety is a filtering inverse limit of smooth k-varieties, with affine transition morphisms: we may therefore extend h^* to $\mathcal{P}_{k(Y)}$ (corresponding to $\mathcal{S}_{k(Y)} := \mathcal{S}m/k(Y)$) by direct limits. This cohomology theory with supports obviously satisfies axiom **COH1**; it also satisfies **COH2** because we assumed the original h^* verified it for all smooth varieties.

We have a commutative diagram with exact rows

$$\begin{array}{ccccccc} 0 & \longrightarrow & H^0(\mathbf{P}^1_Y, \mathcal{H}^i) & \longrightarrow & H^0(\mathbf{P}^1_{k(Y)}, \mathcal{H}^i) & \longrightarrow & \coprod_{y \in (\mathbf{P}^1_Y)^{(1)} - (\mathbf{P}^1_{k(Y)})^{(1)}} h^{i+1}_y(\mathbf{P}^1_Y) \\ & & \uparrow & & \wr\uparrow & & \uparrow \\ 0 & \longrightarrow & H^0(Y, \mathcal{H}^i) & \longrightarrow & h^i(k(Y)) & \longrightarrow & \coprod_{x \in Y^{(1)}} h^{i+1}_x(Y). \end{array}$$

To conclude the proof, it is sufficient to show that the right vertical map in this diagram is injective. But this map factors through

$$\coprod_{x \in Y^{(1)}} h^{i+1}_x(Y) \to \coprod_{x \in Y^{(1)}} h^{i+1}_{\eta_x}(\mathbf{P}^1_Y)$$

where $\eta_x = \operatorname{Spec} k(\mathbf{P}^1_x)$.

Let $x \in Y^{(1)}$: we have to see that the map $h^{i+1}_x(Y) \to h^{i+1}_{\eta_x}(\mathbf{P}^1_Y)$ is injective. We may replace Y by $Y' = \operatorname{Spec} \mathcal{O}_{Y,x}$. By definition, we have

$$h^{i+1}_{\eta_x}(\mathbf{P}^1_{Y'}) = \varinjlim_{Z} h^{i+1}_{\mathbf{P}^1_x - Z}(\mathbf{P}^1_{Y'} - Z)$$

where Z runs through proper closed subsets of $\mathbf{P}^1_x$.

Suppose first that $k(x)$ is infinite. Then $\mathbf{P}^1_x - Z$ contains a $k(x)$-rational point; since $\mathbf{P}^1(Y') \to \mathbf{P}^1(k(x))$ is surjective, we may lift it to a section s of $\mathbf{P}^1_{Y'} \to Y'$, which does not meet Z. Then the composite

$$h^{i+1}_x(Y) \to h^{i+1}_{\mathbf{P}^1_x - Z}(\mathbf{P}^1_{Y'} - Z) \xrightarrow{s^*} h^{i+1}_x(Y)$$

is the identity and $h^{i+1}_x(Y) \to h^{i+1}_{\mathbf{P}^1_x - Z}(\mathbf{P}^1_{Y'} - Z)$ is split injective.

Suppose now that $k(x)$ is finite, hence k is finite. Then $\mathbf{P}^1_x - Z$ contains in any case two closed points x_1, x_2 of coprime degrees d_1, d_2. Extending scalars to the residue fields k_1, k_2 and using axiom **COH6**, we get as above compositions

$$h^{i+1}_x(Y) \to h^{i+1}_{\mathbf{P}^1_x - Z}(\mathbf{P}^1_{Y'} - Z) \to h^{i+1}_{(\mathbf{P}^1_x - Z)_{k_i}}((\mathbf{P}^1_{Y'} - Z)_{k_i})$$

$$\xrightarrow{s_i^*} h^{i+1}_{x_i}(Y_{x_i}) \xrightarrow{\operatorname{Cor}_{k_i/k}} h^{i+1}_x(Y) \qquad (i = 1, 2)$$

with values d_1, d_2. Since these integers are coprime, we get that $h^{i+1}_x(Y) \to h^{i+1}_{\mathbf{P}^1_x - Z}(\mathbf{P}^1_{Y'} - Z)$ is split injective once again. $\square$

Appendix A Etale Cohomology: The Non-Torsion Case

The aim of this appendix is to extend Theorem 2.2.7 to all complexes of sheaves coming from the small étale site of k.

A.1 Proper base change.

The following is a version of the proper base change theorem involving non-torsion coefficients:

Proposition A.1.1 *Let V be an excellent Noetherian scheme of finite Krull dimension and $\tilde{\pi} : X \to V$ be a proper normal morphism* [**EGA4**, (6.8.1)]. *Then, for any complex of étale sheaves $C^{\cdot}$ over V and any geometric point $\overline{x}$ of V, there is a quasi-isomorphism*

$$(R\tilde{\pi}_*\tilde{\pi}^*C^{\cdot})_{\overline{x}} \simeq R\Gamma(X_{\overline{x}}, \tilde{\pi}^*C^{\cdot}).$$

Proof Since $\tilde{\pi}$ is proper and V is Noetherian finite dimensional and excellent, $R\tilde{\pi}_*$ has finite l-cohomological dimension for any prime l, and the argument of [**SGA4-III**, exposé X, proof of Theorem 4.1] shows that its rational cohomological dimension is bounded by $\dim V$. Hence, by comparison of hypercohomology spectral sequences, it is enough to prove this when $C^{\cdot}$ consists of one sheaf C placed in degree 0. We proceed as Deninger in [**10**], first reducing to the case where C is $\mathbf{Z}$-constructible. Using the excellence of V, we can further reduce as in *loc. cit.* to the case where $C = \tau_*F$, where $\tau : V' \to V$ is finite, V' is normal and F is a constant sheaf given by a finitely generated $\mathbf{Z}$-module. Passing to the strict henselization A of V at a geometric point $\overline{x}$, we get a commutative diagram of cartesian squares:

$$\begin{array}{ccccc}
\operatorname{Spec}\kappa' & \xrightarrow{\iota'} & \operatorname{Spec}A' & \longrightarrow & V' \\
{\scriptstyle\tau_\kappa}\downarrow & & {\scriptstyle\tau_A}\downarrow & & {\scriptstyle\tau}\downarrow \\
\operatorname{Spec}\kappa & \xrightarrow{\iota} & \operatorname{Spec}A & \longrightarrow & V
\end{array}$$

Here $\kappa = \kappa(\overline{x})$, $\operatorname{Spec}A' = \operatorname{Spec}A \times_V V'$ and $\operatorname{Spec}\kappa' = \operatorname{Spec}\kappa \times_V V'$, so that $\kappa' = \kappa \otimes_A A'$. Since A' is finite over A, it is a product of strictly henselian local rings and κ' is an Artin local ring with the same residue fields as A'. Since V' is normal, so are A' and any of its connected components. Let B be such a component. Since $\tilde{\pi}$ is proper, it is of finite type, hence $\tilde{\pi}_B$ and $\tilde{\pi}_{\kappa(\overline{y})}$ are normal by [**EGA4**, (6.8.3)], where for any ring R and morphism $\operatorname{Spec}R \to V$, we denote by $\tilde{\pi}_R : X_R \to \operatorname{Spec}R$ the pull-back of $\tilde{\pi}$. In particular, X_B and $X_{\kappa(\overline{y})}$ are normal, where $\overline{y}$ is the closed point of $\operatorname{Spec}B$. Hence, by [**10**, (2.3)], we have:

$$(R^q(\tilde{\pi}_B)_*F)_{\overline{y}} \simeq H^q(X_{\kappa(\overline{y})}, F).$$

On the other hand, letting κ'' be the product of the residue fields of A', the morphism $X_{\kappa''} \to X_{\kappa'}$ is radicial hence induces an isomorphism of étale cohomology [**36**, Chapter II, Remark 3.17]. It follows that

$$\iota'^*(R^q(\tilde{\pi}_{A'})_*F) \simeq R^q(\tilde{\pi}_{\kappa'})_*F.$$

Since τ is finite, $R^q(\tau_A)_* = R^q(\tau_\kappa)_* = 0$ for all $q > 0$ [**36**, Chapter II, Corollary 3.6] and the latter isomorphism implies

$$\iota^*R^q(\tau_A \circ \tilde{\pi}_{A'})_*F \simeq R^q(\tau_\kappa \circ \tilde{\pi}_{\kappa'})_*(\iota'')^*F.$$

$$
\begin{array}{ccccccc}
X_{\kappa''} & \longrightarrow & X_{\kappa'} & \xrightarrow{\iota''} & X_{A'} & \longrightarrow & X_{V'} \\
\Big\downarrow{\scriptstyle \tilde{\pi}_{\kappa''}} & & \Big\downarrow{\scriptstyle \tilde{\pi}_{\kappa'}} & & \Big\downarrow{\scriptstyle \tilde{\pi}_{A'}} & & \Big\downarrow \\
\operatorname{Spec}\kappa'' & \longrightarrow & \operatorname{Spec}\kappa' & \xrightarrow{\iota'} & \operatorname{Spec}A' & \longrightarrow & V' \\
 & & \Big\downarrow{\scriptstyle \tau_\kappa} & & \Big\downarrow{\scriptstyle \tau_A} & & \Big\downarrow{\scriptstyle \tau} \\
 & & \operatorname{Spec}\kappa & \xrightarrow{\iota} & \operatorname{Spec}A & \longrightarrow & V
\end{array}
$$

But using now the finiteness of $X_{V'} \to X_V$ and the induced maps and arguing as in [**10**, proof of (2.4)], we get an isomorphism

$$\iota^* R^q(\tilde{\pi}_A)_*(\tau_* F) \simeq R^q(\tilde{\pi}_\kappa)_*(\iota')^*(\tau_* F),$$

as desired. (See diagram below.)

$$
\begin{array}{ccc}
X_{\kappa'} & \xrightarrow{\iota''} & X_{A'} \\
\Big\downarrow{\scriptstyle \text{finite}} & & \Big\downarrow{\scriptstyle \text{finite}} \\
X_\kappa & \xrightarrow{\iota'} & X_A \\
\Big\downarrow{\scriptstyle \tilde{\pi}_\kappa} & & \Big\downarrow{\scriptstyle \tilde{\pi}_A} \\
\operatorname{Spec}\kappa & \xrightarrow{\iota} & \operatorname{Spec}A
\end{array}
$$

Remark A.1.2 We can use Artin's example in [**SGA4-III**, exposé XII, Section 2] to show that, in general, one cannot extend Proposition A.1.1 to more general complexes of sheaves over X than those of the form $D^{\cdot} = \tilde{\pi}^* \mathcal{C}^{\cdot}$. To be specific, take $V = \operatorname{Spec} R$ where R is a complete discrete valuation ring and $X = \mathbf{P}^1_R$. Let Y be the projective curve over R with equation $zy^2 = x(x-z)(x-\pi z)$, where π is a uniformizing parameter of R, that we view as a two-fold covering $Y \xrightarrow{\tau} X$ via the function x/z. Let $D^{\cdot} = \tau_* \mathbf{Z}[0]$. Then $H^1(X, D^{\cdot}) = H^1(Y, \mathbf{Z}) = 0$ while $H^1(X_0, D^{\cdot}) = H^1(Y_0, \mathbf{Z}) \simeq \mathbf{Z}$ where X_0, Y_0 are the special fibres of X and Y.

A.2 An integral Chern class.

Definition A.2.1 (compare [**13**, proof of Lemma 2]) Let $i \in \mathbf{Z}$.
a) For any prime number p, we denote by $\mathbf{Q}_p/\mathbf{Z}_p(i)$ the extension by 0 of the étale sheaf $\mathbf{Q}_p/\mathbf{Z}_p(i) = \varinjlim \mu_{p^n}^{\otimes i}$ from $\operatorname{Spec} \mathbf{Z}[1/p]$ to $\operatorname{Spec} \mathbf{Z}$. This defines a sheaf over the big étale site of $\operatorname{Spec} \mathbf{Z}$.
b) We define $\mathbf{Q}/\mathbf{Z}(i) = \bigoplus \mathbf{Q}_p/\mathbf{Z}_p(i)$.

Remark. Note that with this definition, $\mathbf{Q}/\mathbf{Z}(0)$ does not in general coincide with $\mathbf{Q}/\mathbf{Z}$!

Let

$$\mathbf{Q}/\mathbf{Z}(0)[-1] \to \mathbf{Z} \tag{A.1}$$

be the morphism in the derived category of big étale sheaves over $\operatorname{Spec} \mathbf{Z}$ defined as follows: for a prime p, we have an exact sequence of sheaves $0 \to \mathbf{Z} \to \mathbf{Z}[1/p] \to \mathbf{Q}_p/\mathbf{Z}_p \to 0$ hence a morphism $\mathbf{Q}_p/\mathbf{Z}_p[-1] \to \mathbf{Z}$ in the derived category. Letting j_p be the open immersion $\operatorname{Spec} \mathbf{Z}[1/p] \to \operatorname{Spec} \mathbf{Z}$, we get a corresponding morphism

$$\mathbf{Q}_p/\mathbf{Z}_p(0)[-1] := (j_p)_!\mathbf{Q}_p/\mathbf{Z}_p[-1] \to (j_p)_!\mathbf{Z} = (j_p)_!(j_p)^*\mathbf{Z}.$$

Composing with the adjunction map $(j_p)_!(j_p)^*\mathbf{Z} \to \mathbf{Z}$, we get a morphism $\mathbf{Q}_p/\mathbf{Z}_p(0)[-1] \to \mathbf{Z}$ over $\operatorname{Spec} \mathbf{Z}$. The desired morphism is the sum of these morphisms for all primes p.

Definition A.2.2 (compare [**13**, Appendix B]) Let X be a scheme and L a line bundle on X. To L we associate a morphism in $D(X_{ét})$

$$\mathbf{Q}/\mathbf{Z}(-1)[-3] \xrightarrow{C_1(L)} \mathbf{Z}$$

as follows. The class of L in $H^1(X_{ét}, \mathbb{G}_m)$ corresponds to a morphism in $D(X_{ét})$

$$\mathbf{Z} \xrightarrow{[L]} \mathbb{G}_m[1].$$

On the other hand, for all primes p, the sheaf $\mathbb{G}_m$ is p-divisible away from the locus where p is not invertible. This yields for all $n \geq 1$ an isomorphism:

$$\mathbb{G}_m \overset{L}{\otimes} \mathbf{Z}/n(0) \xrightarrow{\sim} \mathbf{Z}/n(1)[1].$$

These fit together to give a "Kummer" isomorphism

$$\mathbb{G}_m \overset{L}{\otimes} \mathbf{Q}/\mathbf{Z}(0) \xrightarrow{\sim} \mathbf{Q}/\mathbf{Z}(1)[1].$$

$C_1(L)$ is then the composition of the morphisms in the sequence

$$\mathbf{Q}/\mathbf{Z}(-1)[-3] \to \mathbb{G}_m \overset{L}{\otimes} \mathbf{Q}/\mathbf{Z}(-1)[-2] \xrightarrow{\sim} \mathbf{Q}/\mathbf{Z}(0)[-1] \to \mathbf{Z}$$

where the first morphism is $[L]$ tensored by $\mathbf{Q}/\mathbf{Z}(-1)[-3]$, the second one is the Kummer isomorphism twisted and shifted and the last morphism is (A.1). For two line bundles L and L' on X, we have $C_1(L \otimes L') = C_1(L) + C_1(L')$; in particular, if L is trivial then $C_1(L) = 0$.

The last claim of the definition is obvious from the construction of C_1 and the additivity of line bundle classes in $H^1(X_{ét}, \mathbb{G}_m)$.

A.3 Cohomology of $\mathbf{P}^1$.

Theorem A.3.1 *(compare* [**SGA5**, exposé VII, Theorem 2.2.1] *and* [**13**], *Lemma 3) Let V be as in Proposition A.1.1 and let $\tilde{\pi} : \mathbf{P}^1_V \to V$ be the natural projection. Then, for any complex of sheaves $C^{\cdot}$ over the small étale site of V, there is a natural isomorphism in $D(V_{ét})$:*

$$C^{\cdot} \oplus C^{\cdot} \overset{L}{\otimes} \mathbf{Q}/\mathbf{Z}(-1)[-3] \xrightarrow{\sim} R\tilde{\pi}_*\tilde{\pi}^*C^{\cdot}.$$

This isomorphism is the adjunction of a morphism

$$\tilde{\pi}^*C^{\cdot} \oplus \tilde{\pi}^*C^{\cdot} \overset{L}{\otimes} \tilde{\pi}^*\mathbf{Q}/\mathbf{Z}(-1)[-3] \to \tilde{\pi}^*C^{\cdot}$$

in which the first component is the identity and the second one is given by tensoring (in the derived sense) the Chern class map $\tilde{\pi}^*\mathbf{Q}/\mathbf{Z}(-1)[-3] \xrightarrow{C_1(\mathcal{O}(1))} \tilde{\pi}^*\mathbf{Z}$ of definition A.2.2 by $\tilde{\pi}^*C^{\cdot}$.

Proof By Proposition A.1.1, we have an isomorphism

$$(R\tilde{\pi}_*\tilde{\pi}^*\mathcal{C}^{\cdot})_{\overline{x}} \simeq R\Gamma(\mathbf{P}^1_{\kappa(\overline{x})}, \tilde{\pi}^*_{\overline{x}}\mathcal{C}^{\cdot})$$

for any geometric point $\overline{x}$ of V. To prove Theorem A.3.1, we may therefore assume that $V = \operatorname{Spec}\kappa$, where κ is a separably closed field. We first remark:

Lemma A.3.2 *Let $\mathcal{C}^{\cdot}$ be a complex of abelian groups and $D^{\cdot}$ be a bounded above complex of étale sheaves over $\mathbf{P}^1_\kappa$. Then the natural morphism in the derived category of abelian groups*

$$\mathcal{C}^{\cdot} \overset{L}{\otimes} R\Gamma(\mathbf{P}^1_\kappa, D^{\cdot}) \to R\Gamma(\mathbf{P}^1_\kappa, \mathcal{C}^{\cdot} \overset{L}{\otimes} D^{\cdot})$$

is an isomorphism.

Proof The argument of [**36**, Chapter VI, Lemma 8.7], which consists of reducing to the case where $\mathcal{C}^{\cdot}$ is a single finitely generated free $\mathbf{Z}$-module placed in degree 0, applies (compare *loc. cit.*, Remark 8.14). Note that it is not necessary to assume that $\mathcal{C}^{\cdot}$ is bounded above, since the Tor-dimension of $\mathbf{Z}$ is finite.

Applying Lemma A.3.2 to $D^{\cdot} = \mathbf{Z}[0]$ ($\mathbf{Z}$ placed in degree 0), it now suffices to prove Theorem A.3.1 in the case $\mathcal{C}^{\cdot} = \mathbf{Z}[0]$. In this case, it follows from

Lemma A.3.3 *a) We have $H^1(\mathbf{P}^1_\kappa, \mathbf{Z}) = 0$ and $H^q(\mathbf{P}^1_\kappa, \mathbf{Q}_p/\mathbf{Z}_p) = 0$ for $q > 0$ if $char(\kappa) = p > 0$.*
b) There is an isomorphism in $D(Ab)$:

$$\kappa^* \oplus \mathbf{Z}[-1] \xrightarrow{\sim} R\Gamma(\mathbf{P}^1_\kappa, \mathbb{G}_m)$$

whose first component is the adjunction of the map $\tilde{\pi}^*\kappa^* \to \mathbb{G}_m$ in $D((\mathbf{P}^1_\kappa)_{ét})$ and the second one is the adjunction of $\tilde{\pi}^*\mathbf{Z} \xrightarrow{[\mathcal{O}(1)]} \mathbb{G}_m$ (compare definition A.2.2).

Proof The vanishing of $H^1(\mathbf{P}^1_\kappa, \mathbf{Z})$ follows from the normality of $\mathbf{P}^1_\kappa$. Suppose $\operatorname{char}(\kappa) = p > 0$. We have

$$H^q(\mathbf{P}^1_\kappa, \mathbb{G}_a) = H^q_{Zar}(\mathbf{P}^1_\kappa, \mathcal{O}_{\mathbf{P}^1_\kappa}) = \begin{cases} \kappa & \text{if } \ q = 0 \\ 0 & \text{if } \ q > 0 \end{cases}$$

[**36**, Chapter II, Proposition 3.7 and Remark 3.8] and [**26**, Chapter III, Theorem 5.1]. Using the Artin-Schreier exact sequence $0 \to \mathbf{Z}/p \to \mathbb{G}_a \xrightarrow{F-1} \mathbb{G}_a \to 1$ [**36**, Chapter II, Example 2.18 (c)], this implies that

$$H^q(\mathbf{P}^1_\kappa, \mathbf{Z}/p) = 0 \quad \text{if} \quad q > 0$$

for p equal to the characteristic of κ. Using the exact sequences

$$0 \to \mathbf{Z}/p^n \to \mathbf{Z}/p^{n+1} \to \mathbf{Z}/p \to 0$$

it follows that $H^q(\mathbf{P}^1_\kappa, \mathbf{Z}/p^n) = 0$ for all n, hence $H^q(\mathbf{P}^1_\kappa, \mathbf{Q}_p/\mathbf{Z}_p) = 0$ for $q > 0$, as claimed. Finally, we have

$$H^q(\mathbf{P}^1_\kappa, \mathbb{G}_m) = \begin{cases} \kappa^* & \text{if } \ q = 0 \\ \operatorname{Pic}\mathbf{P}^1_\kappa = \mathbf{Z} & \text{if } \ q = 1 \\ 0 & \text{if } \ q > 1 \end{cases}$$

[**36**, Chapter III, Example 2.23 (b)], where the second isomorphism is induced by the degree map. Since $\mathcal{O}(1)$ generates $\operatorname{Pic}\mathbf{P}^1_\kappa$, the map $\mathbf{Z} \to \mathbb{G}_m[1]$ defined by its class induces an isomorphism

$$\mathbf{Z} = H^0(\mathbf{P}^1_\kappa, \mathbf{Z}) \longrightarrow H^1(\mathbf{P}^1_\kappa, \mathbb{G}_m).$$

inverse to the former. The last claim of Lemma A.3.3 follows.

We now finish proving Theorem A.3.1 for $\mathcal{C}^{\cdot} = \mathbf{Z}[0]$ in the case $V = \operatorname{Spec} \kappa$. By e.g., [**10**, (2.1)], $H^q(\mathbf{P}^1_\kappa, \mathbf{Q}) = 0$ for $q > 0$. From the exact sequence $0 \to \mathbf{Z} \to \mathbf{Q} \to \mathbf{Q}/\mathbf{Z} \to 0$ we derive

$$H^{q-1}(\mathbf{P}^1_\kappa, \mathbf{Q}/\mathbf{Z}) \xrightarrow{\sim} H^q(\mathbf{P}^1_\kappa, \mathbf{Z}) \text{ for } q > 1.$$

In view of Lemma A.3.3 a), this implies that (A.1) induces isomorphisms

$$H^{q-1}(\mathbf{P}^1_\kappa, \mathbf{Q}/\mathbf{Z}(0)) \xrightarrow{\sim} H^q(\mathbf{P}^1_\kappa, \mathbf{Z}) \text{ for } q > 1.$$

Using the Kummer exact sequence $1 \to \mu_n \to \mathbb{G}_m \xrightarrow{n} \mathbb{G}_m \to 1$ [**36**, ch. II, example 2.18 (b)], Lemma A.3.3 b) implies that $H^q(\mathbf{P}^1_\kappa, \mu_n) = 0$ for $q \neq 0, 2$ and n prime to the characteristic of κ; in particular, $H^q(\mathbf{P}^1_\kappa, \mathbf{Q}/\mathbf{Z}(0)) = 0$ for $q \neq 0, 2$. We therefore have $H^q(\mathbf{P}^1_\kappa, \mathbf{Z}) = 0$ for $q \neq 0, 3$.

Clearly, $\mathbf{Z} \to H^0(\mathbf{P}^1_\kappa, \mathbf{Z})$ is an isomorphism; it remains to see that the adjoint to $C_1(\mathcal{O}(1))$ induces an isomorphism $\mathbf{Q}/\mathbf{Z}(-1) \xrightarrow{\sim} H^3(\mathbf{P}^1_\kappa, \mathbf{Z})$. To do this, we follow the definition of C_1. According to definition A.2.2 and Lemma A.3.2, the map

$$\mathbf{Q}/\mathbf{Z}(-1) \longrightarrow H^3(\mathbf{P}^1_\kappa, \mathbf{Z})$$

can be decomposed as follows:

$$\mathbf{Q}/\mathbf{Z}(-1) \to H^1(R\Gamma(\mathbf{P}^1_\kappa, \mathbb{G}_m) \overset{L}{\otimes} \mathbf{Q}/\mathbf{Z}(-1)) \xrightarrow{\sim} H^1(\mathbf{P}^1_\kappa, \mathbb{G}_m \overset{L}{\otimes} \mathbf{Q}/\mathbf{Z}(-1))$$
$$\xrightarrow{\sim} H^2(\mathbf{P}^1_\kappa, \mathbf{Q}/\mathbf{Z}(0)) \xrightarrow{\sim} H^3(\mathbf{P}^1_\kappa, \mathbf{Z}).$$

By Lemma A.3.2, the second map is an isomorphism. According to definition A.2.2, the third map is an isomorphism and as seen above the fourth one is an isomorphism too. It remains to see that the first map is an isomorphism. But it is obtained by tensoring the isomorphism of Lemma A.3.3 b) by $\mathbf{Q}/\mathbf{Z}(-1)$. □

Appendix B The One-Dimensional Case

In this section, we prove a version of Gersten's conjecture for regular one-dimensional schemes, not necessarily in the presence of a base field. The proof mimics Gillet's in [**16**]. As in Section 5, we shall axiomatize the situation. The axioms necessary to make the proof work turn out to be much more costly than those in section 5.

B.1 Some axioms. Let A be a semi-local principal domain, $S = \operatorname{Spec} A$ and $\mathcal{S}_S$ the category of regular schemes separated and *quasifinite* over S. By Zariski's main Theorem, a connected object $X \to S$ of $\mathcal{S}_S$ is of the form $\operatorname{Spec} B$, where B is either a finite extension of one residue field of S or a localization of a finite extension of A at some maximal ideals. We give ourselves a "cohomology theory"

$$h^i : \mathcal{S}_S \to \mathcal{A} \ (i \in \mathbf{Z}),$$

a collection of contravariant functors to some abelian category $\mathcal{A}$, satisfying the following axioms:

(i) **Additivity.** h^* is additive.
(ii) **Transfers.** For a finite morphism $f : Y \to X$ in $\mathcal{S}_S$, there is given a map $f_* : h^{*-2c}(Y) \to h^*(X)$, where $c = \operatorname{codim}_X Y$; this collection of maps makes h^* a covariant functor.

(iii) **Purity.** For $X \in \mathcal{S}_S$ of dimension 1, $Z \xrightarrow{i} X$ a (reduced) closed subset of dimension 0 and $U \xrightarrow{j} X$ the complementary open subset, there is an exact sequence

$$\cdots \to h^{i-2}(Z) \xrightarrow{i_*} h^i(X) \xrightarrow{j^*} h^i(U) \xrightarrow{\partial} h^{i-1}(Z) \to \cdots$$

Moreover, if $f : X' \to X$ is a finite and flat map, the square

$$\begin{array}{ccc} h^i(U') & \xrightarrow{\partial'} & h^{i-1}(Z') \\ {\scriptstyle f_*}\downarrow & & \downarrow{\scriptstyle f'_*} \\ h^i(U) & \xrightarrow{\partial} & h^{i-1}(Z) \end{array}$$

commutes, where $Z' = f^{-1}(Z)_{\mathrm{red}}$, $U' = X' - Z'$ and $f' : Z' \to Z$ is the map induced by f (no multiplicities!)

(iv) **Action of units.** For any $X \in \mathcal{S}_S$ there is a pairing

$$\Gamma(X, \mathcal{O}_X^*) \times h^*(X) \to h^{*+1}(X)$$

which is contravariant in X and satisfies the projection formula for finite flat maps. Moreover,

- for X, Z, U, ∂ as in (iii), we have, for $(f, \alpha) \in \Gamma(U, \mathcal{O}_X^*) \times h^*(X)$:

$$\partial(f \cdot j^*\alpha) = \sum_{z \in Z} v_z(f) i_z^*(\alpha)$$

where i_z is the inclusion $z \hookrightarrow X$ (here we used the additivity of h^*).
- In the situation of (iii), given $f \in \Gamma(X, \mathcal{O}_X^*)$, the following diagram anticommutes:

$$\begin{array}{ccc} h^i(U) & \xrightarrow{\partial} & h^{i-1}(Z) \\ {\scriptstyle \cdot f}\downarrow & & \downarrow{\scriptstyle \cdot f} \\ h^{i+1}(U) & \xrightarrow{\partial} & h^i(Z) \end{array}$$

(v) **Rigidity.** Let $X \in \mathcal{S}_S$ of dimension 1, $x \in X$ a closed point and X_x^h the henselization of X at x. Let $h^*(X_x^h) := \varinjlim h^*(U)$, where U runs through all étale neighbourhoods of x. Then $h^*(X_x^h) \to h^*(x)$ is an isomorphism.

Examples B.1.1

(1) Algebraic K-theory verifies all axioms except (v); algebraic K-theory with coefficients $\mathbf{Z}/n$, where n is invertible in A, satisfies all axioms including (v) [**47**].

(2) Let $\mathcal{C}$ be a complex of sheaves over the small étale site of S; assume that its cohomology sheaves are all locally constant constructible, torsion invertible in A. Then $h^*(X) = \coprod_{n \in \mathbf{Z}} H^*_{ét}(X, \mathcal{C}(n))$ satisfies all the axioms. Axiom (i) is a general property of étale cohomology. Axioms (ii) and (iii) follow

from cohomological purity in dimension 1 [**SGA5**, exposé I, Theorem 5.1] and the existence of trace maps [**SGA4-III**, exposé XVIII]. Axiom (iv) is folklore: see [**42**, Lemma 3] for a detailed proof. Axiom (v) can be deduced from proper base change as in [**SGA4-III**, exposé XII, cor. 5.5]. Note that cohomological purity and proper base change for complexes of sheaves follow from the same for sheaves plus comparison of hypercohomology spectral sequences.

B.2 The result.

Theorem B.2.1 *Let R be a ring and h^* a cohomology theory with values in R-modules, satisfying axioms (i)–(v). Let Z be the set of closed points of S and η its generic point. Then, for all $i \in \mathbf{Z}$, the sequence*

$$0 \to h^i(S) \to h^i(\eta) \xrightarrow{\partial} h^{i-1}(Z) \to 0$$

is universally exact.

Proof For convenience we use ring-theoretic notation. Let $\mathcal{R}$ be the radical of A and F its field of fractions, so that $Z = V(\mathcal{R})$ and $\eta = \operatorname{Spec} F$ and the sequence of Theorem B.2.1 can be rewritten

$$0 \to h^i(A) \to h^i(F) \xrightarrow{\partial} h^{i-1}(A/\mathcal{R}) \to 0. \tag{B.1}$$

Write $h^{i-1}(A/\mathcal{R})$ as a direct limit of finitely presented R-modules. Let M be such a module. We construct an R-linear map $f_M : M \to h^i(F)$ such that $\partial \circ f_M = \iota$, where ι is the map $M \to h^{i-1}(A/\mathcal{R})$. In view of the long cohomology exact sequence of which (B.1) is part (axiom (iii)), this will show that the restriction of (B.1) to M is a split exact sequence of R-modules.

Let A^h be the henselization of A along $\mathcal{R}$. Then A^h splits as a finite product of henselian discrete valuation rings. By axioms (i) and (v), the natural map

$$h^{i-1}(A^h) \to h^{i-1}(A^h/\mathcal{R}A^h)$$

is an isomorphism.

Consider the commutative diagram

$$\begin{array}{ccc}
\operatorname{Hom}_R(M, h^{i-1}(A^h)) & \xrightarrow{\sim} & \operatorname{Hom}_R(M, h^{i-1}(A^h/\mathcal{R}A^h)) \\
\uparrow & & \uparrow \\
\varinjlim \operatorname{Hom}_R(M, h^{i-1}(A')) & \longrightarrow & \varinjlim \operatorname{Hom}_R(M, h^{i-1}(A'/\mathcal{R}A'))
\end{array}$$

where A' runs through the quasi-finite A-subalgebras of A^h. Since M is finitely presented, the two vertical maps are isomorphisms, hence so is also the bottom horizontal one. Therefore there exists an A' and an R-linear map

$$\sigma : M \to h^{i-1}(A')$$

such that the diagram

$$\begin{array}{ccc} h^{i-1}(A') & \longrightarrow & h^{i-1}(A'/\mathcal{R}A') \\ \sigma\uparrow & & \uparrow \\ M & \xrightarrow{\iota} & h^{i-1}(A/\mathcal{R}) \end{array}$$

commutes.

Let A_1 be the integral closure of A in the total ring of fractions F' of A'. Then $A_1 \subseteq A'$ and A' is a semi-localization of A_1 at some of its maximal ideals. Since A' is étale over A, F'/F is separable and A_1 is finite over A [**46**, Proposition 8, p.24]. Then the diagram

$$\begin{array}{ccc} h^i(F') & \xrightarrow{\partial} & h^{i-1}(A_1/\mathcal{R}A_1) \\ \mathrm{Cor}\downarrow & & \mathrm{Cor}\downarrow \\ h^i(F) & \xrightarrow{\partial} & h^{i-1}(A/\mathcal{R}) \end{array} \tag{B.2}$$

commutes (axiom (iii)). Let $\mathcal{R}_1 = \mathcal{R}A_1$. Write $\mathcal{R}_1 = \mathcal{R}'\mathcal{R}''$, where $\mathcal{R}' + \mathcal{R}'' = A_1$, $\mathcal{R}'A' = \mathcal{R}_1 A'$ and $\mathcal{R}''A' = A'$. Then

$$A_1/\mathcal{R}_1 \cong A_1/\mathcal{R}' \times A_1/\mathcal{R}'' \cong A'/\mathcal{R}A' \times A_1/\mathcal{R}'' \cong A/\mathcal{R} \times A_1/\mathcal{R}''$$

and the composite

$$A/\mathcal{R} \to A_1/\mathcal{R}_1 \cong A/\mathcal{R} \times A_1/\mathcal{R}'' \xrightarrow{pr_1} A/\mathcal{R}$$

is the identity. Accordingly, diagram (B.2) becomes:

$$\begin{array}{ccc} h^i(F') & \xrightarrow{(\partial', \partial'')} & h^{i-1}(A/\mathcal{R}) \oplus h^{i-1}(A_1/\mathcal{R}'') \\ \mathrm{Cor}\downarrow & & \begin{pmatrix} Id \\ ? \end{pmatrix}\downarrow \\ h^i(F) & \xrightarrow{\partial} & h^{i-1}(A/\mathcal{R}) \end{array} \tag{B.3}$$

By the Chinese remainder theorem, choose $f \in E'^*$ such that:

- $f \equiv 1 \pmod{\mathcal{R}''}$;
- f generates $\mathcal{R}'$.

For $\beta \in h^{i-1}(A')$, write $\overline{\beta}$ for its image in $h^{i-1}(A'/\mathcal{R}A') = h^{i-1}(A/\mathcal{R})$. Applying axiom (iv), we get:

Lemma B.2.2 *For all $\beta \in h^{i-1}(A')$, one has*

$$\begin{aligned} \partial'(\beta \cdot (f)) &= \overline{\beta}; \\ \partial''(\beta \cdot (f)) &= 0. \end{aligned}$$

□

Corollary B.2.3 *The map $f_M : M \to h^i(F)$ given by*

$$f_M(\alpha) = \mathrm{Cor}_{F'/F}(\sigma(\alpha) \cdot (f))$$

has the required properties.

Proof This follows from diagram (B.2) and Lemma B.2.2. □

B.3 Corollaries.

Notation B.3.1 For any scheme X, we denote by $\mathcal{LC}^+(X)$ the category of bounded below complexes of sheaves over the small étale site of X, whose cohomology sheaves are locally constant constructible, torsion prime to the residue characteristics of X.

Corollary B.3.2 *With notation as in Theorem B.2.1, let B be a finite, étale, Galois A-algebra, with Galois group G. Let $C^\cdot \in \mathcal{LC}^+(B)$. Let E be the total ring of fractions of B. Then the complex of $\mathbf{Z}[G]$-modules*

$$0 \to h^i(B) \to h^i(E) \xrightarrow{\partial} h^{i-1}(B/\mathcal{R}B) \to 0$$

is universally exact for all $i \in \mathbf{Z}$.

Proof Apply Theorem B.2.1 to $R\pi_* C^\cdot$ and $R = \mathbf{Z}[G]$, where $\pi : \operatorname{Spec} B \to \operatorname{Spec} A$ is the natural map. □

Corollary B.3.3 *With notation as in Theorem B.2.1, let X be a proper and smooth A-scheme, X_F its generic fibre and X_0 its closed fibre. Let $C^\cdot \in \mathcal{LC}^+(X)$. Then the complex*

$$0 \to H^i(X_{\acute{e}t}, C^\cdot) \to H^i(X_{F,\acute{e}t}, C^\cdot) \xrightarrow{\partial} H^{i-1}(X_{0,\acute{e}t}, C^\cdot(-1)) \to 0$$

is universally exact for all $i \in \mathbf{Z}$.

Proof Let $\pi : X \to \operatorname{Spec} A$ be the structure map. By [**36**, Corollary VI.4.2], $R\pi_* C^\cdot$ is in $\mathcal{LC}^+(A)$. It is clear that $H^0(F, R\pi_* C^\cdot) \cong H^0(X_F, C^\cdot)$, since $\operatorname{Spec} F \to \operatorname{Spec} A$ is an open immersion. Moreover, the proper base change theorem [**36**, Corollary VI.2.3] shows that $H^0(A/\mathcal{R}, R\pi_* C^\cdot(-1)) \cong H^0(X_0, C^\cdot(-1))$. Therefore the complex of Corollary B.3.3 can be rewritten

$$0 \longrightarrow H^i_{\acute{e}t}(A, R\pi_* C^\cdot) \to H^i_{\acute{e}t}(F, R\pi_* C^\cdot) \xrightarrow{\partial} H^{i-1}(A/\mathcal{R}, R\pi_* C^\cdot(-1)) \to 0$$

and we can apply Theorem B.2.1. □

Remark B.3.4 We can combine corollaries B.3.2 and B.3.3.

Appendix C Unbounded Complexes

In this appendix, we extend the notion of injective resolution from bounded below to unbounded complexes of objects of a suitable abelian category. These results are similar to those of Spaltenstein [**45**] (see also [**3**]).

C.1 Fibrant complexes. Let $\mathcal{A}$ be an abelian category and let $C(\mathcal{A})$ be the category of complexes of objects of $\mathcal{A}$. We set up the following definition:

Definition C.1.1 a) A morphism

$$C^\cdot \to D^\cdot$$

in $C(\mathcal{A})$ is a *trivial cofibration* if it is both a monomorphism and a quasi-isomorphism (i.e., it induces an isomorphism on cohomology).
b) An object $F^\cdot \in C(\mathcal{A})$ is *fibrant* if it has the following property: given a trivial cofibration $C^\cdot \to D^\cdot$, any morphism from $C^\cdot$ to $F^\cdot$ extends to a morphism from $D^\cdot$ to $F^\cdot$.

Fibrant complexes are closely related with K-injective complexes in the sense of J. Bernstein. It can be shown that the latter are those complexes which are homotopy equivalent to the former (compare [**45**, Proposition 1.5]).

Proposition C.1.2 *a) If $F^\cdot$ is fibrant, then F^n is injective for any $n \in \mathbf{Z}$.*
b) If $F^\cdot$ is bounded below, the converse is true.

Proof a) Let $A \overset{\varphi}{\hookrightarrow} B$ be a monomorphism in $\mathcal{A}$, and let $f : A \to F^n$ be a homomorphism. Let $C^\cdot, D^\cdot$ be the complexes such that $C^i = D^i = 0$ for $i \neq n, n+1$, $C^n = C^{n+1} = A$, $D^n = D^{n+1} = B$ and the differentials $C^n \to C^{n+1}$ and $D^n \to D^{n+1}$ are given by the identity. The monomorphism φ induces an obvious monomorphism of acyclic complexes $C^\cdot \hookrightarrow D^\cdot$, and f induces a morphism of complexes $f' : C^\cdot \to F^\cdot$. Applying the defining property, f' extends to a morphism $\tilde{f}' : D^\cdot \to F^\cdot$, whose restriction to $D^n = B$ defines an extension of f to B.

b) It is convenient to give a lemma:

Lemma C.1.3 *Let $F^\cdot \in C(\mathcal{A})$ be such that F^n is injective for some $n \in \mathbf{Z}$. Let $C^\cdot \hookrightarrow D^\cdot$ be a trivial cofibration, and let $f : C^\cdot \to F^\cdot$ be a homomorphism. Assume that $f^{n-1} : C^{n-1} \to F^{n-1}$ extends to $\tilde{f}^{n-1} : D^{n-1} \to F^{n-1}$ such that:*

(i) *$\tilde{f}^{n-1}(B^{n-1}(D^\cdot)) \subseteq B^{n-1}(F^\cdot)$;*
(ii) *$\tilde{f}^{n-1}(Z^{n-1}(D^\cdot)) \subseteq Z^{n-1}(F^\cdot)$.*

Then there exists $\tilde{f}^n : D^n \to F^n$ extending f^n such that

a) *$\tilde{f}^n d = d\tilde{f}^{n-1}$;*
b) *$\tilde{f}^n(Z^n(D^\cdot)) \subseteq Z^n(F^\cdot)$.*

Proof We define $\tilde{f}^n$ first on $B^n(D^\cdot)$, then on $Z^n(D^\cdot)$ and finally on all of D^n.

- On $B^n(D^\cdot)$, define $\tilde{f}^n$ by $\tilde{f}^n(dy) = d\tilde{f}^{n-1}(y)$ for $y \in D^{n-1}$. By assumption (ii), this does not depend on the choice of y.
- On $Z^n(D^\cdot)$, define $\tilde{f}^n$ as the unique map whose restriction to $B^n(D^\cdot)$ is as above and whose restriction to $Z^n(C^\cdot)$ is f^n. This is well-defined by the quasi-isomorphism assumption.
- On D^n, choose for $\tilde{f}^n$ any extension of the above, applying the injectiveness of F^n.

One readily checks that $\tilde{f}^n$ indeed verifies conditions a) and b). □

Proof of Proposition C.1.2 b). Lemma C.1.3 implies that if its conditions a) and b) are satisfied for some $n \in \mathbf{Z}$ then they are satisfied for $n+1$. In case $F^\cdot$ is bounded below, conditions a) and b) are trivially satisfied for $n << 0$. □

Lemma C.1.4 *a) Let $C^\cdot \in C(\mathcal{A})$ be acyclic and $F^\cdot \in C(\mathcal{A})$ be fibrant. Then any morphism $C^\cdot \xrightarrow{f} F^\cdot$ is homotopic to 0.*
b) Let $F^\cdot$ be fibrant and $C^\cdot$ be arbitrary. Then, any quasi-isomorphism $F^\cdot \xrightarrow{f} C^\cdot$ has a homotopy left inverse. If $C^\cdot$ is itself fibrant, f is a homotopy equivalence.

Proof a) Applying the defining property of "fibrant" to the monomorphism $F^\cdot \hookrightarrow C(f)$, where $C(f)$ is the mapping cone of f, we get that the identity $F^\cdot \to F^\cdot$ extends to a morphism $C(f) \to F^\cdot$ (note that, since $C^\cdot$ is acyclic, $F^\cdot \hookrightarrow C(f)$ is a quasi-isomorphism). Since the composite $C^\cdot \xrightarrow{f} F^\cdot \to C(f)$ is homotopic to 0, we get that f itself is homotopic to 0.

b) (cf. [**24**, proof of Lemma 4.5]) Note that the mapping cone $C(f)$ is acyclic. Applying a), we see that the morphism $C(f) \longrightarrow F^\cdot[1]$ is homotopic to 0, and the conclusions easily follow. □

Lemma C.1.5 *Consider a commutative square of objects of $C(\mathcal{A})$*

$$\begin{array}{ccc} C^\cdot & \xrightarrow{\alpha'} & F'^\cdot \\ {\scriptstyle\psi}\downarrow & & \downarrow{\scriptstyle\varphi} \\ D^\cdot & \xrightarrow{\alpha} & F^\cdot \end{array}$$

in which ψ is a trivial cofibration and $F^\cdot$ is fibrant. Let $\beta : D^\cdot \to F'^\cdot$ be a morphism extending α'. Then the two morphisms

$$\alpha, \varphi \circ \beta : D^\cdot \to F$$

are homotopic.

Proof By assumption, $\alpha - \varphi \circ \beta$ factors through the acyclic complex $D^\cdot/C^\cdot$. The conclusion now follows from Lemma C.1.4 a). □

C.2 Homotopy limits.

Definition C.2.1 Let $(F^\cdot_n, \varphi_{n+1,n})_{n\geq 0}$ be a projective system of objects of $C(\mathcal{A})$. Its *homotopy limit* $\operatorname{holim} F^\cdot_n$ is the total complex associated to the double complex (with vertical length 1)

$$\prod F^\cdot_n \xrightarrow{D} \prod F^\cdot_n$$

where $(\prod F^\cdot_n)^q := \prod F^q_n$, the differential being defined componentwise, and where $D_n : \prod F^q_n \to \prod F^q_n$ is defined as $(-1)^q(Id - \varphi^q_{n+1,n})$.

One readily checks that D anticommutes with the differentials of $\prod F_n$, so that the construction indeed defines a complex. It is clear that holim is a functor. By definition and the two spectral sequences associated to the double complex, we have:

Proposition C.2.2 *a) There is a short exact sequence of complexes*

$$0 \to \varprojlim F^\cdot_n \to \operatorname{holim} F^\cdot_n \to \varprojlim{}^1 F^\cdot_n[1] \to 0$$

hence a long exact sequence of cohomology groups

$$\cdots \to H^q(\varprojlim F^\cdot_n) \to H^q(\operatorname{holim} F^\cdot_n) \to H^{q-1}(\varprojlim{}^1 F^\cdot_n) \to H^{q+1}(\varprojlim F^\cdot_n) \to \cdots$$

b) Assume that countable products are exact in $\mathcal{A}$. Then there are short exact sequences

$$0 \to \varprojlim{}^1 H^{q-1}(F_n^\cdot) \to H^q(\operatorname{holim} F_n^\cdot) \to \varprojlim H^q(F_n^\cdot) \to 0$$

□

Corollary C.2.3 *Assume that countable products are exact in $\mathcal{A}$. Let $(C_n^\cdot)$, $(F_n^\cdot)$ be two projective systems in $C(\mathcal{A})$, and let (f_n) be a morphism from $(C_n^\cdot)$ to $(F_n^\cdot)$. Suppose that each f_n is a quasi-isomorphism (resp. a trivial cofibration). Then* $\operatorname{holim} f_n$ *is a quasi-isomorphism (resp. a trivial cofibration).* □

Lemma C.2.4 *For all $n \geq 0$, let $\varphi_n : \operatorname{holim} F_n^\cdot \to F_n^\cdot$ be given on $(\operatorname{holim} F_n^\cdot)^q = \prod F_n^q \oplus \prod F_n^{q-1}$ by $(p_n, 0)$, where p_n is the n-th projection. Then:*
a) The φ_n are morphisms of complexes.
b) For all $n \geq 0$, φ_n and $\varphi_{n+1,n} \circ \varphi_{n+1}$ are homotopic.

Proof a) is trivial. To see b), check that the map of degree -1

$$S_n : \operatorname{holim} F_n \to F_n$$

given by $S_n^q = (0, (-1)^q p_n^{q-1})$ is a homotopy between φ_n and $\varphi_{n+1,n} \circ \varphi_{n+1}$. □

Lemma C.2.5 *Let $(F_n^\cdot, \varphi_{n+1,n})_{n\geq 0}$ be as in Definition C.2.1, and let $C^\cdot$ be in $C(\mathcal{A})$.*
a) Let $f : C^\cdot \to \operatorname{holim} F_n$ be a morphism. Then $\varphi_n \circ f = 0$ for all n if and only if $f = (0, (g_n))$, where, for all n, g_n is a morphism from $C^\cdot$ to $F_n^\cdot[1]$.
b) Let $f_n : C^\cdot \to F_n^\cdot$ be a family of morphisms such that, for all $n \geq 0$, f_n and $\varphi_{n+1,n} \circ f_{n+1}$ are homotopic. Then there exists a morphism $f : C^\cdot \to \operatorname{holim} F_n^\cdot$ such that, for any $n \geq 0$, $\varphi_n \circ f = f_n$.

Proof a) is a simple computation. b) Choose for all n a homotopy s_n between f_n and $\varphi_{n+1,n} \circ f_{n+1}$. Define f by

$$f^q = ((f_n^q), ((-1)^q s_n^q))).$$

One checks easily that f is a morphism of complexes, and that the identity $\varphi_n \circ f = f_n$ holds. □

Proposition C.2.6 *Let $(F_n^\cdot, \varphi_{n+1,n})$ be an inverse system in $C(\mathcal{A})$. If each $F_n^\cdot$ is fibrant, then* $\operatorname{holim} F_n^\cdot$ *is fibrant.*

Proof Let $C^\cdot \xrightarrow{\alpha} D^\cdot$ be a trivial cofibration, and let $f : C^\cdot \to \operatorname{holim} F_n$ be a morphism. By the fibrancy of F_n, we may extend $\varphi_n \circ f$ to a morphism $\tilde{f}_n$ for each n. Lemma C.1.5 shows that $\tilde{f}_n$ and $\varphi_{n+1,n} \circ \tilde{f}_{n+1}$ are homotopic for any n; therefore, applying Lemma C.2.5 b), we can find a $\tilde{f} : D^\cdot \to \operatorname{holim} F_n^\cdot$ such that $\varphi_n \circ \tilde{f} = f_n$ for all n. We have

$$\varphi_n \circ (f - \tilde{f} \circ \alpha) = 0 \qquad \text{for all } n.$$

By Lemma C.2.5 a), we can write $f - \tilde{f} \circ \alpha = (0, (g_n))$, where each g_n is a morphism from $C^\cdot$ to $F_n^\cdot[1]$. Applying the fibrancy of F_n again, we can extend each g_n to a $\tilde{g}_n : D^\cdot \to F_n^\cdot[1]$. Then $\tilde{f} + (0, (\tilde{g}_n)) : D^\cdot \to \operatorname{holim} F_n$ extends f. □

Corollary C.2.7 *If $f : C^\cdot \to D^\cdot$ is a morphism between fibrant complexes, then the mapping cone of f is fibrant.* □

C.3 Resolutions.

Theorem C.3.1 *Suppose that $\mathcal{A}$ verifies axiom AB5 and has a generator in the sense of* [**23**, 1.5 and 1.6], *and that countable products are exact in $\mathcal{A}$. Then there exists a functor $F : C(\mathcal{A}) \to C(\mathcal{A})$ and a natural transformation $\varepsilon : Id \to F$ such that, for any $C^{\cdot} \in C(\mathcal{A})$,*

(i) *$F(C^{\cdot})$ is fibrant;*
(ii) *$\varepsilon_{C^{\cdot}}$ is a trivial cofibration.*

Proof By [**23**, Theorem 1.10.1], the assumptions imply the existence of a functor $I : \mathcal{A} \to \mathcal{A}$ and a natural transformation $\eta : Id \to I$ such that, for all $A \in \mathcal{A}$

(i) $I(A)$ is injective
(ii) η_A is a monomorphism.

We first construct F on bounded below complexes. If $C^{\cdot}$ is such a complex, the construction of [**24**, proof of Lemma 4.6 1)] embeds $C^{\cdot}$ into a bounded below complex of injectives by a trivial cofibration; by Proposition C.1.2 b), the latter complex is fibrant. We note that we can make this construction functorial by using the functor I above.

Suppose now $C^{\cdot}$ arbitrary. For any $n \in \mathbf{Z}$, let $C^{\cdot} \to C^{\cdot}_{\geq n}$ be the canonical truncation of $C^{\cdot}$ at level n. Recall that $C^{\cdot}_{\geq n}$ is defined by

$$(C^{\cdot}_{\geq n})^q = \begin{cases} C^q & \text{if } q \geq n \\ B^q(C^{\cdot}) & \text{if } q = n-1 \\ 0 & \text{if } q < n-1 \end{cases}$$

with the same differentials as $C^{\cdot}$, except that $B^n(C^{\cdot}) \to C^n$ is the canonical injection. We have $H^q(C^{\cdot}_{\geq n}) = \begin{cases} H^q(C^{\cdot}) & \text{if } q \geq n \\ 0 & \text{if } q < n. \end{cases}$ Note also that, for each $q \in \mathbf{Z}$, the inverse system $(C^{\cdot}_{\geq n})^q$ is stationary, hence $\varprojlim{}^1 C_{\geq n} = 0$ and

$$C^{\cdot} \xrightarrow{\sim} \varprojlim C^{\cdot}_{\geq n} \xrightarrow{\sim} \operatorname{holim} C^{\cdot}_{\geq n}$$

(isomorphisms) by Proposition C.2.2 a).

By the first part of the proof, we have a functorial trivial cofibration $C^{\cdot}_{\geq n} \to F(C^{\cdot}_{\geq n})$ for all n. This gives rise to a chain of morphisms

$$C^{\cdot} \xrightarrow{\sim} \varprojlim C^{\cdot}_{\geq n} \xrightarrow{\sim} \operatorname{holim} C^{\cdot}_{\geq n} \to \operatorname{holim} F(C^{\cdot}_{\geq n}). \tag{C.1}$$

We define $F(C^{\cdot})$ as $\operatorname{holim} F(C^{\cdot}_{\geq n})$ and $\varepsilon_{C^{\cdot}}$ as the composition of this chain. By Proposition C.2.6, $F(C^{\cdot})$ is fibrant and, by Corollary C.2.3, $\varepsilon_{C^{\cdot}}$ is a trivial cofibration, as desired. □

References

[1] Bloch, S. and Lichtenbaum, S. [1994], *A spectral sequence for motivic cohomology*, preprint.
[2] Bloch, S. and Ogus, A. [1974], *Gersten's conjecture and the homology of schemes*, Ann. Sci. Ec. Norm. Sup., 4. sér. **7**, 181–202.
[3] Bökstedt, M. and Neeman, A. [1993], *Homotopy limits in triangulated categories*, Compositio Math. **86**, 209–234.
[4] Bousfield, A. and Friedlander, E. M. [1978], *Homotopy theory of Γ-spaces, spectra, and bisimplicial sets*, Lecture Notes in Mathematics **658**, Springer-Verlag, Berlin, Heidelberg, New York, 80–130.

[5] Brown, K. S. and Gersten, S. M. [1973], *Algebraic K-theory as generalized sheaf cohomology*, Lecture Notes in Mathematics **341**, Springer-Verlag, Berlin, Heidelberg, New York, 266–292.

[6] Bourbaki, N. [1982], Algèbre commutative, Masson, Paris.

[7] Colliot-Thélène, J.-L. [1995], *Birational invariants, purity and the Gersten conjecture*, in: K-theory and algebraic geometry: connections with quadratic forms and division algebras, (W. Jacob and A. Rosenberg, ed.), Proceedings of Symposia in Pure Mathematics **58** (I), 1–64.

[8] Colliot-Thélène, J.-L. and Ojanguren, M. [1992], *Espaces principaux homogènes localement triviaux*, Publ. Math. IHES **75**, 97–122.

[9] Colliot-Thélène, J.-L. and Sansuc, J.-J. [1987], *Principal homogeneous spaces under flasque tori: applications*, J. Algebra **106**, 148–205.

[10] Deninger, C. [1988], *A proper base change theorem for non-torsion sheaves in étale cohomology*, J. Pure Appl. Algebra **50**, 231–235.

[11] Dutta, S. P. [1995], *On Chow groups and intersection multiplicities of modules, II* J. Alg. **171**, 370–382.

[12] Friedlander, E. and Voevodsky, V. [1995], *Bivariant cycle cohomology*, preprint.

[13] Gabber, O. [1993], *An injectivity property for étale cohomology*, Compositio Math. **86**, 1–14.

[14] Gabber, O. [1994], *Gersten's conjecture for some complexes of vanishing cycles*, Manuscripta Math. **85**, 323–343.

[15] Gillet, H. [1981], *Riemann-Roch theorems in higher algebraic K-theory*, Adv. in Math. **40**, 203–289.

[16] Gillet, H. [1986], *Gersten's conjecture for the K-theory with torsion coefficients of a discrete valuation ring*, J. Alg. **103**, 377–380.

[17] Gillet, H. and Soulé, C. [1981], *Filtrations on higher algebraic K-theory*, preprint.

[18] Grayson, D. [1978], *Projections, cycles and algebraic K-theory*, Math. Ann. **234**, 69–72.

[19] Grayson, D. [1985], *Universal exactness in algebraic K-theory*, J. Pure Appl. Algebra **36**, 139–141.

[20] Grivel, P.-P. [1987], *Catégories dérivées et foncteurs dérivés*, Exposé 1 of Algebraic $\mathcal{D}$-modules (A. Borel, ed.), Perspectives in Math. **2**, Academic Press, 1–108.

[21] Gros, M. [1985], *Classes de Chern et classes de cycles en cohomologie de Hodge-Witt logarithmique*, Mém. Soc. Math. France **21**.

[22] Gros, M. and Suwa, N. [1988], *La conjecture de Gersten pour les faisceaux de Hodge-Witt logarithmiques*, Duke Math. Journal, **57**, 615–628.

[23] Grothendieck, A. [1957], *Sur quelques points d'algèbre homologique*, Tohoku Math. J. **9**, 119–221.

[24] Hartshorne, R. [1966], *Residues and duality*, Lect Notes in Math. **20**, Springer-Verlag, Berlin, Heidelberg, New York.

[25] Hartshorne, R. [1975], *On the de Rham cohomology of algebraic varieties*, Publ. Math. IHES **45**, 5–99.

[26] Hartshorne, R. [1977], *Algebraic geometry*, Springer-Verlag, Berlin, Heidelberg, New York.

[27] Hu, S. T. [1959], *Homotopy theory*, Academic Press, New York.

[28] Illusie, L. [1979], *Complexe de de Rham-Witt et cohomologie cristalline*, Ann. Sci. Ec. Norm. Sup. **12**, 501–661.

[29] Iversen, B. [1986], *Cohomology of sheaves*, Springer Verlag, Berlin.

[30] Kahn, B. [1993], *Résultats de "pureté" pour les variétés lisses sur un corps fini* (appendice à un article de J.-L. Colliot-Thélène), Actes du Colloque de K-théorie algébrique de Lake Louise, décembre 1991 (P. G. Goerss, J. F. Jardine, ed.), Algebraic K-theory and algebraic topology, NATO ASI Series, Ser. C. **407**, 57–62.

[31] Kahn, B. [1996], *Applications of weight-two motivic cohomology*, Documenta Math. **1**, 395–416.

[32] Lazard, D. [1969], *Autour de la platitude*, Bull. Soc. math. Fr. **97**, 81–128.

[33] Lichtenbaum, S. [1987], *The construction of weight-two arithmetic cohomology*, Invent. Math. **88**, 183–215.

[34] Matsumura, H. [1986], *Commutative ring theory*, Cambridge Studies in advanced mathematics **8**, Cambridge University Press, Cambridge.

[35] McCleary, J. [1985], *User's Guide to Spectral Sequences*, Publish or Perish Inc., Wilmington, Delaware.

[36] Milne, J. [1980], *Étale cohomology*, Princeton University Press, Princeton, NJ.

[37] Nashier, B. [1983], *Efficient generation of ideals in polynomial rings*, J. Alg. **85**, 287–302.

[38] Nisnevich, Y. [1989], *The completely decomposed topology on schemes and the associated descent spectral sequences in algebraic K-theory*, in Algebraic K-theory: connections with geometry and topology, (J. F. Jardine, V. P. Snaith, eds.), NATO ASI Series, Sec. C **279**, 241–342.

[39] Nisnevich, Y. [1990], *Some new cohomological description of the Chow groups of algebraic cycles and the coniveau filtration*, preprint.

[40] Ojanguren, M. [1980], *Quadratic forms over regular rings*, J. Indian Math. Soc. **44**, 109–116.

[41] Paranjape, K. H. [1996], *Some spectral sequences for filtered complexes and applications*, J. Alg. **186**, 793–806.

[42] Peyre, E. [1993], *Unramified cohomology and rationality problems*, Math. Ann. **296**, 247–268.

[43] Quillen, D. [1973], *Higher algebraic K-theory, I*, Lecture Notes in Mathematics **341**, Springer-Verlag, Berlin, Heidelberg, New York, 77–139.

[44] Rost, M. [1996], *Chow groups with coefficients*, Documenta Math. **1**, 319–393.

[45] Spaltenstein, N. [1988], *Resolutions of unbounded complexes*, Compositio Math. **65**, 121–154.

[46] Serre, J.-P. [1968], Corps locaux, Hermann, Paris.

[47] Suslin, A. [1984], *On the K-theory of local fields*, J. Pure Appl. Algebra **34**, 301–318.

[48] Suslin, A. [1994], *Higher Chow groups of affine varieties and étale cohomology*, preprint.

[49] Suslin, A. and Voevodsky, V. [1995], *Bloch-Kato conjecture and motivic cohomology with finite coefficients*, preprint.

[50] Thomason, R. W. [1985], *Algebraic K-theory and étale cohomology*, Ann. Sci. Ec. Norm. Sup. **18**, 437–552.

[51] Thomason, R. W. and Trobaugh, T. [1990], *Higher algebraic K-theory of schemes and of derived categories*, The Grothendieck Festschrift, vol. III, Progress in Mathematics, Birkhäuser, 247–435.

[52] Voevodsky, V. [1995], *Cohomological theory of presheaves with transfers*, preprint.

[53] Voevodsky, V. [1995], *Triangulated categories of motives over a field*, preprint.

[54] Weibel, C. [1997], *The mathematical enterprises of Robert Thomason*, Bull. Amer. Math. Soc. **34**, 1–13.

[EGA4] Grothendieck, A. and Dieudonné, J. [1964], *Eléments de géométrie algébrique*, ch. IV (2de partie), I.H.E.S. Publ. Math. **24**.

[SGA4] Grothendieck, A., Artin, M. and Verdier, J.-L. [1972], *Théorie des topos et cohomologie étale des schémas* (SGA4), part III, Lecture Notes in Mathematics **271**, Springer-Verlag, Berlin, Heidelberg, New York.

[SGA41/2] Deligne, P. [1977], *Cohomologie étale: les points de départ* (rédigé par J.-F. Boutot), in Séminaire de Géométrie algébrique du Bois-Marie (SGA 4 1/2), Lect. Notes in Math. **569**, Springer-Verlag, Berlin, Heidelberg, New York, 129–153.

[SGA5] Grothendieck, A. et al., *Cohomologie l-adique et fonctions L* (SGA5), Lecture Notes in Mathematics **589**, Springer-Verlag, Berlin, Heidelberg, New York.

[SGA7] Deligne, P. and Katz, N. [1973], *Groupes de monodromie en géométrie algébrique* (SGA7), part II, Lecture Notes in Mathematics **340**, Springer-Verlag, Berlin, Heidelberg, New York.

Fields Institute Communications
Volume **16**, 1997

Milnor's Conjecture and Galois Theory I

Wenfeng Gao
Department of Mathematics
University of Western Ontario
London, Ontario, Canada N6A 5B7
wgao@uwovax.uwo.ca

Ján Mináč
Department of Mathematics
The University of Western Ontario
London, Ontario, Canada N6A 5B7
minac@uwo.ca

Dedicated to the memory of Bob Thomason

Abstract. Merkurjev's well-known theorem on the second cohomology of an absolute Galois group G with coefficients in the field of two elements, is a strong condition in the category of profinite groups. In this paper we investigate the group theoretic meaning of this condition in the case when G is a pro-2-group. We are describing G using generators and relations, as well as homomorphic images of G.

Introduction

In Milnor [1970], Milnor introduced his K^M-theory of fields and maps

$$h_n : k_n^M F \to H^n(F, 2)$$
$$\text{and} \quad s_n : k_n F \to \frac{I^n F}{I^{n+1} F}, n = 1, 2, \dots .$$

Here F is a field of characteristic not 2, $k_n F = \dfrac{K_n^M F}{2K_n^M F}$; $H^n(F, 2)$ is the nth Galois cohomology of F with coefficients $\dfrac{\mathbb{Z}}{2\mathbb{Z}}$ and I^n is the nth power of fundamental ideal I in the Witt ring of quadratic forms over F. In this way Milnor made a fascinating connection between Milnor K-theory, Galois cohomology and quadratic

---

1991 *Mathematics Subject Classification.* Primary 12G05; Secondary 11E04.
The first-named author is partially supported by J. F. Jardine's grant from NSERC.
The second-named author is partially supported by NSERC. He also gratefully acknowledges two UWO NSERC internal competition awards.

forms. Milnor and Tate succeeded in proving that all the h_n, s_n are isomorphisms if F is a global field. The question whether h_n and s_n are isomorphisms for all fields is known as the Milnor conjecture.

One has the hypothetical commutative diagram

$$\begin{array}{ccc} k_nF & \xrightarrow{\ s_n\ } & \dfrac{I^nF}{I^{n+1}F} \\ & \searrow^{h_n} & \downarrow{\scriptstyle e_n} \\ & & H^n(F,2). \end{array}$$

It is clear how e_n should be defined but it is not clear whether the resulting map is well defined. If both s_n and h_n are isomorphisms so is e_n. (In particular it is well defined.)

The current status of this conjecture for all fields of characteristic not 2 is:

$n = 0$: $k_0F, \dfrac{I^0F}{IF}, H^0(F,2)$ are all $\dfrac{\mathbb{Z}}{2\mathbb{Z}}$ and all maps s_0, h_0, e_0 are isomorphisms.

$n = 1$: s_1, h_1, e_1 are isomorphisms. These are "Kummer" isomorphisms.

$n = 2$: Milnor showed that s_2 is an isomorphism and in a famous paper Merkurjev [1981] proved that h_2 is an isomorphism.

$n = 3$: h_3 is an isomorphism (Merkurjev-Suslin [1991] and independently Rost [1986]). On the other hand Arason [1975] showed that e_3 is well defined. Therefore s_3 and e_3 are isomorphisms.

$n = 4$: Jacob-Rost [1989] and independently Szyjewski [1990] showed that e_4 is well defined. Finally, Rost announced that h_4 and therefore also s_4 and e_4 are isomorphisms.

Kato [1982] proved the Milnor conjecture if $\mathrm{char}(F) = 2$ and some interesting results were obtained by Arason-Elman-Jacob and others for special fields. (For details see Arason, Elman and Jacob [1984])

Finally, on March 3, 1996, Voevodsky made an announcement that he can prove the full Milnor conjecture in characteristic zero.

In this paper we are interested in the Galois theoretic meaning of Milnor's conjecture. Suppose that h_n is indeed isomorphism for some $n \in \mathbb{N}$. What properties of Galois groups can one deduce from it?

We shall make this question precise. Let G be any profinite group and $T(G)$ be the symmetric tensor algebra of $H^1(G,2)$ over the field with two elements. Let U be the ideal of $T(G)$ generated by all images of $\psi_1 \otimes \psi_2$, $\psi_i \in H^1(G,2)$ and $\psi_1 \cup \psi_2 = 0$ in $H^2(G,2)$ ($\cup$ denotes the cup product in the cohomology ring $H^2(G,2)$). Then $Mil(G) := T(G)/U$ is a graded commutative algebra over $\mathbb{F}_2$ where the degree of elements in $Mil(G)$ which are images of elements of $H^1(G,2)$ is 1. We can write $Mil(G) = \oplus Mil(G)_n, n \in \mathbb{N} \cup 0$, where $Mil(G)_n$ denote the set of all elements of $Mil(G)$ of degree n.

We have the unique extension of the identity map $H^1(G,2) \to H^1(G,2)$ to the ring homomorphism $j : Mil(G) \to H^*(G,2)$ of $Mil(G)$ into the cohomology ring of G with coefficients $\mathbb{Z}/2\mathbb{Z}$.

Observe that if G is an absolute Galois group of a field F then $Mil(G) = k(F)$, the Milnor $K-ring$ of F modulo two.

Definition *Let G be a profinite group and n be a natural number ≥ 2. We say that G* **satisfies the nth surjectivity condition** *if the induced map $j_n : Mil(G)_n \to H^n(G, 2)$ is surjective. We say that G satisfies the nth injectivity condition if j_n is injective.*

If $n = 2$ we say that G satisfies the **Merkurjev surjectivity** *(resp.* **injectivity***)* **condition** *iff G satisfies the second surjectivity (resp. injectivity) condition.*

Similarly we say that G satisfies the **Merkurjev-Suslin-Rost surjectivity** *(resp. injectivity) condition iff G satisfies the third surjectivity (resp. injectivity) condition.*

Question: *What is a group theoretic meaning of a Milnor n^{th} surjectivity (resp. injectivity) condition?*

In this paper we provide an answer to the question above in the case $n = 2$ and G is a $pro - 2 - group$. In their next paper, Gao and Mináč [1997] give an answer to the question above in the case when G is a $pro - 2 - group$ and G satisfies the third injectivity condition as well as both Merkurjev surjectivity and injectivity conditions.

Our answers use a representation of G via generators and relations as well as homomorphic images of G. We also use auxiliary fields to describe properties of relations inside of a free $pro - 2 - groups$ using field theory. The advantages of this approach lies in the fact that we can use Galois theory and field theoretic tools also for the investigation of relations in our groups.

Actually, in this paper we investigate only pro-2-groups and the case when the algebraic closure is replaced by the quadratic closure F_q of F. However, it is known that in order to establish Milnor's conjecture it is enough to establish Milnor's conjecture in this case.

As a matter of fact, assuming Milnor's conjecture one sees immediately that $H^n(F_q, 2) = \{0\}$ for all $n \geq 1$. Therefore from the extension of the absolute Galois group of F_q using the Galois group of F_q/F and its Hochschild-Serre spectral sequence we see that the inflation map $\inf : H^*(Gal(F_q/F), 2) \to H^*(F, 2)$ is an isomorphism. On the other hand if one assumes that $h_n q : k_n(F) \to H^*(Gal(F_q/F), 2)$ is an isomorphism for all $n \in \mathbb{N}$ and all fields F one can imitate the proof of (Arason, Elman and Jacob [1984], Proposition 5.9 and Remark 5.10) to prove that $k_n(F) \to H^*(Gal(F_q/F); 2)$ is an isomorphism for all fields F. In order to adapt the proof in Arason, Elman and Jacob [1984], which uses the Scharlau transfer for quadratic forms one can use (Fesenko and Vostokov [1993], Theorem 3.8), and the result of Bass, Tate and Kato which establishes the norm map for K-groups of field extensions.

Very little seems to be known about answers to the questions above. In the paper Mináč and Spira [1996], it was shown that there exists a rather precise relationship between Witt rings and certain Galois groups. The injectivity of h_2 in Merkurjev's theorem plays a crucial role in Mináč and Spira [1996]. Independently, in Mináč and Spira [1996], Villegas found a very nice Galois theoretic interpretation of injectivity of the Merkurjev theorem. Villegas' result is explained in (Mináč and Spira [1996], Corollaries 2.17 and 2.18).

In this paper, building on an observation in Mináč and Spira [1996], we offer a full Galois theoretic meaning of Merkurjev's theorem as well as injectivity of h_3 under assumption that Merkurjev's theorem is valid. See Theorems 2,3 and 4 below. We shall prove only Theorems 2 and 3, in this paper, and postpone the proof of

Theorem 4 to a subsequent paper. We concentrate on the Galois field theoretic meaning of Merkurjev's theorem.

Perhaps, more importantly, we develop some new tools which should also be useful for $h_n, n > 3$.

Let F be a field of characteristic not 2 and F_q is the quadratic closure of F. Let G be the Galois group of F_q/F. Then G is a pro-2-group. Let $\{\sigma_i\}_{i\in I}$ be a minimal set of generators of G. As is well known we can assume that I is well ordered. Let S be a free pro-2-group with a minimal set of generators $\{s_i\}_{i\in I}$. By sending s_i to σ_i we obtain a continuous projection $\pi : S \longrightarrow G$. We set $R = Ker(\pi)$. Our first observation is

Theorem 1 *For each nonempty set I there exists a field L such that*

1. *$Gal(L_q/L)$ is free pro-2-group on generators $\{s_i | i \in I\}$,*
2. *$\sqrt{-1} \in L$, i.e., level of L denoted by $s(L)$, is 1.*

Also there exists a field L' such that

1. *$Gal(L'_q/L')$ is free pro-2-group on generators $\{s_i | i \in I\}$,*
2. *$\sqrt{-1} \notin L'$ and $-1 = e_1^2 + e_2^2$ for some $e_1, e_2 \in (\dot{L}')$, i.e., $s(L') = 2$.*

Remarks

(1) In the case $\#I = n$, Theorem 1 was proved by Cordes [1973] in a slightly different form. Namely, instead of condition (1) in Theorem 1, Cordes uses $u(L) = 2$. However, if $u(L) = 2$, then $Gal(L_q/L)$ is a free pro-2-group. This is proved in Lemma 2.2. Recall that $u(F) = sup\{dim(q)$: q is an anisotropic form over $F\}$ and $s(F) = min\{n| -1$ is the sum of n squares in $F\}$. (If -1 is not the sum of squares in F we define $s(F) = \infty$. Similarly $u(F)$ can be ∞.) (See Lam [1973], Chapter 11).

(2) Observe that if $\mathrm{Gal}(L_q/L)$ is a free pro-2-group then necessarily $s(L) \leq 2$. Indeed by Remark (1) above, the u-invariant of L, $u(L)$ is 2. This means, in particular, that each element of L can be written as a sum of two squares in L. In particular -1 is expressible as a sum of two squares. This means $s(L) \leq 2$.

By Theorem 1, we may assume that S is the Galois group of the maximal 2-extension of some base field L. Let E be the fixed field of R. Then E is the Galois extension of L with Galois group G. Let $L = L^{(1)} \subset L^{(2)} \subset \cdots \subset L_q$ be the tower of fields such that $L^{(n)}$ is the compositum of all quadratic extensions of $L^{(n-1)}$ which is Galois over L. We define $E^{(1,L)} = E$ and define $E^{(n,L)}$:=the compositum of $E^{(n-1,L)}(\sqrt{a})$ such that $a \in E^{(n-1,L)}$ and $E^{(n-1,L)}(\sqrt{a})/L$ is Galois for $n \geq 2$. Since S and G have the same cardinality of the minimal set of generators, then $L^{(2)} \subset E$, or $L^{(2)}E = E^{(1,L)}$. In this paper, we show

Theorem 2 *$L^{(3)}E = E^{(2,L)}$ iff G satisfies the Merkurjev surjectivity condition.*

We also show a variation of Villegas' theorem. (See Mináč and Spira [1996], Corollary 2.18.)

Theorem 3 *(Variation of Villegas' theorem) Assume that $Gal(E/L)$ is isomorphic to a Galois group of the maximal 2-extension of some field F. Then G satisfies the Merkurjev injectivity condition iff $E \cap L^{(3)}$ is the compositum of all $\mathbb{Z}/2, \mathbb{Z}/4, D_4$ extensions of L sitting inside E.*

The following theorem is proved in Gao and Mináč [1997] (the second reference).

Theorem 4 *Assume that G satisfies both Merkurjev surjectivity and injectivity condition. Then G satisfies the Merkurjev-Suslin-Rost injectivity condition iff* $L^{(4)}E = E^{(3,L)}$.

By Theorem 2 and Theorem 4, it is natural to ask,

Question 5 *Let E be a Galois extension of L such that $G = Gal(E/L)$ is isomorphic to the Galois group of the maximal 2-extension of some base field. Is $L^{(n+1)}E = E^{(n,L)}$ for all $n \geq 1$?*

We also give the group-theoretic meaning of Question 5. For a pro-2-group H, we set $H^{(n+1)} = (H^{(n)})^2[H^{(n)}, H]$ and $H^{(1)} = H$. This means that $H^{(n+1)}$ is the closed subgroup of H generated by all squares in $H^{(n)}$ and all commutators with slots in $H^{(n)}$ and H. Then $H^{(n)}$ is a closed normal subgroup of H. We denote $H^{[n]}$ the quotient. Observe that $S^{[n]} = Gal(L^{(n)}/L)$. We set $R^{(1,S)} = R$ and $R^{(n+1,S)} = (R^{(n,S)})^2[R^{(n,S)}, S]$ for $n \geq 1$. In (Mináč and Spira [1996], Section 5), it was proved that $R^{(2,S)} = R \cap S^{(3)}$ iff the surjectivity part of Merkurjev's theorem is true. We shall prove that $R^{(n,S)} \subset R \cap S^{(n+1)}$ and Question 5 is equivalent to,

Question 6 *Let E be a Galois extension of L such that $G = Gal(E/L)$ is isomorphic to the Galois group of the maximal 2-extension of some base field. Is $R^{(n,S)} = R \cap S^{(n+1)}$ for all $n \geq 1$?*

By Theorem 4, we obtain

Corollary 7 *Assume that $Gal(E/L)$ is realized as the Galois group of the maximal 2-extension of some field F. Then* $R^{(3,S)} = R \cap S^{(4)}$.

In Gao and Mináč [1997] (the first reference) we shall discuss how to climb from $G^{[n]}$ to $G^{[n+1]}$. This is related to the embedding problem in Galois theory.

1 Preliminaries

We shall make the following abuse of notation: $[a]$ means both an element of $\dot{F}/\dot{F}^2$ and its corresponding element (a) in $H^1(F,2)$ or more generally in $H^1(Gal(T/F),2)$ where T/F is any Galois extension of F which contains $F^{(2)} :=$ compositum of all quadratic extensions of F, and is a subfield of F_q - the quadratic closure of F. This abuse is not a crime because by Kummer theory we have the canonical isomorphism $\dot{F}/\dot{F}^2 \to H^1(T,2)$.

We will work in the category of pro-2-groups and make the usual conventions that by "subgroup" we mean "closed subgroup", by "generated" we mean "topologically generated" and by "morphism" we mean "continuous morphism". We shall always work with mod 2 cohomology except when we explicitly mention other coefficients. For the standard facts for Galois cohomology we refer to Serre [1995] and Shatz [1972].

For an extension $1 \to A \to B \to C \to 1$ of profinite groups and a discrete B-module M we have the corresponding Lyndon-Hochschild-Serre spectral sequences $\{E_r^{p,q}, d_r^{p,q}\}$ where $E_2^{p,q} = H^p(C, H^q(A,M))$ (Hochschild and Serre [1953], Shatz [1972]). We also have the five-term exact sequence

$$0 \to H^1(C, M^A) \xrightarrow{Inf} H^1(B,M) \xrightarrow{Res} H^1(A,M)^C \xrightarrow{d_2^{0,1}} H^2(C,M^A) \xrightarrow{Inf} H^2(B,M)$$

where Inf is the inflation and Res is the restriction map.

Observe that k_nF is isomorphic to the factor group of

$$\dot{F}/\dot{F}^2 \otimes ... \otimes \dot{F}/\dot{F}^2$$

by the subgroup (denoted by M_n) generated by $[a_1] \otimes ... \otimes [a_n]$ where a_1 is a norm in the quadratic extension $F(\sqrt{a_j})$ for some $j = 2, ..., n$ and $[a_i]$ is the class of a_i in F^*/F^{*2} for $i = 1, ..., n$.

The following statement is an easy but useful exercise. (See e.g. Ware [1978].)

Statement *Suppose that K/F is a Galois extension, $G = Gal(K/F)$ is its Galois group and $a \in \dot{K}$. Then $K(\sqrt{a})/F$ is Galois iff $(\sigma a)/a$ is a square in $\dot{K}$.*

In other words the latter conditions say that $[a] \in (\dot{K}/\dot{K}^2)^G$, the fixed elements of $\dot{K}/\dot{K}^2$. Since $(\dot{K}/\dot{K}^2)^G = H^0(G, \dot{K}/\dot{K}^2)$ we have a useful link between Galois theory and cohomology.

Here is another very simple fact which enables us to identify [a] and (a) as above also in the case when both $H^1(F, 2)$ and $\dot{F}/\dot{F}^2$ are considered with some Galois group action.

The reason is nicely explained in Waterhouse [1994]. For the reader's convenience, let us recall this basic fact from Kummer theory. We follow quite closely Waterhouse [1994].

Consider a Galois extension K/F with Galois group $G = Gal(K/F)$. Then there is one-to-one correspondence between Galois extensions L/K which Galois group $Gal(L/K)$ is annihilated by 2 and those subgroups N of multiplicative subgroups of K which contain $\dot{K}^2$.

More precisely given a subgroup N of $\dot{K}$ such that $\dot{K}^2 \subset N$ we set $L = K[N^{1/2}]$ and N is recovered from L as $N = \{x \in \dot{K} \mid \sqrt{x} \in L\}$. The group $H = Gal(L/K)$ is the Pontrjagin dual of $N/\dot{K}^2$. In fact we have the following pairing

$$N/\dot{K}^2 \times Gal(L/K) \to \{\pm 1\},$$
$$([n], \gamma) \to \frac{\gamma(\sqrt{n})}{\sqrt{n}}$$

for all $n \in N$ and $\sigma \in Gal(L/K)$ which is a perfect duality between the compact group $Gal(L/K)$ and the discrete group $N/\dot{K}^2$.

As Waterhouse pointed out, this duality is quite standard at least in the finite case. However, relative versions of Kummer theory are not as familiar but are equally straightforward.

Suppose that L/K is as above. Then L/F is Galois iff the G-conjugates of elements in N are again in N. In that case H is a normal subgroup of $Gal(L/F)$, and hence $Gal(L/F)$ acts on $Gal(L/K)$ by the conjugation in $Gal(L/F)$.

A crucial observation is that the Kummer pairing is consistent with the actions of $Gal(K/F)$ on $N/\dot{K}^2$ and $Gal(L/K)$. Namely let σ be an element of $Gal(K/F)$ and $\widehat{\sigma}$ be any extension of it in $Gal(L/F)$. Then we have for each $\gamma \in Gal(L/K)$ and each $n \in N$;

$$\frac{(\widehat{\sigma}\gamma\widehat{\sigma}^{-1})(\widehat{\sigma}(\sqrt{n}))}{\widehat{\sigma}(\sqrt{n})} = \frac{\widehat{\sigma}\gamma(\sqrt{n})}{\widehat{\sigma}(\sqrt{n})} = \widehat{\sigma}\left(\frac{\gamma(\sqrt{n})}{\sqrt{n}}\right) = \frac{\gamma(\sqrt{n})}{\sqrt{n}}.$$

In other words

$$(\sigma[n], \gamma^{\sigma^{-1}}) = ([n], \gamma)$$

and

$$(\sigma[n], \gamma) = ([n], \gamma^{\sigma}),$$

where $\gamma^{\sigma} = \sigma^{-1}\gamma\sigma$. Observe that we are making a choice $\widehat{\sigma}(\sqrt{n}) = \sqrt{\sigma(n)}$. (This does not matter because if we take $\widehat{\sigma}(\sqrt{n}) = -\sqrt{\sigma(n)}$ then minus signs will cancel each other.) This means that $\sigma[n]$ acts on $Gal(L/K)$ as the corresponding element $\sigma((n)) \in H^1(\dot{K}, 2)$.

In particular we can identify $(\dot{K}/\dot{K}^2)^G$ with $H^1(\dot{K}, 2)^G$. This simple fact makes a direct link between Galois theory and cohomology.

The relations $\psi_1 \cup \psi_2 = 0 \in H^2(G, 2), \psi_1, \psi_2 \in H^1(G, 2)$ have a group theoretic interpretation. In order to explain it set D_4 to be the dihedral group of order 8 and let $\delta : D_4 \to \mathbb{Z}/2\mathbb{Z} \times \mathbb{Z}/2\mathbb{Z}$ and $\theta : \mathbb{Z}/4\mathbb{Z} \to \mathbb{Z}/2\mathbb{Z}$ be the natural projections of D_4 and $\mathbb{Z}/4\mathbb{Z}$ on their quotients by central elements of order 2.

Let D_4 be generated by involutions σ_1 and σ_2 and we write $\mathbb{Z}/2\mathbb{Z} \times \mathbb{Z}/2\mathbb{Z} = \langle g_1 \rangle \times \langle g_2 \rangle$.

We require that $\delta(\sigma_1) = (g_1, 1)$ and $\delta(\sigma_2) = (1, g_2)$.

Fact 1 *Let $\psi \in H^1(G, 2)$ be considered as a surjective homomorphism $G \to \mathbb{Z}/2\mathbb{Z}$. Then there exists a surjective homomorphism $\widetilde{\psi} : G \to \mathbb{Z}/4\mathbb{Z}$ such that the diagram*

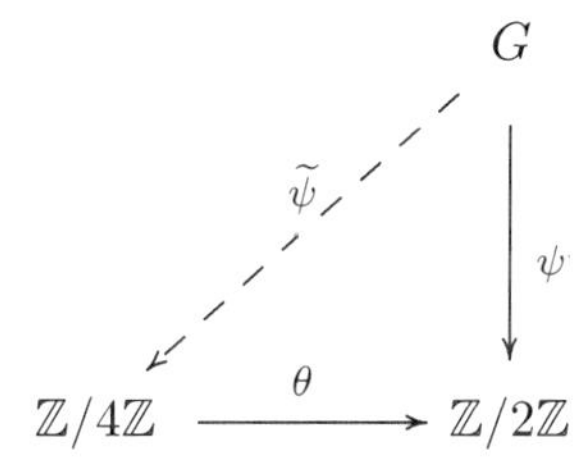

is commutative iff $\psi \cup \psi = 0$ in $H^1(G, 2)$.

Fact 2 *Let $\psi_1, \psi_2 \in H^1(G, 2)$ such that $0 \neq \psi_1 \neq \psi_2 \neq 0$. Set $\psi : G \to \mathbb{Z}/2\mathbb{Z} \times \mathbb{Z}/2\mathbb{Z} = \langle g_1 \rangle \times \langle g_2 \rangle$ be the group homomorphism defined by $\psi(g) = (\psi_1(g), \psi_2(g))$ for each $g \in G$. Then there exists a surjective homomorphism $\widetilde{\psi} : G \to D_4$ such that the following diagram is commutative:*

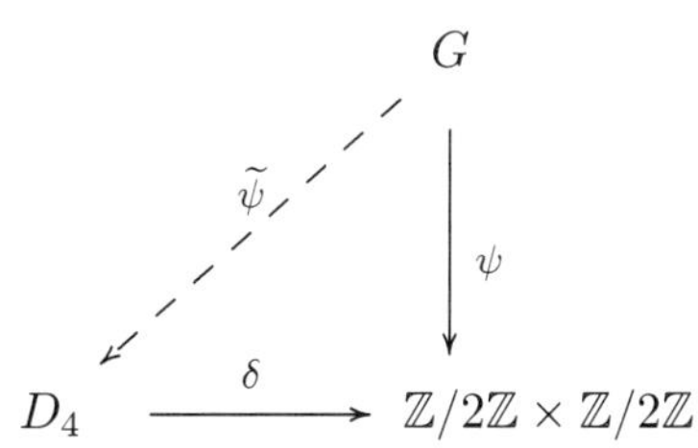

iff $\psi_1 \cup \psi_2 = 0$ in $H^2(G, 2)$.

Both of these two facts above follow from 1.1 and 2.3 in Hoechsmann's paper (Hoechsman [1968]). Translating these facts into field theory we obtain the following well known embedding criteria.

Fact 3 *Suppose that $F(\sqrt{a})/F$ is a quadratic extension ($a \notin \dot{F}^2$). Then there exists a Galois extension L/F such that $F(\sqrt{a}) \subset L$ and $Gal(L/F) = \mathbb{Z}/4\mathbb{Z}$ iff $[a] \cup [a] = 0$ in $H^2(F, 2)$.*

Fact 4 *Let* $[a],[b] \in \dot{F}/\dot{F}^2$ *are linearly independent in the vector space* $\dot{F}/\dot{F}^2$ *over* $\mathbb{F}_2$. *(Of course* $\mathbb{F}_2$ *is the field with two elements.) Then there exists a Galois extension* L/F *such that* $F(\sqrt{a},\sqrt{b}) \subset L$, *Galois group* $Gal(L/F)$ *is* D_4 *and* $Gal(L/F(\sqrt{ab}))$ *is the cyclic group of order 4 iff* $[a] \cup [b] = 0$ *in* $H^2(F,2)$.

These last two facts are both a special case of a more general theorem. (See e.g. Fröhlich [1985], Section 7.)

2 The Proof of Theorem 1

Proof of Theorem 1 For any non-empty set I there exists a field F_0 and a subset $\{[a_i]\}_{i\in I}$ of $\dot{F}_0/\dot{F}_0^2$ such that elements $[a_i], i \in I$, are linearly independent over $\mathbb{F}_2$, the field with two elements. Set, for example, $F_0 = F_0'(x_i)_{i\in I}$, a field with transcedence basis $\{x_i, i \in I\})$ over F_0'. Here F_0' is any field of characteristic not 2.

In the case when we want to manufacture a field L with $s(L) = 1$ we simply choose F_0 such that $s(F_0) = 1$.

In the case when we desire to construct a field L with $s(L) = 2$ we slightly modify our zero field F_0. We choose any field F_0' (of characteristic not 2) such that $s(F_0') \neq 1$. Then we set $x_1 = -1$ and $(x_i)_{i\in I, i\neq 1}$ such that $F_0 = F_0'(x_i)_{i\in I,i\neq 1}$ has transcedence basis $\{x_i, i \in I, i \neq 1\}$ over F_0'.

It is a fairly straightforward exercise to check that $\{[a_i] = [x_i], i \in I\}$ in both cases considered above are linearly independent over $\mathbb{F}_2$.

For each $a, b \in F_0$ consider the function field of conic $ax^2 + by^2 - z^2 = 0$, say: $F_{a,b} = F_0(t)(\sqrt{b(t^2-a)})$ and set F_1 to be the free compositum of $F_{a,b}$ for all $a, b \in \dot{F}_0$. Then, for each $a, b \in \dot{F}_0$ we have $[a] \cup [b] = 0 \in H^2(F_1)$. Also F_1/F_0 is known to be a regular field extension. This means that it contains no algebraic extension of F_0. In particular, $\dot{F}_1^2 \cap F_0 = \dot{F}_0^2$. Hence $\{[a_i]\}_{i\in I}$ remain linearly independent in $H^1(F_1)$.

Similarly set $F_2/F_1, F_3/F_2, ..., F_0 \subset F_1 \subset F_2 \subset F_3 \subset ..., F = \cup_{n=0}^{\infty} F_n$. Then $u(F) = 2$ and $\{[a_i]_F\}_{i\in I}$ is linearly independent in $H^1(F)$.

Consider $F \subset L \subset F_q$, where L is the maximal field extension of F in F_q such that all elements $\{[a_i]_L\}_{i\in I}$ remain linearly independent over L. (Use Zorn's Lemma to show the existence of L.)

Claim 1 $\{[a_i]\}_{i\in I}$ *forms a basis of* $H^1(L)$.

Proof of Claim 1 Because $\{[a_i]\}_{i\in I}$ is linearly independent over $H^1(L)$, it is enough to show that it generates $H^1(L)$. Suppose that it is not true. Then there exists $b \in \dot{L}$ such that $[b] \notin < [a_i]_L, i \in I >$. Consider $L(\sqrt{b}) \subset L_q$. Then, since $[b] \notin < [a_i]_L, i \in I >$, we have an exact sequence [L, page 202]:

$$1 \to \{[1]_L, [b]_L\} \to H^1(L) \to H^1(L(\sqrt{b})),$$

we ee that if, say $[a_1]_{L(\sqrt{b})}[a_2]_{L(\sqrt{b})}...[a_k]_{L(\sqrt{b})} = [1]_{L(\sqrt{b})}$, then $[a_1]_L[a_2]_L...[a_k]_L = [b]_L$. A contradiction of our hypothesis: $[b] \notin < [a_i]_L, i \in I >$.

□

Claim 2 $Gal(L_q/L)$ *is free pro-2-group on* $\{s_i\}_{i\in I}$.

Proof of Claim 2 Because $\{[a_i]_L\}_{i\in I}$ is linearly independent in $H^1(L)$, from the Kummer theory, we see $Gal(L_q/L) =< s_i | i \in I >$. Here $\{s_i \mid i \in I\}$ is a minimal set of generators of $Gal(L_q/L)$. To show that $Gal(L_q/L)$ is free pro-2-group, we recall that $u(F) = 2$ implies that $Gal(F_q/F)$ is a free pro-2-group. (See Lemma 2.2 below.) However $Gal(L_q/L) = Gal(F_q/L)$ is closed subgroup of $Gal(F_q/F)$. Therefore, from Serre [1995] we conclude $cd(L_q/L) \leq cd(F_q/F) = 1$. This means that $Gal(L_q/L)$ is a free pro-2-group.

□

The proof of the theorem is complete.

□

Corollary 2.1 [Waterhouse [1974] *For each pro-2-group G there exists a Galois extension L/F such that $Gal(L/F) \cong G$.*

For our reader's convenience we include the following well known result.

Lemma 2.2 $u(L) = 2$ *iff $Gal(L_q/L)$ is a free nontrivial pro-2-group.*

Proof Assume $u(L) = 2$. To show that $Gal(L_q/L)$ is a free pro-2-group, it is enough to show that $H^2(Gal(L_q/L)) = 0$. Using the surjectivity part of Merkurjev's theorem, it is enough to show that each class $[a] \cup [b] = 0 \in H^2(Gal(L_q/L))$, where $a, b \in \dot{L}$. However it is well known, Lam [1973], that $[a] \cup [b] = 0$ iff $aX^2 + bY^2 - Z^2 = 0$ has nontrivial solution $(x, y, z) \in L \times L \times L$. Since $u(L) = 2$, we see that $< a, b, -1 >$ is isotropic. Therefore, indeed $[a] \cup [b] = (0)$ for each $a, b \in \dot{L}$ and consequently $Gal(L_q/L)$ is a free pro-2-group.

Conversely suppose that $Gal(L_q/L)$ is a free nontrivial pro-2-group. Then each ternary quadratic form $a'x^2 + b'y^2 + c'z^2, a', b', c' \in \dot{L}$ is isotropic over L iff a suitable scaled ternary form $ax^2 + by^2 = z^2$ is isotropic over L. But this last quadratic form is indeed isotropic because $[a] \cup [b] = 0$ in $H^2(Gal(L_q/L))$. (See Lam [1973].) Hence we see $u(L) \leq 2$. However $u(L) = 1$ iff L is a quadratically closed field. (This means $\dot{L}^2 = \dot{L}$.) Since we assume that $Gal(L_q/L)$ is nontrivial this is not the case. Hence $u(L) = 2$ as we claimed.

□

3 Reformulations of the Surjectivity Part of Merkurjev's Theorem

Since S and G have the same cardinality of the minimal set of generators, then $R \subset S^{(2)}$.

Lemma 3.1 $R^{(n,S)} \subset R \cap S^{(n+1)}$.

Proof Our inclusion is true if $n = 1$. Assume that it is true for $k \leq n$. Then $R^{(n+1,S)} = (R^{(n,S)})^2[R^{(n,S)}, S] \subset (S^{(n+1)})^2[S^{(n+1)}, S] = S^{(n+2)}$. So $R^{(n+1,S)} \subset R \cap S^{(n+2)}$ as we desired.

□

Observe that $L^{(n+1)}E$ is the fixed field of $R \cap S^{(n+1)}$ and $E^{(n,L)}$ is the fixed field of $R^{(n,S)}$. Then we obtain

Corollary 3.2 $L^{(n+1)}E \subset E^{(n,L)}$ *for* $n \geq 1$.

Corollary 3.3 $L^{(n+1)}E = E^{(n,L)}$ *iff* $R^{(n,S)} = R \cap S^{(n+1)}$.

Proof of Theorem 2 Observe that $L^{(3)}E$ is the fixed field of $R \cap S^{(3)}$ which corresponds to quaternion algebras, $E^{(2,L)}$ is the fixed field of $R^2[R,S]$ which corresponds to $H^2(G)$. As is pointed out in (Minac and Spira [1996], Section 5), $R^2[R,S] = R \cap S^{(3)}$ iff the surjectivity part of Merkurjev's theorem is true. Our inclusion follows.

□

Remark We offer a detailed proof of a generalization of Corollary 3.3 in Theorem 2′ and Proposition 3.5 below.

Corollary 3.4 *$H^2(G)$ is dual to $Gal(EL^{(3)}/E)$.*

Recall that $H_n(G,2)$, the nth homology group of profinite group G with coefficients in the field with two elements, is a compact abelian group. As a matter of fact, for each $n \in N$ the group $H_n(G,2)$ is dual to $H^n(G,2)$. If we denote the Pontrjagin dual via an asterisk we have in particular: $(H^2(G))^* = H_2(G)$. This leads immediately to the following reformulation of Corollary 3.4. (See e.g. (Kuz'min [1969], Section 3).)

Corollary 3.4′ *$Gal(EL^{(3)}/E) \cong H_2(G)$.*

So that we can interpret certain abelian Galois groups as homology groups, one can try to "redefine" homology of profinite groups such that one obtains even some nonabelian Galois groups as new "homology groups". We hope to address these questions in the case of pro-2-groups in a subsequent paper.

We shall extend Theorem 2 to the case when the presentation $1 \to R \to S \to G \to 1$ is not necessarily minimal. This is important for translating the Merkurjev surjectivity property to each subgroup of G. In order to do this, set

$$\begin{aligned}(L, \text{rel } E)^{(3)} = &\text{ compositum of all quadratic extensions}\\ &(L^{(2)} \cap E)(\sqrt{\gamma}), \gamma \in L^{(2)} \cap \dot{E}, \text{ which are Galois over } L.\end{aligned}$$

Observe that if our representation above is minimal then $(L, \text{rel } E)^{(3)} = L^{(3)}$. Under the hypothesis above we have:

Theorem 2′ *$(L, rel\ E)^{(3)}E = E^{(2,L)}$ iff the Merkurjev surjectivity condition for the group G is true.*

Proof For the sake of convenience let us rename $L^{(2)} \cap E$ as $E^{(2)}$. Then

$$\begin{aligned}(L, rel\ E)^{(3)} = &\text{ compositum of all quadratic extensions } E^{(2)}(\sqrt{\gamma}), \gamma \in \dot{E}^{(2)},\\ &\text{such that the square class of } \gamma \text{ is fixed under Galois action of}\\ &Gal(E^{(2)}/L).\end{aligned}$$

Observe that $Gal(E^{(2)}/L) = G^{[2]}$. Therefore it is enough to show that the group G satisfies Merkurjev surjectivity condition iff the map

$$\left(\frac{\dot{E}^{(2)}}{(\dot{E}^{(2)})^2}\right)^{G^{[2]}} \longrightarrow \left(\frac{\dot{E}}{\dot{E}^2}\right)^{G}.$$

is surjective.

Our fundamental sequence applied to the group extension

$$1 \to Gal\left(\frac{L_q}{E^{(2)}}\right) \to S \to G^{[2]} \to 1$$

gives us

$$0 \to H^1(G^{[2]}) \to H^1(S) \to H^1\left(\frac{L_q}{E^{(2)}}\right)^{G^{[2]}} \to H^2(G^{[2]}) \to 0.$$

Here we use the fact that S is a free pro-2-group and therefore $H^2(S) = (0)$. Observe that we have canonical isomorphisms

$$\begin{array}{ccc} H^1\left(\frac{L_q}{E^{(2)}}\right)^{G^{[2]}} & \xrightarrow{\cong} & \left(\frac{\dot{E}^{(2)}}{(\dot{E}^{(2)})^2}\right)^{G^{[2]}} \\ {\scriptstyle Res}\downarrow & & \downarrow{\scriptstyle Res} \\ H^1(R)^G & \xrightarrow{\cong} & \left(\frac{\dot{E}}{\dot{E}^2}\right)^G \end{array}$$

such that the diagram above is commutative. Therefore it is enough to show that

$$Res : H^1\left(\frac{L_q}{E^{(2)}}\right)^{G^{[2]}} \longrightarrow H^1(R)^G$$

is surjective.

Let us compare our fundamental sequences for extensions below:

$$\begin{array}{ccccccccc} 1 & \longrightarrow & R & \longrightarrow & S & \longrightarrow & G & \longrightarrow & 1 \\ & & \downarrow & & \downarrow & & \downarrow & & \\ 1 & \longrightarrow & Gal\left(\frac{L_q}{E^{(2)}}\right) & \longrightarrow & S & \longrightarrow & G^{[2]} & \longrightarrow & 1. \end{array}$$

We obtain the commutative diagram

$$\begin{array}{ccccccc} H^1(S) & \longrightarrow & H^1(R)^G & \xrightarrow{tra} & H^2(G) & \longrightarrow & 0 \\ \| & & \uparrow{\scriptstyle Res} & & \uparrow{\scriptstyle inf} & & \\ H^1(S) & \longrightarrow & H^1\left(\frac{L_q}{E^{(2)}}\right)^{G^{[2]}} & \xrightarrow{tra} & H^2(G^{[2]}) & \longrightarrow & 0. \end{array}$$

Because $H^2(G^{[2]})$ is the degree 2 part of symmetric tensor algebra on $H^1(G^{[2]})$ we see that G satisfies Merkurjev surjectivity condition iff $H^2(G^{[2]}) \to H^2(G)$ is surjective.

Therefore by chasing the diagram above we see that the surjectivity part of Merkurjev's theorem is equivalent to the surjectivity of

$$H^1\left(\frac{L_q}{E^{(2)}}\right)^{G^{[2]}} \to H^1(R)^G.$$

Our proof is completed.

□

Theorem 2′ allows us to formulate conditions for the Merkurjev surjectivity condition for each subgroup N of G.

In order to formulate this result we set E^N - the fixed field of N, $E_N = E^{N^{(2)}}$ - the fixed field of $N^{(2)}$. Hence $\frac{E_N}{E^N}$ is the maximal multiquadratic extension of E^N in E.

Proposition 3.5 *N satisfies Merkurjev surjectivity condition iff the homomorphism*

$$Res : \left(\frac{\dot{E}_N}{\dot{E}_N^2}\right)^{N^{[2]}} \longrightarrow \left(\frac{\dot{E}}{\dot{E}^2}\right)^{N}$$

is surjective.

Proof Let $1 \to R \to S \xrightarrow{\pi} G \to 1$ be our presentation of G, and N be a subgroup of G. Then let $S(N)$ be the inverse image of N under the canonical map π. Then we obtain a presentation $1 \to R \to S(N) \to N \to 1$. Observe that $S(N)$ is a subgroup of S and therefore $S(N)$ is again a free pro-2-group. Then set $L' = E^N$, the fixed field of E. We can now apply Theorem 2′ for a field L'. We obtain $(L', rel\, E)^{(3)}E = E^{(2,L')}$ and $Res : \left(\frac{\dot{E}_N}{\dot{E}_N^2}\right)^{N^{[2]}} \to \left(\frac{\dot{E}}{\dot{E}^2}\right)^{N}$ is surjective.

□

Consider once again a presentation $1 \to R \to S \xrightarrow{\pi} G \to 1$ which is not necessarily minimal. Set $S(2)$ to be the inverse image of $G^{(2)}$ under π. Observe that if the presentation above is minimal then $S^{(2)} = S(2)$. Indeed because $G^{(2)}$ is the subgroup of G generated by squares we see that $\pi(S^{(2)}) = G^{(2)}$. On the other hand since we assume that our presentation is minimal we have $R \subset S^{(2)}$. Therefore $\pi^{-1}(G^{(2)}) = S^{(2)}$ as was claimed.

Set $D := (L, rel\, E)^{(3)}$ and $\mathcal{D} = Gal\left(\frac{L_q}{D}\right)$.

Lemma 3.6 $\mathcal{D} = (S(2))^2[S(2), S]$.

Proof Because $Gal(D/L^{(2)} \cap E)$ is a group of exponent 2 and since $S(2) = Gal\left(\frac{L_q}{L^{(2)} \cap E}\right)$ we see that $(S(2))^2 \subset \mathcal{D}$.

Because D is compositum of all quadratic extensions $(L^{(2)} \cap E)(\sqrt{\gamma}), \gamma \in L^{(2)} \cap E$, such that $(L^{(2)} \cap E)(\sqrt{\gamma})$ is Galois over L we see that each Galois group $Gal(L^{(2)} \cap \dot{E})(\sqrt{\gamma})/L$ is a central extension of $Gal((L^{(2)} \cap E)(\sqrt{\gamma})/(L^{(2)} \cap E))$ by $Gal((L^{(2)} \cap E)/L)$. On the other hand because $S(2)$ is the normal subgroup of S we see $[S(2), S] \subset S(2)$. Therefore the restriction of each commutator $[\sigma, \tau], \sigma \in S(2), \tau \in S$, to D is the identity. Hence $[S(2), S] \subset \mathcal{D}$ as well.

Therefore $(S(2))^2[S(2), S] \subset \mathcal{D}$. To show inclusion $\mathcal{D} \subset (S(2))^2[S(2), S]$ it is enough to show that the fixed field C of $(S(2))^2[S(2), S]$ is a subfield of D. Because $(S(2))^2 \subset (S(2))^{(2)}[S(2), S]$ we see that C is a compositum of some quadratic extensions of $E \cap L^{(2)}$. On the other hand because $[S(2), S] \subset (S(2))^2[S(2), S]$ we see that $Gal(C/(E \cap L^{(2)}))$ is in the center of $Gal(C/L)$. In particular each quadratic extension $(E \cap L^{(2)})(\sqrt{\gamma}) \subset C$, $\gamma \in E \cap L^{(2)}$ is Galois over L. Therefore $C \subset \mathcal{D}$. Our proof is now completed.

□

We pointed out above that if

$$1 \to R \to S \to G \to 1$$

is a minimal presentation, then the Merkurjev surjectivity condition for $G \to 1$ is equivalent to the equality $R^2[R,S] = R \cap S^{(3)}$. Now we shall extend this criterion when we do not assume that the presentation above is minimal. This result is a Corollary of Theorem 2′ and Lemma 3.6 above.

Corollary 3.7 $R \cap (S(2))^2[S(2),S] = R^2[R,S]$ *is true iff the Merkurjev surjectivity condition for* G *is satisfied.*

Proof First observe that since $R \subset S(2)$ we have $R^2[R,S] \subset R \cap (S(2))^2[S(2),S]$. The equivalence of $R^2[R,S] = R \cap (S(2))^2[S(2),S]$ with surjectivity of Merkurjev's theorem follows from the following facts.
Fact A: $E^{(2,L)}$ is the fixed field of $R^2[R,S]$.
Fact B: $(L, \mathit{Rel}\, E)^{(3)} E = E^{(2,L)}$ by Theorem 2′.
Fact C: $Gal\left(\dfrac{L_q}{(L, \mathit{Rel}\, E)^{(3)}}\right) = (S(2))^2[S(2),S]$. Indeed

$$R^2[R,S] = Gal\left(\frac{L_q}{E^{(2,L)}}\right) = R \cap (S(2))^2[S(2),S].$$

□

For a reader's convenience, we illustrate the proof above with the following diagram of fields:

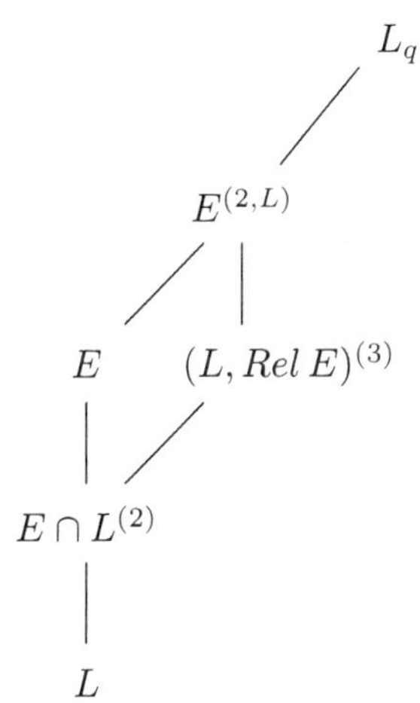

Here $Gal\left(\dfrac{L_q}{E}\right) = R; Gal\left(\dfrac{L_q}{E^{(2,L)}}\right) = R^2[R,S]; Gal\left(\dfrac{L_q}{(L, \mathit{Rel}\, E)^{(3)}}\right) = (S(2))^2[S(2),S]$.

Example 3.8 Let $G = D_4$, the dihedral group of order 8. It is well known that $H^*(D_4) \cong \mathbb{Z}/2[x_1, x_2, w]/(x_1(x_1 + x_2))$, where x_1, x_2 are generators of $H^1(D_4)$ and $w \neq 0 \in H^2(D_4)$ is the Stiefel-Whitney class of degree 2. We have $EL^{(3)} = L^{(3)}$

and $[L^{(3)} : L^{(2)}] = 8$. Indeed $Gal(L^{(3)}/L)$ is the quotient of the free pro-2-group on two generators subject to only relations which are consequences of the fact that $Gal(L^{(3)}/L^{(2)})$ is the central 2-elementary subgroup of $Gal(L^{(3)}/L)$. Therefore one can pick generators σ_1, σ_2 of $Gal(L^{(3)}/L^{(2)})$ such that

$$\begin{aligned} Gal(L^{(3)}/L^{(2)}) &\cong \langle \sigma_1^2 \rangle \times \langle \sigma_2^2 \rangle \times \langle [\sigma_1, \sigma_2] \rangle \\ &\cong \mathbb{Z}/2\mathbb{Z} \times \mathbb{Z}/2\mathbb{Z} \times \mathbb{Z}/2\mathbb{Z}. \end{aligned}$$

On the other hand, $dim_{\mathbb{Z}/2} H^2(G) = 3$ and $H^2(S) = 0$, then $[E^{(2,L)} : E] = 8$ and $[E^{(2,L)} : L^{(2)}] = 16$. So $EL^{(3)} \neq E^{(2,L)}$ and the quadratic extension $E^{(2,L)}/EL^{(3)}$ corresponds to the Stiefel-Whitney class w.

Of course D_4 is never a Galois group of a maximal 2-closure of some field F. Observe that in general $Gal(E^{(2,L)}/EL^{(3)})$ measures the failure of the Merkurjev surjectivity property of G.

4 The Proof of Theorem 3

We assume again that S and G have the same cardinality of the minimal set of generators, then $R \subset S^{(2)}$ and we have an extension $1 \to R \to S^{(2)} \to G^{(2)} \to 1$. By applying the fundamental exact sequence, we obtain

$$0 \to H^1(G^{(2)}) \to H^1(S^{(2)}) \to H^1(R)^{G^{(2)}} \to \dots.$$

This is an exact sequence of $G^{[2]} (= S^{[2]})$-modules. We then obtain an exact sequence

$$0 \to H^1(G^{(2)})^{G^{[2]}} \to H^1(S^{(2)})^{G^{[2]}} \to H^1(R)^G.$$

Let $a \in \dot{L}^{(2)}$. Observe that $L^{(2)}(\sqrt{a})$ is a Galois extension of L iff the class $[a]$ of a is in $H^1(S^{(2)})^{S^{[2]}}$. $L^{(3)}$ is the compositum of all the extensions $\dot{L}^{(2)}(\sqrt{a})$ of $L^{(2)}$ such that $a \in L^{(2)}$ and $[a] \in H^1(S^{(2)})^{S^{[2]}}$. Observe that $[a] \in H^1(G^{(2)})^{G^{[2]}}$ iff $[a]|_R = 0$ iff $\sqrt{a} \in E$. Then $E \cap L^{(3)}$ is the compositum of all the extensions $L^{(2)}(\sqrt{a})$ of $L^{(2)}$ such that $a \in L^{(\dot{2})}$ and $[a] \in H^1(G^{(2)})^{G^{[2]}}$. In particular, $E \cap L^{(3)}$ contains the compositum of all $\mathbb{Z}/2, \mathbb{Z}/4, D_4$ extensions of L sitting inside E (denoted by $E_{(3,L)}$).

Now we shall prove that $E \cap L^{(3)} = E_{(3,L)}$. We proved $E_{(3,L)} \subset L^{(3)} \cap E$; therefore it is enough to show that $E \cap L^{(3)} \subset E_{(3,L)}$. Consider the extension $G^{(2)} \to G \to G^{[2]}$. Since $inf : H^1(G^{[2]}) \to H^1(G)$ is an isomorphism, by applying the fundamental exact sequence, we obtain an exact sequence

$$0 \to H^1(G^{(2)})^{G^{[2]}} \xrightarrow{d_2^{0,1}} H^2(G^{[2]}) \xrightarrow{inf} H^2(G).$$

Let M_2 be the subgroup of $\dot{F}/\dot{F}^2 \otimes_{\mathbb{Z}/2} \dot{F}/\dot{F}^2$ generated by Steinberg's relations and B_2 the subgroup of $\dot{F}/\dot{F}^2 \otimes_{\mathbb{Z}/2} \dot{F}/\dot{F}^2$ generated by $x \otimes y + y \otimes x$, where $x, y \in \dot{F}/\dot{F}^2$. It is well known that $B_2 \subset M_2$ and $H^2(G^{[2]}) \cong (\dot{F}/\dot{F}^2 \otimes_{\mathbb{Z}/2} \dot{F}/\dot{F}^2)/B_2$. Here we use the isomorphism $H^1(G^{[2]}) \xrightarrow{inf} H^1(G) \xrightarrow{\sim} \dot{F}/\dot{F}^2$. Then we have a commutative diagram

$$\begin{array}{ccccccccc} 0 & \longrightarrow & M_2/B_2 & \longrightarrow & H^2(G^{[2]}) & \longrightarrow & k_2F & \longrightarrow & 0 \\ & & \Big\downarrow{\scriptstyle m_2} & & \Big\| & & \Big\downarrow{\scriptstyle t_2} & & \\ 0 & \longrightarrow & H^1(G^{(2)})^{G^{[2]}} & \xrightarrow{d_2^{0,1}} & H^2(G^{[2]}) & \xrightarrow{inf} & H^2(G) & \longrightarrow & 0 \end{array}$$

where $m_2 : M_2/B_2 \longrightarrow H^1(G^{(2)})^{G^{[2]}}$ is the induced homomorphism. Observe that we obtain $0 \longrightarrow ker\, t_2 \longrightarrow coker\, m_2 \longrightarrow 0$. Hence t_2 is injective iff m_2 is surjective. Therefore $Im(d_2^{0,1})$ is generated by $[c] \cup [d]$, where $cx^2 + dy^2 = 1$ for some $x, y \in \dot{F}$.

We have $[c] \cup [d] \in Im(d_2^{0,1})$ iff there exists $a \in L^{(2)}$ such that $\sqrt{a} \in E$ and $d_2^{0,1}([a]) = [c] \cup [d]$, where $c, d \in \dot{L}$. For each nontrivial class $[c] \cup [d] \neq 0$ as above we set $T = L(\sqrt{c}), b \in \dot{T}$ such that $N_{T/L}(b) = d$. Then $L(\sqrt{c}, \sqrt{d})(\sqrt{b})$ is a $\mathbb{Z}/4\mathbb{Z}$ or D_4-extension. (This means that $Gal(L(\sqrt{c}, \sqrt{d})(\sqrt{b})/L)$ is either $\mathbb{Z}/4\mathbb{Z}$ or D_4.) Moreover $d_2^{0,1}([b]) = [c] \cup [d]$ also! Therefore $d_2^{0,1}([a]) = d_2^{0,1}([b])$. Because $d_2^{0,1} : H^1(S^{(2)})^{S^{[2]}} \longrightarrow H^2(S^{[2]})$ is injective we see $[a] = [b]$. Hence $b = al^2$ with $l \in \dot{L}^{(2)}$. Then $\sqrt{b} = \pm\sqrt{a}l \in E$. Thus finally we conclude $L(\sqrt{c}, \sqrt{d})(\sqrt{b}) \subset E$. So $E \cap L^{(3)} \subset E_{(3,L)}$. □

Example 4.1 Let $G = \mathbb{Q}_8$ be the quaternion group. It is well known that $H^*(G) \cong \mathbb{Z}/2[x_1, x_2, w]/(x_1^2 + x_1x_2 + x_2^2, x_1^2x_2 + x_1x_2^2)$, where x_1, x_2 are generators of $H^1(G)$ and $w \neq 0 \in H^4(G)$ is the Pontrjagin class. We have $E \subset L^{(3)}$. So $E \cap L^{(3)} = E$. On the other hand, $E_{(3,L)} = L^{(2)} \neq E$.

Remark 4.2 Observe that $H^1(G^{(2)})^{G^{[2]}}$ is dual to $Gal(E \cap L^{(3)}/L^{(2)})$. Then k_2F is dual to $Gal(L^{(3)}/E \cap L^{(3)})$. On the other hand, $H^2(G) \cong H^1(R)^G$ is dual to $Gal(E^{(2,L)}/E)$. So Merkurjev's theorem implies that these two Galois groups are isomorphic. One can think that the Galois cohomology is upstairs (above E) and Milnor's k-theory (mod 2) is downstairs between E and L.

Acknowledgements

This paper is a part of the first-named author's Ph.D. thesis under the supervision of Professor J. F. Jardine. The first-named author is very grateful to him for his supervision and support, especially for his recommendation to be directed under and work with the second-named author. Both authors appreciate the useful comments and suggestions of both referees.

References

Arason, J. K., Elman, R. and Jacob, B. [1984] *Graded Witt rings and Galois cohomology I*, Can. Math. Soc. Conf. Proc., Vol. 4, AMS, Providence, RI, pp. 17-50.

Arason, J. Kr. [1975] *Cohomologische invarianten quadratischer formen* J. of Alg. **36**, 448-491.

Cordes, C. [1973] *The Witt group and the equivalence of fields with respect to quadratic forms* J. of Alg. **26**, 400-421.

Elman, R. and Lam, T.-Y. [1972] *Pfister forms and K-theory of fields* Jour. Algebra **23**, 181-213.

Fesenko, I. B. and Vostokov, S. V. [1993] *Local fields and their extensions. A constructive approach* Translations of Mathematical Monographs, AMS, Vol 121.

Fröhlich, A. [1985] *Orthogonal representations of Galois groups, Stiefel-Whitney classes and Hasse-Witt invariants* J. Reine Angew. Math. **360**, 83-123.

Gao, W. and Mináč, J. [1997] *Milnor's conjecture and embedding problems*, in preparation.

Gao, W. and Mináč, J. [1997] *Milnor's conjecture and Galois theory II*, in preparation.

Hoechsman, K. [1968] *Zum Einbettungsproblem* J. Reine Angew. Math. **229**, 81-106.

Hochschild, G. and Serre, J.-P. [1953] *Cohomology of group extensions* Trans. AMS **74**, 110-134.

Jacob, B. and Rost, M. [1989] *Degree four cohomological invariants for quadratic forms* Invent. Math. **96**, 551-580.

Jacob, B. and Ware, R. [1989] *A recursive description of the maximal pro-2-group via Witt rings* Math. Z. **200**, 379-396.

Kato, K. [1982] *Symmetric bilinear forms, quadratic forms, and Milnor K-theory in characteristic two* Invent. Math. **66**, 493-510.

Koch, H. [1978] *On p-extensions with given ramification*, Appendix 1 in K. Haberland, Galois cohomology of algebraic number fields, VEB Deutscher Verlag der Wissenschaften, Springer-Verlag.

Koch, H. [1970] *Galoische theorie der p-Erweiterungen*, Grundlehren der mathematische Wissenschaften, Springer-Verlag.

Kuz'min, L. V. [1969] *Homology of profinite groups, Schur multipliers and class field theory* Math. USSR Izvestija, Vol. **3**, no. 6, 1149-1181, English translation from Russian.

Lam, T. Y. [1973] *The algebraic theory of quadratic forms*, Benjamin/Addison-Wesley, Reading, Mass..

Merkurjev, A. S. [1981] *On the norm residue symbol of degree 2* Dokl. Akad. Nauk. USSR. **261**, 542-547; English transl. inSoviet. Math. Dokl. **24** (1982), 546-551.

Merkurjev, A. S. and Suslin, A. A. [1991] *The norm residue homomorphism of degree three* Math. USSR. Izv. **36**, 349-367.

Milnor, J. [1970] *Algebraic K-theory and quadratic forms* Invent. Math. **9**, 318-344.

Mináč, J. and Spira, M. [1996] *Witt rings and Galois groups* Ann. of Math. **144**, 35-60.

Rost, M. [1986] *Hilbert 90 for K_3 for degree two extensions*, preprint.

Šafarevič, I. R. [1955] *On p-extensions* Izv. Akad. Nauk. Math. Ser **19**, 445-470, (or see his collected mathematical papers, Springer-Verlag (1989)).

Scharlau, W. [1985] *Quadratic and Hermitian forms*, Grundlehren mat. Wiss. Vol. 270, Springer-Verlag, Berlin.

Serre, J.-P. [1995] *Cohomologie Galoisienne, Fifth Edition*, Lecture Notes in Mathematics, Vol. 5, Springer Verlag.

Shatz, S. [1972] *Profinite groups, arithmetic, and geometry*, Ann. of Math. Studies, Vol. 67, Princeton Univ. Press and Univ. of Tokyo Press.

Szyjewski, M. [1990] *The fifth invariant of quadratic forms*, English translation Algebra i Analiz **2**, 179-198.

Waterhouse, W. C. [1974] *Profinite groups are Galois groups* Proc. AMS **42**, 639-640.

Waterhouse, W. C. [1994] *The normal closures of certain Kummer extensions* Can. Math. Bull. **37**, 133-139.

Ware, R. [1978] *When are Witt rings group rings? II* Proc. J. Math. **76**, 541-564.

Fields Institute Communications
Volume **16**, 1997

Ultraproducts and the Discrete Cohomology of Algebraic Groups

J. F. Jardine
Mathematics Department
University of Western Ontario
London, Ontario, Canada N6A 5B7
jardine@uwo.ca

Abstract. We give a criterion, in terms of whether or not the ultraproduct construction for simplicial sets preserves certain homology localizations, for the isomorphism conjecture relating the discrete and étale cohomology of the special linear group Sl_n. Ultraproducts are characterized here as stalks of direct image sheaves, for a topology of finite covers on the subsets of an infinite set.

Introduction

Suppose that G is a reductive algebraic group over an algebraically closed field k, and that ℓ is a prime not equal to the characteristic of k. The generalized isomorphism conjecture of Friedlander and Milnor asserts that the comparison map

$$H^*_{et}(BG_k, \mathbb{Z}/\ell) \xrightarrow{\epsilon^*} H^*(BG(k), \mathbb{Z}/\ell),$$

relating étale and discrete mod ℓ cohomology, is an isomorphism. The positive evidence for this conjecture consists of the Friedlander-Mislin theorem [**6**], [**12**], which asserts that the map ϵ^* is an isomorphism if k is the algebraic closure of a finite field [**6**], and the Gabber rigidity theorem [**8**], which implies that ϵ^* is an isomorphism if G is the general linear group Gl, for all algebraically closed fields k [**11**]. As is well known, the result for the general linear group Gl leads to a calculation of the mod ℓ K-groups of all algebraically closed fields. It can also be bootstrapped to a result for both the infinite orthogonal and infinite symplectic groups, leading to calculations of the mod ℓ Karoubi L-groups for all algebraically closed fields [**14**].

The argument for the Gabber rigidity theorem has been shown to be essentially a formal consequence of a series of axioms [**18**], which are satisfied by mod ℓ K-theory, and involve both the transfer and the homotopy property. There has been some hope that one could demonstrate analogs of these axioms for the mod ℓ homology sheaves $\tilde{H}_*(BG, \mathbb{Z}/\ell)$ for the simplicial sheaf BG on some big étale site

---

1991 *Mathematics Subject Classification.* Primary 20G10; Secondary 18G30, 03C20.
This research was supported by NSERC.

for the algebraically closed field k, to show that these sheaves are constant. The assertion that the sheaves $\tilde{H}_*(BG, \mathbb{Z}/\ell)$ are constant is a rigidity statement for the simplicial sheaf BG, and would lead to a proof of the isomorphism conjecture for G in the same way that the Gabber rigidity theorem implies the conjecture for the general linear group Gl. The entire unstable case has, however, stoutly resisted all such attempts in the intervening years: the sticking point seems to be the homotopy property.

The purpose of this paper is to present a new attack on the problem. The conjecture is not proved here, not even in an interesting special case. It is not attacked in full generality either, but it is shown that the isomorphism conjecture for the special linear group Sl_n is equivalent to a problem about the homology of ultraproducts of spaces of the form $BSl_n(\overline{\mathbb{F}}_p)$ (Theorem 17 below), and that this problem can be reduced to the corresponding problem for ultraproducts of spaces $BSl_n(F)$ arising from finite fields (Corollary 19). The special linear group Sl_n was chosen to fix ideas: one expects the same arguments to work for arbitrary simple simply connected algebraic groups. One expects further that a proof of the isomorphism conjecture for simple simply connected algebraic groups would lead to a proof of the statement in full generality.

The ideas that appear here are old as well as new. This work was originally motivated by a desire to understand what classical model theory has to say about a Lefschetz principle for the homology groups $H_*(BSl_n(k), \mathbb{Z}/\ell)$. A short answer is quickly apparent: nothing, because homology is not a first order construction. At the same time, the circle of ideas around the Lefschetz principle for first order statements about algebraically closed fields is extremely suggestive.

The idea, roughly, is that there are two different methods of constructing a very large algebraically closed extension L of a given algebraically closed field k, and both are useful. The first of these is to construct a big transcendental extension of k and then to close it up, while the second is the classical ultraproduct construction. The utility of attaching transcendentals in the context of the isomorphism conjecture is already well understood: the experts all know, and it is proved here using an argument of Suslin [**16**], that any inclusion $k \subset L$ of algebraically closed fields induces a monomorphism $H_*(BSl_n(k), \mathbb{Z}/\ell) \hookrightarrow H_*(BSl_n(L), \mathbb{Z}/\ell)$. What it means (see the argument for Theorem 17) is that, if you want to show that the canonical map

$$H^*_{et}(BSl_{n,k}, \mathbb{Z}/\ell) \xrightarrow{\epsilon^*} H^*(BSl_n(k), \mathbb{Z}/\ell)$$

is an isomorphism, then it's enough to find an algebraically closed field L containing k such that the map

$$H^*_{et}(BSl_{n,L}, \mathbb{Z}/\ell) \xrightarrow{\epsilon^*} H^*(BSl_n(L), \mathbb{Z}/\ell)$$

is an isomorphism. The production of such extension fields L is the role of the ultraproduct construction.

In order to talk about ultraproducts of simplicial sets sensibly, it is necessary to reinterpret classical ultraproducts as stalks of certain direct image sheaves, where the direct image arises from a comparison of the discrete topology with the topology of finite coverings on the set of subsets of a fixed infinite set. As such, ultraproducts of simplicial sets become stalks of simplicial sheaves, so that the homotopy theory of simplicial sheaves can be brought into play. The classical approach remains important, because we always want to be sure that our constructions produce simplicial sets $BSl_n(L)$ as ultraproducts, for large enough algebraically closed fields L. The usual facts about ultrafilters and the topos theoretic interpretation of them that are required here are surveyed in the first two sections of this paper. I defer to the book of Bell and Slomson [**1**] for all of the more delicate cardinality calculations, but these are standard theorems of model theory.

The third and last section contains the main results, along with most of the homotopy theory. The main obstruction to proving all theorems in this context (true or otherwise) is that the infinite product construction does not necessarily preserve weak equivalences, much less homology isomorphisms. It's not true, for example, that an infinite product of perfect groups is perfect[1], [**9**]. This is not an issue for us here, because the stalks (or ultraproducts) of the family of groups $\{G(k_i)\}$ of k_i-points of a simply-connected algebraic group G at algebraically closed fields k_i have the form $G(L)$ for some algebraically closed field L. This is typical of the sort of distinction that is wiped out by passing to ultraproducts, rather than confronting infinite products head on. The homotopy theory of ultraproducts of simplicial sets, or rather the homotopy theory of simplicial objects in Grothendieck toposes arising from topologies of finite coverings of infinite sets, may prove to be a good vehicle for discussing homotopy problems arising from infinite products. One expects, in particular, a strong relation between this theory and the recent work of Bousfield [**2**] and Goerss [**10**] on the homology of infinite products of nilpotent spaces.

On a final technical note, the proof of Theorem 17 makes essential use of the Friedlander-Mislin theorem and standard results of étale homotopy theory [**5**] to identify the homology sheaves of a certain sheaf of completed spaces. I believe that this is the first time that the homotopy theory of simplicial sheaves and étale homotopy theory have appeared in naturally complementary roles.

1 Ultrafilters and Sheaves

Suppose that F_i, $i \in I$ is a collection of fields indexed by an infinite set I, and consider the infinite product ring

$$R = \prod_{i \in I} F_i.$$

The Zariski spectrum $Sp(R)$ has an interesting structure, in that the prime ideals $\mathcal{P}$ of the ring R correspond to the "ultrafilters" of the set I.

Explicitly, suppose that (a_i) is an element of R. The *zero set* $Z(a_i)$ of this element is defined by

$$Z(a_i) = \{i \in I |\ a_i = 0\}.$$

Conversely, any subset $X \subset I$ gives rise to a canonical element $e_X = (e_{X,i}) \in R$, with

$$e_{X,i} = \begin{cases} 0 & \text{if } i \in X, \\ 1 & \text{if } i \notin X. \end{cases}$$

Suppose that $\mathcal{P}$ is a prime ideal of R. The set $\mathcal{U}(\mathcal{P})$ consists of the set of subsets $X \subset I$ such that $e_X \in \mathcal{P}$. Equivalently, $\mathcal{U}(\mathcal{P})$ is the set of zero sets of elements of $\mathcal{P}$. The set $\mathcal{U}(\mathcal{P})$ satisfies the following properties:

(1) The empty set $\emptyset$ is not a member of $\mathcal{U}(\mathcal{P})$.

(2) If $X \subset Y$ and $X \in \mathcal{U}(\mathcal{P})$, then $Y \in \mathcal{U}(\mathcal{P})$.

[1] I would like to thank Crichton Ogle for supplying this reference.

(3) $\mathcal{U}(\mathcal{P})$ is closed under finite intersections.

(4) If $X \sqcup Y$ is in $\mathcal{U}(\mathcal{P})$ then $X \in \mathcal{U}(\mathcal{P})$ or $Y \in \mathcal{U}(\mathcal{P})$.

Item (1) follows from the fact that $e_\emptyset = 1 \notin \mathcal{P}$. Item (2) follows from the relation $e_Y e_X = e_Y$. Item (3) follows from item (2) and the relation

$$e_{Y \cup X^c} + e_X = e_{X \cap Y}.$$

Item (4) is a consequence of $e_X e_Y = e_{X \cup Y}$.

Conversely, if $\mathcal{U}$ is a set of subsets of I satisfying properties (1) – (4), then the set $\mathcal{P}(\mathcal{U})$ of all elements of R having zero sets in $\mathcal{U}$ is a prime ideal of R. In effect, $Z(a_i) \cap Z(b_i) \subset Z((a_i)+(b_i))$, so $\mathcal{P}(\mathcal{U})$ is closed under addition. Also $Z(b_i) \subset Z((a_i)(b_i))$, so $\mathcal{P}(\mathcal{U})$ is closed under multiplication by elements of R. Trivially, $\mathcal{P}(\mathcal{U})$ contains 0 but does not contain 1. Finally,

$$Z((a_i)(b_i)) = Z(a_i) \cup Z(b_i) = Z(a_i) \sqcup (Z(b_i) \cap Z(a_i)^c).$$

Thus, if $(a_i)(b_i) \in \mathcal{P}(\mathcal{U})$, then $(a_i) \in \mathcal{P}(\mathcal{U})$ or $(b_i) \in \mathcal{P}(\mathcal{U})$.

A subset $\mathcal{U}$ of the set of subsets of I which satisfies properties (1) – (4) above is said to be an *ultrafilter* of the set I. We have just seen that the ultrafilters of I are in one-to-one correspondence with the prime ideals of R. In particular, the kernel $\mathcal{M}_i$ of the projection function

$$pr_i : R = \prod_{i \in I} F_i \to F_i$$

is a maximal ideal of R which corresponds to the ultrafilter $\mathcal{U}_i = \mathcal{U}(\mathcal{M}_i)$ that is generated by the element $e_{\{i\}}$. Ultrafilters of the form $\mathcal{U}_i$ are said to be *principal*.

Set theorists know that there are lots of non-principal ultrafilters about: it is a theorem of Tarski that the set of non-principal ultrafilters on a set I has cardinality 2^{2^α} if the cardinality of the set I is α [**1**, p.108]. The collection of non-principal ultrafilters on I coincides with the set of ultrafilters which contain the set of all complements of finite subsets of the set I.

More generally, a *filter* of I is a collection $\mathcal{F}$ of subsets of I which satisfies the following properties:

(1) The empty set $\emptyset$ is not a member of $\mathcal{F}$.

(2) If $X \subset Y$ and $X \in \mathcal{F}$, then $Y \in \mathcal{F}$.

(3) $\mathcal{F}$ is closed under intersection.

Note that the correspondence

$$\{\text{ultrafilters of } I\} \cong \{\text{prime ideals of } R\}$$

is the restriction of an analogously defined (order preserving) correspondence

$$\{\text{filters of } I\} \cong \{\text{proper ideals of } R\}.$$

In effect, if $\mathcal{I}$ is an ideal of R, then the corresponding filter $\mathcal{F}(\mathcal{I})$ is the set of zero sets of elements of $\mathcal{I}$.

Every ultrafilter $\mathcal{U}$ of I is a maximal filter. To see this, note that every maximal ideal of R is prime, so that every maximal filter is an ultrafilter. Then, if $\mathcal{F}$ is a maximal filter of I which properly contains $\mathcal{U}$, then $\mathcal{F}$ has an element X which is not in $\mathcal{U}$, and then $X^c \in \mathcal{U}$ implies that $\emptyset = X \cap X^c \in \mathcal{F}$. In particular, every prime ideal of R is maximal.

Now suppose that $\mathcal{P}$ is a prime ideal of R, and take an element $f = (f_i) \in R - \mathcal{P}$. Then the set $X_f \subset I$ defined by

$$X_f = Z(f_i)^c = \{i \in I | \ f_i \neq 0\}$$

is a member of the ultrafilter $\mathcal{U}(\mathcal{P})$, since $Z(f_i)$ is not a member of $\mathcal{U}(\mathcal{P})$. There is a surjective projection homomorphism

$$\prod_{i \in I} F_i \xrightarrow{pr} \prod_{i \in X_f} F_i.$$

The element $pr(f)$ is a unit of $\prod_{i \in X_f} F_i$, and this projection induces an isomorphism of rings

$$(\prod_{i \in I} F_i)_f \cong \prod_{i \in X_f} F_i.$$

Traditionally, the *ultraproduct* of the fields F_i corresponding to an ultrafilter $\mathcal{U}$ is the set

$$(\prod_{i \in I} F_i)/ \sim,$$

where the equivalence relation $\sim$ is defined by decreeing that $(a_i) \sim (b_i)$ if there is an $X \in \mathcal{U}$ such that $a_i = b_i$ for all $i \in X$. Observe that $(a_i) \sim (b_i)$ if and only if there is an element $f \in R - \mathcal{P}(\mathcal{U})$ (in this case, e_{X^c}) such that $pr(a_i) = pr(b_i)$ for the projection

$$pr : \prod_{i \in I} F_i \to \prod_{i \in X_f} F_i.$$

It follows that the ultraproduct $(\prod_{i \in I} F_i)/ \sim$ arising from the ultrafilter $\mathcal{U}$ is exactly the local ring $(\prod_{i \in I} F_i)_{\mathcal{P}(\mathcal{U})}$ associated to the corresponding prime ideal. This local ring is in fact a field, which will be denoted by $F_{\mathcal{U}}$.

More generally, if Y_i, $i \in I$ is a collection of sets, then there is a sheaf j_*Y on the Zariski site for $Sp(R)$, which is given on basic open subsets X_f by

$$j_*Y(X_f) = \prod_{i \in X_f} Y_i.$$

Then, what is traditionally called the ultraproduct of the sets Y_i at the ultrafilter $\mathcal{U}$ is none other than the stalk $j_*Y_{\mathcal{P}(\mathcal{U})}$ of the sheaf j_*Y at the corresponding point $\mathcal{P}(\mathcal{U})$.

The direct image notation reflects the fact that there is a canonical morphism of schemes

$$j : \bigsqcup_{i \in I} Sp(F_i) \to Sp(R),$$

and that the collection of sets Y_i can be identified with a sheaf Y for the Zariski topology on $\bigsqcup_{i \in I} Sp(F_i)$. Then j_*Y means exactly what it is supposed to.

The scheme homomorphism j is in fact an open immersion. If $\mathcal{F}$ is the filter generated by all complements of finite subsets in I, then it's straightforward to show that an ultrafilter $\mathcal{U}$ contains $\mathcal{F}$ if and only if $\mathcal{U}$ is non-principal. The collection of non-principal ultrafilters therefore forms a closed subset of the scheme $Sp(R)$.

Recall that sheaves on the Zariski site $Zar|_{Sp(R)}$ are completely determined by their restriction to the subcategory $Basic|_{Sp(R)}$ of basic affine open subschemes. In particular, a sheaf on $Zar|_{Sp(R)}$ can be identified with a contravariant functor

F defined $Basic|_{Sp(R)}$ which satisfies the finite patching property: given a finite covering $V = U_1 \cup \cdots \cup U_n$ in $Basic|_{Sp(R)}$, there is an equalizer diagram

$$F(V) \to \prod_i F(U_i) \rightrightarrows \prod_{i,j} F(U_i \cap U_j).$$

The equalizer makes sense, since $Sp(R_f) \cap Sp(R_g) = Sp(R_h)$, where h is an element of R such that $h_i \neq 0$ if and only if $f_i \neq 0$ and $g_i \neq 0$, so that the collection of basic affine open subschemes of R is closed under finite intersections. The standard patching property reduces to a property involving finite covers in this case, because all of the objects in $Basic|_{Sp(R)}$ are affine, and hence quasi-compact.

An open subscheme $U \subset Sp(R)$ is a basic affine open subscheme if and only if U consists of a set of prime ideals $\mathcal{P}$ not containing some $f \in R$. A prime ideal $\mathcal{P}$ doesn't contain f if and only if $\mathcal{P}$ contains the element $e_{Z(f)^c}$. The collection of all primes not containing f coincides with the collection of all primes not containing g provided that $e_{Z(f)^c} \in \mathcal{P}$ if and only if $e_{Z(g)^c} \in \mathcal{P}$. But by restricting attention to the principal ideals $\mathcal{M}_i$, this means that $e_{Z(f)^c}(i) = 0$ if and only if $e_{Zg^c}(i) = 0$, which means that $f_i = 0$ if and only if $g_i = 0$. Thus, two ring elements f and g define the same basic affine open subscheme if and only if their "non-zero sets" $Z(f)^c$ and $Z(g)^c$ coincide. Observe finally that $Sp(R_f) \subset Sp(R_g)$ if and only if $Z(f)^c \subset Z(g)^c$. We can therefore deduce the following:

Lemma 1 *Let $\mathcal{P}(I)$ denote the poset of subsets of the index set I. Then there is an isomorphism of categories*

$$Basic|_{Sp(R)} \cong \mathcal{P}(I),$$

which is defined by $Sp(R_f) \mapsto Z(f)^c$. This isomorphism gives $\mathcal{P}(I)$ the structure of a Grothendieck site, where the corresponding topology is induced by the pretopology of finite coverings.

The *pretopology of finite coverings* on the poset $\mathcal{P}(I)$ is the pretopology given by finite collections U_i, $i = 1, \ldots, n$ of subsets of $V \in \mathcal{P}(I)$ such that $\bigcup_{i=1}^n U_i = V$. I shall say that the corresponding Grothendieck topology on $\mathcal{P}(I)$ is the *topology of finite covers.* Write $\mathcal{P}(I)_{fin}$ for the category of subsets of I, equipped with the topology of finite covers.

Proof of Lemma 1 The statement about the induced topology is a consequence of the observation that $Sp(R) = Sp(R_{f_1}) \cup \ldots Sp(R_{f_n})$ if and only if 1 is in the ideal $(f_1, \ldots, f_n)$ in R, which holds if and only if $I = Z(f_1)^c \cup \cdots \cup Z(f_k)^c$. □

The sheaves on $\mathcal{P}(I)_{fin}$ are easily identified:

Lemma 2 *A contravariant functor $F : \mathcal{P}(I)^{op} \to$ **Sets** is a sheaf on $\mathcal{P}(I)_{fin}$ if and only if F takes finite disjoint unions to products.*

Proof First of all, note that F is a sheaf if and only if all finite collections U_i, $1 \leq i \leq n$, give rise to equalizers

$$F(U_1 \cup \cdots \cup U_n) \to \prod_{i=1}^n F(U_i) \rightrightarrows \prod_{i,j} F(U_i \cap U_j).$$

In particular, F takes binary patching data to pullbacks in the sense that all induced diagrams

$$\begin{array}{ccc} F(U \cup V) & \longrightarrow & F(U) \\ \downarrow & & \downarrow{\scriptstyle i_*} \\ F(V) & \longrightarrow & F(U \cap V) \end{array} \tag{1}$$

are pullbacks.

It's a standard argument to show the converse, so that a presheaf F is a sheaf if and only if it takes binary patching data to pullbacks and satisfies $F(\emptyset) = *$, where $*$ denotes a set with one element.

It's obvious now that if F is a sheaf, and U and V are subsets of I such that $U \cap V = \emptyset$, then the canonical map $F(U \cup V) \to F(U) \times F(V)$ is an isomorphism. Conversely, suppose that F takes disjoint unions to products in this sense. Then $U \cup V = U \sqcup (V - U)$ and $V = (U \cap V) \sqcup (V - U)$, and the diagram

$$\begin{array}{ccc} F(U) \times F(V - U) & \xrightarrow{pr} & F(U) \\ {\scriptstyle i^* \times 1}\downarrow & & \downarrow{\scriptstyle i^*} \\ F(U \cap V) \times F(V - U) & \xrightarrow[pr]{} & F(U \cap V) \end{array}$$

is a pullback, where pr denotes projection. This diagram is isomorphic to the diagram (1), so that F takes binary patching data to pullbacks. $\square$

The topological space underlying the scheme $\bigsqcup_{i \in I} Sp(F_i)$ is the set I, equipped with the discrete topology. Write $\mathcal{P}(I)_{disc}$ for the corresponding Grothendieck site: its objects are all subsets of the set I, and the covering families of an object U consist of all collections of subsets $U_j \subset U$, $j \in J$, such that $\bigcup_{j \in J} U_j = U$. Note that there is no size requirement for the index sets J.

The map of schemes

$$j : \bigsqcup_{i \in I} Sp(F_i) \to Sp(R)$$

determines a site morphism

$$j : \mathcal{P}(I)_{disc} \to \mathcal{P}(I)_{fin}$$

which is induced by the identity functor on $\mathcal{P}(I)$. A sheaf on the site $\mathcal{P}(I)_{disc}$ is a contravariant functor $F : \mathcal{P}(I)^{op} \to$ **Sets** such that the canonical map

$$F(U) \cong \prod_{i \in U} F(\{i\})$$

is an isomorphism for each subset U of I. A sheaf F is therefore plainly determined by the objects $F_i = F(\{i\})$, $i \in I$, so the topos of sheaves on the site $\mathcal{P}(I)_{disc}$ is equivalent to the category of I-indexed families of sets. It's clear, as well, that the direct image sheaf j_*F on $\mathcal{P}_{fin}$ is defined for $U \subset I$ by

$$j_*F(U) = \prod_{i \in U} F_i.$$

2 Cardinality issues

Suppose that k_i, $i \in I$ is a collection of algebraically closed fields indexed by the infinite set I, and let $\mathcal{U}$ be a non-principal ultrafilter on the set I. The field $k_{\mathcal{U}}$ is algebraically closed; the proof is standard model theory, amounting to the observation that any polynomial $P(x) \in k_{\mathcal{U}}[x]$ is in the image of some polynomial $Q(x)$ defined over a ring $\prod_{i\in Z} k_i$, where $Z \in \mathcal{U}$. Then any polynomial in $(\prod_{i\in Z} k_i)[x]$ has a root in $\prod_{i\in Z} k_i$ since all of the fields k_i are algebraically closed, so $Q(x)$ has a root in $\prod_{i\in Z} k_i$ and so $P(x)$ has a root in $k_{\mathcal{U}}$.

Questions about the size of $k_{\mathcal{U}}$ will be of interest here, but are more subtle, and devolve to general questions about the cardinality of ultraproducts $j_*X_{\mathcal{U}}$ arising from families of sets X_i, $i \in I$.

The cardinality of the ultraproduct $j_*X_{\mathcal{U}}$ depends on some rather delicate structural information about the ultrafilter $\mathcal{U}$, as well as on the cardinality of the sets X_i and the cardinality of the index set I. Since I is presumed to be infinite, the cardinality of the set $Fin(I)$ of all finite subsets of I coincides with that of I, so there is a bijection

$$f : I \xrightarrow{\cong} Fin(I).$$

For each $i \in I$, there is a subset $E_i \subset I$ defined by

$$E_i = \{j \in I |\ i \in f(j)\}.$$

Any finite intersection of the sets E_i is non-empty, so that the set of all such subsets E_i, $i \in I$ generates a proper filter and is therefore contained in some ultrafilter $\mathcal{F}$. In effect, suppose that $F = \{i_1, \dots, i_n\}$ is a finite subset of I. Then $F = f(i_0)$ for some element i_0 of I, and $i_0 \in E_{i_1} \cap \dots \cap E_{i_n}$. Any ultrafilter $\mathcal{U}$ which contains all the E_i *for some choice of bijection* f is said to be *regular* (see [**1**, p.114]). It is known, for example, that all non-principal ultrafilters on a countable infinite set I are regular [**1**, p.116].

An ultrafilter $\mathcal{U}$ on I is said to be *uniform* if all the subsets of I which appear in $\mathcal{U}$ have the same cardinality as the index set I. Observe, for example, that if $\mathcal{U}$ is a non-principal ultrafilter on a countable set I, then $\mathcal{U}$ is uniform. More generally, the following is well known:

Lemma 3 *Every regular ultrafilter $\mathcal{U}$ is uniform.*

Proof Suppose that X is a subset of I, and that $card(X) < card(I)$. Then

$$card(\bigcup_{i\in X} f(i)) < card(I),$$

so there is some $j_0 \in I - \bigcup_{i\in X} f(i)$. Furthermore

$$\{i \in I |\ j_0 \in f(i)\} \subset I - X,$$

so that $I - X \in \mathcal{U}$, so $X \notin \mathcal{U}$. □

This last proof was shamelessly lifted from [**1**, p.115].

Without further ado, here's the main result for big cardinals:

Theorem 4 *Suppose that the sets X_i, $i \in I$ are infinite sets of cardinality α, and that the (infinite) index set I has cardinality γ. Suppose that $\mathcal{U}$ is a regular ultrafilter on the set I. Then the ultraproduct $j_*X_{\mathcal{U}}$ has cardinality α^γ. It therefore has cardinality greater than γ.*

Proof See [**1**, p.132]. □

Bell and Slomson [**1**] also give a cardinality calculation for the smallest possible infinite cardinals:

Theorem 5 *Suppose that the sets X_i, $i \in I$ are countable infinite sets, and that the (infinite) index set I is also countable. Suppose that $\mathcal{U}$ is a non-principal ultrafilter on the set I. Then the ultraproduct $j_*X_{\mathcal{U}}$ has cardinality $2^{\aleph_0}$.*

Proof See [**1**, p.129]. □

Remark 6 The point of the Theorem 4, for us, will be that it's possible to build ultraproducts of arbitrarily high cardinality by using an index set I of suitable size. For example, suppose that k is an algebraically closed field and let I be a set of infinite cardinality γ. Let $k_i = k$ for $i \in I$, and let $\mathcal{U}$ be a regular ultrafilter on I. Then the "ultrapower" $j_*k_{\mathcal{U}}$ is an algebraically closed field of cardinality strictly greater than γ. Observe as well that the characteristic of $j_*k_{\mathcal{U}}$ coincides with the characteristic of the field k.

Remark 7 Theorem 5 gives a method for using the algebraic closures of the finite fields to construct the complex numbers. In effect, let I be the set of primes, and let $\overline{\mathbb{F}}$ denote the collection of algebraic closures $\overline{\mathbb{F}}_p$ of the finite fields $\mathbb{F}_p$ of the various prime orders, thought of as a sheaf on the site $\mathcal{P}(I)_{disc}$. Let $\mathcal{U}$ be any non-principal ultrafilter on the countable set I. Then the algebraically closed field $j_*\overline{\mathbb{F}}_{\mathcal{U}}$ has cardinality $2^{\aleph_0}$, by the second theorem. It also has characteristic 0, since $n \cdot (1) \neq (0)$ in all of the rings $\prod_{p \in X} \overline{\mathbb{F}}_p$, where X is infinite. It follows, by comparing cardinalities of transcendence bases over the algebraic numbers $\overline{\mathbb{Q}}$, that $j_*\overline{\mathbb{F}}_{\mathcal{U}}$ is isomorphic to the field of complex numbers $\mathbb{C}$. This last observation is commonly attributed to Steinitz (1910).

In cases of uncountable index sets there are special types of regular ultrafilters that one wants to study. In particular, suppose that I and J are infinite sets, suppose that $\mathcal{U}$ and $\mathcal{V}$ are ultrafilters on I and J, respectively, and suppose that both ultrafilters are regular. Write I_x for the copy of I in the product set $J \times I$ which corresponds to $\{x\} \times I$, for $x \in J$. Let $\mathcal{W}$ be the set of all subsets W of $J \times I$ such that the subset

$$J(W) = \{x \in J | \ W \cap I_x \in \mathcal{U}\}$$

of J is an element of of $\mathcal{V}$.

Lemma 8 *The set $\mathcal{W}$ is a regular ultrafilter of $J \times I$.*

Proof To see that $\mathcal{W}$ is closed under finite intersection, suppose that $W_i \in \mathcal{W}$, $i = 1, 2$. Then $J(W_1) \cap J(W_2) \subset J(W_1 \cap W_2)$ since $W_i \cap I_x \in \mathcal{U}$, $i = 1, 2$ means that

$$W_1 \cap W_2 \cap I_x = (W_1 \cap I_x) \cap (W_2 \cap I_x) \in \mathcal{U}.$$

But then $J(W_1) \cap J(W_2) \in \mathcal{V}$ implies that $J(W_1 \cap W_2) \in \mathcal{V}$.

To finish the proof of the assertion that $\mathcal{W}$ is an ultrafilter, it's enough to show that if $W \notin \mathcal{W}$, then $W^c \notin \mathcal{W}$. If $J(W) = \{x \in J | \ W \cap I_x \in \mathcal{U}\} \notin \mathcal{V}$, then $\{x \in J | \ W \cap I_x \notin \mathcal{U}\} \in \mathcal{V}$. But

$$\{x \in J | \ W \cap I_x \notin \mathcal{U}\} = \{x \in J | \ W^c \cap I_x \in \mathcal{U}\} = J(W^c),$$

so the complement W^c of W in $J \cdot I$ is a member of $\mathcal{W}$.

The regularity of the ultrafilters $\mathcal{U}$ and $\mathcal{V}$ is defined by bijections $f : I \cong Fin(I)$ and $g : J \cong Fin(J)$, respectively. The fibre Fib_F over a finite subset F of J of the map $pr_* : Fin(J \times I) \to Fin(J)$ which is induced by the projection $pr : J \times I \to J$ can be identified up to bijection with a set $Fin(I^n)$, where n is the number of elements of the finite set F. The set I^n has the cardinality of I, so one is entitled to a composite bijection

$$I \xrightarrow[\cong]{f} Fin(I) \cong Fin(Fib_F) \tag{2}$$

for each $F \in Fin(J)$. The product set $J \times I$ is a disjoint union of the subset I_x, $x \in J$, so that there is a bijection

$$J \times I \xrightarrow[\cong]{(f,g)_*} Fin(J \times I)$$

which is defined by bijections (2) in each fibre, and makes the diagram

$$\begin{array}{ccc} J \times I & \xrightarrow[\cong]{(f,g)_*} & Fin(J \times I) \\ {\scriptstyle pr}\downarrow & & \downarrow{\scriptstyle pr_*} \\ J & \xrightarrow[g]{\cong} & Fin(J) \end{array}$$

commute. From this diagram, one sees that $pr(E_{(x,i)}) = E_x$, so that $E_x = \{y \in J|\ I_y \cap E_{(x,i)} \neq \emptyset\}$. But then, for $y \in E_x$,

$$\begin{aligned} I_y \cap E_{(x,i)} &\cong \{H \in Fin(Fib_{g(y)})|\ (x,i) \in H\} \\ &\cong \{H \in Fin(I)|\ i' \in H\} \end{aligned}$$

where $i' \mapsto (x,i)$ under the bijection $I \cong Fib_{g(y)}$. Since $\mathcal{U}$ is regular (for the map f), it follows that $I_y \cap E_{(x,i)}$ is a member of $\mathcal{U}$ for all $y \in E_x$. The set E_x is a member of $\mathcal{V}$, since $\mathcal{V}$ is regular for the map g. It follows that $E_{(x,i)}$ is a member of $\mathcal{W}$; this is true for all pairs (x,i), so $\mathcal{W}$ is regular.

□

Write $\mathcal{W} = \mathcal{V} \cdot \mathcal{U}$. The point of this construction, for us, is that it implies the following:

Corollary 9 *Suppose that I is a countable infinite set, and that J is an infinite set. Let $\mathcal{U}$ be a non-principal ultrafilter on I, and let J be a regular ultrafilter on J. Then the ultrafilter $\mathcal{W} = \mathcal{V} \cdot \mathcal{U}$ is a regular ultrafilter on $J \times I$.*

We are now in position to demonstrate

Lemma 10 *Characteristic 0 algebraically closed fields of arbitrarily high cardinality may be constructed as ultraproducts of algebraically closed fields having finite characteristics.*

Proof Let I be the set of all prime numbers, and let J be an infinite set of cardinality α, and choose a non-principal ultrafilter $\mathcal{U}$ on I and a regular ultrafilter $\mathcal{V}$ on J.

The projection function $pr : J \times I \to I$ defines a family of fields $\overline{\mathbb{F}}_{pr(x,p)} = \overline{\mathbb{F}}_p$, $(x,p) \in J \times I$. Let $\overline{\mathbb{F}}_{pr}$ denote the corresponding sheaf on the site $\mathcal{P}(J \times I)_{disc}$. Let $\mathcal{W} = \mathcal{V} \cdot \mathcal{U}$ be the regular ultrafilter on $J \times I$ of Corollary 9. Then the algebraically closed field

$$j_*(\overline{\mathbb{F}}_{pr})_{\mathcal{W}}$$

has cardinality greater than α, by Theorem 4.

Suppose that p is a prime and that the element

$$(p) \in \prod_{(x,p)\in J\times I} \overline{\mathbb{F}}_p$$

represents 0 in $j_*(J \cdot \mathbb{F})_{\mathcal{W}}$. Then (p) vanishes on some element $W \in \mathcal{W}$, so that, for some x, (p) vanishes on some element $W \cap I_x$ of the ultrafilter $\mathcal{U}$ on I. This is a contradiction, since $W \cap I_x$ contains infinitely many elements. □

A rather different and much easier question arises when one wants to analyze the cardinality of ultraproducts $j_*X_{\mathcal{U}}$, where X is a family of finite sets X_i, $i \in I$, all having the same number of elements, namely n. It's well known that $j_*X_{\mathcal{U}}$ must also have n elements [**1**, p.126], but I want to express it in the following way:

Lemma 11 *Suppose that F is a finite set, and let Γ^*F denote the associated constant sheaf on the site $\mathcal{P}(I)_{fin}$. Then the canonical adjunction map*

$$\eta : \Gamma^*F \to j_*j^*\Gamma^*F$$

is an isomorphism.

Proof The map η is a map of associated sheaves which is defined on the presheaf level, at a subset U of I, by the diagonal map

$$F \xrightarrow{\Delta} \prod_{i\in U} F.$$

The induced map of stalks for this presheaf map, at a principal ultrafilter $\mathcal{U}_i$, is the composite

$$F \xrightarrow{\Delta} \prod_{i\in U} F \xrightarrow{pr_i} F,$$

which is the identity map on F and is therefore a bijection.

Suppose that $F = \{x_1, \ldots, x_n\}$, and take an element $(y_i) \in \prod_{i\in U} F$, where U is a member of a non-principal ultrafilter $\mathcal{U}$. Let U_j be the subset of U defined by $U_j = \{i \in U |\ y_i = x_j\}$. Then $U = U_1 \sqcup \cdots \sqcup U_n$ and $U \in \mathcal{U}$, so there is a j such that $U_j \in \mathcal{U}$. But then the restriction of (y_i) to U_j is constant on the element x_j, and is therefore the image of x_j under the diagonal map

$$F \xrightarrow{\Delta} \prod_{i\in U_j} F,$$

so that the induced map of stalks

$$F \xrightarrow{\eta_*} j_*j^*\Gamma^*F_{\mathcal{U}}$$

is surjective. The map of sheaves is already monic, since it's induced by a monomorphism of presheaves. □

3 The Cohomology of $BSl_n(k)$

The *isomorphism conjecture* for BSl_n [**6**], [**12**] says that, if k is an algebraically closed field of characteristic not equal to a prime ℓ, then the comparison map

$$H^*_{et}(BSl_{n,k}, \mathbb{Z}/\ell) \xrightarrow{\epsilon^*} H^*(BSl_n(k), \mathbb{Z}/\ell) \tag{3}$$

is an isomorphism. The map ϵ^* can variously be seen as induced by the canonical map

$$\epsilon : \Gamma^*BSl_n(k) \to BSl_{n,k}$$

of simplicial sheaves on the big étale site $(Sch|_k)_{et}$ for the algebraically closed field k, or by a map of simplicial schemes

$$\epsilon : Sp(k) \otimes BSl_n(k) \to BSl_{n,k}.$$

Here, $\Gamma^* BSl_n(k)$ denotes the constant simplicial sheaf on the simplicial set $BSl_n(k)$ of global sections of the simplicial sheaf $BSL_{n,k}$, and ϵ is the adjunction map. Also, $Sp(k) \otimes BSl_n(k)$ is the simplicial scheme having m-simplices given by

$$\bigsqcup_{BSl_n(k)_m} Sp(k).$$

The simplicial k-scheme $Sp(k) \otimes BSl_n(k)$ represents $\Gamma^* BSL_n(k)$ in standard simplicial sheaf categories, so the two versions of the map ϵ are effectively the same.

The theorem of Friedlander and Mislin [**6**] says that the comparison map ϵ^* in (3) is an isomorphism if k is the algebraic closure $\overline{\mathbb{F}}_p$ of a finite field. The point of this section (and of this paper) is to give a homological criterion for the full isomorphism conjecture, in terms of suitable ultraproducts.

We're going to need the following two results:

Lemma 12 *Suppose that* Y *is a connected Kan complex having finite homotopy groups. Let* $\Gamma^* Y$ *denote the associated constant simplicial sheaf on* $\mathcal{P}(I)_{fin}$, *for some infinite index set* I. *Then the canonical map* $\eta : \Gamma^* Y \to j_* j^* \Gamma^* Y$ *is a stalkwise weak equivalence.*

Proof Let $\mathcal{U}$ be an ultrafilter on I, and take $U \in \mathcal{U}$. The simplicial set Y is a connected Kan complex, so there is a path

$$\Delta^1 \to \prod_{i \in U} Y$$

between any two 0-simplices (x_i) and (y_i) of $\prod_{i \in U} Y$, so the stalk $j_* j^* \Gamma^* Y_{\mathcal{U}}$ is connected. Recall that the map η is induced by the map of simplicial presheaves defined in sections corresponding to $U \in \mathcal{U}$ by the diagonal map

$$Y \xrightarrow{\Delta} \prod_{i \in U} Y.$$

This map induces a homomorphism

$$\pi_n Y \xrightarrow{\Delta} \prod_{i \in U} \pi_n Y. \tag{4}$$

of homotopy group presheaves, for any $n \geq 1$, and any choice of base point of the Kan complex Y, which in turn induces the map of sheaves of abelian groups

$$\pi_n \Gamma^* Y \xrightarrow{\eta_*} \pi_n j_* j^* \Gamma^* Y.$$

The presheaf homomorphism (4) induces an isomorphism of associated sheaves, by Lemma 11. □

Lemma 13 *Suppose that* Y *is a simply connected Kan complex such that all of its mod* ℓ *homotopy groups* $\pi_i(Y, \mathbb{Z}/\ell)$ *are finite. Then the canonical map*

$$\eta : \Gamma^* Y \to j_* j^* \Gamma^* Y$$

induces isomorphisms of all associated mod ℓ *homology sheaves on* $\mathcal{P}(I)_{fin}$.

Proof Suppose that the simplicial presheaf F is defined in sections by the fibre sequence

$$F(U) \to Y \xrightarrow{\Delta} \prod_{i \in U} Y.$$

The mod ℓ homotopy groups of a Kan complex Y are defined to be the homotopy groups of pointed function complexes $\mathbf{hom}_*(P^2_\ell, Y)$, where P^2_ℓ is a finite simplicial set representing the Moore space given by the cofibre of multiplication by ℓ on the circle S^1. The homotopy groups of the Kan complex $\mathbf{hom}_*(P^2_\ell, Y)$ are finite by assumption, so that the presheaf map

$$\mathbf{hom}_*(P^2_\ell, Y) \xrightarrow{\Delta} \prod_{i \in U} \mathbf{hom}_*(P^2_\ell, Y)$$

is a stalkwise weak equivalence, by the previous result. It follows that the stalks of the simplicial presheaf $\mathbf{hom}_*(P^2_\ell, F)$ are contractible, so that the stalks $F_{\mathcal{U}}$ of the simplicial presheaf F are simple spaces with uniquely ℓ-divisible homotopy groups. But then $\tilde{H}_*(F_{\mathcal{U}}, \mathbb{Z}/\ell) \cong 0$, and the spaces $j_*j^*\Gamma^*Y_{\mathcal{U}}$ are simply connected, so that the induced maps of stalks

$$Y \to j_*j^*\Gamma^*Y_{\mathcal{U}}$$

are mod ℓ homology isomorphisms. □

We'll also need to know various technical things about the completion $BSl_n(F)^\wedge_\ell$ of classifying spaces for the special linear groups $Sl_n(F)$ over fields F of characteristic distinct from ℓ. The first of these is that $Sl_n(F)$ is perfect, so that $Sl_n(F)$ is "ℓ-good" in the sense of Bousfield and Kan [**3**]; in particular, the canonical map $BSl_n(F) \to BSl_n(F)^\wedge_\ell$ is a mod ℓ-homology isomorphism, and the space $BSl_n(F)^\wedge_\ell$ is ℓ-complete.

Secondly, the integral homology group $H_2(BSl_n(k), \mathbb{Z})$ is isomorphic to the group $K_2(k)$ in the case where k is an algebraically closed field, by a well known theorem of Matsumoto. The abelian group $K_2(k)$ is divisible, on account of the usual presentation, since the group of units k^* of the algebraically closed field is divisible. It follows that $H_2(BSl_n(k), \mathbb{Z}/\ell) = 0$, so that the completion $BSl_n(k)^\wedge_\ell$ is 2-connected.

We shall also require the following basic homotopy theoretic result:

Lemma 14 *Suppose that $f : X \to Y$ is a map of spaces such that Y is 2-connected and X is simply connected. Then f induces an isomorphism in mod ℓ homology if and only if f induces an isomorphism in mod ℓ homotopy groups.*

Proof The homotopy fibre F of the map f is a simply connected space. A Serre spectral sequence argument can be used to show that f induces an isomorphism in mod ℓ homology if and only if the fibre F is also mod ℓ acyclic in the sense that $\tilde{H}_*(F, \mathbb{Z}/\ell) = 0$. Finally, one can use the mod ℓ Hurewicz theorem (see [**15**], for example) to show – or argue directly (see also the proof of Lemma 13) – that since F is simply connected, it is mod ℓ acyclic if and only if the mod ℓ homotopy groups $\pi_*(F, \mathbb{Z}/\ell)$ vanish. □

The theorem of Friedlander and Mislin says that the comparison map

$$\epsilon^* : H^*_{et}(BSl_{n,\overline{\mathbb{F}}_p}, \mathbb{Z}/\ell) \to H^*(BSl_n(\overline{\mathbb{F}}_p), \mathbb{Z}/\ell)$$

relating étale to discrete cohomology over the algebraic closure $\overline{\mathbb{F}}_p$ of the finite field $\mathbb{F}_p$ is an isomorphism. This result means that the machinery of étale homotopy theory [**5**] can be invoked to show that there is a weak equivalence

$$(BSU_n)_\ell^\wedge \xrightarrow{\simeq} BSl_n(\overline{\mathbb{F}}_p)_\ell^\wedge$$

for each prime $p \neq \ell$. This map is induced, really, by a sequence of maps of simplicial schemes which are mod ℓ étale cohomology isomorphisms, and each such map induces a weak equivalence of spaces that arise as homotopy inverse limits of ℓ-completions of the respective étale homotopy types.

Observe, as well, that the map

$$BSU_n \to (BSU_n)_\ell^\wedge$$

is a mod ℓ homology isomorphism that satisfies the conditions of Lemma 14. It follows that the mod ℓ homotopy groups $\pi_*((BSU_n)_\ell^\wedge, \mathbb{Z}/\ell)$ are all finite.

Lemma 15 *Suppose that $i : k \hookrightarrow L$ is an inclusion of algebraically closed fields. Then the induced map*

$$i_* : H_*(BSl_n(k), \mathbb{Z}/\ell) \to H_*(BSl_n(L), \mathbb{Z}/\ell)$$

is a monomorphism.

Proof Suppose that L/k has transcendance degree one. Then L is a filtered colimit

$$L = \varinjlim_{X=Sp(A)} A,$$

where X runs through the set of all affine curves over k. All such curves have closed points, which can be identified with k-algebra maps $A \to k$. It follows that each map

$$BSl_n(k) \to BSl_n(A)$$

has a left inverse, so that each induced map

$$H_*(BSl_n(k), \mathbb{Z}/\ell) \to H_*(BSl_n(A), \mathbb{Z}/\ell)$$

in mod ℓ homology groups is a monomorphism. Then

$$H_*(BSl_n(L), \mathbb{Z}/\ell) \cong \varinjlim_{X=Sp(A)} H_*(BSl_n(A), \mathbb{Z}/\ell).$$

The inclusion $k \subset L$ therefore induces a monomorphism

$$H_*(BSl_n(k), \mathbb{Z}/\ell) \to H_*(BSl_n(L), \mathbb{Z}/\ell).$$

The proof for the general case is a Zorn's Lemma argument. □

Now here's a conjecture:

Conjecture 16 *Let I be the set of prime numbers, and suppose that $\mathbf{P} : J \to I$ is a function, where J is some infinite set. Let $BSl_n(\overline{\mathbb{F}}_{\mathbf{P}})$ denote the sheaf on $\mathcal{P}(J)_{disc}$ corresponding to the family of simplicial sets $\{BSl_n(\overline{\mathbb{F}}_{\mathbf{P}(j)})\}$, and let $BSl_n(\overline{\mathbb{F}}_{\mathbf{P}})_\ell^\wedge$ be the sheaf on $\mathcal{P}(J)_{disc}$ associated to the family of ℓ-completions $\{BSl_n(\overline{\mathbb{F}}_{\mathbf{P}(j)})_\ell^\wedge\}$. Suppose that $\mathcal{U}$ is an ultrafilter on the set J such that the set $\{j \in J \mid \mathbf{P}(j) = \ell\}$ is not a member of $\mathcal{U}$. Then the induced map*

$$j_*(BSl_n(\overline{\mathbb{F}}_{\mathbf{P}}))_{\mathcal{U}} \to j_*(BSl_n(\overline{\mathbb{F}}_{\mathbf{P}})_\ell^\wedge)_{\mathcal{U}}$$

is a mod ℓ homology isomorphism.

This conjecture is obviously true if the ultrafilter $\mathcal{U}$ in the statement is principal, but nothing is yet known about non-principal ultrafilters.

The following is the main result of this paper.

Theorem 17 *Conjecture 16 is equivalent to the isomorphism conjecture for* Sl_n.

Proof Suppose that α is an infinite cardinal. The constructions of Remark 6 and Lemma 10 imply that there is a function $\mathbf{P} : J \to I$ and a choice of regular ultrafilter $\mathcal{U}$ on J such that the algebraically closed field $j_*(\overline{\mathbb{F}}_{\mathbf{P}})_{\mathcal{U}}$ has cardinality greater than α and can have arbitrary characteristic.

The weak equivalences

$$f : (BSU_n)^\wedge_\ell \xrightarrow{\simeq} BSl_n(\overline{\mathbb{F}}_{\mathbf{P}(j)})^\wedge_\ell,$$

$j \in J$, resulting from the Friedlander-Mislin theorem induce a pointwise weak equivalence

$$j_* j^* \Gamma^* (BSU_n)^\wedge_\ell \to j_*(BSl_n(\overline{\mathbb{F}}_{\mathbf{P}})^\wedge_\ell)$$

of simplicial sheaves on $\mathcal{P}(J)_{fin}$, and hence a weak equivalence

$$j_* j^* \Gamma^* ((BSU_n)^\wedge_\ell)_{\mathcal{U}} \to j_*(BSl_n(\overline{\mathbb{F}}_{\mathbf{P}})^\wedge_\ell)_{\mathcal{U}}.$$

The mod ℓ homotopy groups of BSU_n are finite, so Lemma 13 implies that the induced map

$$f_* : (BSU_n)^\wedge_\ell \to j_* j^* \Gamma^* ((BSU_n)^\wedge_\ell)_{\mathcal{U}}$$

is a mod ℓ homology isomorphism. Conjecture 16 therefore implies that there is an induced isomorphism

$$H_*(BSU_n, \mathbb{Z}/\ell) \cong H_*(BSl_n(j_*(\overline{\mathbb{F}}_{\mathbf{P}})_{\mathcal{U}}), \mathbb{Z}/\ell).$$

Within any fixed characteristic, we can therefore find an algebraically closed field L of arbitrarily high cardinality such that

$$H_m(BSl_n(L), \mathbb{Z}/\ell) \cong H_m(BSU_n, \mathbb{Z}/\ell)$$

for all $m \geq 0$. It follows, for such fields L, that there is an abstract isomorphism

$$H^m_{et}(BSl_{n,L}, \mathbb{Z}/\ell) \cong H^m(BSl_n(L), \mathbb{Z}/\ell)$$

of finite dimensional vector spaces for all $m \geq 0$. The canonical comparison map

$$H^m_{et}(BSl_{n,k}, \mathbb{Z}/\ell) \xrightarrow{\epsilon^*} H^m(BSl_n(k), \mathbb{Z}/\ell)$$

is known to be a monomorphism for all algebraically closed fields k. The map

$$H^m_{et}(BSl_{n,L}, \mathbb{Z}/\ell) \xrightarrow{\epsilon^*} H^m(BSl_n(L), \mathbb{Z}/\ell) \tag{5}$$

is therefore a monomorphism of vector spaces over $\mathbb{Z}/\ell$-vector spaces, both of which have the same finite dimension. It follows that the morphism (5) is an isomorphism for all $m \geq 0$.

Suppose that k is now an arbitrary algebraically closed field. The preceding argument implies that there is an algebraically closed field L such that $k \subset L$ and such that the comparison map

$$H^*_{et}(BSl_{n,L}, \mathbb{Z}/\ell) \xrightarrow{\epsilon^*} H^*(BSl_n(L), \mathbb{Z}/\ell)$$

is an isomorphism. There is a commutative diagram

$$\begin{array}{ccccc} & & Sp(k) \otimes BSl_n(k) & \xrightarrow{\epsilon} & BSl_{n,k} \\ & \overset{i_*}{\nearrow} & & & \uparrow i_* \\ Sp(L) \otimes BSl_n(k) & & & & \\ & \searrow & & & \\ & & Sp(L) \otimes BSl_n(L) & \xrightarrow[\epsilon]{} & BSl_{n,L} \end{array}$$

of simplicial scheme homomorphisms, where the maps denoted by i_* are induced by the scheme homomorphism $i : Sp(L) \to Sp(k)$ associated to the inclusion $k \subset L$. There is an induced diagram in cohomology of the form

$$\begin{array}{ccc} H^*_{et}(BSl_{n,k}, \mathbb{Z}/\ell) & \xrightarrow{\epsilon^*} & H^*(BSl_n(k), \mathbb{Z}/\ell) \\ \cong \big\downarrow i^* & & \big\uparrow i_* \\ H^*_{et}(BSl_{n,L}, \mathbb{Z}/\ell) & \xrightarrow[\epsilon^*]{} & H^*(BSl_n(L), \mathbb{Z}/\ell) \end{array}$$

The map $i_* : H^*(BSl_n(L), \mathbb{Z}/\ell) \to H^*(BSl_n(k), \mathbb{Z}/\ell)$ is the map induced by $k \subset L$, and is surjective, by Lemma 15. The map i^* is known to be an isomorphism, by the Lefschetz principle for the étale cohomology of BSl_n, while $\epsilon^* : H^*_{et}(BSl_{n,L}, \mathbb{Z}/\ell) \to H^*(BSl_n(L), \mathbb{Z}/\ell)$ is an isomorphism by assumption, so the monomorphism $\epsilon^* : H^*_{et}(BSl_{n,k}, \mathbb{Z}/\ell) \to H^*(BSl_n(k), \mathbb{Z}/\ell)$ must also be surjective.

To see that the isomorphism conjecture for Sl_n implies Conjecture 16, let H_n denote the diagonal subgroup of Sl_n, and let $T_n \subset SU_n$ be the corresponding maximal compact subgroup of $H_n(\mathbb{C})$. The homotopy groups $\pi_*(BA)^{\wedge}_{\ell}$ are of the form $Ext^*(\mathbb{Z}_{\ell^\infty}, A)$ [**3**, VI.2] for abelian groups A, so that the ℓ-completion functor preserves all products of classifying spaces of abelian groups. It follows that the completion maps

$$BH_n(\overline{\mathbb{F}}_{\mathbf{P}(j)}) \to BH_n(\overline{\mathbb{F}}_{\mathbf{P}(j)})^{\wedge}_{\ell}$$

induce $H_*(\ , \mathbb{Z}/\ell)$-isomorphisms

$$\prod_{j \in V} BH_n(\overline{\mathbb{F}}_{\mathbf{P}(j)}) \to \prod_{j \in V} BH_n(\overline{\mathbb{F}}_{\mathbf{P}(j)})^{\wedge}_{\ell}$$

for all members V of the ultrafilter $\mathcal{U}$, and hence induce an $H_*(\ , \mathbb{Z}/\ell)$-isomorphism

$$BH_n(j_*(\overline{\mathbb{F}}_{\mathbf{P}})_{\mathcal{U}}) \cong j_* BH_n(\overline{\mathbb{F}}_{\mathbf{P}})_{\mathcal{U}} \to j_*(BH_n(\overline{\mathbb{F}}_{\mathbf{P}})^{\wedge}_{\ell})_{\mathcal{U}},$$

where the simplicial sheaves $BH_n(\overline{\mathbb{F}}_{\mathbf{P}})$ and $BH_n(\overline{\mathbb{F}}_{\mathbf{P}})^{\wedge}_{\ell}$ are defined on $\mathcal{P}(J)_{disc}$ by analogy with $BSl_n(\overline{\mathbb{F}}_{\mathbf{P}})$ and $BSl_n(\overline{\mathbb{F}}_{\mathbf{P}})^{\wedge}_{\ell}$.

The weak equivalences

$$f : (BSU_n)^{\wedge}_{\ell} \to BSl_n(\overline{\mathbb{F}}_p)^{\wedge}_{\ell}$$

arising from the Friedlander-Mislin theorem restrict to weak equivalences

$$f : (BT_n)^{\wedge}_{\ell} \to BH_n(\overline{\mathbb{F}}_p)^{\wedge}_{\ell}$$

and hence determine a mod ℓ homology isomorphism

$$f_* : (BT_n)^{\wedge}_{\ell} \to j_*(BH_n(\overline{\mathbb{F}}_{\mathbf{P}})^{\wedge}_{\ell})_{\mathcal{U}}$$

which fits into a commutative diagram

$$\begin{array}{ccccc}
BH_n(j_*(\overline{\mathbb{F}}_{\mathbf{P}})_{\mathcal{U}}) & \longrightarrow & j_*(BH_n(\overline{\mathbb{F}}_{\mathbf{P}})^{\wedge}_{\ell})_{\mathcal{U}} & \xleftarrow{\ f_*\ } & (BT_n)^{\wedge}_{\ell} \\
\downarrow{\scriptstyle j_*} & & \downarrow{\scriptstyle j_*} & & \downarrow{\scriptstyle j_*} \\
BSl_n(j_*(\overline{\mathbb{F}}_{\mathbf{P}})_{\mathcal{U}}) & \longrightarrow & j_*(BSl_n(\overline{\mathbb{F}}_{\mathbf{P}})^{\wedge}_{\ell})_{\mathcal{U}} & \xleftarrow[\ f_*\ ]{} & (BSU_n)^{\wedge}_{\ell}
\end{array}$$

where all vertical maps are induced by the inclusion $j : H_n \hookrightarrow Sl_n$. The induced morphism

$$j_* : H_*(BT_n, \mathbb{Z}/\ell) \to H_*(BSU_n, \mathbb{Z}/\ell)$$

is well known to be surjective in all degrees. If the isomorphism conjecture for Sl_n holds, then the map

$$j_* : H_*(BH_n(j_*(\overline{\mathbb{F}}_{\mathbf{P}})_{\mathcal{U}}), \mathbb{Z}/\ell) \to H_*(BSl_n(j_*(\overline{\mathbb{F}}_{\mathbf{P}})_{\mathcal{U}}), \mathbb{Z}/\ell)$$

is surjective in all degrees, because the dual map would be isomorphic to the degreewise monomorphism

$$j^* : H^*(BSU_n, \mathbb{Z}/\ell) \to H^*(BT_n, \mathbb{Z}/\ell).$$

It would then follow that the induced map

$$H_*(BSl_n(j_*(\overline{\mathbb{F}}_{\mathbf{P}})_{\mathcal{U}}), \mathbb{Z}/\ell) \to H_*(j_*(BSl_n(\overline{\mathbb{F}}_{\mathbf{P}})^{\wedge}_{\ell})_{\mathcal{U}}, \mathbb{Z}/\ell)$$

is, in each degree, a surjective homomorphism of finite dimensional $\mathbb{Z}/\ell$-vector spaces having the same dimension. This map would therefore be an isomorphism. □

We shall now show that Conjecture 16 is implied by a statement involving only finite fields.

Observe, first of all, that if $X_j : I_j \to$ **Sets**, $\in J$, is a collection of functors defined on filtered small categories I_j, then the product category $I = \prod_{j \in J} I_j$ is filtered, and the canonical map

$$\varinjlim_{I} \prod_{j \in J} X_j \to \prod_{j \in J} \varinjlim_{I_j} X_j$$

is a bijection. Here, $\prod_{j \in J} X_j$ is the functor on $I = \prod_{j \in J} I_j$ defined by

$$(i_j) \mapsto \prod_{j \in J} X_j(i_j).$$

In particular, the space $\prod_{j \in X} BSl_n(\overline{\mathbb{F}}_{\mathbf{P}(j)})$ is a filtered colimit of spaces of the form

$$\prod_{j \in X} BSl_n(\mathbb{F}_{\mathbf{P}(j)^{m(j)}})$$

where $\mathbb{F}_{\mathbf{P}(j)^{m(j)}}$ is a finite field of characteristic $\mathbf{P}(j)$ and of order $\mathbf{P}(j)^{m(j)}$. It also follows that the simplicial sheaf $j_* BSl_n(\overline{\mathbb{F}}_{\mathbf{P}})$ is a filtered colimit in the simplicial presheaf category, in the sense that there is an isomorphism

$$\varinjlim_{m : J \to \mathbb{N}} j_* BSl_n(\mathbb{F}_{\mathbf{P}^m}) \cong j_* BSl_n(\overline{\mathbb{F}}_{\mathbf{P}}),$$

where $BSl_n(\mathbb{F}_{\mathbf{P}^m})$ is the family of simplicial sets $\{BSl_n(\mathbb{F}_{\mathbf{P}(j)^{m(j)}})\}$. Note that if $m_1, m_2 : J \to \mathbb{N}$, then $m_1 \leq m_2$ in the poset $\mathbb{N}^J = \prod_{j \in J} \mathbb{N}$ if and only if

$m_1(j)$ divides $m_2(j)$ for all $j \in J$. There are corresponding simplicial sheaves $j_*BSl_n(\mathbb{F}_{\mathbf{P}^m})^\wedge_\ell$ on $\mathcal{P}(J)_{fin}$ defined at subsets X of J by

$$j_*BSl_n(\mathbb{F}_{\mathbf{P}^m})^\wedge_\ell(X) = \prod_{j\in X} BSl_n(\mathbb{F}_{\mathbf{P}(\mathbf{j})^{m(j)}})^\wedge_\ell,$$

and the inclusions $\mathbb{F}_{\mathbf{P}^{m(j)}} \subset \overline{\mathbb{F}}_{\mathbf{P}(j)}$ and completion maps together induce a commutative diagram of simplicial presheaf maps of the form

$$\begin{array}{ccc} \varinjlim_{m:J\to\mathbb{N}} j_*BSl_n(\mathbb{F}_{\mathbf{P}^m}) & \xrightarrow{\cong} & j_*BSl_n(\overline{\mathbb{F}}_{\mathbf{P}}) \\ \downarrow & & \downarrow \\ \varinjlim_{m:J\to\mathbb{N}} j_*BSl_n(\mathbb{F}_{\mathbf{P}^m})^\wedge_\ell & \longrightarrow & j_*BSl_n(\overline{\mathbb{F}}_{\mathbf{P}})^\wedge_\ell. \end{array}$$

Lemma 18 *Subject to the definitions above, the map of simplicial presheaves*

$$\varinjlim_{m:J\to\mathbb{N}} j_*BSl_n(\mathbb{F}_{\mathbf{P}^m})^\wedge_\ell \to j_*BSl_n(\overline{\mathbb{F}}_{\mathbf{P}})^\wedge_\ell$$

is a pointwise mod ℓ homology isomorphism.

Proof In sections corresponding to $X \subset J$, this map can be identified with the map

$$\prod_{j\in X} \varinjlim_{m(j)\in\mathbb{N}} BSl_n(\mathbb{F}_{\mathbf{P}(j)^{m(j)}})^\wedge_\ell \to \prod_{j\in X} BSl_n(\overline{\mathbb{F}}_{\mathbf{P}(j)})^\wedge_\ell \tag{6}$$

which is the product of the maps

$$\varinjlim_{m(j)\in\mathbb{N}} BSl_n(\mathbb{F}_{\mathbf{P}(j)^{m(j)}})^\wedge_\ell \to BSl_n(\overline{\mathbb{F}}_{\mathbf{P}(j)})^\wedge_\ell. \tag{7}$$

Each of the maps (7) is a mod ℓ homology isomorphism from a simply connected space to a 2-connected space, and therefore induces an isomorphism in mod ℓ homotopy groups. The map (6) is thus a map from a 1-connected space to a 2-connected space which induces an isomorphism in mod ℓ homotopy groups. It is therefore a mod ℓ homology isomorphism, as required. □

Corollary 19 *In order to prove Conjecture 16, it suffices to show that, for each function $m : J \to \mathbb{N}$, and each ultrafilter $\mathcal{U}$ on the set J such that $\{j \in J | \mathbf{P}(j) = \ell\}$ is not a member of $\mathcal{U}$, the induced map*

$$j_*BSl_n(\mathbb{F}_{\mathbf{P}^m})_{\mathcal{U}} \to j_*(BSl_n(\mathbb{F}_{\mathbf{P}^m})^\wedge_\ell)_{\mathcal{U}}$$

is a mod ℓ homology isomorphism.

The isomorphism conjecture for the mod ℓ cohomology of $BSl_n(k)$ over arbitrary algebraically closed fields k thus becomes a problem about finite fields. Morally, for a subset $U \subset J$, one expects that the $H_*(\ ,\mathbb{Z}/\ell)$-isomorphisms

$$BSl_n(\mathbb{F}_{\mathbf{P}^{m(j)}}) \to BSl_n(\mathbb{F}_{\mathbf{P}^{m(j)}})^\wedge_\ell$$

will together induce a homology isomorphism

$$\prod_{j\in U} BSl_n(\mathbb{F}_{\mathbf{P}^{m(j)}}) \to \prod_{j\in U} BSl_n(\mathbb{F}_{\mathbf{P}^{m(j)}})^\wedge_\ell$$

provided that ℓ does not appear as one of the characteristics of the finite fields $\mathbb{F}_{\mathbf{P}^{m(j)}}$, $j \in U$. If it were true, this claim would imply Conjecture 16, but such a pointwise statement seems like a lot to expect. The techniques of [**2**] and [**10**] are not available, in particular, because the spaces $BSl_n(\mathbb{F}_{\mathbf{P}^{m(j)}})$ are not nilpotent. It is probably much more reasonable to try to prove a statement, as in Corollary 19, on the ultraproduct level.

References

[1] Bell, J. L. and Slomson, A. B. [1974], *Models and Ultraproducts: an Introduction*, 3rd revised printing, North-Holland, Amsterdam.

[2] Bousfield, A. K. [1987], *On the homology spectral sequence of a cosimplicial space*, Amer. J. Math. **109**, 361–394.

[3] Bousfield, A. K. and Kan, D. M. [1987], *Homotopy Limits, Completions and Localizations*, 2nd corrected printing, Springer Lecture Notes in Math., Vol. 304, Springer-Verlag, Berlin-Heidelberg-New York.

[4] Carlsson, G. [1994], *On the Algebraic K-theory of Infinite Product Categories*, preprint.

[5] Friedlander, E. M. [1982], *Etale Homotopy of Simplicial Schemes*, Annals of Math. Studies, No. 104, Princeton University Press, Princeton.

[6] Friedlander, E. and Mislin, G. [1984], *Cohomology of classifying spaces of complex Lie groups and related discrete subgroups*, Comment. Math. Helv. **59**, 347–361.

[7] Friedlander, E. M. and Parshall, B. [1981], *Etale cohomology of reductive groups*, Springer Lecture Notes in Math., Vol. 854 , 127–140.

[8] Gabber, O. [1992], *K-theory of Henselian local rings and Henselian pairs*, Contemporary Math. **126**, 59–70.

[9] Gobel, R. and Richter, M. [1972], *Cartesian closed classes of perfect groups*, J. Algebra **23**, 370–381.

[10] Goerss, P. G. [1994], *The Homology of Homotopy Inverse Limits*, preprint, to appear in J. Pure Appl. Algebra.

[11] Jardine, J. F. [1986], *Simplicial objects in a Grothendieck topos*, Contemporary Math. **55**, 193–239.

[12] Jardine, J. F. [1986], *Cup products in sheaf cohomology*, Canad. Math. Bull. **29**, 469–477.

[13] Jardine, J. F. [1987], *Simplicial presheaves*, J. Pure Applied Algebra **47**, 35–87.

[14] Karoubi, M. [1984], *Relations between algebraic K-theory and hermitian K-theory*, J. Pure Appl. Algebra **34**, 259–263.

[15] Neisendorfer, J. [1980], *Primary homotopy theory*, Memoirs AMS **232**.

[16] Suslin, A. [1983], *On the K-theory of algebraically closed fields*, Invent. Math. **73**, 241–245.

[17] Suslin, A. [1984], *On the K-theory of local fields*, J. Pure Appl. Algebra **34**, 301–318.

[18] Suslin, A. [1987], *Algebraic K-Theory of Fields*, Proceedings of the International Congress, Berkeley, 1986, AMS, Providence, 222–244.

Fields Institute Communications
Volume **16**, 1997

Lambda-operations, K-Theory and Motivic Cohomology

Marc Levine
Department of Mathematics
Northeastern University
Boston, MA USA 02115
marc@neu.edu

Dedicated to the memory of Robert Thomason

Abstract. We give a construction of functorial lambda operations, satisfying the special lambda ring identities, for relative K-theory with supports. As an applicaton, we show that Bloch's higher Chow group $\mathrm{CH}^q(X,p)$ of a smooth, quasi-projective variety X of dimension d over a field of arbitrary characteristic agrees with $\mathrm{gr}^q_\gamma K_p(X)$, after inverting $(p+d-1)!$.

Introduction

This paper is in two parts: in part one, our main object is to give a construction of natural λ-operations for relative K-theory with supports, satisfying the special λ-ring identities. In the second part, we give an application to a comparison of the relation of motivic cohomology and algebraic K-theory of a smooth quasi-projective variety over a field.

The main idea in the construction of the λ-operations is to redo the constructions of Quillen (Hiller gives an exposition of Quillen's construction in [12]) and Kratzer [15] for the K-theory of a commutative ring, in the setting of the K-theory of an I-*diagram of commutative rings*, i.e., a functor

$$A\colon I \to \mathbf{Rings}$$

with I a small category. This approach can contrasted with the construction of Soulé [24], where use is made of the closed model structure on the category of sheaves of simplicial sets. Our point of view is to look at the "discrete" setting, viewing for example, a scheme as a finite diagram of affine schemes, and replacing the use of topology and sheaf theory with the Mayer-Vietoris property of K-theory provided by the Thomason-Trobaugh theorem [26], together with the remark that the operations we construct are natural with respect to refinements.

1991 *Mathematics Subject Classification.* Primary 19D55, 19E20; Secondary 14F20, 18F20.

Partially supported by the NSF Grant no. DMS-9401164, and the DFG Forschergruppe "Arithmetik und Geometrie".

We rely on the general machinery of closed simplicial model categories, especially that of I-diagrams of simplicial sets: the category of functors

$$X: I \to \mathcal{S},$$

where I is a fixed small category, and $\mathcal{S}$ is the category of simplicial sets. Much of the general theory, as well as the fundamental properties, of Quillen's +-construction carry over without change to the setting of I-diagram of commutative rings. However, at some point we need to work with an analog of finite CW complexes, which restricts our theory to *finite* categories, i.e., a category I whose nerve has only finitely many non-degenerate simplices. Fortunately, the diagrams of affine schemes needed to express the relative K-theory of a noetherian scheme, with supports in a closed subset, are all finite.

Besides the works mentioned above, there is a recent construction of Lecomte [16], building on previous work of Schechtman [22], giving Adams operations for the K-groups of schemes, which in turn gives a special λ-ring structure for the rational K-groups. It seems that her construction would also give natural Adams operations for relative K-theory with supports, and for other groups expressable as finite diagrams of schemes, or perhaps even for arbitrary small diagrams. There is also the unpublished work of Gillet and Soulé [7], which uses the sheaf-theoretic point of view. Together with the results of Jardine [13], the arguments of [7] would give a special λ-ring structure for the K-groups of an arbitrary small diagram of schemes satisfying a certain cohomological finiteness (see e.g. [25]). Grayson ([8], [9]) has constructed λ-operations and Adams operations on the simplicial level, but it is not clear that these operations satisfy special λ-ring identities. Our justification for producing this work is that there is currently nothing in the literature which constructs λ-operations, satisfying the special λ-ring identities, in the generality required for relative K-theory with supports.

The argument given here consists of reworking the basic ingredients of the construction of Quillen in the setting of a diagram of rings. These ingredients are:

i) the special λ-ring structure on the representation ring of a group
ii) the universal property of the plus construction, and the structure of an H-space on BGL^+
iii) the construction of a "universal" natural transformation from the representation ring of $\pi_1(X)$ to the set of homotopy classes of maps $[X, \mathrm{BGL}^+]$.

The representation theory is quite straightforward and is discussed in Section 1. The generalization of (ii) is done in Section 2. For (iii), we require the finiteness of the parameter category; this is discussed in Section 4. We give the applications to the construction of the special λ-ring structure for relative K-theory with supports in Section 5. For the reader's convenience, we have included two appendices, the first recalling the basic notions from the theory of closed simplicial model categories, and the second reviewing basic properties of homotopy limits, and relations with Hom-complexes in the model category of functors to simplicial sets.

The application in the second part is a refinement of our earlier rational computation of motivic cohomology in [17], namely, for X a smooth quasi-projective variety over a field k, there is a natural isomorphism of the motivic cohomology, defined as Bloch's higher Chow groups [3], with the weight-graded pieces of K-theory, after tensoring with $\mathbb{Q}$

$$\mathrm{CH}^q(X,p)_{\mathbb{Q}} \cong K_p(X)_Q^{(q)}. \tag{1}$$

As the γ-filtration on the K-group $K_p(X)$ splits into eigenspaces for the Adams operations ψ^k after inverting $(d+p-1)!$ (at least for $p \geq 1$; for $p = 0$ one needs to restrict to the subgroup $F^2_\gamma K_0(X)$), where d is the dimension of X over k, it is natural to ask if there is an isomorphism as in (1), after inverting only $(d+p-1)!$. This turns out to be the case: we show here that there are natural isomorphisms for all p and q (see Corollary 8.2)

$$\mathrm{CH}^q(X,p)\left[\tfrac{1}{(d+p-1)!}\right] \cong \mathrm{gr}^q_\gamma K_p(X)\left[\tfrac{1}{(d+p-1)!}\right]. \tag{2}$$

There is in fact a somewhat finer result; we refer the reader to Theorem 8.1 for details. Soulé [25] has proved a similar result for $X = \mathrm{Spec}(F)$, where F is a field of characteristic zero, using the "Atiyah-Hirzebruch" spectral sequence of Bloch and Lichtenbaum [4].

The main idea of the proof is to note that, although the γ-filtration does not in general behave well with respect to exact sequences, a decomposition into characteristic subspaces for the Adams operations does. This remark is systematized in Section 6. In Section 7, we give the applications of the Adams decomposition to the case of relative K-theory with supports, and we prove the main result in Section 8.

The construction of λ-operations given here is a somewhat refined version of an earlier construction I gave in a course on algebraic K-theory and motivic cohomology at MIT in the winter of 1993. I would like to thank all the participants in that course for their comments and suggestions, especially Thomas Geisser and Lorenzo Ramero; I would like to thank Thomas Geisser doubly for preparing an excellent set of notes. The systematic use of techniques of closed model categories that appears here is largely due to a comment of Rick Jardine, who pointed out to me the use of closed model categories in solving coherence problems. Finally, I would like to thank the University of Essen, especially Hélène Esnault, Eckart Viehweg, and the DFG Forschergruppe "Arithmetik und Geometrie", for their hospitality and support.

Part I: Lambda-Operations

1 Representation Theory Over a Category

For a small category I, and a category $\mathcal{C}$, we let $\mathcal{C}^I$ denote the category of functors $F\colon I \to \mathcal{C}$. Let **Rings** be the category of commutative rings (with unit); for a commutative ring S, we let $\mathbf{Rings}_S$ be the subcategory of **Rings** consisting of S-algebras and S-algebra maps.

Let I be a small category, and let

$$A\colon I \to \mathbf{Rings}_S; \qquad G\colon I \to \mathbf{Groups}$$

be functors. We have the category $\mathbf{Mod}_A$ of A-modules, defined as the category of pairs (M,μ), with M in $\mathbf{Ab}^I$, and $\mu\colon A \times M \to M$ a map in $\mathbf{Sets}^I$ such that $\mu(i)$ makes $M(i)$ into an $A(i)$-module for each $i \in I$. Maps of A-modules are defined similarly. We have the full subcategory $\mathcal{P}_{A/S}$ of $\mathbf{Mod}_A$ with objects M isomorphic to an A-module of the form $P \otimes_S A$:

$$i \mapsto P \otimes_S A(i)$$

for some finitely generated projective S-module P.

If M is in $\mathbf{Mod}_A$, a *representation* ρ *of* G *on* M is a collection of homomorphisms

$$\rho(i)\colon G(i) \to \mathrm{Aut}_{A(i)}(M(i)); \qquad i \in I,$$

such that, for each morphism $s\colon i \to j$ in I, we have

$$M(s)(\rho(i)(g)(m)) = \rho(j)(G(s)(g))(M(s)(m))$$

for each $m \in M(i)$ and $g \in G(i)$. We let $\mathrm{Rep}_S(G; A)$ denote the category of pairs (M, ρ), with M in $\mathcal{P}_{A/S}$, and ρ a representation of G on M, with the obvious notion of morphism. If we choose an isomorphism of M with $P \otimes_S A$, then a representation ρ of G on M is the same as a homomorphism

$$\rho\colon G \to \mathrm{Aut}_A(P \otimes_S A)$$

where $\mathrm{Aut}_A(P \otimes_S A)$ is the group over I given by

$$i \mapsto \mathrm{Aut}_{A(i)}(P \otimes_S A(i)).$$

We make $\mathrm{Rep}_S(G; A)$ into an exact category by taking the sequences in $\mathrm{Rep}_S(G; A)$

$$0 \to (M', \rho') \to (M, \rho) \to (M'', \rho'') \to 0$$

such that

$$0 \to M' \to M \to M'' \to 0$$

is *split* exact in $\mathbf{Mod}_A$. We let $R_S(G; A)$ denote the Grothendieck group of the exact category $\mathrm{Rep}_S(G; A)$, and define $K_0(A/S)$ by

$$K_0(A/S) := R_S(\{\mathrm{id}\}; A).$$

The inclusion id $\to G$ gives the augmentation

$$\epsilon_{G,A}\colon R_S(G; A) \to K_0(A/S). \tag{1.1}$$

Sending a finitely generated projective S-module P to the A-module $P \otimes_S A$ determines the surjection

$$K_0(S) \to K_0(A/S). \tag{1.2}$$

Taking the point-wise tensor product of representations gives $R_S(G; A)$ the natural structure of a commutative ring; point-wise exterior product gives the natural operations

$$\lambda^k\colon R_S(G; A) \to R_S(G; A) \tag{1.3}$$

We recall from [1], [11] or [2] the notions of a λ-ring and a special λ-ring.

Propositon 1.1 *The ring $R_S(G; A)$ with the operations (1.3) is a special λ-ring; the augmentation (1.1) is a map of λ-rings.*

Proof The second assertion follows directly from the first, and the naturality of the operations (1.3).

For a finitely generated projective S-module P, let $Aut_{P/S}$ denote the group scheme over S representing the functor

$$R \mapsto \mathrm{Aut}_R(P \otimes_S R).$$

For $P = S^n$, we write $\mathcal{GL}_{n/S}$ for $Aut_{P/S}$. Let $\mathcal{G}$ be a group scheme over S. We have the exact category $\mathrm{Rep}_S(\mathcal{G})$ with objects consisting of pairs (P, ρ), P a finitely generated projective S-module, and

$$\rho\colon \mathcal{G} \to Aut_{P/S} \tag{1.4}$$

a homomorphism of group schemes over S.

Let $R_S(\mathcal{G})$ denote the Grothendieck group of $\mathrm{Rep}_S(\mathcal{G})$. The operation of tensor product of representations gives $R_S(\mathcal{G})$ the structure of a ring, and the operation sending ρ to $\Lambda^k\rho$ extends to give $R_S(\mathcal{G})$ the structure of a λ ring. Using the fact that the K_0-class of a representation of $\mathcal{GL}_{n/S}$ on a free S-module is determined by its character (see e.g. [23]), it is easy to show that $R_S(\mathcal{GL}_{n/S})$ is a special λ-ring. This easily implies that $R_S(\mathcal{G})$ is a special λ-ring for $\mathcal{G}$ of the form

$$\mathcal{G} = \prod_{i=1}^{N} \mathcal{GL}_{n_i/S},$$

as

$$R_S(\prod_{i=1}^{N} \mathcal{GL}_{n_i/S}) \cong \bigotimes_{i=1}^{N} R_S(\mathcal{GL}_{n_i/S}).$$

As an arbitrary homomorphism (1.4) is stably equivalent to a homomorphism with $P = S^n$ for some n, this implies that $R_S(\mathcal{G})$ has the natural structure of a special λ-ring for all algebraic group schemes over S.

If we now have a finite collection of representations ρ_i of some $G \in \mathbf{Groups}^I$ on A-modules $P_i \otimes A$, these representations factor through the identity representation of

$$\prod_i \mathrm{Aut}_S(P_i \otimes A).$$

As this latter object of $\mathbf{Groups}^I$ is the functor gotten by composing A with

$$\mathcal{G}(-) := \prod_i Aut_{P_i/S}(-),$$

the special λ-ring identities in $R_S(\mathcal{G})$ imply the special λ-ring identities in the sub-λ-ring of $R_S(G; A)$ generated by the ρ_i. As $R_S(G; A)$ is a direct limit of its finitely generated sub-λ-rings, it follows that $R_S(G; A)$ is a special λ-ring. □

2 The Classifying Space, and the Classifying Map

We refer the reader to Appendix A for the basic notions from the theory of closed model categories, including the closed model structure on the category $\mathcal{S}$ of simplicial sets, the category **Top** of topological spaces, and the pointed versons $\mathcal{S}^*$ and $\mathbf{Top}^*$.

For a group G, we have the simplicial set $\mathrm{E}G$ with free G-action, and the quotient simplicial set $\mathrm{B}G := \mathrm{E}G/G$. The geometric realization $|\mathrm{B}G|$ is a functorial model for the classifying space of G, i.e., $|\mathrm{B}G|$ is pointed and connected, $\pi_1(|\mathrm{B}G|, *) = G$ and $|\mathrm{B}G|$ has no higher homotopy groups. In particular, if $(X, *)$ is a pointed simplicial set with $|X|$ connected, sending a map $f\colon (X, *) \to (\mathrm{B}G, *)$ to the induced map on π_1 gives an isomorphism

$$\phi_X\colon [(X, *), (\mathrm{B}G, *)] \to \mathrm{Hom}_{\mathbf{Groups}}(\pi_1(|X|, *), G).$$

(Here $[-, -]$ is the set of pointed homotopy classes of pointed maps). Indeed, since $\mathrm{B}G$ is a simplicial group, $\mathrm{B}G$ is fibrant, hence

$$[(X, *), (\mathrm{B}G, *)] \cong [(|X|, *), (|\mathrm{B}G|, *)],$$

and the result follows from classical obstruction theory.

Now let I be a small category, and $G\colon I \to \mathbf{Groups}$ a functor. We let $\tilde{B}G : I \to S$ denote a bifibrant model of the functor

$$i \mapsto \mathrm{B}G(i).$$

For a cofibrant $(Y,*)$ in $\mathcal{S}^{*I}$, let $\pi_1(Y,*)$ denote the object of $\mathbf{Groups}^I$

$$i \mapsto \pi_1(|Y(i)|, *(i)).$$

Propositon 2.1 *Let $(X,*)$ be a cofibrant object of $\mathcal{S}^{*I}$, and let $G: I \to \mathbf{Groups}$ be a functor. Suppose $|X(i)|$ is connected for each $i \in I$. Then sending a map $f\colon (X,*) \to (\tilde{\mathrm{B}}G,*)$ to the induced map on π_1 defines an isomorphism*

$$\phi_{X,G}\colon [(X,*),(\tilde{\mathrm{B}}G,*)] \to \mathrm{Hom}_{\mathbf{Groups}^I}(\pi_1(X,*),G).$$

Proof For each i and j in I, the space

$$\mathcal{H}om_{\mathcal{S}^*}((X(i),*),(\tilde{\mathrm{B}}G(j),*))$$

has

$$\pi_n = \begin{cases} \mathrm{Hom}_{\mathbf{Groups}}(\pi_1(X(i),*),G(j)) & \text{for } n = 0 \\ 0 & \text{for } n \geq 1. \end{cases}$$

Thus, sending $\mathcal{H}om_{\mathcal{S}^*}((X(i),*),(\tilde{\mathrm{B}}G(j),*))$ to the constant simplicial set

$$\pi_0\mathcal{H}om_{\mathcal{S}^*}((X(i),*),(\tilde{\mathrm{B}}G(j),*))$$

defines a weak equivalence in $\mathcal{S}^{*aI}$ (see Appendix B, following Theorem B.1, for the definition of the *twisted arrow category* aI):

$$\Pi_0\colon \mathcal{H}om((X,*),(\tilde{\mathrm{B}}G,*)) \to \pi_0\mathcal{H}om((X,*),(\tilde{\mathrm{B}}G,*)).$$

Since $(X,*)$ is cofibrant and $(\tilde{\mathrm{B}}G,*)$ is fibrant, this is a trivial fibration of fibrant objects in $\mathcal{S}^{*aI}$. Thus, by ([5], XI, Section 5, Lemma 5.6) Π_0 induces a weak equivalence

$$\operatorname*{holim}_{aI} \mathcal{H}om((X,*),(\tilde{\mathrm{B}}G,*)) \to \operatorname*{holim}_{aI} \pi_0\mathcal{H}om((X,*),(\tilde{\mathrm{B}}G,*)). \tag{2.1}$$

It is easy to see that the natural map

$$\mathrm{Hom}_{\mathbf{Groups}^I}(\pi_1(X,*),G) = \varprojlim_{aI} \pi_0\mathcal{H}om((X,*),(\tilde{\mathrm{B}}G,*))$$

$$\to \pi_0(\operatorname*{holim}_{aI} \pi_0\mathcal{H}om((X,*),(\tilde{\mathrm{B}}G,*)))$$

is an isomorphism; by Theorem B.2, the weak equivalence (2.1) induces the isomorphism

$$[(X,*),(\tilde{\mathrm{B}}G,*)] \cong \pi_0 \operatorname*{holim}_{aI} \mathcal{H}om((X,*),(\tilde{\mathrm{B}}G,*)) \cong \mathrm{Hom}_{\mathbf{Groups}^I}(\pi_1(X,*),G).$$

□

Let G and H be in $\mathbf{Groups}^I$. Taking $X = \tilde{\mathrm{B}}H$, Proposition 2.1 gives the isomorphism

$$\Phi_{H,G}\colon \mathrm{Hom}_{\mathbf{Groups}^I}(H,G) \to [(\tilde{\mathrm{B}}H,*),(\tilde{\mathrm{B}}G,*)]. \tag{2.2}$$

In particular, if $g\colon I \to G$ is a section, the automorphism of G given by conjugation by g determines the pointed map

$$C_g\colon \tilde{\mathrm{B}}G \to \tilde{\mathrm{B}}G, \tag{2.3}$$

uniquely up to pointed homotopy.

Let

$$q_Y\colon (Y,*) \to \tilde{\mathrm{B}}\pi_1(Y,*) \tag{2.4}$$

be the map $\phi_{Y,\pi_1(Y,*)}(\mathrm{id}_{\pi_1(Y,*)})$.

3 BGL and BGL$^+$

For $R \in$ **Rings**, n a positive integer, we have the functorial map of the permutation group S_n into $\mathrm{GL}_n(R)$ as the group of permutation matrices; this gives the functorial map $A_5 \to \mathrm{GL}_n(R)$. We let

$$\mathrm{B}A_5 \to \mathrm{B}A_5^+$$

be a cofibration of simplicial sets whose geometric realization is Quillen's +-construction applied to the classifying space $|\mathrm{B}A_5|$. Define the simplicial set $\mathrm{BGL}_n(R)^+$, $n \geq 5$, by

$$\mathrm{BGL}_n(R)^+ = \mathrm{BGL}_n(R) \cup_{\mathrm{B}A_5} \mathrm{B}A_5^+. \tag{3.1}$$

This gives the natural transformation of functors

$$\mathrm{BGL}_n(-) \to \mathrm{BGL}_n(-)^+$$

from **Rings** to $\mathcal{S}^*$.

It is shown in [12] that $|\mathrm{BGL}_n(R)^+|$ is a model for the +-construction on $|\mathrm{BGL}_n(R)|$; i.e., the cofibration $i\colon |\mathrm{BGL}_n(R)| \to |\mathrm{BGL}_n(R)^+|$ satisfies

i) $i_*\colon \pi_1(|\mathrm{BGL}_n(R)|, *) = \mathrm{GL}_n(R) \to \pi_1(|\mathrm{BGL}_n(R)^+|, *)$ is surjective, with kernel $\mathrm{E}_n(R)$

ii) i is a homology isomorphism for all local systems on $|\mathrm{BGL}_n(R)^+|$.

Let I be a small category, and $A\colon I \to$ **Rings** a functor. We may apply the functors $\mathrm{BGL}_n(-)$ and $\mathrm{BGL}_n(-)^+$, giving the objects $\mathrm{BGL}_n(A)$ and $\mathrm{BGL}_n(A)^+$ of $\mathcal{S}^{*I}$, and the map

$$i_n\colon \mathrm{BGL}_n(A) \to \mathrm{BGL}_n(A)^+ \tag{3.2}$$

in $\mathcal{S}^{*I}$. We have as well the stabilization maps

$$\begin{aligned} \iota_{n,n+1}\colon \mathrm{BGL}_n(A) &\to \mathrm{BGL}_{n+1}(A) \\ \iota^+_{n,n+1}\colon \mathrm{BGL}_n(A)^+ &\to \mathrm{BGL}_{n+1}(A)^+ \end{aligned} \tag{3.3}$$

induced by the stabilization maps $\mathrm{GL}_n \to \mathrm{GL}_{n+1}$. Passing to the direct limit over n gives the functors

$$\begin{aligned} \mathrm{BGL}(A)\colon I &\to \mathcal{S}^* \\ \mathrm{BGL}(A)^+\colon I &\to \mathcal{S}^*. \end{aligned} \tag{3.4}$$

and the maps

$$\begin{aligned} \iota_n\colon \mathrm{BGL}_n(A) &\to \mathrm{BGL}(A) \\ \iota_n^+\colon \mathrm{BGL}_n(A)^+ &\to \mathrm{BGL}(A)^+ \\ i\colon \mathrm{BGL}(A) &\to \mathrm{BGL}(A)^+. \end{aligned} \tag{3.5}$$

We sometimes denote $\mathrm{BGL}(A)$ and $\mathrm{BGL}(A)^+$ by $\mathrm{BGL}_\infty(A)$ and $\mathrm{BGL}_\infty(A)^+$.

If we apply the functor $\mathrm{Sin}|-|$ (singular complex of the geometric realization), we have the canonical weak equivalences with fibrant objects

$$\mathrm{BGL}_n(A) \to \mathrm{Sin}|\mathrm{BGL}_n(A)|$$

$$\mathrm{BGL}_n(A)^+ \to \mathrm{Sin}|\mathrm{BGL}_n(A)^+|$$

as well as for $n = \infty$.

We may find commutative diagrams in $\mathcal{S}^{*I}$:

$$\begin{array}{ccc} \tilde{\mathrm{BGL}}_n(A) & \xrightarrow{\tilde{\imath}_n} & \tilde{\mathrm{BGL}}_n(A)^+ \\ \downarrow & & \downarrow \\ \mathrm{Sin}|\mathrm{BGL}_n(A)| & \xrightarrow{\mathrm{Sin}|i_n|} & \mathrm{Sin}|\mathrm{BGL}_n(A)^+| \end{array}$$

$n = 5, 6, \ldots$, and $n = \infty$, with the vertical arrows trivial fibrations, and $\tilde{\imath}_n$ a cofibration of bifibrant objects of $\mathcal{S}^{*I}$. We may assume that there are stabilization maps

$$\begin{aligned} \tilde{\iota}_{n,n+1} &\colon \tilde{\mathrm{BGL}}_n(A) \to \tilde{\mathrm{BGL}}_{n+1}(A) \\ \tilde{\iota}^+_{n,n+1} &\colon \tilde{\mathrm{BGL}}_n(A)^+ \to \tilde{\mathrm{BGL}}_{n+1}(A)^+ \end{aligned} \tag{3.6}$$

extending the stabilization maps $\mathrm{Sin}|\iota_{n,n+1}|$ and $\mathrm{Sin}|\iota^+_{n,n+1}|$, that the maps (3.6) are cofibrations, and that there are cofibrations

$$\begin{aligned} \tilde{\iota}_n &\colon \tilde{\mathrm{BGL}}_n(A) \to \tilde{\mathrm{BGL}}(A) \\ \tilde{\iota}^+_n &\colon \tilde{\mathrm{BGL}}_n(A)^+ \to \tilde{\mathrm{BGL}}(A)^+ \end{aligned} \tag{3.7}$$

extending the maps $\mathrm{Sin}|\iota_n|$ and $\mathrm{Sin}|\iota^+_n|$. In addition, we may assume that this fits together so that all the obvious diagrams commute: the maps $\tilde{\imath}_*$, $\tilde{\iota}_{*,*+1}$, $\tilde{\iota}_*$ $\tilde{\iota}^+_{*,*+1}$ and $\tilde{\iota}^+_*$ all commute, $* = 5, 6, \ldots, \infty$, and this data commutes with the corresponding data for $\mathrm{Sin}|\mathrm{BGL}_n(A)|$ and $\mathrm{Sin}|\mathrm{BGL}_n(A)^+|$. The construction of this data satisfying the required commutativities is an easy exercise using elementary properties of closed model categories (see Appendix A, axioms CM1-CM5, and properties following the axioms).

Lemma 3.1 *Let $(Z, *)$ be a fibrant object of $\mathcal{S}^{*I}$, such that $\pi_1(|Z(i)|, *)$ has no non-trivial perfect subgroup for each $i \in I$. Let*

$$f\colon X \to X^+$$

be a pointed map of cofibrant objects of $\mathcal{S}^{*I}$, such that, for each $i \in I$,

a) *$|X(i)|$ and $|X^+(i)|$ are connected*

b) *the map $f(i)$ is acyclic ($f(i)$ induces an isomorphism in homology for all local systems on $|X(i)^+|$).*

Then the map

$$f^*\colon \mathcal{H}om_{\mathcal{S}^{*I}}(X^+, Z) \to \mathcal{H}om_{\mathcal{S}^{*I}}(X, Z)$$

is a weak equivalence. In particular, for each $n \geq 5$, including $n = \infty$, the map

$$\tilde{\imath}_n\colon \tilde{\mathrm{BGL}}_n(A) \to \tilde{\mathrm{BGL}}_n(A)^+$$

induces an isomorphism

$$\tilde{\imath}_n\colon [\tilde{\mathrm{BGL}}_n(A)^+, *), (Z, *)] \to [(\tilde{\mathrm{BGL}}_n(A), *), (Z, *)].$$

Proof It follows from classical obstruction theory that the map

$$f^*(i)\colon \mathcal{H}om_{\mathcal{S}^*}((X^+(i), *), (Z(j), *)) \to \mathcal{H}om_{\mathcal{S}^*}((X(i), *), (Z(j), *))$$

is a weak equivalence for each $i, j \in I$. The result then follows from Corollary B.4. □

Let $g \in \Gamma(\mathcal{C}, \mathrm{GL}_n(A))$ be a section. By Lemma 3.1, the map (2.3)

$$C_g\colon \mathrm{B\tilde{G}L}_n(A) \to \mathrm{B\tilde{G}L}_n(A)$$

extends to the pointed map

$$C_g^+\colon \mathrm{B\tilde{G}L}_n(A)^+ \to \mathrm{B\tilde{G}L}_n(A)^+$$

uniquely up to (pointed) homotopy.

Lemma 3.2 *Let*

$$\tilde{\iota}^+_{n,2n}\colon \mathrm{B\tilde{G}L}_n(A)^+ \to \mathrm{B\tilde{G}L}_{2n}(A)^+$$

be the stabilization map. Then $\tilde{\iota}^+_{n,2n}$ and $\tilde{\iota}^+_{n,2n} \circ C_g^+$ induce homotopic maps

$$\mathrm{B\tilde{G}L}_n(A)^+ \to \mathrm{B\tilde{G}L}_{2n}(A)^+,$$

for $n \geq 5$.

Proof By the usual trick of writing the matrix

$$\begin{pmatrix} m & 0 \\ 0 & m^{-1} \end{pmatrix}$$

as an element in $\mathrm{E}_{2n}(R)$:

$$\begin{pmatrix} m & 0 \\ 0 & m^{-1} \end{pmatrix} = \begin{pmatrix} 1 & m \\ 0 & 1 \end{pmatrix}\begin{pmatrix} 1 & 0 \\ -m^{-1} & 1 \end{pmatrix}\begin{pmatrix} 1 & m \\ 0 & 1 \end{pmatrix}\begin{pmatrix} 1 & -1 \\ 0 & 1 \end{pmatrix}\begin{pmatrix} 1 & 0 \\ 1 & 1 \end{pmatrix}\begin{pmatrix} 1 & -1 \\ 0 & 1 \end{pmatrix}.$$

we need only show that C_g^+ is homotopically trivial for each section g of the form $g = e^s_{ij}$, where s is a section of A over I.

Let $(X, *)$ be a pointed simplicial set. We have the canonical homomorphism

$$\nu_{|X|}\colon \pi_1(|X|, *) \to [(|X|, *), (|X|, *)]$$

giving the action of $\pi_1(|X|, *)$ on X. Applying Sin, we have the map

$$\nu_X\colon \pi_1(|X|, *) \to [(\mathrm{Sin}|X|, *), (\mathrm{Sin}|X|, *)].$$

For $(X, *)$ in $\mathcal{S}^{*I}$, define $\pi_1^I(|X|, *)$ as the homotopy classes of pointed maps in $\mathbf{Top}^I$

$$\sigma\colon (S^1, *) \times I \to (|X|, *),$$

where a homotopy between two pointed maps $f, g\colon (Y, *) \to (Z, *)$ is given in the obvious way by a map

$$h\colon (Y \times [0,1], * \times [0,1]) \to (Z, *).$$

It is then easy to see that there is an action of $\pi_1^I(|X|, *)$ on $|X|$, given as a homomorphism

$$\nu_{|X|}\colon \pi_1^I(|X|, *) \to [(|X|, *), (|X|, *)].$$

The main point to note is that, taking the barycentric subdivision of each $X(i)$, the resulting closed star neighborhood of $*$ in each $|X(i)|$ gives a functorial system of closed neighborhoods of $*$, functorially contractible to $*$. Taking the singular complex gives the homomorphism

$$\nu_X\colon \pi_1^I(|X|, *) \to [(\widetilde{\mathrm{Sin}|X|}, *), (\widetilde{\mathrm{Sin}|X|}, *)]$$

where $\tilde{}$ denotes bifibrant model. When evaluated at a point $i \in I$, these give the maps $\nu_{|X(i)|}$ and $\nu_{X(i)}$, respectively.

If we apply this to the case $X = \mathrm{B}G$, and $\sigma := \sigma_g \in \pi_1$ is given by a section g of G over I, it follows from Proposition 2.1 that $\nu_X(\sigma_g)$ is homotopic to the map

$$\mathrm{Sin}|\tilde{\mathrm{B}G}| \to \mathrm{Sin}|\tilde{\mathrm{B}G}|$$

given by conjugation by g. Thus, to show that $C_{e^s_{ij}}$ is homotopically trivial, it suffices to show that the map

$$|i_n| \circ \sigma_{e^s_{ij}} : (S^1 \times I, * \times I) \to (|\mathrm{BGL}_n(A)^+|, *)$$

is homotopically trivial. If we take A to be the constant ring $\mathbb{Z}[t]$, and s the constant section $s(i) = t$, the homotopy of $\sigma_{e^t_{ij}}$ to $*$ in $|\mathrm{BGL}_n(\mathbb{Z}[t])^+|$ gives the desired homotopy over I; in general, a choice of a section s of A over I defines the map

$$\tilde{s}\colon \mathbb{Z}[t] \times I \to A$$

$$\tilde{s}(i)(t) = s(i)$$

and the homotopy for e^s_{ij} is gotten by applying $\tilde{s}$ to the homotopy for e^t_{ij}. □

For a ring R, the shuffle sum of matrices:

$$((g) \oplus (h))_{ij} = \begin{cases} g_{i'j'} & \text{if } i = 2i' - 1, j = 2j' - 1, \\ h_{i'j'} & \text{if } i = 2i', j = 2j', \\ 0 & \text{otherwise} \end{cases}$$

defines the map

$$\oplus\colon \mathrm{BGL}(R) \times \mathrm{BGL}(R) \to \mathrm{BGL}(R)$$

and extends, uniquely up to homotopy, to the map

$$\oplus\colon \mathrm{BGL}(R)^+ \times \mathrm{BGL}(R)^+ \to \mathrm{BGL}(R)^+.$$

This gives $\mathrm{BGL}(R)^+$ the structure of a commutative H-group, i.e., an H-space which is homotopy associative and homotopy commutative, and with homotopy inverse.

Let $\mathcal{A}$ be a closed model category. An *H-space in $\mathcal{A}$*, (X, e, μ), is a monoid object in the homotopy category $\mathrm{Ho}\mathcal{A}$ (see Appendix A); an *H-group in $\mathcal{A}$* is an H-space in $\mathcal{A}$ which is a group object in $\mathrm{Ho}\mathcal{A}$.

Theorem 3.3 *The point-wise shuffle sum gives the bifibrant model* $\mathrm{B\tilde{G}L}(A)^+$ *the structure of a commutative H-group in $\mathcal{S}^I$.*

Proof An injection

$$\alpha\colon \mathbb{N} \to \mathbb{N}$$

induces the map

$$\tilde{\mathrm{B}}\alpha^+\colon \mathrm{B\tilde{G}L}(A)^+ \to \mathrm{B\tilde{G}L}(A)^+$$

by rearranging the matrix entries. To show that that $(\mathrm{B\tilde{G}L}(A)^+, \oplus, *)$ is a commutative H-space, it suffices to show that each $\tilde{\mathrm{B}}\alpha^+$ is homotopic to the identity.

It is well-known that, for each i, the map

$$\tilde{\mathrm{B}}\alpha^+(i)\colon \mathrm{B\tilde{G}L}(A(i))^+ \to \mathrm{B\tilde{G}L}(A(i))^+$$

is homotopic to the identity; in particular, a weak equivalence. Thus each $\tilde{\mathrm{B}}\alpha^+$ is a weak equivalence. Since $\mathrm{B\tilde{G}L}(A)^+$ is bifibrant, each $\tilde{\mathrm{B}}\alpha^+$ is a homotopy equivalence. Sending α to $\tilde{\mathrm{B}}\alpha^+$ defines a homomorphism of the monoid $\mathcal{M}$ of injective self-maps of $\mathbb{N}$ (under composition) to the monoid of homotopy classes of self-maps of $\mathrm{B\tilde{G}L}(A)^+$, which thus factors through the group completion of $\mathcal{M}$. As this group completion is the trivial group, all the $\tilde{\mathrm{B}}\alpha^+$ are homotopic to the identity. We now show that $\mathrm{B}\tilde{\mathrm{GL}}(A)^+$ is an H-group. It suffices to prove,

Claim Let (X, e, μ) be an H-space in $\mathcal{S}^{*I}$, with X bifibrant. Suppose that $|X(i)|$ is connected for each $i \in I$. Then there is a map

$$\iota \colon X \to X$$

which is a homotopy inverse for μ.

Proof of Claim $X \times X$ is fibrant. Let

$$p \colon X \times X \to X \times X$$

be the map

$$p = (\mu, p_2).$$

Applying the classical result for connected H-spaces (in **Top**), the map

$$|p(i)| \colon |X(i)| \times |X(i)| \to |X(i)| \times |X(i)|$$

is a weak equivalence for each $i \in I$, hence p is a weak equivalence in $\mathcal{S}^{*I}$.

Since X is cofibrant and $X \times X$ is fibrant there is an up-to-homotopy lifting

$$\phi \colon X \to X \times X$$

in the diagram

$$\begin{array}{ccc} & & X \times X \\ & & \big\downarrow p \\ X = * \times X & \xrightarrow{e \times \mathrm{id}_X} & X \times X. \end{array}$$

Then $\pi \circ \phi$ is homotopic to (ι, id_X), for some map $\iota \colon X \to X$, and ι is thus a left homotopy inverse for μ. As μ is homotopy associative, ι is the desired homotopy inverse. □

For a ring R, let $\mathrm{GL}_{n,m}(R)$ be the subgroup of $\mathrm{GL}_{n+m}(R)$ consisting of matrices of the form

$$\begin{pmatrix} g_n & a_{n,m} \\ 0 & g_m \end{pmatrix}$$

with $g_n \in \mathrm{GL}_n(R)$, $g_m \in \mathrm{GL}_m(R)$, and $a_{n,m}$ an $n \times m$ matrix. The shuffle sum

$$\mathrm{GL}_n(R) \times \mathrm{GL}_m(R) \to \mathrm{GL}(R)$$

extends in the obvious way to the inclusion

$$s_{n,m} \colon \mathrm{GL}_{n,m}(R) \to \mathrm{GL}(R).$$

Classifying, applying Lemma 3.1, and taking $+$, we have the map

$$Bs^+_{n,m} \colon \mathrm{BGL}_{n,m}(R) \to \mathrm{BGL}(R)^+.$$

Similarly, the stabilization map

$$\mathrm{GL}_n(R) \times \mathrm{GL}_m(R) \to \mathrm{GL}_{n+1}(R) \times \mathrm{GL}_{m+1}(R)$$

extends to

$$\mathrm{BGL}_{n,m}(R) \to \mathrm{BGL}_{n+1,m+1}(R);$$

let $\mathrm{GL}_{*,*}(R)$ be the direct limit of the $\mathrm{GL}_{n,m}(R)$. We let $j_{n,m}:\mathrm{GL}_n(R)\times\mathrm{GL}_m(R)\to\mathrm{GL}_{n,m}(R)$ denote the inclusion,

$$r_{n,m}\colon \mathrm{GL}_{n,m}(R)\to \mathrm{GL}_n(R)\times\mathrm{GL}_m(R)$$

$$\begin{pmatrix} g_n & a_{n,m}\\ 0 & g_m\end{pmatrix}\mapsto (g_n, g_m)$$

the projection.

If $A\colon I\to$ **Rings** is a functor, the above constructions generalize to define the groups $\mathrm{GL}_{n,m}(A)$, maps

$$s_{n,m}\colon \mathrm{GL}_{n,m}(A)\to \mathrm{GL}(A),$$

etc. We may take as in (3.3)-(3.7) a system of bifibrant models $\mathrm{B\tilde{G}L}_{n,m}(A)$ of $\mathrm{BGL}_{n,m}(A)$, together with a compatible system of cofibrations ($n'\geq n$, $m'\geq m$)

$$\mathrm{B\tilde{G}L}_{n,m}(A)\to\mathrm{B\tilde{G}L}_{n',m'}(A)\to\ldots\to\mathrm{B\tilde{G}L}_{*,*}(A)$$

$$\mathrm{BGL}_n(A)\tilde{\times}\mathrm{BGL}_m(A)\to\mathrm{BGL}_{n'}(A)\tilde{\times}\mathrm{BGL}_{m'}(A)\to\ldots$$

and compatible maps

$$\mathrm{BGL}_n(A)\tilde{\times}\mathrm{BGL}_m(A)\xrightarrow{\tilde{\mathrm{B}}j_{n,m}}\mathrm{B\tilde{G}L}_{n,m}(A)$$

$$\mathrm{BGL}(A)\tilde{\times}\mathrm{BGL}(A)\xrightarrow{\tilde{\mathrm{B}}j_{*,*}}\mathrm{B\tilde{G}L}_{*,*}(A)$$

and

$$\mathrm{B\tilde{G}L}_{n,m}(A)\xrightarrow{\tilde{\mathrm{B}}r_{n,m}}\mathrm{BGL}_n(A)\tilde{\times}\mathrm{BGL}_m(A);$$

as well as a commutative diagram

$$\begin{array}{ccc}\mathrm{B\tilde{G}L}_{n,m}(A) & \xrightarrow{\tilde{\mathrm{B}}s^+_{n,m}} & \mathrm{B\tilde{G}L}(A)^+\\ \downarrow & & \downarrow\\ \mathrm{B\tilde{G}L}_{*,*}(A) & \xrightarrow{\tilde{\mathrm{B}}s^+_{*,*}} & \mathrm{B\tilde{G}L}(A)^+\end{array}$$

Lemma 3.4 *The maps*

$$\tilde{\mathrm{B}}s^+_{n,m}\colon \mathrm{B\tilde{G}L}_{n,m}(A)\to\mathrm{B\tilde{G}L}(A)^+$$

and

$$\tilde{\mathrm{B}}s^+_{n,m}\circ\tilde{\mathrm{B}}j_{n,m}\circ\tilde{\mathrm{B}}r_{n,m}\colon \mathrm{B\tilde{G}L}_{n,m}(A)\to\mathrm{B\tilde{G}L}(A)^+$$

are homotopic for all m,n.

Proof From Quillen [21], we know that the map

$$\mathrm{B}j_{*,*}\colon \mathrm{BGL}(R)\times\mathrm{BGL}(R)\to\mathrm{BGL}_{*,*}(R)$$

induces a weak equivalence

$$\mathrm{B}j^*_{*,*}\colon \mathcal{H}om_{\mathcal{S}^*}(\mathrm{BGL}_{*,*}(R),\mathrm{B\tilde{G}L}(R')^+)\to\mathcal{H}om_{\mathcal{S}^*}(\mathrm{BGL}(R)\times\mathrm{BGL}(R),\mathrm{B\tilde{G}L}(R')^+)$$

for all rings R and R', where $\mathrm{B\tilde{G}L}(R')^+$ is a fibrant model. We may apply Corollary B.4, which implies that

$$\tilde{\mathrm{B}}j^*_{*,*}\colon \mathcal{H}om_{\mathcal{S}^{*I}}(\mathrm{B\tilde{G}L}_{*,*}(A),\mathrm{B\tilde{G}L}(A)^+)\to\mathcal{H}om_{\mathcal{S}^{*I}}(\mathrm{BGL}(A)\tilde{\times}\mathrm{BGL}(A),\mathrm{B\tilde{G}L}(A)^+)$$

is a weak equivalence. As the maps $\tilde{\mathrm{B}}s_{n,m}$ factor through $\mathrm{B}\tilde{\mathrm{GL}}_{*,*}(A)$ this, together with the commutative diagrams given above, completes the proof. □

4 λ-operations

A simplicial set X is called *finite* if X has only finitely many non-degenerate simplices. We have the full subcategories $\mathcal{S}_{\mathrm{fin}}$ and $\mathcal{S}_{\mathrm{c}}$ of $\mathcal{S}$, consisting respectively of the finite and the connected simplicial sets; we let $\mathcal{S}_{\mathrm{cfin}}$ denote the intersection $\mathcal{S}_{\mathrm{fin}} \cap \mathcal{S}_{\mathrm{c}}$ This gives us the functor category $\mathcal{S}_{\mathrm{c}}^I$, which is a full subcategory of $\mathcal{S}^I$.

Call an object X of $\mathcal{S}^I$ *finite* if $X(i)$ is finite for each $i \in I$. We let $\mathcal{S}_{\mathrm{fin}}^I$ denote the full subcategory of $\mathcal{S}^I$ consisting of finite cofibrant objects, and let $\mathcal{S}_{\mathrm{bfin}}^I$ denote the full subcategory of $\mathcal{S}^I$ with objects X such that

i) X is bifibrant

ii) there is a weak equivalence $f\colon X_0 \to X$, with X_0 in $\mathcal{S}_{\mathrm{fin}}^I$.

We note that (ii) implies the map

$$f^*\colon \mathcal{H}om(X, Z) \to \mathcal{H}om(X_0, Z)$$

is a weak equivalence for all fibrant Z. Indeed, factor f as uv with

$$u\colon X_0 \to X'$$

a trivial cofibration, and

$$v\colon X' \to X$$

a trivial fibration. Then X' is also bifibrant, hence v is a homotopy equivalence. Thus, we may assume that f is a trivial cofibration; as Z is fibrant, f^* is a trivial fibration by axiom SM7.

We let $\mathcal{S}_c^I$, $\mathcal{S}_{\mathrm{cfin}}^I$ and $\mathcal{S}_{\mathrm{cbfin}}^I$ be the full subcategories of $\mathcal{S}^I$, $\mathcal{S}_{\mathrm{fin}}^I$ and $\mathcal{S}_{\mathrm{bfin}}^I$ of X such that $|X(i)|$ is connected for each $i \in I$. We have the pointed versions defined similarly.

We now suppose I is a *finite* category, i.e., that the nerve of I is a finite simplicial set. This implies that I is *direct* (in the sense of [6]), i.e., that the nerve $\mathcal{N}(I/i)$ of the over category I/i is finite dimensional for each $i \in I$. Concretely, the condition that I is finite is simply that I has finitely many objects, finitely many morphisms, and there is an N such that each sequence of morphisms

$$i_0 \xrightarrow{f_0} i_1 \xrightarrow{f_1} \ldots \xrightarrow{f_N} i_{N+1}$$

contains an identity morphism.

For a direct category J, we let J_n be the full subcategory of objects i such that

$$\dim \mathcal{N}(J/i) \leq n.$$

and J^n the set of objects with

$$\dim \mathcal{N}(J/i) = n.$$

If X is in $\mathcal{S}^J$, and i is in J^n, have the canonical map

$$\lim_{\substack{\rightarrow \\ (J/i)_{n-1}}} X/i \to X(i),$$

where

$$X/i\colon J/i \to \mathcal{S}$$

is the functor

$$X(s\colon j \to i) = X(j).$$

We write the inductive limit as

$$X(J//i) := \varinjlim_{(J/i)_{n-1}} X/i.$$

Lemma 4.1 *Each cofibrant object X in $\mathcal{S}^I$ is an inductive limit:*

$$X = \varinjlim_{\alpha \in A} Z_\alpha; \qquad Z_\alpha \in \mathcal{S}^I_{\mathrm{fin}},$$

with A a totally ordered set, and

$$\begin{array}{ccc} Z_\alpha & \longrightarrow & Z_\beta \\ & {\scriptstyle i_\alpha}\searrow \quad \swarrow{\scriptstyle i_\beta} & \\ & X & \end{array}$$

a commutative diagram of cofibrations in $\mathcal{S}^I$ for $\alpha \leq \beta \in A$. If $(X, *)$ is a cofibrant object of $\mathcal{S}^{*I}$, and each $X(i)$ is connected, then the above holds for $(X, *)$, where we may take the inductive limit in $\mathcal{S}^{*I}_{\mathrm{cfin}}$.

Proof We proceed by induction on the number of objects of I. If $I = I_0$, then there are no non-identity maps in I, hence the result follows from the analogous statement in $\mathcal{S}$. In general $I = I_N$ for some minimal N; take some $i \in I^N$, and consider the category $J := I \backslash \{i\}$.

If j is in J, there are no morphisms $i \to j$ in I; if there were, this would put j in I^M for some $M > N$. The restriction functor

$$\mathrm{res}^I_J \colon \mathcal{S}^I \to \mathcal{S}^J$$

has the left adjoint

$$\mathrm{cor}^J_I \colon \mathcal{S}^J \to \mathcal{S}^I$$

defined by

$$\mathrm{res}^I_J \mathrm{cor}^J_I = \mathrm{id}$$

$$\mathrm{cor}^J_I(X)(i) = X(I//i).$$

As res^I_J preserves fibrations and weak equivalences, cor^J_I sends cofibrations to cofibrations. It is easy to see that, if $f \colon B \to C$ is a map in $\mathcal{S}^I$, then f is a cofibration if and only if

i) $\mathrm{res}^I_J f$ is a cofibration in $\mathcal{S}^J$, and
ii) the natural map

$$C(I//i) \cup_{B(I//i)} B(i) \to C(i)$$

is a cofibration in $\mathcal{S}$.

In addition, the natural map

$$f(I//i) \colon B(I//i) \to C(I//i)$$

is a cofibration in $\mathcal{S}$.

If now X is a cofibrant object of $\mathcal{S}^I$, we may write $\mathrm{res}^I_J X$ as the nested inductive limit of cofibrations

$$i_\alpha \colon Z^J_\alpha \to \mathrm{res}^I_J X; \alpha \in A_J$$

with Z^J_α in $\mathcal{S}^J_{\mathrm{fin}}$, which satisfy the various conditions of the lemma. Since I is finite, the objects $\mathrm{cor}^J_I Z^J_\alpha$ are in $\mathcal{S}^I_{\mathrm{fin}}$. Thus $\mathrm{cor}^J_I \mathrm{res}^I_J X$ is the inductive limit of the cofibrations

$$\mathrm{cor}^J_I Z^J_\alpha \to \mathrm{cor}^J_I \mathrm{res}^I_J X.$$

Evaluating at i, this writes the subcomplex $X(I//i)$ of $X(i)$ as the nested inductive limit of subcomplexes

$$X(I//i) = \lim_{\overrightarrow{\alpha}} Z^J_\alpha(I//i). \tag{4.1}$$

We may write $X(i)$ as a nested inductive limit of finite subcomplexes

$$X(i) = \lim_{\overrightarrow{\beta}} Z^i_\beta$$

indexed by a totally ordered set B. For each $\beta \in B$, we may find an $\alpha_\beta \in A_J$ such that the map

$$X(I//i) \cup_{Z^J_\alpha(I//i)} [Z^J_\alpha(I//i) \cup Z^i_\beta] \to X(i)$$

is a cofibration for all $\alpha \geq \alpha_\beta$. Let $A_J B$ be the partially ordered set consisting of (α, β) with $\alpha \geq \alpha_\beta$; for $(\alpha, \beta) \in A_J B$, let $Z_{\alpha\beta}$ be the object of $\mathcal{S}^I$ given by

$$Z_{\alpha\beta}(j) = \begin{cases} Z^J_\alpha(j) & \text{for } j \neq i, \\ Z^J_\alpha(I//i) \cup Z^i_\beta & \text{for } j = i, \end{cases}$$

with the obvious maps. We thus have X written as the inductive limit

$$X = \lim_{\overrightarrow{(\alpha,\beta) \in A_J B}} Z_{\alpha\beta},$$

via commutative diagrams of cofibrations in $\mathcal{S}^I$

$$\begin{array}{ccc} Z_{\alpha\beta} & \longrightarrow & Z_{\alpha'\beta'} \\ & \searrow \quad \swarrow & \\ & X & \end{array}$$

for $\alpha\beta \leq \alpha'\beta'$ in $A_J B$. By an obvious diagonalization, we may find a totally ordered subset A of $A_J B$, which gives a cofinal subfamily of the family $\{Z_{\alpha\beta}\}$.

The proof for $(X, *)$ in $\mathcal{S}^{*I}_{\mathrm{c}}$ is similar, and is left to the reader. □

We now proceed to make minor modifications in the argument of Quillen [12] to extend the universality statement of ([12], Corollary 2.4) to our setting. Let F denote, as in ([12], p. 245), the two-skeleton of the simplicial set BA_5, and $F \to F^+$ a simplicial version of the $+$-construction on F: $F \to F^+$ is a cofibration of finite pointed connected simplicial sets. Applying the functor (B.3)(ii)

$$(-) \wedge \mathcal{N}(I/-) \colon \mathcal{S}^* \to \mathcal{S}^{*I}$$

to $F \to F^+$, we have the cofibration

$$i \colon \mathcal{F} \to \mathcal{F}^+$$

of cofibrant objects of $\mathcal{S}^{*I}_{\mathrm{c}}$, which is weakly equivalent to the map

$$F \times I \to F^+ \times I.$$

As I is finite dimensional, i is a map in $\mathcal{S}^{*I}_{\rm cfin}$. The natural maps

$$F \to \mathrm{BGL}(R)$$

gives rise to map

$$j\colon \mathcal{F} \to \mathrm{BGL}(A).$$

Replacing $\mathrm{BGL}(A)$ with a weakly equivalent model, and changing notation, we may assume that j is a cofibration; in particular, that $\mathrm{BGL}(A)$ is cofibrant. By ([12], p. 245), the push-out

$$\mathrm{BGL}(A) \to \mathrm{BGL}(A) \cup_{\mathcal{F}} \mathcal{F}^{+} \tag{4.2}$$

is (point-wise) weakly equivalent to the map

$$\mathrm{BGL}(A) \to \mathrm{BGL}(A)^{+}.$$

By Lemma 3.1, a bifibrant model of (4.2) is weakly equivalent to

$$\tilde{i}\colon \mathrm{B\tilde{G}L}(A) \to \mathrm{B\tilde{G}L}(A)^{+}.$$

We now apply Lemma 4.1, writing $\mathrm{BGL}(A)$ as the nested inductive limit of cofibrations

$$j_\alpha\colon \mathcal{F}_\alpha \to \mathrm{BGL}(A);$$

with $\mathcal{F}_\alpha$ in $\mathcal{S}^{*I}_{\rm cfin}$; we may assume that j factors (uniquely) through each j_α.

The geometric realization functor

$$|-|\colon \mathcal{S}^I \to \mathbf{Top}^I$$

has the singular complex functor

$$\mathrm{Sin}\colon \mathbf{Top}^I \to \mathcal{S}^I$$

as right adjoint. For each $T \in \mathbf{Top}^I$, the object $\mathrm{Sin}T$ is fibrant in $\mathcal{S}^I$. The functor $|-|$ commutes with direct limits. This is not in general the case for the functor Sin, however, if a simplicial complex X is written as an nested inductive limit of subcomplexes

$$X = \cup_{\alpha\in A} Z_\alpha$$

$$Z_\alpha \subset \ldots \subset Z_\beta \subset \ldots \subset X; \quad \alpha \le \beta \in A,$$

indexed by a totally ordered set A, then we have

$$\mathrm{Sin}|X| = \varinjlim_{\alpha\in A} \mathrm{Sin}|Z_\alpha|.$$

Take B in $\mathcal{S}^I_{\rm fin}$, and let $\tilde{B}$ be a bifibrant model. Let

$$\tilde{\mathcal{F}}_\alpha \to \mathrm{Sin}|\mathcal{F}_\alpha|$$

$$\mathcal{F}_\alpha \,\tilde{\cup}_{\mathcal{F}}\, \mathcal{F}^{+} \to \mathrm{Sin}|\mathcal{F}_\alpha \cup_{\mathcal{F}} \mathcal{F}^{+}|$$

be trivial fibrations, with $\tilde{\mathcal{F}}_\alpha$ and $\mathcal{F}_\alpha \tilde{\cup}_{\mathcal{F}} \mathcal{F}^+$ bifibrant. Then we have the isomorphisms (for both the pointed and unpointed homotopy classes)

$$\begin{aligned}
[\tilde{B}, \mathrm{B}\tilde{\mathrm{GL}}(A)^+] &\cong [B, \mathrm{B}\tilde{\mathrm{GL}}(A)^+] \\
&\cong [B, \mathrm{Sin}|\mathrm{BGL}(A) \cup_{\mathcal{F}} \mathcal{F}^+|] \\
&\cong \varinjlim_{\alpha} [B, \mathrm{Sin}|\mathcal{F}_\alpha \cup_{\mathcal{F}} \mathcal{F}^+|] \\
&\cong \varinjlim_{\alpha} [B, \mathcal{F}_\alpha \tilde{\cup}_{\mathcal{F}} \mathcal{F}^+] \\
&\cong \varinjlim_{\alpha} [\tilde{B}, \mathcal{F}_\alpha \tilde{\cup}_{\mathcal{F}} \mathcal{F}^+].
\end{aligned} \tag{4.3}$$

Similarly, we have the isomorphism

$$[\tilde{B}, \mathrm{B}\tilde{\mathrm{GL}}(A)] \cong \varinjlim_{\alpha} [\tilde{B}, \tilde{\mathcal{F}}_\alpha]. \tag{4.4}$$

Let G be in $\mathbf{Groups}^I$, P a finitely generated projective S-module, and

$$\rho \colon G \to \mathrm{Aut}(P \otimes_S A)$$

be a representation of G on the A-module $P \otimes_S A$. Let Q be a finitely generated projective S-module with $P \oplus Q$ isomorphic to S^N, and choose an isomorphism $P \oplus Q \to S^N$. The representation $\rho \oplus \mathrm{id}_Q$ then determines the homomorphism

$$\tilde{\rho}(A) \colon G \to \mathrm{GL}_N(A).$$

Classifying, applying Lemma 3.1, and taking $+$, we have the map in $\mathcal{S}^{*I}$

$$\tilde{\mathrm{B}}\tilde{\rho}^+ \colon \tilde{\mathrm{B}}G \to \mathrm{B}\tilde{\mathrm{GL}}_N(A)^+, \tag{4.5}$$

uniquely determined up to homotopy by $\tilde{\rho}$. We may then compose with

$$\tilde{\iota}_N \colon \mathrm{B}\tilde{\mathrm{GL}}_N(A)^+ \to \mathrm{B}\tilde{\mathrm{GL}}(A)^+.$$

Lemma 4.2 *The (pointed) homotopy class of $\tilde{\iota}_N \circ \tilde{\mathrm{B}}\tilde{\rho}^+$ is uniquely determined by ρ.*

Proof If Q' is another choice of inverse to P, we have

$$Q \oplus S^m \cong Q' \oplus S^{m'}$$

for certain m and m'. If ρ_1 and ρ_2 are isomorphic representations

$$\rho_i \colon G \to \mathrm{GL}_{M/S},$$

then they differ by conjugation by a matrix in $\mathrm{GL}_M(S)$. Thus, the choices made alter $\tilde{\mathrm{B}}\tilde{\rho}^+$ by stabilizing, and by conjugation by some section g of $\mathrm{GL}_M(A)$ over I. Applying Lemma 3.2 completes the proof. □

Let $\mathcal{C}$ be the category of pairs (Z, B_Z), with Z a fibrant object in $\mathcal{S}^I$, B_Z a set of base points

$$s \colon * \to Z; s \in B_Z,$$

such that $\pi_1(|Z(i)|, s(i))$ has no non-trivial perfect subgroup for each i in I. Maps $(Z, B_Z) \to (W, B_W)$ are given by pairs of maps $h \colon B_Z \to B_W$, $g \colon Z \to W$ such that g defines a pointed map $(Z, s) \to (W, h(s))$ for each $s \in B_Z$.

For $(X,*)$ in $\mathcal{S}^{*I}$, set

$$[X,(Z,B_Z)]_* = \coprod_{s\in B_Z} [(X,*),(Z,s)].$$

We usually omit the B_Z from the notation. The set of base-points $K_0(A/S)\times *$ for $K_0(A/S)\times \mathrm{B\tilde{G}L}(A)^+$ makes $K_0(A/S)\times \mathrm{B\tilde{G}L}(A)^+$ an object of $\mathcal{C}$.

Let $(X,*)$ be in $\mathcal{S}^{*I}_{\mathrm{cfin}}$. Define the map

$$\mathrm{Rep}_S(\pi_1(X,*);A) \to [X, K_0(A/S)\times \mathrm{B\tilde{G}L}(A)^+]_*$$

by sending

$$\rho\colon \pi_1(X,*) \to \mathrm{Aut}(P\otimes_S A)$$

to the map (see (2.3), (3.7) and (4.4))

$$([P], \tilde{\iota}_N \circ \mathrm{B}\tilde{\rho}^+ \circ q_X).$$

This determines the natural transformation of functors

$$q^s\colon \mathrm{Rep}_S(\pi_1(-);A) \to [-, K_0(A/S)\times \mathrm{B\tilde{G}L}(A)^+]_* \tag{4.6}$$

from $[\mathcal{S}^{*I}_{\mathrm{cfin}}]^{\mathrm{op}}$ to **Sets**.

We let $\mathrm{Ho}\mathcal{S}^{*I}_{\mathrm{cbfin}}$ denote the full subcategory of $\mathrm{Ho}\mathcal{S}^{*I}$ with objects the objects of $\mathcal{S}^{*I}_{\mathrm{cbfin}}$.

Theorem 4.3 *The natural transformation (4.6) extends to the natural transformation*

$$q^s\colon R_S(\pi_1(-);A) \to [-, K_0(A/S)\times \mathrm{B\tilde{G}L}(A)^+]_* \tag{4.7}$$

of contravariant functors from $\mathrm{Ho}\mathcal{S}^{*I}_{\mathrm{cbfin}}$ *to* **Sets**, *which is universal relative to* $\mathcal{C}$*: for each natural transformation*

$$g\colon R_S(\pi_1(-);A) \to [-,Z]_*$$

of contravariant functors from the pointed homotopy category $\mathrm{Ho}\mathcal{S}^{*I}_{\mathrm{cbfin}}$ *to* **Sets**, *with* Z *in* $\mathcal{C}$*, there is a unique natural transformation*

$$h\colon [-, K_0(A/S)\times \mathrm{B\tilde{G}L}(A)^+]_* \to [-,Z]_*$$

with $g = h\circ q^s$ *(see (2.4), (3.7) and (4.5)).*

Proof The arguments of ([12], Section 1, especially Proposition 1.1), combined with Theorem 3.3 and Lemma 3.4, give the extension of the functor (4.6) to the functor (4.7). The arguments of [12] (Theorem 2.2, Corollary 2.3 and Corollory 2.4), together with the results of Section 3 mentioned above, as well as Lemma 4.1 and Lemma 4.2, prove the universality: we replace Quillen's categories $\mathcal{C}_1$ and $\mathcal{C}_2$ with

$$\begin{aligned} \mathcal{C}_1 &:= \{(Z,*) \text{ in } \mathcal{S}^{*I} \mid Z \text{ is fibrant and } \pi_1(|Z(i)|,*(i)) \text{ has no} \\ &\qquad\qquad \text{non-trivial perfect subgroup for each } i\in I\} \\ \mathcal{C}_2 &:= \mathcal{C}_1, \end{aligned}$$

and replace $\mathcal{C}_3$ with our $\mathcal{C}$. In Quillen's proof of Theorem 2.2 of [12], we replace his F_α and $F_\alpha \cup_F F^+$ with our $\tilde{\mathcal{F}}_\alpha$ and $\mathcal{F}_\alpha \tilde{\cup}_{\mathcal{F}} \mathcal{F}^+$, use the isomorphisms (4.3) and (4.4), and replace his use of obstruction theory with Lemma 3.1. We use pointed maps and pointed homotopy classes of maps throughout.

The same argument shows that the natural transformation

$$[-,(\mathrm{B\tilde{G}L}(A))^n]_* \to [-,(\mathrm{B\tilde{G}L}(A)^+)^n]_* \tag{4.8}$$

is universal relative to $\mathcal{C}_1$, for all $n\geq 1$.

One notes that, for X in $\mathcal{S}^{*I}_{\text{cbfin}}$, we have

$$
\begin{aligned}
[X, \mathrm{B}\tilde{\mathrm{GL}}(A)]_* &\cong \mathrm{Hom}_{\mathbf{Groups}^I}(\pi_1(X), \mathrm{GL}) \\
&= \varinjlim_n \mathrm{Hom}_{\mathbf{Groups}^I}(\pi_1(X), \mathrm{GL}_n) \\
&= \varinjlim_n [X, \mathrm{B}\tilde{\mathrm{GL}}_n(A)]_*,
\end{aligned}
$$

since $\pi_1(X(i))$ is finitely generated for each i, and I is finite.

In Quillen's proof of ([12], Corollary 2.3), we replace his $\tilde{K}(-, A)$ with $[-, \mathrm{B}\tilde{\mathrm{GL}}(A)^+]_*$, and replace his $\tilde{R}(-, A)$ with the subgroup $\tilde{R}_S(-; A)$ of $R_S(-; A)$ generated by representations into A-modules of the form A^n. We also replace his statement that "products of universal transformations relative to $\mathcal{C}_2$ are universal relative to $\mathcal{C}_2$" with the universality of (4.8) for all n. The argument of ([12], Corollary 2.3), together with Lemma 3.2, then shows that the natural transformation

$$
\tilde{R}_S(\pi_1(-); A) \to [-, \mathrm{B}\tilde{\mathrm{GL}}(A)^+]_* \tag{4.9}
$$

defined as the composition

$$
\tilde{R}_S(\pi_1(-); A) \to R_S(\pi_1(-); A) \to [-, K_0(A/S) \times \mathrm{B}\tilde{\mathrm{GL}}(A)^+]_* \xrightarrow{p_{2*}} [-, \mathrm{B}\tilde{\mathrm{GL}}(A)^+]_*,
$$

is universal relative to $\mathcal{C}_2$; the change in the argument mentioned above accounts for the difference in our $\mathcal{C}_2$ from his.

Now suppose we have a natural transformation

$$
\eta\colon R_S(\pi_1(-); A) \to [-, (Z, B_Z)]_*
$$

with (Z, B_Z) in $\mathcal{C}$. Applying η to $(\mathcal{N}(I/-), *)$ (which is point-wise contractible) gives the map

$$
\eta(*)\colon K_0(A/S) = R_S(\mathrm{id}; A) \to B_Z.
$$

This, together with the universality of (4.9), shows that (4.7) is universal relative to $\mathcal{C}$. □

Theorem 4.4 *The functor*

$$
[-, K_0(A/S) \times \mathrm{B}\tilde{\mathrm{GL}}(A)^+]_*\colon \mathcal{S}^{*I}_{\text{cbfin}} \to \mathbf{Sets} \tag{4.10}
$$

admits a canonical structure of a functor from $\mathcal{S}^{*I}_{\text{cbfin}}$ to the category of special λ-rings such that the natural transformation (4.7) is a natural transformation of functors to the category of special λ-rings. In addition, this functor is natural in the pair (I, A).

Proof The first assertion follows directly from Theorem 4.3 and the arguments of *loc. cit.* The second follows similarly from Theorem 4.3, and the naturality of the functor

$$
R_S(-; A)\colon \mathbf{Groups}^I \to \mathbf{Rings}
$$

in (I, A). □

We now pass from the pointed category to the unpointed category.

Theorem 4.5 *The functor*

$$
[-, K_0(A/S) \times \mathrm{B}\tilde{\mathrm{GL}}(A)^+]\colon \mathcal{S}^{I}_{\text{fin}} \to \mathbf{Sets} \tag{4.11}
$$

admits a canonical structure of a functor from $\mathcal{S}^{I}_{\text{fin}}$ to the category of special λ-rings. In addition, this functor is natural in the pair (I, A).

Proof We extend the functor $[-,-]_*$ to finite disjoint unions of objects in $\mathcal{S}^{*I}_{\rm cbfin}$ by

$$\left[\coprod_{i=1}^N (X_i,*),-\right]_* = \prod_{i=1}^N [(X_i,*),-]_*.$$

The functor (4.10) then extends to a functor to special λ-rings in the obvious way.

Referring to (4.3), let

$$\tilde{W}_\alpha = \mathcal{F}_\alpha \tilde{\cup}_{\mathcal{F}} \mathcal{F}^+.$$

From (4.3) we have the canonical isomorphisms of functors from $\mathrm{Ho}\mathcal{S}^{*I}_{\rm cbfin}$ to pointed sets (resp. from $\mathcal{S}_{\rm fin}$ to **Sets**):

$$[-,\mathrm{B\tilde{GL}}(A)^+]_* \cong \lim_{\overrightarrow{\alpha}}[-,\tilde{W}_\alpha]_*$$

$$[-,\mathrm{B\tilde{GL}}(A)^+] \cong \lim_{\overrightarrow{\alpha}}[-,\tilde{W}_\alpha]. \tag{4.12}$$

Write the set $K_0(A/S)$ as a nested union of finite subsets S_α. The pointed version in (4.12) gives the isomorphisms

$$\mathrm{Nat}\left([-,K_0(A/S)\times \mathrm{B\tilde{GL}}(A)^+]_*,[-,K_0(A/S)\times \mathrm{B\tilde{GL}}(A)^+]_*\right)$$

$$\cong \lim_{\overleftarrow{\alpha}}\mathrm{Nat}([-,S_\alpha\times\tilde{W}_\alpha]_*,[-,K_0(A/S)\times \mathrm{B\tilde{GL}}(A)^+]_*)$$

$$\cong \lim_{\overleftarrow{\alpha}}[S_\alpha\times\tilde{W}_\alpha,K_0(A/S)\times \mathrm{B\tilde{GL}}(A)^+]_*$$

Thus, by Theorem 4.4, the set

$$\lim_{\overleftarrow{\alpha}}[S_\alpha\times\tilde{W}_\alpha,K_0(A/S)\times \mathrm{B\tilde{GL}}(A)^+]_* \tag{4.13}$$

has the natural structure of a special λ-ring. Now, if B is in $\mathcal{S}^I_{\rm fin}$, the isomorphism (4.12) (in the unpointed case), the canonical pairing

$$\lim_{\overrightarrow{\alpha}}[B,S_\alpha\times\tilde{W}_\alpha]\times\lim_{\overleftarrow{\alpha}}[S_\alpha\times\tilde{W}_\alpha,K_0(A/S)\times\mathrm{B\tilde{GL}}(A)^+]_*\to[-,K_0(A/S)\times\mathrm{B\tilde{GL}}(A)^+]_*,$$

together with the natural λ-ring structure on (4.13), gives the natural special λ-ring structure on the unpointed homotopy set

$$[B,K_0(A/S)\times \mathrm{B\tilde{GL}}(A)^+].$$

□

Remark 4.6

i) We will usually use take $S=\mathbb{Z}$, in which case $K_0(A/\mathbb{Z})$ is $\mathbb{Z}$ by rank.

ii) Using the fact that the functor

$$\operatorname*{holim}_I\colon \mathcal{S}^I\to\mathcal{S}$$

is right adjoint to the functor (B.1)

$$(-)\times\mathcal{N}(I/-)\colon\mathcal{S}\to\mathcal{S}^I,$$

and noting that this latter functor sends $\mathcal{S}_{\rm fin}$ to $\mathcal{S}^I_{\rm fin}$ in case I is finite, Theorem 4.5 gives a natural special λ-ring structure to the functor

$$[-,\operatorname*{holim}_I K_0(A/S)\times \mathrm{B\tilde{GL}}(A)^+]\colon\mathcal{S}_{\rm fin}\to\mathbf{Sets}.$$

iii) One may replace GL_n with SL_n, and get analogous results, with minor changes. Indeed, for a group G and a commutative ring R, we let $\mathrm{Rep}^0(G;R)$ denote the set of conjugacy classes (by $\mathrm{GL}_n(R)$) of homomorphisms

$$\rho\colon G \to \mathrm{SL}_n(R)$$

and $R^0(G;R)$ the Grothendieck group of the $\mathrm{Rep}^0(G;R)$. The same argument as in this and the preceding section gives $\mathrm{B\tilde{S}L}(A)^+$ the structure of an H-group, and we have the universal natural transformation

$$q_0^s : R^0(\pi_1(-);A) \to [-,\mathbb{Z}\times \mathrm{B\tilde{S}L}(A)^+] \tag{4.14}$$

which gives a natural special λ-ring structure to the functor $[-,\mathbb{Z}\times\mathrm{B\tilde{S}L}(A)^+]$. The map

$$[-,\mathbb{Z}\times \mathrm{B\tilde{S}L}(A)^+] \to [-,\mathbb{Z}\times \mathrm{B\tilde{G}L}(A)^+]$$

is a map of functors from $\mathcal{S}_{\mathrm{fin}}$ to special λ-rings.

□

5 λ-operations on the K-theory of Schemes and Variants

Let X be a (noetherian, separated) scheme over a commutative noetherian ring S. We let $K^Q(X)$ denote the Quillen K-theory space of X:

$$K^Q(X) = \Omega\mathrm{BQ}\mathcal{P}_X$$

where $\mathcal{P}_X$ is the category of locally free coherent sheaves on X. If we have a small category I, and a functor

$$X\colon I \to \mathbf{Sch}$$

one can make $X \mapsto \mathcal{P}_X$ into a functor, giving the functor

$$\mathrm{BQ}\mathcal{P}_X : I \to \mathcal{S}.$$

We let

$$K^Q(X)\colon I \to \mathcal{S}$$

be a bifibrant model of the loop space functor applied to a fibrant model of $\mathrm{BQ}\mathcal{P}_X$, and let $K_0(A)$ be the functor

$$i \mapsto K_0(A(i)).$$

Lemma 5.1 *Let I be a small category, $A\colon I \to$ **Rings** a functor. Let $\mathrm{B\tilde{G}L}_n(A)^+$, $\mathrm{B\tilde{G}L}(A)^+$ be bifibrant models of $\mathrm{BGL}_n(A)^+$, $\mathrm{BGL}(A)^+$, resp., with the sequence of cofibrations*

$$\begin{array}{ccccc} \mathrm{B\tilde{G}L}_n(A)^+ & \longrightarrow & \mathrm{B\tilde{G}L}_{n+1}(A)^+ & \longrightarrow & \cdots \\ & \searrow & \downarrow & & \\ & & \mathrm{B\tilde{G}L}(A)^+ & & \end{array} \tag{5.1}$$

as in (3.3)-(3.7). Then there is a weak equivalence

$$\psi_A\colon K_0(A)\times \mathrm{B\tilde{G}L}(A)^+ \to K^Q(\mathrm{Spec}(A))$$

ψ_A is not necessarily natural in A, but the composition

$$K_0(A) \times \tilde{\mathrm{BGL}}_n(A)^+ \to K_0(A) \times \tilde{\mathrm{BGL}}(A)^+ \to K^Q(\mathrm{Spec}(A))$$

is natural in A, up to homotopy, for each n.

Proof The definition of the map ψ_A follows essentially from Grayson's article [8] relating $\Omega K^Q(\mathrm{Spec}R)$ and $\mathrm{BGL}(R)^+$ for a commutative ring R, together with some techniques from the theory of the simplicial closed model categories applied to $\mathcal{S}^I$. Indeed, for a small category I, and a functor

$$A\colon I \to \mathbf{Rings}$$

we have the functors

$$\begin{array}{ll} P(A)\colon I \to \mathbf{cat}; & S(A)\colon I \to \mathbf{cat} \\ QP(A)\colon I \to \mathbf{cat}; & E(A)\colon I \to \mathbf{cat} \end{array}$$

where, for a ring R, $P(R)$ is the category of finitely generated projective R-modules, $S(R)$ is the category $\mathrm{Iso}(P(R))$, with the same objects as $P(R)$, with only isomorphisms for morphisms, $QP(R)$ is Quillen's Q-construction on $P(R)$, and $E(R)$ is the extension construction on $S(R)$ ([8], p. 226). One needs to fiddle a bit to make P into a functor, but this is easily done. We have the commutative diagram of functors from I to **cat** ([8], p. 228):

$$\begin{array}{ccc} S^{-1}S(A) & \longrightarrow & S^{-1}E(A) \\ \downarrow & & \downarrow \\ * & \longrightarrow & QP(A) \end{array}$$

taking nerves gives the resulting diagram in $\mathcal{S}^{*I}$, which we write the same way. If we pass to a commuting diagram of bifibrant models in $\mathcal{S}^{*I}$:

$$\begin{array}{ccc} \widetilde{S^{-1}S(A)} & \longrightarrow & \widetilde{S^{-1}E(A)} \\ \downarrow & & \downarrow \\ * & \longrightarrow & \widetilde{QP(A)} \end{array} \tag{5.2}$$

the first two theorems of ([8], p. 228) imply that the diagram (5.2) is homotopy cartesian. Applying the results of ([5], XI, 5.6), taking homotopy limits results in a weak equivalence

$$\beta(A)\colon \operatorname{holim}_I \widetilde{S^{-1}S(A)} \to \operatorname{holim}_I \Omega\widetilde{\widetilde{QP}(A)} = \operatorname{holim}_I K^Q(A).$$

On the other hand, for a ring R, there is, for each n, a canonical map

$$\mathrm{BGL}_n(R) \to S^{-1}S(R)_0 \tag{5.3}$$

(see [8], p. 224), which is natural, up to homotopy, with respect to the stabilization maps

$$\mathrm{BGL}_n(R) \to \mathrm{BGL}_{n+1}(R).$$

Taking a functor $A\colon I \to \mathbf{Rings}$ as above, the maps (5.3) define maps

$$\alpha_n\colon K_0(A) \times \mathrm{BGL}_n(A) \to S^{-1}S(A);$$

if we pass to the sequence (5.1) of bifibrant models, the maps α_n induce (uniquely up to homotopy for each n) a sequence of maps

$$\tilde{\alpha}_n\colon K_0(A) \times \mathrm{B\tilde{G}L}_n(A) \to \widetilde{S^{-1}S}(A),$$

which strictly commute with the stabilization maps

$$\mathrm{B\tilde{G}L}_n(A) \to \mathrm{B\tilde{G}L}_{n+1}(A).$$

This gives a well-defined map on the inductive limit

$$\alpha_\infty\colon K_0(A) \times \lim_{\vec{n}} \mathrm{B\tilde{G}L}_n(A) \to \widetilde{S^{-1}S}(A).$$

In addition, it is easy to see that $\lim_{\vec{n}}\mathrm{B\tilde{G}L}_n(A)$ is cofibrant, and that the canonical map

$$\lim_{\vec{n}} \mathrm{B\tilde{G}L}_n(A) \to \mathrm{B\tilde{G}L}(A)$$

is a weak equivalence. This gives the map

$$\alpha\colon K_0(A) \times \mathrm{B\tilde{G}L}(A) \to \widetilde{S^{-1}S}(A),$$

which restricted to $K_0(A) \times \mathrm{B\tilde{G}L}_n(A)$ is homotopic to $\tilde{\alpha}_n$.

As $\widetilde{S^{-1}S}(A)$ is an H-space, the map α extends canonically (up to homotopy) to

$$\alpha^+(A)\colon K_0(A) \times \mathrm{B\tilde{G}L}(A)^+ \to \widetilde{S^{-1}S}(A),$$

by Lemma 3.1. Applying the theorem of ([8], p. 224), α^+ is a weak equivalence. Composing $\alpha^+(A)$ with $\beta(A)$ gives the desired weak equivalence.

For the naturality, the map β is canonical. It follows from the various uniqueness statements of Section 5, that the restriction of α^+ to $K_0(A) \times \mathrm{B\tilde{G}L}_n(A)$ is canonical, up to homotopy, for each n. Suppose we have a functor

$$f\colon J \to I$$

of small categories, a functor

$$B\colon J \to \mathbf{Rings}$$

and a natural transformation

$$\omega\colon B \to A \circ f.$$

Let $(f\colon J \to I)$ denote the category with objects $\mathrm{Obj}I \coprod \mathrm{Obj}J$, such that I and J are embedded as full subcategories, there are no maps $i \to j$ for $i \in I$ and $j \in J$, and where

$$\mathrm{Hom}(j,i) = \mathrm{Hom}_I(f(j),i).$$

The data (A,B,ω) determines the functor

$$(A,B,\omega)\colon (f\colon J \to I) \to \mathbf{Rings}.$$

The weak equivalence $\beta(A,B,\omega)\alpha^+(A,B,\omega)$ then gives the naturality of $\beta(-)\alpha^+(-)$. $\square$

Lemma 5.2 *Let I be a finite category, $A\colon I \to$ **Rings** a functor. For each B in $\mathcal{S}^I_{\text{fin}}$, there is an isomorphism*

$$\psi_{A,B}\colon [B, K_0(A) \times \tilde{\mathrm{B}}\mathrm{GL}(A)^+] \to [B, K^Q(A)]$$

*which is natural in B, and in functors $A\colon I \to$ **Rings** for finite categories I. In particular, for each finite simplicial complex B, there is an isomorphism*

$$\Psi_{A,B}\colon [B, \operatorname*{holim}_I K_0(A) \times \mathrm{BGL}(A)^+] \to [B, \operatorname*{holim}_I K^Q(A)]$$

*which is natural in B, and in functors $A\colon I \to$ **Rings** for finite categories I.*

Proof The first assertion follows from Lemma 5.1, noting that the natural map

$$\varinjlim_n [X, K_0(A) \times \mathrm{BGL}_n(A)^+] \to [X, K_0(A) \times \mathrm{BGL}(A)^+]$$

is an isomorphism. For the second, we note that the functor $\operatorname*{holim}_I$ is right adjoint to the functor (see Appendix B)

$$(-) \times \mathcal{N}(I/-)\colon \mathcal{S} \to \mathcal{S}^I$$

and that this functor sends $\mathcal{S}_{\text{fin}}$ to $\mathcal{S}^I_{\text{fin}}$. □

For a set T, we let $\mathcal{M}(T)$ denote the category with objects the elements $a_* := a_1 + \ldots + a_n$, $n = 0, 1, \ldots$, of the free abelian monoid $M(T)$ on T, with a unique morphism $a_* \to b_*$ if and only if there is a $c_* \in M(T)$ with

$$b_* = a_* + c_*.$$

We let $<T>$ denote the full subcategory of $\mathcal{M}(T)$ with objects the sums $a_1 + \ldots + a_n$ with *distinct* a_i in T (and the empty sum).

Let $\mathfrak{U} = \{U_\alpha \mid \alpha \in T\}$ be an affine cover of X; $U_\alpha = \mathrm{Spec}(R_\alpha)$. Let $<\mathfrak{U}>$ be the full subcategory of $<T>$ consisting of the non-empty sums $\alpha_1 + \ldots + \alpha_n$. Let $R_\mathfrak{U}$ be the S-algebra over $<\mathfrak{U}>$ defined by

$$R_\mathfrak{U}(\alpha_1 + \ldots + \alpha_n) = \Gamma(\cap_j U_{\alpha_j}, \mathcal{O}_X)$$

(we make the convention that $\Gamma(\emptyset, \mathcal{O}_X)$ is the zero ring, and that $K_0(S) \times \mathrm{BGL}(0)^+$ is the one-point space).

Theorem 5.3 *Let X be a noetherian scheme over S which admits an ample family of line bundles, and let $\mathfrak{U}$ be a finite affine open cover of X. Then*

i) *There is a map*

$$\psi_\mathfrak{U}\colon \operatorname*{holim}_{<\mathfrak{U}>}(K_0(A/S) \times \mathrm{BGL}(R_\mathfrak{U})^+) \to K^Q(X).$$

For each finite simplicial complex B, the induced map on $[B, -]$ is natural in the triple $(B, \mathfrak{U}, X)$.

ii) *$\psi_\mathfrak{U}$ induces an isomorphism on homotopy groups π_n for $n > 0$ and an injection for $n = 0$; the image of $\psi_\mathfrak{U}$ on π_0 is the subgroup of $K_0(X)$ generated by locally free sheaves P on X such that $[P \otimes_{\mathcal{O}_X} \mathcal{O}_{U_\alpha}]$ is in the image of $K_0(S) \to K_0(U_\alpha)$ for each $\alpha \in A$. Furthermore, $\psi_\mathfrak{U}$ induces an isomorphism*

$$\varinjlim \pi_0(\operatorname*{holim}_{<\mathfrak{U}>}(K_0(A/S) \times \mathrm{BGL}(R_\mathfrak{U})^+)) \to K_0(X)$$

where the limit is over finite affine covers $\mathfrak{U}$ of X.

Proof We note that the category $<\mathfrak{U}>$ is finite. By the two previous lemmas, we need only show that the natural map

$$K^Q(X) \to \underset{I}{\operatorname{holim}}\, K^Q(R_{\mathfrak{U}}) \tag{5.4}$$

is a weak equivalence, and compute the difference between

$$\pi_0 \underset{<\mathfrak{U}>}{\operatorname{holim}} K_0(R_{\mathfrak{U}}/S) \times \mathrm{BGL}(R_{\mathfrak{U}})^+$$

and

$$\pi_0 \underset{<\mathfrak{U}>}{\operatorname{holim}} K_0(R_{\mathfrak{U}}) \times \mathrm{BGL}(R_{\mathfrak{U}})^+.$$

In the case of regular X, (5.4) being a weak equivalence is a direct consequence of the Mayer-Vietoris spectral sequence arising from Quillen's localization theorem [20], which one compares with the Bousfield-Kan spectral sequence (Theorem B.1) for $\underset{<\mathfrak{U}>}{\overset{*}{\operatorname{holim}}}$. In fact, for a pre-sheaf $\mathcal{T}$ of simplicial spaces on X, we have the space $\mathcal{T}(\mathfrak{U})$ over $<\mathfrak{U}>$, gotten by taking global sections over the various intersections of elements of $\mathfrak{U}$. The Bousfield-Kan spectral sequence for

$$\underset{<\mathfrak{U}>}{\operatorname{holim}} \mathcal{T}(\mathfrak{U})$$

is exactly the Mayer-Vietoris spectral sequence for $\mathcal{T}$ with respect to the cover $\mathfrak{U}$.

In general, a similar argument using the Thomason-Trobaugh extension [26] of Quillen's localization theorem to the case of arbitrary schemes gives the proof; although the theorem of Thomason-Trobaugh uses spectra with (possibly) negative homotopy groups, the homotopy groups in non-negative degree agree with the Quillen K-groups, which implies that the natural map

$$K^Q(X) \to \underset{<\mathfrak{U}>}{\operatorname{holim}}(K^Q(-))$$

is a weak equivalence.

The difference in two π_0's is also readily computable from the Mayer-Vietoris spectral sequence, with the result as claimed. □

Corollary 5.4 Let X be a noetherian scheme over a noetherian ring S, admitting an ample family of line bundles. Then there is a special $K_0(S)$-λ-algebra structure on the K-groups of X over S, which is natural in X.

Proof By Remark 4.6(ii), there is a natural special $K_0(S)$-λ-algebra structure for the functor

$$[-, \underset{<\mathfrak{U}>}{\operatorname{holim}}\, (K_0(R_{\mathfrak{U}}/S) \tilde{\times} \mathrm{BGL}(R)^+)] \colon \mathcal{S}_{\mathrm{fin}} \to \mathbf{Sets}.$$

In particular, this gives a natural special $K_0(S)$-λ-algebra structure on the homotopy groups of the H-space

$$\underset{<\mathfrak{U}>}{\operatorname{holim}}\, (K_0(R_{\mathfrak{U}}/S) \tilde{\times} \mathrm{BGL}(R)^+).$$

The result then follows from Theorem 5.3. □

If U is an open subset of a scheme X, we have the K-theory space with supports in $W := X \setminus U$, $K^W(X)$, defined as the homotopy fiber of the natural map

$$i_U^* \colon K(X) \to K(U).$$

If we have an affine open cover of X as above, we have the ring $R_{\mathfrak{U}}$ over $<\mathfrak{U}>$, and similarly the ring $R_{U,\mathfrak{U}}$ over $<\mathfrak{U}>$ gotten by restricting the open cover to U.

We have the category $[0,1]$ associated to the ordered set $0 < 1$, and for each $n = 1, 2, \dots$ the n-cube $[0,1]^n$; we may identify $[0,1]^n$ with the category associated to the partially ordered set of subsets of $\{1, \dots, n\}$, ordered under inclusion.

The map $i_U^*\colon R_{\mathfrak{U}} \to R_{U,\mathfrak{U}}$ defines the ring $R_{\mathfrak{U}}^U$ over $<\mathfrak{U}> \times [0,1]$, where

$$(R \to R_U)_{|<\mathfrak{U}>\times 0} \quad = R_{\mathfrak{U}}$$

$$(R \to R_U)_{|<\mathfrak{U}>\times 1} \quad = R_{U,\mathfrak{U}}$$

As above, we may compute the component of the identity of $K^W(X)$ is given as the component of the identity of

$$\overset{*}{\operatorname{holim}}_{<\mathfrak{U}>\times[0,1]} K_0(R_{\mathfrak{U}}^U/S) \tilde{\times} \operatorname{BGL}(R_{\mathfrak{U}}^U)^+,$$

and $K_0^W(X)$ may be computed as a limit over finite open covers. Here $\overset{*}{\operatorname{holim}}$ is the pointed version of holim which gives the desired homotopy fiber. This gives

Corollary 5.5 Let X be a noetherian scheme over a noetherian ring S, admitting an ample family of line bundles, W a closed subset. There is a special $K_0(S)$-λ-algebra structure on the $K_p^W(X)$, which is natural in the pair (X, W). In addition, the long exact sequence

$$\dots \to K_p^W(X) \to K_p(X) \to K_p(X \backslash W) \to K_{p-1}^W(X) \to \dots \tag{5.5}$$

associated to the fiber sequence

$$K^W(X) \to K(X) \to K(U)$$

is a sequence of $K_0(S)$-λ-algebras.

Proof We may apply the special $K_0(S)$-λ-algebra structure on the functor

$$[-, \overset{*}{\operatorname{holim}}_{<\mathfrak{U}>\times[0,1]} K_0(R_{\mathfrak{U}}^U/S) \tilde{\times} \operatorname{BGL}(R_{\mathfrak{U}}^U)^+)],$$

given by Theorem 4.5 and Remark 4.6(ii), to give a natural special $K_0(S)$- λ-algebra structure on the homotopy groups of

$$\overset{*}{\operatorname{holim}}_{<\mathfrak{U}>\times[0,1]} K_0(R_{\mathfrak{U}}^U/S) \tilde{\times} \operatorname{BGL}(R_{\mathfrak{U}}^U)^+.$$

The first assertion then follows from Theorem 5.3.

For the second, the naturality portion of Theorem 4.5 implies that the sequence of functors on finite simplicial sets

$$[-, K^W(X)] \to [-, K(X)] \to [-, K(U)]$$

is a sequence of functors from finite simplicial sets to $K_0(S)$- λ-algebras. This shows that the long exact sequence (5.5) is a sequence of $K_0(S)$-λ- algebras. □

Finally, if $Y_1, \dots, Y_n$ are closed subschemes of X, we may form the n-cube of spaces $K(X; Y_1, \dots, Y_n)_*$, with

$$K(X; Y_1, \dots, Y_n)_I = K(\cap_{i\in I} Y_i); \qquad I \subset \{1, \dots, n\}.$$

The relative K-theory space $K(X; Y_1, \dots, Y_n)$ defined as

$$K(X; Y_1, \dots, Y_n) = \overset{*}{\operatorname{holim}}_{[0,1]^n}(K(X; Y_1, \dots, Y_n)_*).$$

If U is an open subset of X, with complement W, we have the relative K-theory space with supports, $K^W(X; Y_1, \ldots, Y_n)$, defined as the homotopy fiber of

$$K(X; Y_1, \ldots, Y_n) \to K(U; Y_1 \cap U, \ldots, Y_n \cap U);$$

we may also consider $K^W(X; Y_1, \ldots, Y_n)$ as a (pointed) homotopy limit over the obvious $n+1$-cube. If we have an affine open cover of X as above, we may restrict this cover to each intersection $\cap_{i \in I} Y_i$ or $\cap_{i \in I} Y_i \cap U$, forming the ring

$$R^U_{\mathfrak{U}}(Y_1, \ldots, Y_n)$$

over $<\mathfrak{U}> \times [0,1]^n \times [0,1] = <\mathfrak{U}> \times [0,1]^{n+1}$. As above, the component of the identity of $K^W(X; Y_1, \ldots, Y_n)$ is given as the component of the identity of

$$\overset{*}{\operatorname{holim}}_{<\mathfrak{U}> \times [0,1]^{n+1}} K_0(R^U_{\mathfrak{U}}(Y_1, \ldots, Y_n)/S) \tilde{\times} \mathrm{BGL}(R^U_{\mathfrak{U}}(Y_1, \ldots, Y_n))^+,$$

and $K_0^W(X; Y_1, \ldots, Y_n)$ may be computed as a limit over finite open covers. This gives as before,

Corollary 5.6 Let X be a noetherian scheme over a noetherian ring S, admitting an ample family of line bundles. There is a special $K_0(S)$-λ-algebra structure for relative K-theory with supports:

$$K^W_p(X; Y_1, \ldots, Y_n),$$

which is natural in the tuple $(X, Y_1, \ldots, Y_n, W)$. The long exact relativization sequence

$$\ldots \to K^W_p(X; Y_1, \ldots, Y_n) \to K^W_p(X; Y_1, \ldots, Y_{n-1})$$
$$\to K^{W \cap Y_n}_p(X \cap Y_n; Y_1 \cap Y_n, \ldots, Y_{n-1} \cap Y_n) \to \ldots$$

and the long exact localization sequence

$$\ldots \to K^W_p(X; Y_1, \ldots, Y_n) \to K_p(X; Y_1, \ldots, Y_n)$$
$$\to K_p(X \backslash W; Y_1 \backslash W, \ldots, Y_n \backslash W) \to \ldots$$

are sequences of special $K_0(S)$-λ-algebras.

□

Part II: Motivic Cohomology

6 Adams Decompositions

For a set S of primes in $\mathbb{Z}$, let $\mathbb{Z}_S$ be the subring of $\mathbb{Q}$

$$\mathbb{Z}_S = \mathbb{Z}[\{\frac{1}{l} \mid l \in S\}].$$

Let $\mathcal{A}_S$ denote the category of $\mathbb{Z}_S$-modules M, together with a collection of endomorphisms

$$\psi^k_M : M \to M; \qquad k = 1, 2, \ldots$$

satisfying

$$\psi^k_M \circ \psi^l_M = \psi^{kl}_M$$

for all k and l. Morphisms $(M, \psi_M^*) \to (N, \psi_N^*)$ are $\mathbb{Z}_S$-module maps $\phi\colon M \to N$ with

$$\phi \circ \psi_M^k = \psi_N^k \circ \phi$$

for all k. Clearly, $\mathcal{A}_S$ is an abelian category, with forgetful functor to the abelian category of $\mathbb{Z}_S$-modules.

If N is an integer, we let $S(N)$ denote the set of primes p which divide N, and write $\mathcal{A}_N$ for $\mathcal{A}_{S(N)}$. Similarly, if T is a set of primes, we write (T) for the set of primes not in T, giving the category $\mathcal{A}_{(T)}$, and the category of $\mathbb{Z}_{(T)}$-modules. We write $\mathbb{Z}_{(l)}$ for $\mathbb{Z}_{(\{l\})}$ when l is a prime. We let ∞ denote the set of all primes.

Let M be a $\mathbb{Z}_S$-module, $\phi\colon M \to M$ a locally nilpotent endomorphism (for each $x \in M$, there is an N such that $\phi^N(x) = 0$). We have the increasing filtration on M

$$N_s(M, \phi) = \{m \in M \mid \phi^s(m) = 0\}$$

which is a finite filtration if and only if ϕ is nilpotent for all k.

Let (M, ψ_M^*) be in $\mathcal{A}_S$. An *Adams decomposition* of (M, ψ_M^*) is a finite direct sum decomposition of (M, ψ_M^*) into subobjects in $\mathcal{A}_S$:

$$(M, \psi_M^*) = \oplus_{i=a}^b (M_i, \psi_{M_i}^*); \qquad 0 \le a \le b,$$

such that

i) $\psi_{M_i}^k - k^i \mathrm{id}_{M_i}$ is nilpotent for all k.

ii) There is a finite, exhaustive, increasing filtration N_*M $(* \ge 0)$ on M such that $N_sM \cap M_i \subset N_s(M_i, \psi_{M_i}^k - k^i \mathrm{id}_{M_i})$ for all k.

We call $b - a + 1$ the *length* of the Adams decomposition; we call the summand $(M_i, \psi_{M_i}^*)$ the *summand of weight* i in the Adams decomposition, and say a decomposition of the form above has weights $a, \ldots, b$.

Let (M, ψ_M^*) be an object of $\mathcal{A}_S$. If $S' \supset S$, then an Adams decomposition of (M, ψ_M^*) induces an Adams decomposition of $(M \otimes \mathbb{Z}_{S'}, \psi_M^* \otimes \mathrm{id})$ by extension of scalars.

If (M, ψ_M^*) in $\mathcal{A}_S$ has an Adams decomposition as above, then the endomorphisms

$$\Psi_M^k(a, b) := \prod_{i=a}^b \psi_M^k - k^i \mathrm{id}_M$$

are all nilpotent, and $N_sM \subset N_s(M, \Psi_M^k(a, b))$ for all k. If we set $N_s^{\mathrm{can}}M$ equal to the intersection

$$N_s^{\mathrm{can}}M := \cap_k N_s(M, \Psi_M^k(a, b)),$$

then we may take $N_*M = N_*^{\mathrm{can}}M$ in (ii) above. We call $N_*^{\mathrm{can}}M$ the *canonical filtration* of (M, ψ_M^*), associated to the Adams decomposition $\oplus_{i=a}^b (M_i, \psi_{M_i}^*)$.

Lemma 6.1

(i) *Let (M, ψ_M^*) be in $\mathcal{A}_\infty$. An Adams decomposition of (M, ψ_M^*) is unique.*

ii) *Let l be a prime, and let (M, ψ_M^*) be in $\mathcal{A}_{(l)}$. Fix $0 \le a \le b$ with $b - a + 1 < l$. An Adams decomposition of (M, ψ_M^*) of the form*

$$(M, \psi_M^*) = \oplus_{i=a}^b (M_i, \psi_{M_i}^*)$$

is unique.

Proof (i) Take any $k \geq 2$. If (M, ψ^*_M) has an Adams decomposition

$$(M, \psi^*_M) = \oplus_{i=a}^b (M_i, \psi^*_{M_i})$$

then $\psi^k_M - k^i \mathrm{id}_M$ is nilpotent on M_i and invertible on M_j for $j \neq i$. Thus

$$M_i = \cup_{n=1}^\infty \ker[(\psi^k_M - k^i \mathrm{id}_M)^n]$$

This characterizes M_i.

(ii) The group of units in $\mathbb{Z}/l\mathbb{Z}$ is cyclic of order $l-1$, hence there is an integer k prime to l such that the powers k^i, $i = a, a+1, \dots, b$ are distinct modulo l. Thus, if we have an Adams decomposition

$$(M, \psi^*_M) = \oplus_{i=a}^b (M_i, \psi^*_{M_i})$$

then $\psi^k_M - k^i \mathrm{id}$ acts nilpotently on M_i and invertibly on M_j for $j \neq i$. We may then characterize M_i as above. □

Lemma 6.2 *Let (M, ψ^*_M) be in $\mathcal{A}_S$, and suppose*

i) $(M \otimes \mathbb{Z}_{(l)}, \psi^*_M \otimes \mathrm{id})$ *has an Adams decomposition*

$$(M \otimes \mathbb{Z}_{(l)}, \psi^*_M \otimes \mathrm{id}) = \oplus_{i=a}^b (M^l_i, \psi^*_{M^l_i})$$

for each prime l outside S, where a and b are independent of l.

ii) *There is a finite increasing filtration $N_* M$ $(* = 1, 2 \dots)$ on M such that*

$$N_s M \otimes \mathbb{Z}_{(l)} \subset N_s(M^l_i, \psi^k_{M^l_i} - k^i \mathrm{id}_{M^l_i}) \ \text{ for all } l \text{ and } k.$$
$$M = \cup_{s=1}^\infty N_s M.$$

Then there is a unique Adams decomposition of M:

$$(M, \psi^*_M) = \oplus_{i=a}^b (M_i, \psi^*_{M_i})$$

such that

$$M_i \otimes \mathbb{Z}_{(l)} = M^l_i.$$

Proof If $S = (l)$ for some prime l, or if $S = \infty$, there is nothing to prove, so we may assume there are at least two primes not in S.

For $a \leq i \leq b$, let M_i be the subset of M defined by

$$M_i = \{m \in M \mid m \otimes 1 \in M^l_i \text{ for all } l \notin S\}.$$

Then M_i is a $\mathbb{Z}_S$-submodule of M, stable under ψ^k_M for all k. Let p and q be distinct primes outside of S. By Lemma 6.1, the $\mathbb{Q}$-subspaces $M^p_i \otimes \mathbb{Q}$ and $M^q_i \otimes \mathbb{Q}$ of $M \otimes \mathbb{Q}$ agree. Thus, by an elementary argument, we have

$$M_i \otimes \mathbb{Z}_{(l)} = M^l_i \tag{6.1}$$

for all $l \notin S$, and

$$M = \oplus_i M_i. \tag{6.2}$$

This gives the direct sum decomposition in $\mathcal{A}_S$

$$(M, \psi^*_M) = \oplus_{i=a}^b (M_i, \psi^*_{M_i}). \tag{6.3}$$

A similar argument shows that the direct sum decomposition (6.3) is the unique such which induces the given Adams decomposition of $M \otimes \mathbb{Z}_{(l)}$ for each l not in S. The assumption (ii) readily implies that $\psi^k_M - k^i \mathrm{id}$ is nilpotent on M_i for each k, and that (6.3) defines an Adams decomposition of (M, ψ^*). The uniqueness follows from the uniqueness of the direct sum decomposition (6.3). □

Lemma 6.3 *Let N be a positive integer, let S be a set of primes containing all primes $p \leq N$, and let (M, ψ_M^*) be in $\mathcal{A}_S$. Let $0 \leq a \leq b$ be integers with $b-a < N$. Then an Adams decomposition of (M, ψ_M^*) as*

$$(M, \psi_M^*) = \oplus_{i=a}^b (M_i, \psi_{M_i}^*)$$

is unique.

Proof This follows from Lemma 6.1(ii) and Lemma 6.2. □

In particular, for an Adams decomposition as in Lemma 6.3, the canonical filtration $N_*^{\text{can}} M$ is an invariant of M.

Propositon 6.4 *Let $0 \leq a \leq b$ be integers, let S be a set of primes containing all primes $p \leq b-a+1$, and let*

$$(A, \psi_A^*) \xrightarrow{\iota} (M, \psi_M^*) \xrightarrow{\pi} (B, \psi_B^*)$$

be an exact sequence in $\mathcal{A}_S$. Suppose we have Adams decompositions

$$(A, \psi_A^*) = \oplus_{i=a}^b (A_i, \psi_{A_i}^*); \qquad (B, \psi_B^*) = \oplus_{i=a}^b (B_i, \psi_{B_i}^*). \tag{6.4}$$

Then there is a unique Adams decomposition of (M, ψ_M^*) of the form

$$(M, \psi_M^*) = \oplus_{i=a}^b (M_i, \psi_{M_i}^*).$$

In addition, the sequence (6.4) induces exact sequences in $\mathcal{A}_S$

$$(A_i, \psi_{A_i}^*) \to (M_i, \psi_{M_i}^*) \to (B_i, \psi_{B_i}^*) \tag{6.5}$$

for each i.

Proof The uniqueness follows from Lemma 6.3. For existence, suppose $N_r A = A$. Set

$$N_s M = \begin{cases} \iota(N_s A) & \text{for } s = 1, \dots, r \\ \pi^{-1}(N_{s-r} B) & \text{for } s = r+1, \dots \end{cases}$$

Take a prime $l > N$. Write M^l for $M \otimes \mathbb{Z}_{(l)}$, $\psi_{M^l}^k$ for the endomorphism induced by ψ_M^k, and similarly for A and B. By Lemma 6.2, it suffices to construct an Adams decomposition for $M \otimes \mathbb{Z}_{(l)}$ of the form

$$(M^l, \psi_{M^l}^*) = \oplus_{i=a}^b (M_i^l, \psi_{M_i}^*),$$

with the filtration $N_* M^l := N_* M \otimes \mathbb{Z}_{(l)}$ being finer than the canonical filtration.

For this, choose an integer k prime to l which gives a generator for $(\mathbb{Z}/l)^\times$, as in the proof of Lemma 6.1(ii). Let M_i^l be given by

$$M_i^l = \cup_{n=1}^\infty \ker[(\psi_{M^l}^k - k^i \mathrm{id})^n].$$

The endomorphism $\psi_{A^l}^k - k^i\mathrm{id}$ is nilpotent on A_i^l, invertible on on A_j^l for $i \neq j$, and similarly for B^l. Thus, we have the exact sequence

$$A_i^l \to M_i^l \to B_i^l \tag{6.6}$$

and inclusion

$$N_s M_i^l \subset N_s(M_i^l, \psi_{M^l}^k - k^i) \tag{6.7}$$

for each $i = a, \dots, b$.

We claim that

$$B_i^l \cap \pi(M^l) = \pi(M_i^l). \tag{6.8}$$

The inclusion $\supset$ is clear. To prove the inclusion $\subset$, take m in M^l, and suppose $\pi(m)$ is in B^l_i. There is an integer $N > 0$ such that $(\psi^k_{A^l} - k^j)^N$ is zero on A^l_j and $(\psi^k_{B^l} - k^j)^N$ is zero on B^l_j for all j. Let $P_N(T)$ be the polynomial

$$P_N(T) = T\sum_{i=0}^{N-1}(-1)^i(T-1)^i.$$

Letting

$$p_i^B := \prod_{j\neq i,\ a\leq j\leq b} P_N\left(\frac{(\psi^k_{B^l} - k^j)^N}{(k^i - k^j)^N}\right) : B^l \to B^l,$$

p_i^B is the projection of B^l onto B^l_i. We have a similar formula for the projection p_i^A of A^l onto A^l_i. Let

$$p_i^M := \prod_{j\neq i,\ a\leq j\leq b} P_N\left(\frac{(\psi^k_{M^l} - k^j)^N}{(k^i - k^j)^N}\right) : M^l \to M^l.$$

Then p_i^M commutes with $(\psi^k_{M^l} - k^i)$,

$$\pi\left(p_i^M(m)\right) = p_i^B(\pi(m)) = \pi(m),$$

and

$$\pi\left((\psi^k_{M^l} - k^i)^N(p_i^M(m))\right) = \left(\psi^k_{M^l} - k^i\right)^N\left(p_i^B(\pi(m))\right) = 0.$$

Thus $(\psi^k_{M^l} - k^i)^N(p_i^M(m))$ is in the image of A^l:

$$\left(\psi^k_{M^l} - k^i\right)^N\left(p_i^M(m)\right) = \iota(x); \quad x \in A^l.$$

We have the same identities with p_i^M replaced by $[p_i^M]^2$; in addition

$$\begin{aligned}(\psi^k_{M^l} - k^i)^{2N}([p_i^M]^2(m)) &= [(\psi^k_{M^l} - k^i)^N \circ p_i^M] \circ [(\psi^k_{M^l} - k^i)^N \circ p_i^M](m)\\ &= \iota((\psi^k_{A^l} - k^i)^N(p_i^A(x)))\\ &= 0.\end{aligned}$$

Thus $[p_i^M]^2(m)$ is an element of M^l_i lifting $\pi(m)$, proving (6.8).

By (6.8) and the exactness of (6.6), it follows that we have the direct sum decomposition of M^l as a $\mathbb{Z}_{(l)}[\psi^k_{M^l}]$-module

$$M^l = \oplus_{i=a}^b M^i_i. \tag{6.9}$$

Since $\psi^k_{M^l}$ and $\psi^t_{M^l}$ commute for all t and k, the decomposition (6.9) is a direct sum decompostion of $(M^l, \psi^*_{M^l})$ in $\mathcal{A}_{(l)}$, and we have the exact sequences in $\mathcal{A}_{(l)}$,

$$A^l_i \to M^l_i \to B^l_i,$$

for all $i = a, \ldots, b$. Since $\psi^t_{A^l} - t^i$ is nilpotent on A^l_i, and $\psi^t_{B^l} - t^i$ is nilpotent on B^l_i, it follows that $\psi^t_{M^l} - t^i$ is nilpotent on M^l_i; similarly, we have

$$N_s M^l_i \subset N_s(M^l_i, \psi^t_{M^l} - t^i)$$

for all s, t and i. Thus (6.9) gives the desired Adams decomposition of M^l. □

We let $\mathcal{A}_S(a,b)$ denote the full subcategory of $\mathcal{A}_S$ consisting of (M,ψ^*_M) which admit an Adams decomposition with weights $a,\ldots,b$.

Theorem 6.5 *Let $0 \le a \le b$ be integers, and let S be a set of primes, containing all primes $p \le b-a+1$. Then*

1) *The category $\mathcal{A}_S(a,b)$ is a strictly full abelian subcategory of $\mathcal{A}_S$.*
2) *The category $\mathcal{A}_S(a,b)$ is closed under extensions in $\mathcal{A}_S$.*
3) *Each object of $\mathcal{A}_S(a,b)$ has a unique Adams decomposition with weights $a,\ldots,b$.*
4) *Each morphism $f\colon (M,\psi^*_M) \to (N,\psi^*_N)$ in $\mathcal{A}_S(a,b)$ is strict, i.e.,*

$$f(M_i) \subset N_i$$

for all i, where M_i and N_i are the weight i summands of M and N, respectively.

Proof (3) follows from Lemma 6.1 and Lemma 6.2, and (2) follows from Proposition 6.4. For (4), it suffices to show that

$$f(M_i \otimes \mathbb{Z}_{(l)}) \subset N_i \otimes \mathbb{Z}_{(l)}$$

for all primes l not in S, i.e., we may assume that $S = (l)$. In this case, for each object (A,ψ^*_A) of $\mathcal{A}_S(a,b)$, we have the projector p_i^A of A onto the weight i summand A_i, defined in the proof of Proposition 6.4:

$$p_i^A := \prod_{j\neq i,\ a\le j\le b} P_N\left(\frac{(\psi^k_{A^l} - k^j)^N}{(k^i-k^j)^N}\right),$$

where N is any sufficiently large integer. Clearly, we have

$$p_i^N \circ f = f \circ p_i^M$$

for each morphism $f\colon (M,\psi^*_M) \to (N,\psi^*_N)$ in $\mathcal{A}_S(a,b)$, whence (4).

(1) follows easily from (4): If we have a morphism $f\colon (M,\psi^*_M) \to (N,\psi^*_N)$ in $\mathcal{A}_S(a,b)$, with kernel (A,ψ^*_A) and cokernel (B,ψ^*_B) in $\mathcal{A}_S$, from (3) we have

$$(A,\psi^*_A) = \oplus_{i=a}^b (A\cap M_i, \psi_{A\cap M_i})$$

$$(B,\psi^*_B) = \oplus_{i=a}^b (\bar{N}_i, \psi_{\bar{N}_i})$$

where M_i and N_i are the weight i summands of M and N, and $\bar{N}_i$ is the image of N_i in B.

7 Adams Operations on Relative K-theory with Support

Let Y be a scheme, W, $D_1,\ldots,D_s$ closed subschemes. By Corollary 5.6, we have natural λ-operations on the relative K-groups with support $K_*^W(Y;D_1,\ldots,D_s)$, which satisfy the special λ-ring identities. The resulting Adams operations ψ^k thus define the object

$$(K_p^W(Y;D_1,\ldots,D_s),\psi^*)$$

of $\mathcal{A}$. All the maps of groups of the form $K_*^W(Y;D_1,\ldots,D_s)$, arising from pull-back by morphisms of schemes, or from maps in the various long exact localization sequences or relativization sequences, define maps in $\mathcal{A}$.

Let Y be a scheme of finite type over a field k. We say that subschemes $D_1, \dots, D_s$ define a *reduced normal crossing subscheme of* Y if each D_i has pure codimension, say $d_i > 0$ in Y, and each scheme-theoretic intersection

$$D_{i_1} \cap \dots \cap D_{i_t}$$

(including the empty intersection) is smooth over k, and of pure codimension $\sum_j d_{i_j}$ on Y. For the multi-index $I = (i_1, \dots, i_t)$, we let D_I denote the above intersection, and d_I the sum

$$d_I := \sum_j d_{i_j}.$$

We let $\mathcal{Z}^q(Y)$ denote the group of codimension q algebraic cycles on Y; for W a closed subset of Y, we let $\mathcal{Z}^q_W(Y)$ denote the subgroup of W consisting of cycles with support contained in W.

Lemma 7.1 *Let Y be a scheme finite type over a field k, W is a closed subset of Y of pure codimension q. Let $D_1, \dots, D_s$ define a reduced normal crossing subscheme of Y, and suppose that*

$$W \cap D_I$$

has pure codimension q on D_I for each multi-index I. Suppose that W has pure dimension d over k. Then

i) *the image of $K^W_p(Y; D_1, \dots, D_s)$ in $\mathcal{A}_{(d+p+1-\epsilon)!}$ has an Adams decomposition with weights $q+\epsilon, \dots, q+d+p$, where*

$$\epsilon = \begin{cases} 0 & \text{for } p = 0 \\ 1 & \text{for } p = 1 \\ 2 & \text{for } p \geq 2. \end{cases}$$

ii) *Letting*

$$K^W_p(Y; D_1, \dots, D_s)^{(a)}$$

denote the weight a-submodule of $K^W_p(Y; D_1, \dots, D_s)[1/(d+p+1-\epsilon)!]$, we have the exact sequence of $\mathbb{Z}[1/(d+1)!]$-modules

$$0 \to K^W_0(Y; D_1, \dots, D_s)^{(q)} \to K^W_0(Y)^{(q)} \to \oplus_{i=1}^s K_0^{W \cap D_i}(D_i)^{(q)}.$$

iii) *Let w denote the set of closed points in the semi-local scheme $\mathrm{Spec}(\mathcal{O}_{Y,W})$; sending an irreducible component W_i of W to the K_0-class of $\mathcal{O}_{W_i}$ gives the identification*

$$K^w_0(\mathrm{Spec}(\mathcal{O}_{Y,W})) \cong \mathcal{Z}^q_W(Y).$$

Then the restriction map

$$K^W_0(Y) \to K^w_0(\mathrm{Spec}(\mathcal{O}_{Y,W}))$$

induces isomorphisms

$$K^W_0(Y)^{(q)} \cong K^w_0(\mathrm{Spec}(\mathcal{O}_{Y,W}))^{(q)} \cong \mathcal{Z}^q_W(Y)[1/(d+1)!].$$

of $\mathbb{Z}[1/(d+1)!]$-modules, and identifies the restriction maps in (2)

$$K^W_0(Y)^{(q)} \to K_0^{W \cap D_i}(D_i)^{(q)}$$

with the cycle intersection

$$Z \mapsto Z \cdot D_i.$$

Proof (i) If $W = Y$, and $s = 0$, this is well-known (see e.g. [24]). More generally, suppose W is regular, and a codimension q complete intersection in Y, such that the inclusion

$$i\colon W \to Y$$

is split by a morphism

$$p\colon Y \to W.$$

The structure sheaf $\mathcal{O}_W$ defines canonically a class $[\mathcal{O}_W]$ in $K_0^W(Y)$; as W is a complete intersection of codimension q, a direct computation shows that

$$\psi^k([\mathcal{O}_W]) = k^q[\mathcal{O}_W].$$

As the map

$$([\mathcal{O}_W] \cup (-)) \circ p^*\colon K_p(W) \to K_p^W(Y)$$

is an isomorphism, and the Adams operations are multiplicative, the result follows in this case as well. In general, W has a dense open subset W_0 such that there is an étale map

$$g\colon (Y', W_0) \to (Y, W)$$

which on W_0 is the inclusion $W_0 \to W$, with W_0 a closed codimension q split complete intersection in a smooth Y'. Let $\bar{W}$ be the complement $W \setminus W_0$. As g induces an isomorphism in $\mathcal{A}$

$$K_p^{W_0}(Y \backslash \bar{W}) \cong K_p^{W_0}(Y')$$

we have the result for $K_p^{W_0}(Y\backslash\bar{W})$. The result for general W and $s = 0$ follows by induction on dimension, using the exact sequence

$$K_p^{\bar{W}}(Y) \to K_p^W(Y) \to K_p^{W_0}(Y\backslash\bar{W})$$

and Theorem 6.5. This completes the proof for $s = 0$.

For general s, we proceed by induction. We have the exact sequence

$$K_{p+1}^{W\cap D_s}(D_s; D_1 \cap D_s, \dots, D_{s-1} \cap D_s) \\ \to K_p^W(Y; D_1, \dots, D_s) \to K_p^W(Y; D_1, \dots, D_{s-1}).$$

Since

$$\dim(W \cap D_s) = \dim W - d_s \leq d - 1$$

we may apply our induction hypothesis and Theorem 6.5 to complete the proof of (i).

For (ii), we have the exact sequence

$$K_1^{W\cap D_s}(D_s; D_1 \cap D_s, \dots, D_{s-1} \cap D_s) \\ \to K_0^W(Y; D_1, \dots, D_s) \to K_0^W(Y; D_1, \dots, D_{s-1})$$

After inverting $(d+1)!$, this gives by (i) the exact sequence

$$0 = K_1^{W\cap D_s}(D_s; D_1 \cap D_s, \dots, D_{s-1} \cap D_s)^{(q)} \to K_0^W(Y; D_1, \dots, D_s)^{(q)} \\ \to K_0^W(Y; D_1, \dots, D_{s-1})^{(q)} \to K_0^{W\cap D_s}(D_s; D_1 \cap D_s, \dots, D_{s-1} \cap D_s)^{(q)}.$$

Thus, by induction, the map

$$K_0^W(Y; D_1, \dots, D_s)^{(q)} \to K_0^W(Y)^{(q)}$$

is injective. This, together with the similar injectivity for the restrictions to D_I, proves (ii).

For (iii), we may apply the results of (i) to $K_0^w(\mathrm{Spec}(\mathcal{O}_{Y,W}))$ by taking a direct limit. As the kernel of the surjection

$$K_0^W(Y) \to K_0^w(\mathrm{Spec}(\mathcal{O}_{Y,W}))$$

comes from groups $K_0^{\bar{W}}(Y)$, with

$$\mathrm{codim}_Y(\bar{W}) > \mathrm{codim}_Y(W),$$

the map on the weight q submodule is thus an isomorphism. The final assertion in (iii) follows from Serre's intersection formula, and the fact that the appropriate higher Tor's vanish in this situation. □

Let A be a λ-ring, and let $a > 2$ be an integer. Form the operator

$$\phi^k_{\geq a} := \prod_{i=2}^{a-1} (\psi^k - k^i \mathrm{id}).$$

Lemma 7.2 *Let Y be a scheme, x an element of $F^2_\gamma K_0(Y)$. Then*

$$\phi^k_{\geq a}(x) \in F^a_\gamma K_0(Y).$$

Proof This follows from the fact that $\psi^k - k^i\mathrm{id}$ maps $F^i_\gamma(K_0(Y))$ to $F^{i+1}_\gamma(K_0(Y))$. □

Lemma 7.3 *Let Y be a quasi-projective scheme over an infinite field k, $q \geq 1$ an integer, x an element of $F^q_\gamma K_0(Y)$, and $F_1, \dots, F_s$ closed irreducible subsets of Y. Then there is a closed subset W of Y, of pure codimension q, such that*

i) *$W \cap F_i$ has codimension q on F_i, for each $i = 1, \dots, s$.*
ii) *There is an element $x_W \in K_0^W(Y)$ with image x in $K_0(Y)$.*

Proof Let $E \to Y$ be a vector bundle. Since Y is quasi-projective, there is a map $f\colon Y \to H$, into a homogeneous space H for a linear reductive algebraic group G (G can be taken to be a product of general linear groups), and a vector bundle E_H of rank n on H such that

$$E \cong f^*(E_H).$$

Taking products of such maps, we see that the same holds true for each finite set of vector bundles on Y. This implies that there is a map $f\colon Y \to H$ as above, and an element y of $F^q_\gamma(H)$ with

$$f^*(y) = x.$$

As H is smooth over k, the result of ([11], Chapter II, Section 5, Theorem 2.16) implies that y is in the image of $K_0^T(H)$ for some pure codimension q closed subset T of H:

$$y = i_{T*}(y_T); \quad y_T \in K_0^T(H).$$

As G acts trivially on $K_0(H)$, we may assume ([14], Theorem 2) that $W := f^{-1}(T)$ has pure codimension q on Y, and intersects each F_i in codimension q. Taking

$$x_W = f^*(y_T) \in K_0^W(Y)$$

completes the proof. □

For a noetherian scheme Y, define the groups $\tilde{S}K_p(Y)$ as follows: For a ring R, we have the simplicial set $\mathrm{BSL}(R)^+$ (which we make into a functor as for $\mathrm{BGL}(R)^+$). Take a finite affine cover

$$\mathfrak{U} := \{\mathrm{Spec}(R_i) \mid i \in I\}$$

of Y, and set

$$\mathrm{B\tilde{S}L}(\mathfrak{U})^+ := \operatorname*{holim}_I \mathrm{B\tilde{S}L}(-)^+,$$

where ˜ means bifibrant model, as in Section 3. A refinement $\mathfrak{V} \to \mathfrak{U}$ induces a map on the homotopy groups

$$\pi_*(\mathrm{B\tilde{S}L}(\mathfrak{U})^+) \to \pi_*(\mathrm{B\tilde{S}L}(\mathfrak{V})^+)$$

which is independent of the choice of refinement mapping. We then set

$$\tilde{S}K_p(Y) := \lim_{\vec{\mathfrak{U}}} \pi_*(\mathrm{B\tilde{S}L}(\mathfrak{U})^+).$$

The generalization to groups with support and the relative version are defined similarly, by taking compatible collections of open covers of the various schemes involved. The natural H-group structure on $\mathrm{BSL}(-)$ gives $\mathrm{B\tilde{S}L}(\mathfrak{U})^+$ the structure of an H-group, which in turn makes $\tilde{S}K_p(-)$ into an abelian group-valued functor, for all $p \geq 0$.

We define the functor $K^{(1)}(-)$ similarly, by using $\mathrm{B}\mathbb{G}_m = \mathrm{B}\mathbb{G}_m^+$ instead of $\mathrm{BSL}(-)^+$. The sequence of group schemes

$$\mathrm{SL}_n \to \mathrm{GL}_n \to \mathbb{G}_m$$

gives rise to the sequence of functors

$$\tilde{S}K_p(-) \to \tilde{K}_p(-) \to K_p^{(1)}(-),$$

where

$$\tilde{K}_p(-) = \begin{cases} K_p(-) & \text{for } p > 0 \\ \ker[\mathrm{rnk}\colon K_0(-) \to H^0(-,\mathbb{Z})] & \text{for } p = 0. \end{cases}$$

Theorem 7.4 *Let Y be a k-scheme of finite type, W and $D_1, \ldots, D_s$ closed subschemes.*

i) *The sequence of abelian groups*

$$0 \to \tilde{S}K_p^W(Y; D_1, \ldots, D_s) \to \tilde{K}_p^W(Y; D_1, \ldots, D_s) \to K_p^{(1)W}(Y; D_1, \ldots, D_s) \to 0$$

is exact for all $p \geq 0$, and split for $p \geq 1$.

ii) *Let*

$$i\colon D := \cup_{i=1}^s D_i \to Y$$

be the inclusion, $j\colon U \to Y$ the inclusion of the complement

$$U := Y \backslash D,$$

and let $j_!\mathcal{O}_Y^*$ be the subsheaf of $\mathcal{O}_Y^*$ on Y given by the exactness of

$$0 \to j_!\mathcal{O}_Y^* \to \mathcal{O}_Y^* \to i_*\mathcal{O}_D^* \to 0.$$

Then $K_p^{(1)W}(Y; D_1, \ldots, D_s)$ is 0 for $p \geq 2$; for $p \leq 1$, we have

$$K_1^{(1)}(Y; D_1, \ldots, D_s) = \ker[\Gamma(Y, \mathcal{O}_Y^*) \to \textstyle\prod_{i=1}^s \Gamma(D_i, \mathcal{O}_{D_i}^*)]$$

$$K_0^{(1)}(Y; D_1, \ldots, D_s) = H^1_{\mathrm{Zar}}(Y, j_!\mathcal{O}_Y^*).$$

Proof Let $\mathrm{E}(R)$ be the group of elementary matrices. It follows from the definition of the + contruction that the sequence

$$\mathrm{BE}(R)^+ \to \mathrm{BSL}(R)^+ \to \mathrm{BGL}(R)^+$$

is homotopy equivalent to a sequence of covering spaces of $\mathrm{BGL}(R)^+$, corresponding to the sequence of subgroups

$$\{0\} \subset \mathrm{SL}(R)/\mathrm{E}(R) \subset K_1(R) = \mathrm{GL}(R)/\mathrm{E}(R).$$

In particular, the sequence

$$\mathrm{BSL}(R)^+ \to \mathrm{BGL}(R)^+ \xrightarrow{\det^+} \mathrm{B}\mathbb{G}_m(R) \tag{7.1}$$

is a homotopy fiber sequence.

The stabilization map

$$\mathrm{B}\mathbb{G}_m(R) \to \mathrm{BGL}(R) \tag{7.2}$$

gives a natural section s to the sequence (7.1); since $\mathrm{BGL}(R)^+$ is an H-group, we have the homotopy equivalence

$$\mathrm{BGL}(R)^+ \sim \mathrm{BSL}(R)^+ \times \mathrm{B}\mathbb{G}_m(R).$$

Let

$$A\colon I \to \mathbf{Rings}$$

be a functor, giving the functors

$$\mathrm{BSL}(A)^+\colon I \to \mathcal{S}^*, \quad \mathrm{BGL}(A)^+\colon I \to \mathcal{S}^*, \quad \mathrm{B}\mathbb{G}_m(A)\colon I \to \mathcal{S}^*,$$

and the sequence

$$\mathrm{BSL}(A)^+ \to \mathrm{BGL}(A)^+ \to \mathrm{B}\mathbb{G}_m(A) \tag{7.3}$$

of functors. The sequence (7.3) is weakly equivalent to a homotopy fiber sequence. The splitting (7.2) gives a splitting of (7.3); this and the H-group structure on $\mathrm{BGL}(A)^+$ defines a weak equivalence of bifibrant objects

$$\mathrm{B}\tilde{\mathrm{GL}}(A)^+ \to \mathrm{B}\tilde{\mathrm{SL}}(A)^+ \times \tilde{\mathrm{B}}\mathbb{G}_m(A). \tag{7.4}$$

Taking the homotopy limit over I gives ([5], XI, Section 5, Lemma 5.6) the weak equivalence

$$\operatorname*{holim}_I \mathrm{B}\tilde{\mathrm{SL}}(A)^+ \times \operatorname*{holim}_I \tilde{\mathrm{B}}\mathbb{G}_m(A) \sim \operatorname*{holim}_I \mathrm{B}\tilde{\mathrm{GL}}(A)^+;$$

this proves (i).

For (ii), we note that the groups $K_p^{(1)}(-)$ are given by

$$\begin{aligned} K_1^{(1)}(Y) &= \Gamma(Y, \mathcal{O}_Y^*) \\ K_0^{(1)}(Y) &= H^1(Y, \mathcal{O}_Y^*) \end{aligned} \tag{7.5}$$

and $K_p^{(1)}(Y) = 0$ for $p > 1$. Indeed, the last equality follows from the spectral sequence of Theorem B.1, which degenerates at E_2 (we avoid the fringe effects by taking deloopings of $\mathrm{B}\mathbb{G}_m$). The same spectral sequence gives

$$\pi_1 \operatorname*{holim}_I \mathrm{B}\mathbb{G}_m(A) = \lim_{\overleftarrow{I}} \mathbb{G}_m(A)$$

$$\pi_0 \operatorname*{holim}_I \mathrm{B}\mathbb{G}_m(A) = \lim_{\overleftarrow{I}}{}^1 \mathbb{G}_m(A).$$

For A the functor associated to an affine open cover $\mathfrak{U}$ of Y, the lim term is $H^0(\mathfrak{U}, \mathcal{O}_Y^*)$, and the $\lim^1$ term is $H^1(\mathfrak{U}, \mathcal{O}_Y^*)$, where $H^0(\mathfrak{U}, -)$ is the Čech cohomology with respect to the cover $\mathfrak{U}$. As the Zariski cohomology of a sheaf of abelian groups is computable via Čech cohomology, for degrees ≤ 1, we arrive at the canonical identites (7.5). The computation of relative groups $K_p^{(1)}(Y; D_1, \dots, D_s)$ follows from the case $s = 0$ by noting that the inclusion

$$j_!\mathcal{O}_Y^* \to \mathcal{O}_Y^*$$

induces a natural map

$$H^*_{\text{Čech}}(Y, j_!\mathcal{O}_Y^*) \to K_{1-*}^{(1)}(Y; D_1, \dots, D_s)$$

and using the respective long exact relativization sequences. □

Corollary 7.5 Let $D_1, \dots, D_s$ define a reduced normal crossing subscheme of a finite type k-scheme Y, and let W be a closed subscheme of Y. Then

i) the natural map

$$\tilde{S}K_*^W(Y; D_1, \dots, D_s) \to \tilde{S}K_*^{W\times\mathbb{A}^1}(Y \times \mathbb{A}^1; D_1 \times \mathbb{A}^1, \dots, D_s \times \mathbb{A}^1)$$

is an isomorphism.

ii) for all $p \geq 0$, there is a natural isomorphism of λ-rings (without unit)

$$\tilde{S}K_p^{W\times\mathbb{A}^1}(Y \times \mathbb{A}^1; D_1 \times \mathbb{A}^1, \dots, D_s \times \mathbb{A}^1, Y \times \{0,1\}) \to \tilde{S}K_{p+1}^W(Y; D_1, \dots, D_s).$$

Proof The second assertion follows from the first by the usual long exact relativization sequence, and the naturality of the λ-ring operations. The first assertion follows from the homotopy invariance of Quillen K-theory and homotopy invariance for the functors Pic and $\Gamma(-, \mathcal{O}^*)$, for regular schemes, together with Theorem 7.4. □

Lemma 7.6 *Let $D_1, \dots, D_s$ define a reduced normal crossing subscheme of an irreducible finite type k-scheme Y, and let W be a closed subscheme of Y. Suppose that W has pure codimension $q \geq 2$ on Y, and that $W \cap D_I$ has codimension q on D_I for each multi-index I. Let U be the complement of W in Y, and $E_i = D_i \cap U$. Let x be in kernel of the map*

$$K_0^W(Y; D_1, \dots, D_s) \to K_0(Y; D_1, \dots, D_s).$$

Then there is an element η in $\tilde{S}K_1(U; E_1, \dots, E_s)$ mapping to x via the boundary map in the long exact localization sequence

$$\to K_1(U; E_1, \dots, E_s) \to K_0^W(Y; D_1, \dots, D_s) \to K_0(Y; D_1, \dots, D_s).$$

Proof Since Y is regular and irreducible, and W has codimension at least 2, the map

$$\Gamma(Y, \mathcal{O}_Y^*) \to \Gamma(U, \mathcal{O}_U^*)$$

is an isomorphism. This implies that the map

$$K_1^{(1)}(Y; D_1, \dots, D_s) \to K_1^{(1)}(U; E_1, \dots, E_s)$$

is an isomorphism as well. Similarly, we have the identities

$$\tilde{K}_0^W(Y; D_1, \dots, D_s) = K_0^W(Y; D_1, \dots, D_s);$$

$$\tilde{K}_1(U; E_1, \dots, E_s) = K_1(U; E_1, \dots, E_s).$$

The lemma then follows from the commutative diagram with exact rows and split exact columns

$$
\begin{array}{ccccc}
0 & & 0 & & 0 \\
\downarrow & & \downarrow & & \downarrow \\
\tilde{S}K_1(Y; D_1, \ldots, D_s) & \longrightarrow & \tilde{S}K_1(U; E_1, \ldots, E_s) & \longrightarrow & \tilde{S}K_0^W(Y; D_1, \ldots, D_s) \\
\downarrow & & \downarrow & & \downarrow \\
\tilde{K}_1(Y; D_1, \ldots, D_s) & \longrightarrow & \tilde{K}_1(U; E_1, \ldots, E_s) & \longrightarrow & \tilde{K}_0^W(Y; D_1, \ldots, D_s) \\
\downarrow & & \downarrow & & \downarrow \\
K_1^{(1)}(Y; D_1, \ldots, D_s) & \longrightarrow & K_1^{(1)}(U; E_1, \ldots, E_s) & \longrightarrow & K_0^{(1)W}(Y; D_1, \ldots, D_s) \\
\downarrow & & \downarrow & & \downarrow \\
0 & & 0 & & 0
\end{array}
$$

□

It follows from Remark 4.6(iii) that the functor $\tilde{S}K_0(-)$ has the natural structure of a functor from schemes to λ-rings (without unit), and that the map

$$\tilde{S}K_p(-) \to \tilde{K}_p(-)$$

is a map of functors to λ-rings (without unit). Similarly, the identity

$$\det(\lambda^k(\rho - \mathrm{id})) = \lambda^k(\det(\rho)),$$

for a representation

$$\rho\colon G \to \mathrm{GL}_n(R)$$

implies that the map

$$\tilde{K}_p(-) \to K_p^{(1)}(-)$$

is a map of functors to λ-rings without unit, where we give $K_p^{(1)}(-)$ the operations

$$\lambda^k = \begin{cases} \mathrm{id} & \text{for } k = 1, \\ 0 & \text{for } k > 1. \end{cases}$$

Lemma 7.7 *Let Y be a k-scheme of finite type. Then*

$$\tilde{S}K_p(Y) = F_\gamma^2 K_p(Y)$$

for all p.

Proof The identity

$$\gamma_1 + \ldots + \gamma_{n+1} = 0$$

in the representation ring of SL_n, together with Remark 4.6(iii), implies the inclusion

$$\tilde{S}K_p(Y) \subset F_\gamma^2 K_p(Y)$$

for all p. Conversely, since $\lambda^k = 0$ on $K_p^{(1)}(Y)$ for all $k > 1$, it follows that

$$F_\gamma^2 K_p^{(1)}(Y) = 0,$$

hence

$$F^2_\gamma K_p(Y) \subset \ker(\tilde{K}_p(Y) \to K_p^{(1)}(Y)) = \tilde{S}K_p(Y).$$

□

8 K-theory and Motivic Cohomology

We conclude with an application of the results of the two previous sections to a comparison of K-theory and motivic cohomology, the latter defined as Bloch's higher Chow groups $\mathrm{CH}^q(X,p)$. In [17], we have shown that there is a natural isomorphism

$$\mathrm{CH}^q(X,p)_\mathbb{Q} \cong K_p(X)^{(q)}_\mathbb{Q} \tag{8.1}$$

for X smooth and quasi-projective over a field.

The isomorphism (8.1) uses a cubical version $\mathcal{Z}^q(X,*)_c$ of Bloch's simplicial complex $\mathcal{Z}^q(X,*)$. In ([17], Section 3) it is shown that there is a natural isomorphism

$$\mathcal{Z}^q(X,*)_c \to \mathcal{Z}^q(X,*).$$

in the derived category, giving the natural isomorphism

$$H_*(\mathcal{Z}^q(X,*)_c) \cong \mathrm{CH}^q(X,p)$$

We let $Z_p(\mathcal{Z}^q(X,*)_c)$ denote the (homological) cycles of dimension p in the complex $\mathcal{Z}^q(X,*)_c$.

We refine the isomorphism (8.1) to the following result:

Theorem 8.1 *Let X be a smooth, quasi-projective variety over a field k, of dimension d over k. Let $(p \geq 0, q \geq 0)$ be a pair of integers, with $(p,q) \neq (0,0),(0,1)$.*

i) *Let*

$$M_{p,q,d} = \begin{cases} d+p-q+1 & \text{for } q > 1 \\ 1 & \text{for } q = 0,1; p > 0. \end{cases}$$

Then there are natural maps

$$\mathrm{cl}^{q,p}\colon \mathrm{CH}^q(X,p)\left[\frac{1}{M_{p,q,d}!}\right] \to K_p(X)\left[\frac{1}{M_{p,q,d}!}\right].$$

ii) *Let*

$$N_{p,d} = \begin{cases} \max(d+p-q+1, q-1) & \text{for } q > 1 \\ 1 & \text{for } q = 0,1; p > 0. \end{cases}$$

Then

$$\mathrm{cl}^{q,p}\left(\mathrm{CH}^q(X,p)\left[\frac{1}{N_{p,q,d}!}\right]\right) \subset F^q_\gamma K_p(X)\left[\frac{1}{N_{p,q,d}!}\right],$$

and the induced map

$$\mathrm{cl}^{q,p}\colon \mathrm{CH}^q(X,p)\left[\frac{1}{N_{p,q,d}!}\right] \to \mathrm{gr}^q_\gamma K_p(X)\left[\frac{1}{N_{p,q,d}!}\right]$$

is an isomorphism.

□

As immediate corollary, we have

Corollary 8.2 Let X be a smooth, quasi-projective variety of dimension d over a field k. Then there are natural graded isomorphisms

$$\oplus_q \mathrm{CH}^q(X,p)\left[\tfrac{1}{(d+p-1)!}\right] \to \oplus_q \mathrm{gr}^q_\gamma K_p(X)\left[\tfrac{1}{(d+p-1)!}\right].$$

Proof For $(p,q) \neq (0,0),(0,1)$, this follows from Theorem 8.1. From Theorem 7.4 and Lemma 7.7, we have the natural isomorphism (this is of course well known)

$$\mathrm{gr}^1_\gamma K_0(X) \cong \mathrm{Pic}(X);$$

by definition of F^1, we have

$$\mathrm{gr}^0_\gamma K_0(X) \cong H^0(X,\mathbb{Z}).$$

The result then follows from the natural isomorphisms

$$\mathrm{CH}^1(X,0) = \mathrm{CH}^1(X) \cong \mathrm{Pic}(X); \quad \mathrm{CH}^0(X,0) \cong H^0(X,\mathbb{Z}).$$

□

The proof of Theorem 8.1 is essentially the same as the proof of the rational result in [17]; we merely use the results of Section 6 and Section 7 to make the refinement.

For the convenience of the reader, we give a sketch of the proof of the isomorphism (8.1):

Let $\square^n$ be the affine space $\mathbb{A}^n_k$, let $\partial\square^n$ be the collection of divisors

$$D^\epsilon_i : t_i = \epsilon; i = 1, \ldots, n; \epsilon = 0, 1,$$

and let

$$\partial_0\square^n = \partial\square^n \setminus D^0_n.$$

The group $\mathcal{Z}^q(X,n)_c$ is the subgroup of the group of codimension q cycles on $X \times \square^n$, consisting of those cycles Z such that

i) each irreducible component W of Z intersectes each "face":

$$X \times [D^{\epsilon_1}_{i_1} \cap \ldots \cap D^{\epsilon_s}_{i_s}]$$

properly.

ii)

$$Z \cdot D^\epsilon_i = 0$$

for $\epsilon = 1$, $i = 1, \ldots, n$ and for $\epsilon = 0$, $i = 1, \ldots, n-1$.

Identifying $\square^{n-1}$ with D^0_n in the obvious way, the restriction to D^0_n defines the map

$$d_{n-1} \colon \mathcal{Z}^q(X,n)_c \to \mathcal{Z}^q(X,n-1)_c,$$

giving the complex $(\mathcal{Z}^q(X,*)_c, d)$. We denote the homology $H_p(\mathcal{Z}^q(X,*)_c)$ by $\mathrm{CH}^q(X,p)_c$. As mentioned above, there is a canonical isomorphism

$$\mathrm{CH}^q(X,p)_c \cong \mathrm{CH}^q(X,p). \tag{8.2}$$

Step 1: Let T be a smooth k-scheme, $D_1, \ldots, D_n$ subschemes forming a reduced normal crossing divisor, and W a closed, codimension q subset which intersects each D_I properly. By Lemma 7.1, we have the isomorphism

$$K^W_0(T)^{(q)}_{\mathbb{Q}} \cong \mathcal{Z}^q_W(T)_{\mathbb{Q}}$$

and the exact sequence

$$0 \to K^W_0(T; D_1 \ldots, D_n)^{(q)}_{\mathbb{Q}} \to \mathcal{Z}^q_W(T)_{\mathbb{Q}} \xrightarrow{\cap} \oplus_{i=1}^n \mathcal{Z}^q_{W\cap D_i}(D_i)_{\mathbb{Q}}, \tag{8.3}$$

where $\cap$ is the map

$$Z \mapsto \sum_{i=1}^{n} Z \cdot D_i.$$

Let $K_r^{[q]}(X \times \square^p; \partial\square^p)$ be the direct limit

$$K_r^{[q]}(X \times \square^p; X \times \partial\square^p) := \lim_{\rightarrow} K_r^W(X \times \square^p; X \times \partial\square^p)$$

over closed subsets W of $X \times \square^p$ of codimension q, such that each component of W intersects each face of $X \times \square^p$ in codimension q. Similarly, let $K_r^{[q]}(X \times \square^p; \partial_0\square^p)$ be the direct limit

$$K_r^{[q]}(X \times \square^p; X \times \partial_0\square^p) := \lim_{\rightarrow} K_r^W(X \times \square^p; X \times \partial_0\square^p)$$

over the same set of W. By (8.3), we have the canonical isomorphisms

$$\begin{aligned} K_0^{[q]}(X \times \square^p; X \times \partial\square^p)_{\mathbb{Q}}^{(q)} &\cong Z_p(\mathcal{Z}^q(X, *)_{c\mathbb{Q}}), \\ K_0^{[q]}(X \times \square^{p+1}; X \times \partial_0\square^{p+1})_{\mathbb{Q}}^{(q)} &\cong \mathcal{Z}^q(X, p+1)_{c\mathbb{Q}}. \end{aligned} \tag{8.4}$$

In particular, the natural map

$$K_0^{[q]}(X \times \square^p; X \times \partial\square^p)_{\mathbb{Q}}^{(q)} \to K_0(X \times \square^p; X \times \partial\square^p)_{\mathbb{Q}}^{(q)}$$

combined with (8.4) defines the map

$$\mathrm{cyc}^{q,p} \colon Z_p(\mathcal{Z}^q(X, *)_{c\mathbb{Q}}) \to K_0(X \times \square^p; X \times \partial\square^p)_{\mathbb{Q}}^{(q)}. \tag{8.5}$$

As

$$K_0(X \times \square^{p+1}; X \times \partial_0\square^{p+1}) = 0,$$

$\mathrm{cyc}^{q,p}$ descends to

$$\mathrm{cl}^{q,p} \colon \mathrm{CH}^q(X, *)_{c\mathbb{Q}} \to K_0(X \times \square^p; X \times \partial\square^p)_{\mathbb{Q}}^{(q)} \cong K_p(X)_{\mathbb{Q}}^{(q)}. \tag{8.6}$$

Step 2: Let T be a smooth k-scheme, $D_1, \ldots, D_n$ a reduced normal crossing divisor on T. Form the *n-fold double of T along* $D_1, \ldots, D_n$ by gluing together 2^n copies of T (indexed by the subsets of $\{1, \ldots, n\}$) by identifying T_I with T_J along the common subset $\cap_{j \in J \setminus I} D_i$, for all $I \subset J$. We denote the n-fold double by $\mathcal{D}(T; D_1, \ldots, D_n)$. Let $\mathcal{D}^i$ denote the subscheme of $\mathcal{D}(T; D_1, \ldots, D_n)$ formed as the union of the T_I with $i \in I$. Then restriction to the component $T_\emptyset$ of $\mathcal{D}$ defines the map of relative K-groups

$$K_p(\mathcal{D}; \mathcal{D}^1, \ldots, \mathcal{D}^n) \to K_p(T; D_1, \ldots, D_n).$$

In addition, the map

$$K_p(\mathcal{D}; \mathcal{D}^1, \ldots, \mathcal{D}^n) \to K_p(\mathcal{D}) \tag{8.7}$$

is a split injection for all p, with splitting given by a certain sequence of splittings of maps of schemes (see [17], Section 1). In particular, the splitting (8.7) defines a splitting of the natural maps

$$\tilde{S}K_p(\mathcal{D}; \mathcal{D}^1, \ldots, \mathcal{D}^n) \to \tilde{S}K_p(\mathcal{D}).$$

By ([17], Corollary 1.11) the restriction map

$$K_0(\mathcal{D}; \mathcal{D}^1, \ldots, \mathcal{D}^n) \to K_0(T; D_1, \ldots, D_n) \tag{8.8}$$

is an isomorphism, if there is a map

$$T \to S$$

to a smooth affine k-scheme S, such that the divisors $D_1, \ldots, D_n$ are pull backs of reduced normal crossing divisors $E_1, \ldots, E_n$ on S, and the collection of divisors $E_1, \ldots, E_n$ is *locally split* on S (see the statement of [17], Corollary 1.11 for a precise definition). In this case, one has the following result similar to Lemma 7.3 ([17], Theorem 2.3),

Lemma *Let x be an element of $K_0(T; D_1, \ldots, D_n)^{(q)}_{\mathbb{Q}}$. Then there is a closed, codimension q subset W of T, intersecting each D_I properly, such that x is in the image*

$$K_0^W(T; D_1, \ldots, D_n)^{(q)}_{\mathbb{Q}} \to K_0(T; D_1, \ldots, D_n)^{(q)}_{\mathbb{Q}}.$$

□

From this lemma, one easily shows that the map (8.6) is surjective.

Step 3: To show that (8.6) is injective, suppose

$$\mathrm{cyc}^{q,p}(Z) = 0$$

for some Z. Denote $X \times \square^n$ by $\square^n_X$, and $X \times \partial\square^n$ by $\partial\square^n_X$. Let $W \subset \square^n_X$ be the support of Z, and let $U = \square^n_X \setminus W$. By the localization sequence

$$K_1(U; U \cap \partial\square^n_X)^{(q)}_{\mathbb{Q}} \to \delta K_0^W(X \times \square^n, X \times \partial\square^n)^{(q)}_{\mathbb{Q}} \to K_0(X \times \square^n, X \times \partial\square^n)^{(q)}_{\mathbb{Q}},$$

there is an element $\eta \in K_1(U; U \cap \partial\square^n_X)^{(q)}_{\mathbb{Q}}$ with

$$\delta(\eta) = \mathrm{cyc}^{q,p}(Z). \tag{8.9}$$

By the homotopy property, we identify the above K_1 with

$$K_0(U \times \mathbb{A}^1; (U \cap \partial\square^n_X) \times \mathbb{A}^1, U \times \{0,1\})^{(q)}_{\mathbb{Q}}$$

giving the element $\tilde{\eta}$ corresponding to η. We then use the lemma of Step 2 to find a closed, codimension q subset W_U of $U \times \mathbb{A}^1$, intersection all the "faces" in $U \cap \partial\square^n_X) \times \mathbb{A}^1, U \times \{0,1\}$ properly, and an element b of

$$K_0^{W_0}(U \times \mathbb{A}^1; (U \cap \partial\square^n_X), \times\mathbb{A}^1, U \times \{0,1\})^{(q)}_{\mathbb{Q}}$$

lifting $\tilde{\eta}$. One checks that the closure W of W_U in $X \times \square^{n+1}$ intersects all faces in $X \times \partial\square^{p+1}$ properly. The element b thus defines, via the isomorphism (8.4), an element B of $\mathcal{Z}^q(X, n+1)_{c\mathbb{Q}}$. The relation (8.9) implies (see [17], Lemma 2.6)

$$d_p(B) = Z,$$

which gives the injectivity of the map (8.6). □

We now show how one modifies this argument to prove Theorem 8.1. The cases $q = 0, 1$ are well known (see [3], Section 6), so we assume that $q \geq 2$.

First of all, let $(T; D_1, \ldots, D_n)$ be as in Step 1, and let $d_T = \dim_k(T)$. By Lemma 7.1, we have the isomorphism

$$K_0^W(T)\left[\frac{1}{(d_T-q+1)!}\right]^{(q)} \cong \mathcal{Z}^q_W(T)\left[\frac{1}{(d_T-q+1)!}\right]$$

and the exact sequence

$$0 \to K_0^W(T; D_1 \ldots, D_n)\left[\frac{1}{(d_T-q+1)!}\right]^{(q)} \to \mathcal{Z}^q_W(T)\left[\frac{1}{(d_T-q+1)!}\right]$$
$$\xrightarrow{\cap} \oplus_{i=1}^n \mathcal{Z}^q_{W\cap D_i}(D_i)\left[\frac{1}{(d_T-q+1)!}\right],$$

as in (8.3). As in Step 1, this gives us the isomorphism

$$K_0^{[q]}(X\times\square^p; X\times\partial\square^p)\left[\tfrac{1}{(d+p-q+1)!}\right]^{(q)} \cong Z_p\left(\mathcal{Z}^q(X,*)_c\left[\tfrac{1}{(d+p-q+1)!}\right]\right) \tag{8.4$'$}$$

and thus the map

$$\begin{aligned}\mathrm{cyc}'^{q,p}: Z_p\left(\mathcal{Z}^q(X,*)_c\left[\tfrac{1}{(d+p-q+1)!}\right]\right) &\to \\ K_0(X\times\square^p; X\times\partial\square^p)\left[\tfrac{1}{(d+p-q+1)!}\right]^{(q)}&.\end{aligned} \tag{8.5$'$}$$

Next, we note the following

Lemma 8.3 *For all n, p, q and X, the map*

$$K_p^{[q]}(X\times\square^n; X\times\partial_0\square^n)\to K_p^{[q]}(X\times\square^n)$$

is injective.

Proof Let $E_1,\dots,E_{2n-1}$ be the divisors

$$E_i=\begin{cases} D_i^1 & i=1,\dots,n\\ D_{n-i}^0 & i=n+1,\dots 2n-1.\end{cases}$$

It suffices to show that the maps

$$K_p^{[q]}(X\times\square^n; X\times\{E_1,\dots,E_{i+1}\})\to K_p^{[q]}(X\times\square^n; X\times\{E_1,\dots,E_i\})$$

is injective for all i.

Suppose first that $i\le n$. The inclusion

$$\iota_i^1:\square^{n-1}\cong D_i^1\to\square^n$$

is split by the projection p_i:

$$(x_1,\dots,x_n)\mapsto(x_1,\dots,x_{i-1},1,x_{i+1},\dots,x_n).$$

As p_i sends D_j^1 to $D_i^1\cap D_j^1$, and p_i is flat, p_i defines a splitting in the relativization sequence

$$\begin{aligned}\dots\to K_{p+1}^{[q]}&(X\times D_i^1; X\times\{D_i^1\cap D_1^1,\dots,D_i^1\cap D_{i-1}^1\})\\ &\to K_p^{[q]}(X\times\square^n; X\times\{D_1^1,\dots,D_i^1\})\\ &\to K_p^{[q]}(X\times\square^n; X\times\{D_1^1,\dots,D_{i-1}^1\})\to\dots\end{aligned}$$

giving the injectivity in this case. If $n<i\le 2n-1$, then the inclusion

$$D_i^0\to\square^n$$

is split by the flat map q_i:

$$(x_1,\dots,x_n)\mapsto(x_1,\dots,x_{i-1},0,(1-x_i)(x_{i+1}-1)+1,x_{i+2},\dots,x_n).$$

Then

$$q_i(D_j^1)=\begin{cases} D_i^0\cap D_j^1 & \text{for } j\ne i,\\ D_i^0\cap D_{i+1}^i & \text{for } j=i,\end{cases}$$

$$q_i(D_j^0)=D_i^0\cap D_j^1 \quad\text{for } 1\le j<i.$$

Thus q_i defines a splitting in the relativization sequence

$$\dots \to K^{[q]}_{p+1}\left(X \times E_{n+i}; X \times \{E_{n+i} \cap E_1, \dots, E_{n+i} \cap E_{n+i-1}\}\right)$$

$$\to K^{[q]}_p(X \times \square^n; X \times \{E_1, \dots, E_{n+i}\})$$

$$\to K^{[q]}_p(X \times \square^n; X \times \{E_1, \dots, E_{n+i-1}\}) \to \dots$$

□

Lemma 8.4 *The map (8.5)′ descends to the map*

$$\mathrm{cl}'^{q,p}\colon \mathrm{CH}^q(X,*)_c\left[\tfrac{1}{(d+p-q+1)!}\right] \to K_0(X \times \square^p; X \times \partial\square^p)\left[\tfrac{1}{(d+p-q+1)!}\right]^{(q)}. \quad (8.10)$$

Proof Fix a prime $l > d+p-q+1$, and let k be a generator of $\mathbb{Z}/l^\times$. Using Lemma 8.3, the map sending a subvariety Z to the K_0-class of $\mathcal{O}_Z$ defines the map

$$\alpha^{q,p}\colon \mathcal{Z}^q(X,p+1)_c \to K^{[q]}_0(X \times \square^{p+1}; X \times \partial_0\square^{p+1}).$$

We let

$$\beta^{q,p+1,l}\colon \mathcal{Z}^q(X,p+1)_c \otimes \mathbb{Z}_{(l)} \to K^{[q]}_0(X \times \square^{p+1}; X \times \partial_0\square^{p+1}) \otimes \mathbb{Z}_{(l)}$$

be the map

$$\prod_{i=q+1}^{d+p} \frac{\psi^k - k^i\mathrm{id}}{k^q - k^i} \circ \alpha^{q,p}.$$

Then the image of $\beta^{q,p+1,l}(W)$ in $K^{[q]}_0(X \times \square^p; X \times \partial\square^p) \otimes \mathbb{Z}_{(l)}$ under the restriction to the face $t_n = 0$, followed by the canonical map

$$K^{[q]}_0(X \times \square^p; X \times \partial\square^p) \otimes \mathbb{Z}_{(l)} \to K_0(X \times \square^p; X \times \partial\square^p) \otimes \mathbb{Z}_{(l)}$$

is equal to $\mathrm{cyc}'^{q,p}(d_p(W))$. As

$$K_0(X \times \square^{p+1}; X \times \partial_0\square^{p+1}) = 0,$$

we have

$$\mathrm{cyc}'^{q,p}(d_p(W)) = 0.$$

□

This proves the first part of Theorem 8.1. To prove the injectivity of the map $\mathrm{cl}'^{q,p}$, we proceed as in Step 3. Let l be a prime with

$$l \geq N_{p,q,d};$$

we tensor everything with $\mathbb{Z}_{(l)}$ and fix a generator k of $\mathbb{Z}/l^\times$. Suppose

$$\mathrm{cyc}'^{q,p}(Z) = 0.$$

Take W, U and $\eta \in K_1(U; U \cap \partial\square^n_X) \otimes \mathbb{Z}_{(l)}$ as in Step 3, with

$$\delta(\eta) = \mathrm{cyc}'^{q,p}(Z), \quad (8.9')$$

in the obvious modification of the localization sequence of Step 3.

Passing to an infinite prime to l extension of k and changing notation, we may assume that k is infinite.

As W has codimension at least two, we may assume, by Lemma 7.6, that η is in $\tilde{S}K_1(U; U \cap \partial\square^n_X) \otimes \mathbb{Z}_{(l)}$. By Corollary 7.5, this gives us the corresponding element

$$\eta \in \tilde{S}K_0(U \times \mathbb{A}^1; (U \cap \partial\square^n_X) \times \mathbb{A}^1, U \times \{0,1\}) \otimes \mathbb{Z}_{(l)}.$$

Let $\mathcal{D}$ be the iterated double of $U\times\mathbb{A}^1$ with respect to the divisors $U\cap X\partial\square^n\times\mathbb{A}^1$, $U\times\{0,1\}$. Using the isomorphism (8.8), and the splitting (8.7), we have an element

$$\tau\in\tilde{S}K_0(\mathcal{D}),$$

mapping to $\tilde{\eta}$. By Lemma 7.7, τ is in $F^2_\gamma K_0(\mathcal{D})$.

Let τ_q be the element of $K_0(\mathcal{D})$:

$$\tau_q:=\prod_{i=q+1}^{d+p}\prod_{i=2}^{q-1}\frac{\psi^k-k^i\mathrm{id}}{k^q-k^i}(\tilde{\eta}).$$

By Lemma 7.2, τ_q in in $F^q_\gamma K_0(\mathcal{D})$.

By Lemma 7.1, we have the injective map in $\mathcal{A}_{(l)}$

$$(K_0^{[q]}(X\times\square^p;X\times\partial\square^p)\otimes\mathbb{Z}_{(l)})^{(q)}\to\oplus_{w\in(X\times\square^p)^{(q)}}(K_0^w(\mathcal{O}_{X\times\square^p,w})\otimes\mathbb{Z}_{(l)})^{(q)}.$$

As ψ^k acts as $k^q\mathrm{id}$ on $(K_0^w(\mathcal{O}_{X\times\square^p,w})\otimes\mathbb{Z}_{(l)})^{(q)}$, the same holds true for $(K_0^{[q]}(X\times\square^p;X\times\partial\square^p)\otimes\mathbb{Z}_{(l)})^{(q)}$. In particular, $\mathrm{cyc}'^{q,p}(Z)$ satisfies

$$\prod_{i=q+1}^{d+p+q}\prod_{i=2}^{q-1}\frac{\psi^k-k^i\mathrm{id}}{k^q-k^i}(\mathrm{cyc}'^{q,p}(Z))=\mathrm{cyc}'^{q,p}(Z).$$

Thus, we may replace η with

$$\prod_{i=q+1}^{d+p}\prod_{i=2}^{q-1}\frac{\psi^k-k^i\mathrm{id}}{k^q-k^i}(\eta),$$

and assume from the start that $\tau=\tau_q$.

Using Lemma 7.3, we find a closed, codimension q subset $\mathcal{T}$ of $\mathcal{D}$, intersecting all the iterated doubles of the "faces" in $(U\cap\partial\square^n_X)\times\mathbb{A}^1,U\times\{0,1\}$ properly, such that τ is in the image of

$$K_0^{\mathcal{T}_U}(U\times\mathbb{A}^1:(U\cap\partial\square^n_X)\times\mathbb{A}^1,U\times\{0,1\})\otimes\mathbb{Z}_{(1)}$$

Projecting back to $U\times\mathbb{A}^1$ by the splitting (8.7) (see [17], proof of Theorem 2.3, for details), we find a closed, codimension q subset T_U of $U\times\mathbb{A}^1$, intersecting all "faces" in $U\cap X\partial\square^n\times\mathbb{A}^1,U\times\{0,1\}$ properly, and an element $\tilde{\eta}^{T_U}$ of

$$K_0^{T_U}(U\times\mathbb{A}^1;(U\cap\partial\square^n_X)\times\mathbb{A}^1,U\times 1)\otimes\mathbb{Z}_{(l)}$$

lifting $\tilde{\eta}$.

Let T denote the closure of T_U in $X\times\square^{p+1}$. It is shown in ([17], proof of Theorem 2.7, p. 274) that T intersects all faces of $X\times\partial\square^{p+1}$ properly. By using the splitting of Lemma 8.3, as well as Lemma 8.3 itself, (see [17], p. 274) we may extend the image of $\tilde{\eta}^{T_U}$ in

$$K_0^{T_U}(U\times\mathbb{A}^1;(U\cap\partial\square^n_X)\times\mathbb{A}^1,U\times 1)\otimes\mathbb{Z}_{(l)}$$

to an element ρ of

$$K_0^T(X\times\square^{p+1};X\times\partial\square^{p+1})\otimes\mathbb{Z}_{(l)}.$$

Let T^0 be the set of generic points of T. Taking the image of ρ in

$$K_0^{T^0}(\mathcal{O}_{X\times\square^{p+1},T})\otimes\mathbb{Z}_{(l)}\cong\mathcal{Z}^q_{T_0}(T)\otimes\mathbb{Z}_{(l)}$$

we get a cycle B on $X\times\square^{p+1}$, with $\mathbb{Z}_{(l)}$ coefficients. One easily sees that B is in fact in $\mathcal{Z}^q(X,n+1)\otimes\mathbb{Z}_{(l)}$. By ([17], Lemma 2.6), the relation (8.9)′ implies that

$$d_p(B)=Z,$$

yielding the injectivity of the map $(8.6)'$.

To show that the image of $\mathrm{cl}'^{q,p}$ is contained in $F^q_\gamma K_p(X)$, first suppose that $p > 1$. Then the γ-filtration on

$$K_p(X)/F^q_\gamma = F^2_\gamma K_p(X)/F^q_\gamma$$

is split into eigenspaces for the Adams operators, after inverting $(q-1)!$. As the image of $\mathrm{cl}'^{q,p}$ in $K_p(X)/F^q_\gamma$ must land in the weight q eigenspace by construction, this forces the image in this quotient to be zero. For $p = 0, 1$, $q \geq 2$, it follows easily from our construction, together with Theorem 7.4 and Lemma 7.7, that the image of $\mathrm{cl}'^{q,p}$ is contained in $F^2_\gamma K_p(X)$; the proof then proceeds as above. For $p = 1$, $q = 1$, we have $K_1(X) = F^1_\gamma K_1(X)$, whence the result.

Similarly, the restriction of gamma filtration to $F^q_\gamma K_p(X)$ is split into eigenspaces for the ψ^k, after inverting $(d+p-q+1)!$. By Lemma 7.3, and the argument of Step 2, $\mathrm{cl}'^{q,p}$ is surjective; as we have just proved the injectivity, the proof of Theorem 8.1 is complete. □

9 Appendix A. Closed Simplicial Model Categories

Suppose we are given a class of morphisms $\mathcal{F}$ in a category $\mathcal{C}$. A morphism $q\colon A \to B$ has the *left lifting property* (LLP) with respect to $\mathcal{F}$ if there is a lifting $h\colon B \to Z$ for each commutative diagram

$$\begin{array}{ccc} A & \longrightarrow & Z \\ {\scriptstyle q}\downarrow & & \downarrow{\scriptstyle p} \\ B & \longrightarrow & W \end{array}$$

with $p\colon Z \to W$ in $\mathcal{F}$. A morphism $p\colon Z \to W$ has the *right lifting property* (RLP) with respect to $\mathcal{F}$ if there is a lifting $h\colon B \to Z$ for each commutative diagram

$$\begin{array}{ccc} A & \longrightarrow & Z \\ {\scriptstyle q}\downarrow & & \downarrow{\scriptstyle p} \\ B & \longrightarrow & W \end{array}$$

with $q\colon A \to B$ in $\mathcal{F}$.

We refer to [18] and [19] for the notions and basic properties of closed model categories; we recall that a closed model category is a category $\mathcal{C}$ together with three distinguished classes of morphisms: fibrations, cofibrations and weak equivalences; a morphism which is a (co)fibration and a weak equivalence is called a *trivial* (co)fibration. In addition, these classes satisfy the axioms CM1-CM5 of ([19], p. 233):

CM1: $\mathcal{C}$ is closed under finite projective and injective limits.

CM2: Let $X \xrightarrow{f} Y \xrightarrow{g} Z$ be maps in $\mathcal{C}$. Then if two of the maps f, g and gf are weak equivalences, so is the third.

CM3: If f is a retract of g (i.e., there are maps $a\colon f \to g$, $b\colon g \to f$ in the category of maps such that $ba = \mathrm{id}_f$) and g is a weak equivalence, a fibration, or a cofibration, then so is f.

CM4: Cofibrations have the LLP with respect to trivial fibrations, and fibrations have the RLP with respect to trivial cofibrations.

CM5: Any map f may be factored in two ways:

i) as pq, where p is a fibration, q is a cofibration, and p is a weak equivalence.

ii) as pq, where p is a fibration, q is a cofibration, and q is a weak equivalence.

These five axioms imply

i) The class of cofibrations (resp. trivial cofibrations) is closed under composition and co-base change, and contains all isomorphisms.

ii) The class of fibrations (resp. trivial fibrations) is closed under composition and base change, and contains all isomorphisms.

iii) (Axiom M6 of [18]) The fibrations, cofibrations and weak equivalences determine each other by

a) a map f is a fibration $\Leftrightarrow$ f has the RLP with respect to trivial cofibrations,

b) a map f is a cofibration $\Leftrightarrow$ f has the LLP with respect to trivial fibrations,

c) a map f is a weak equivalence $\Leftrightarrow$ f has a factorization $f = uv$ with u a trivial fibration and v a trivial cofibration.

We have the category Δ with objects the ordered sets $[0], \ldots, [n], \ldots$,

$$[n] := \{0 < 1 < \ldots < n\},$$

and with morphisms being order-preserving maps; we let $\mathcal{S}$ denote the category of simplicial sets: the category of functors

$$\Delta^{\mathrm{op}} \to \mathbf{Sets}.$$

We have the object

$$\Delta_n := \mathrm{Hom}_\Delta(-, [n])\colon \Delta^{\mathrm{op}} \to \mathbf{Sets}$$

of $\mathcal{S}$. Similarly, we consider the functor

$$\mathrm{Hom}_\Delta(-,-)\colon \Delta^{\mathrm{op}} \times \Delta \to \mathbf{Sets}$$

as a functor

$$\begin{aligned} \tilde{\Delta}\colon \Delta &\to \mathcal{S} \\ \tilde{\Delta}([n]) &= \Delta_n. \end{aligned}$$

We have the category of topological spaces **Top**, and the geometric realization functor

$$|-|\colon \mathcal{S} \to \mathbf{Top}.$$

We recall that the categories $\mathcal{S}$ and **Top** are both closed model categories, with the following notions of fibrations, cofibrations and weak equivalences:

Fibrations: In **Top**, we have the classical notion of a fibration being a map $f\colon X \to Y$ with the (right) homotopy lifting property. In $\mathcal{S}$, we have the notion of a *Kan fibration*: for each n and k, $0 \le k \le n$, the sub-simplicial set $\wedge_{n,k}$ of Δ_n, where $f\colon [i] \to [n]$ in Δ is in $\wedge_{n,k}$ if $f([i])$ does not contain the subset $[n]\backslash\{k\}$ of $[n]$. A map $f\colon X \to Y$ in $\mathcal{S}$ is a Kan fibration if f has the right lifting property with respect to the map $\wedge_{n,k} \to \Delta_n$, for each n and k.

Weak equivalences: In **Top**, a map f is a weak equivalence if f induces an isomorphism on π_n for each n. In $\mathcal{S}$, a map f is a weak equivalence if the geometric realization $|f|$ is a weak equivalence in **Top**.

Cofibrations: In both **Top** and $\mathcal{S}$, cofibrations are defined by requiring the left lifting property with respect to maps which are both a fibration and a weak equivalence. In $\mathcal{S}$, a map $f\colon X \to Y$ is a cofibration if and only if the map on n-simplices, $f_n\colon X_n \to Y_n$, is injective for each n.

We recall that a *simplicial category* ([18], II, Section 1) $\mathcal{C}$ is a category enriched in simplicial sets; i.e, a category $\mathcal{C}$ together with

i) a functor $(X,Y) \mapsto \mathcal{H}om_{\mathcal{C}}(X,Y)$ from $\mathcal{C}^{\mathrm{op}} \times \mathcal{C}$ to $\mathcal{S}$,

ii) an associative "composition law" in $\mathcal{S}$

$$\mathcal{H}om_{\mathcal{C}}(X,Y) \times \mathcal{H}om_{\mathcal{C}}(Y,Z) \to \mathcal{H}om_{\mathcal{C}}(X,Z),$$

iii) an identification

$$u \mapsto \tilde{u}$$

of $\mathrm{Hom}_{\mathcal{C}}(-,-)$ with the 0-simplices $\mathcal{H}om_{\mathcal{C}}(-,-)_0$ of $\mathcal{H}om_{\mathcal{C}}(-,-)$, such that

$$f \circ s_0^n(\tilde{u}) = \mathcal{H}om_{\mathcal{C}}(u,Z)_n(f); \qquad s_0^n(\tilde{u}) \circ g = \mathcal{H}om_{\mathcal{C}}(W,u)_n(g)$$

for $u \in \mathrm{Hom}_{\mathcal{C}}(Y,Z)$, $f \in \mathcal{H}om_{\mathcal{C}}(X,Y)_n$ and $g \in \mathcal{H}om_{\mathcal{C}}(Y,W)_n$, where s_0^n is the map induced by the unique surjection $[n] \to [0]$.

The category $\mathcal{S}$ is a simplicial category, with

$$\mathcal{H}om_{\mathcal{S}}(X,Y) = \mathrm{Hom}_{\mathcal{S}}(X \times \tilde{\Delta}, Y).$$

Similarly, category **Top** is a simplicial category, with

$$\mathcal{H}om_{\mathbf{Top}}(X,Y) = \mathrm{Hom}_{\mathbf{Top}}(X \times |\tilde{\Delta}|, Y).$$

If $\mathcal{C}$ is a simplicial category, X an object of $\mathcal{C}$ and K a simplicial set, we say that $X \otimes K$ exists if there is a pair $(X \otimes K, \alpha)$, with $X \otimes K$ an object of $\mathcal{C}$, and α a map (evaluation)

$$\alpha\colon K \to \mathcal{H}om_{\mathcal{C}}(X, X \otimes K),$$

yielding an isomorphism

$$\mathcal{H}om_{\mathcal{C}}(X \otimes K, -) \to \mathcal{H}om_{\mathcal{S}}(K, \mathcal{H}om_{\mathcal{C}}(X,-))$$

by the obvious composition; the functor $\mathcal{H}om_{\mathcal{S}}(K, \mathcal{H}om_{\mathcal{C}}(X,-))$ is thus represented by $(X \otimes K, \alpha)$. We say X^K exists if the functor $\mathcal{H}om_{\mathcal{S}}(K, \mathcal{H}om_{\mathcal{C}}(-,X))$ is similarly represented by an object X^K and an evaluation map

$$\beta\colon K \to \mathcal{H}om_{\mathcal{C}}(X^K, X).$$

We recall from ([18], II, Section 2) that a *closed simplicial model category* is a closed model category $\mathcal{C}$ which is also a simplicial category, and satisfies

SM0: If X is an object of $\mathcal{C}$, then the objects $X \otimes K$ and X^K exist for all finite simplicial sets K.

SM7: If $q\colon A \to B$ is a cofibration, and $p\colon X \to Y$ is a fibration, then

$$\mathcal{H}om(B,X) \xrightarrow{(q^*,p_*)} \mathcal{H}om(A,X) \times_{\mathcal{H}om(A,Y)} \mathcal{H}om(B,Y)$$

is a fibration in $\mathcal{S}$, which is trivial if either q or p is trivial.

□

The closed model categories $\mathcal{S}$ and **Top** are closed simplicial model categories ([18], II, Section 3); making the obvious modifications in the definitions, the pointed categories $\mathcal{S}^*$ and **Top*** are closed simplicial model categories as well.

In ([18], II, Section 4), Quillen gives a sufficient criterion for the category $s\mathcal{A}$ of simplicial objects in a category $\mathcal{A}$ to be a simplicial closed model category, where a map f in $s\mathcal{A}$ is a fibration (resp. weak equivalence) if $\mathcal{H}om(P, f)$ is a fibration (resp. weak equivalence) for all projective objects of $\mathcal{A}$, and cofibrations are determined by the LLP with respect to trivial fibrations.

For a category $\mathcal{C}$ and a small category I, we let $\mathcal{C}^I$ denote the category of functors

$$X: I \to \mathcal{C}.$$

As an application of the criterion of Quillen mentioned above, Bousfield and Kan ([5], XI, Section 8, proof of Proposition 8.1) prove the following result:

Theorem A.1 *Let I be a small category. Then the simplicial category $\mathcal{S}^I$ is a closed simplicial model category with the following notions of fibration, weak equivalence, and cofibration:*

Fibrations and weak equivalences: a map $f: X \to Y$ is a fibration (resp. weak equivalence) in $\mathcal{S}^I$ if and only if the map $f(i): X(i) \to Y(i)$ is a fibration (resp. weak equivalence) in $\mathcal{S}$ for each $i \in I$.

Cofibrations: A map $f: X \to Y$ is a cofibration in $\mathcal{S}^I$ if and only if f has the left lifting property with respect to maps g which are a fibration and a weak equivalence.

□

We conclude with a few additional results which we need in the body of the paper.

Let $\mathcal{A}$ be a closed model category, with initial object $\emptyset$ and final object $*$. An object X of $\mathcal{A}$ is called *cofibrant* (resp. fibrant) if the canonical map $\emptyset \to X$ (resp. $X \to *$) is a cofibration (resp. fibration). Call X *bifibrant* if X is fibrant and cofibrant. If A is an object of $\mathcal{A}$, a *bifibrant model of A* is a diagram of weak equivalences

$$A = A_0 \xrightarrow{i_0} B_1 \xleftarrow{j_1} A_1 \xrightarrow{i_1} \ldots \xleftarrow{j_n} A_n = A',$$

with A' bifibrant. Given A, we can always find a bifibrant model, e.g., first by factoring the canonical map $A \to *$ as

$$A \xrightarrow{i} B \xrightarrow{p} *$$

with i a trivial cofibration and p a fibration, then factoring the canonical map $\emptyset \to B$ as

$$\emptyset \xrightarrow{q} A' \xrightarrow{j} B$$

with q a cofibration and j a trivial fibration.

For a closed model category $\mathcal{A}$, the *homotopy category of $\mathcal{A}$*, Ho$\mathcal{A}$ is the category gotten from $\mathcal{A}$ by localizing with respect to the weak equivalences; this category exists ([18], I, Section 1 Theorem 1). If $\mathcal{A}$ is a closed simplicial model category, the relation of homotopy equivalence of maps is defined by setting

$$[A, B] := \pi_0(|\mathcal{H}om_{\mathcal{A}}(A, B)|).$$

This gives the notion of a homotopy equivalence $f: A \to B$, and a strong deformation retract, as in the case of topological spaces.

Lemma A.2 *Let A be a closed simplicial model categroy. A trivial cofibration between fibrant objects, and a trivial fibration between cofibrant objects are strong* *deformation retracts. A weak equivalence between bifibrant objects is a homotopy equivalence.*

Proof The first pair of assertions is a part of ([18], II, Section 2, Corollary to Proposition 4). If $f\colon X \to Y$ is a weak equivalence of bifibrant objects, factor f as $f = uv$, with $v\colon X \to Z$ a trivial cofibration, and $u\colon Z \to Y$ a trivial fibration. Then Z is also bifibrant, hence u and v are both homotopy equivalences. □

Remark A. 3 We give a list of basic properties of homotopy classes of maps, and the relation with maps in the homotopy category, for a closed simplicial model category $\mathcal{A}$:

i) If A is cofibrant, and B is fibrant, then $[A, B]$ is the same as the set of maps from A to B in the homotopy category $\mathrm{Ho}\mathcal{A}$.

ii) A homotopy equivalence $f\colon A \to B$ is a weak equivalence; if A and B are bifibrant, then a weak equivalence $f\colon A \to B$ is a homotopy equivalence. In particular, if A, X and Y are bifibrant, and $f\colon X \to Y$ is a weak equivalence, then

$$f_*\colon [A, X] \to [A, Y]; \qquad f^*\colon [X, A] \to [Y, A]$$

are isomorphisms.

iii) $\mathrm{Ho}\mathcal{A}$ is equivalent to the category with objects the bifibrant objects of $\mathcal{A}$, with morphisms

$$\mathrm{Hom}(A, B) := [A, B].$$

10 Appendix B. Homotopy Limits

Let I be a small category. For an object i of I, we have the *over category* I/i of maps $j \to i$ in I, as well as the *nerve* $\mathcal{N}(I/i)$. Each map $s\colon i \to i'$ gives the functor

$$s_*\colon I/i \to I/i';$$

taking the nerve gives the functor

$$\mathcal{N}(I/-)\colon I \to \mathcal{S}. \tag{B.1}$$

Bousfield and Kan have defined homotopy the limit of an object X of $\mathcal{S}^I$ as

$$\operatorname*{holim}_I X = \mathcal{H}om_{\mathcal{S}^I}(\mathcal{N}(I/-), X). \tag{B.2}$$

This gives the functor

$$\operatorname*{holim}_I\colon \mathcal{S}^I \to \mathcal{S}$$

having the following properties:

(B.3i) Let $f\colon X \to Y$ be a fibration (resp. weak equivalence). Then

$$\operatorname*{holim}_I f\colon \operatorname*{holim}_I X \to \operatorname*{holim}_I Y$$

is a fibration (resp. weak equivalence).

ii) $\operatorname*{holim}_I$ is right adjoint to the functor

$$\begin{aligned} (-) \times \mathcal{N}(I/-)\colon \mathcal{S} &\to \mathcal{S}^I \\ A &\mapsto A \times \mathcal{N}(I/-) \end{aligned}$$

and hence preserves projective limits.

iii) If $f\colon J \to I$, $Y\colon J \to \mathcal{C}$ and $X\colon I \to \mathcal{C}$ are functors, and

$$\omega\colon X \circ f \to Y$$

is a natural transformation, we have the natural map

$$\omega_* \circ f^*\colon \operatorname{holim}_I X \to \operatorname{holim}_J Y.$$

In addition, there is the basic construction of ([5], XI, Section 7.6):

Theorem B.1 *Let X be a fibrant object of $\mathcal{S}^{*I}$. Then there is a "fringed" spectral sequence*

$$E_2^{s,t} := \lim_{\leftarrow}{}^s \pi_{-t} X; \quad 0 \le s \le t \Longrightarrow \pi_{-s-t} \operatorname{holim}_I X,$$

which converges if e.g. there is for each s an integer $N(s)$, $s \le N(s) < \infty$ with

$$E_r^{s,s+i} = E_{N(s)}^{s,s+i}$$

for all $r \ge N(s)$.

□

This is extended by Dwyer and Kan to function complexes. Let aI be the *twisted arrow category* with objects maps $i \to j$ in I, and maps $(i \to j) \to (i' \to j')$ commutative diagrams

$$\begin{array}{ccc} i & \longrightarrow & i' \\ \big\downarrow & & \big\downarrow \\ j & \longrightarrow & j' \end{array}$$

For X and Y in $\mathcal{S}^I$, we have the functor

$$\mathcal{H}om(X,Y)\colon aI \to \mathcal{S}$$
$$\mathcal{H}om(X,Y)(i \to j) = \mathcal{H}om_{\mathcal{S}}(X(i), Y(j))$$

with the obvious maps. In addition, there is the obvious isomorphism

$$\mathcal{H}om_{\mathcal{S}^I}(X,Y) \cong \lim_{\substack{\leftarrow \\ aI}} \mathcal{H}om(X,Y). \tag{B.4}$$

The main result of [6] (Theorem 3.3) is

Theorem B.2 *Suppose that X is cofibrant and Y is fibrant. Then the natural map*

$$\lim_{\substack{\leftarrow \\ aI}} \mathcal{H}om(X,Y) \to \operatorname{holim}_{aI} \mathcal{H}om(X,Y)$$

is a weak equivalence. The same holds in the pointed setting. □

Actually, the pointed case is not considered in [6], but the same proof gives the result. As consequence, we have

Corollary B.3 *Let X be a cofibrant object of $\mathcal{S}^I$, and Y a pointed fibrant object of $\mathcal{S}^I$. There is a "fringed" spectral sequence*

$$E_2^{s,t} := \lim_{\substack{\leftarrow \\ aI}}{}^s \pi_{-t}\mathcal{H}om(X,Y); \quad 0 \le s \le t \Longrightarrow \pi_{-s-t}\mathcal{H}om_{\mathcal{S}^I}(X,Y).$$

This spectral sequence converges if e.g. the condition of Theorem B.1 are satisfied (for the object $\mathcal{H}om(X,Y)$ of $\mathcal{S}^{aI}$). If X is pointed as well, the same holds for the pointed $\mathcal{H}om$.

Proof This follows directly from the isomorphism (B.4), Theorem B.2, and Theorem B.1, applied to the fibrant object $\mathcal{H}om(X,Y)$ of $\mathcal{S}^{aI}$; the proof in the pointed case is the same. □

Corollary B.4 *Let $f\colon X \to X'$ be a map of cofibrant objects of $\mathcal{S}^I$, $g\colon Y' \to Y$ a map of fibrant objects of $\mathcal{S}^I$. If the maps*

$$f(i)^*g(j)_*\colon \mathcal{H}om_{\mathcal{S}}(X'(i),Y'(j)) \to \mathcal{H}om_{\mathcal{S}}(X(i),Y(j))$$

are weak equivalences for all pairs $i,j \in I$ for which there is a map $i \to j$ in I, then

$$f^*g_*\colon \mathcal{H}om_{\mathcal{S}}(X',Y') \to \mathcal{H}om_{\mathcal{S}}(X,Y)$$

is a weak equivalence. The same holds in the pointed case as well.

Proof The map in $\mathcal{S}^{aI}$

$$f^*g_*\colon \mathcal{H}om(X',Y') \to \mathcal{H}om(X,Y)$$

is a weak equivalence of fibrant objects of $\mathcal{S}^{aI}$. By ([5], XI, 5.5, 5.6), the map

$$\operatorname*{holim}_{aI} \overset{*}{f} g_*\colon \operatorname*{holim}_{aI} \mathcal{H}om(X',Y') \to \operatorname*{holim}_{aI} \mathcal{H}om(X,Y)$$

is a weak equivalence of fibrant objects of $\mathcal{S}$. By Theorem B.2, the natural maps

$$\mathcal{H}om_{\mathcal{S}^I}(X',Y') \cong \lim_{\substack{\leftarrow \\ aI}} \mathcal{H}om(X',Y') \to \operatorname*{holim}_{aI} \mathcal{H}om(X',Y')$$
$$\mathcal{H}om_{\mathcal{S}^I}(X,Y) \cong \lim_{\substack{\leftarrow \\ aI}} \mathcal{H}om(X,Y) \to \operatorname*{holim}_{aI} \mathcal{H}om(X,Y)$$

are weak equivalences, whence the result. □

References

[1] Atiyah, M. and Tall, D. [1969], *Group representations, λ-rings and the J-homomorphism*, Topology **8**, 253-297.

[2] Berthelot, P. [1971], *Généralités sur les λ-anneaux*, Exposé V, SGA 6, Lecture Notes in Math. **225**, 297-364.

[3] Bloch, S. [1986], *Algebraic cycles and higher K-theory*, Adv. in Math. **61** No. 3, 267-304.

[4] Bloch, S. and Lichtenbaum, S. [1994], *A spectral sequence for motivic cohomology*, preprint.

[5] Bousfield, A. K. and Kan, D. M. [1972], *Homotopy limits, Completions and Localizations*, Lecture Notes in Math. **304**, Springer.

[6] Dwyer, W. G. and Kan, D. M. [1983], *Function complexes for diagrams of simplicial sets*, Nederl. Akad. Wetensch. Proc. Ser. A **86(2)**, 139-146 , Indag. Math. **45** (1983).

[7] Gillet, H. and Soulé, C. *Filtrations on higher algebraic K-theory*, unpublished manuscript.

[8] Grayson, D. [1976], *Higher algebraic K-theory II (after Daniel Quillen)*, in Algebraic $\boldsymbol{K}$-Theory, Lecture Notes in Math. **551**, Springer.

[9] Grayson, D. [1989], *Exterior power operations on algebraic K-theory*, K-theory **3**, 247-260.

[10] Grayson, D. [1992], *Adams operations on higher K-theory*, K-theory **6**, 97-111.

[11] Grothendieck, A. [1971], *Classes de faisceaux et théorème de Riemann-Roch*, Exposé 0, SGA 6, Lecture Notes in Math. **225**, 297-364.

[12] Hiller, H. [1981], *λ-rings and algebraic K-theory*, JPAA **20**, 241-266.

[13] Jardine, J. F. [1987], *Simplicial presheaves*, JPAA **47**, 35-87.

[14] Kleiman, S. [1974], *The transversality of a general translate*, Comp. Math. **28**, 287-297.

[15] Kratzer, C. [1980], *λ-structure en K-théorie algébrique*, Com. Math. Helv. **55** No. 2, 233-254.

[16] Lecomte, F. [1996], *Opération d'Adams en K-théorie algébrique*, preprint.

[17] Levine, M. [1994], *Bloch's higher Chow groups revisited*, in K-theory Strasbourg 1992 (C. Kassel and J. L. Loday, ed.), Astérisque vol. 226, N. Schappacher, Soc. Math. de France, 235-320.

[18] Quillen, D. [1967], *Homotopical Algebra*, Lecture Notes in Math. **43**, Springer.

[19] Quillen, D. [1969], *Rational homotopy theory*, Ann. Math. **90**, 205-295.
[20] Quillen, D. [1973], *Higher algebraic K-theory I*, in Algebraic $\boldsymbol{K}$-Theory I, Lecture Notes in Math. **341**, Springer.
[21] Quillen, D. [1976], *Charateristic classes of representations*, in Algebraic $\boldsymbol{K}$-Theory, Lecture Notes in Math. **551**, Springer.
[22] Schechtmann, V., *On the delooping of Chern character and Adams operations*, in K-theory, Arithmetic, and Geometry (Yu. Manin, ed.) LNM 1289, 265-319.
[23] Serre, J.-P. [1968], *Groupes de Grothendieck des schémas en groupes réductifs déployés*, Publ. Math. I.H.E.S. **34**, 37-52.
[24] Soulé, C. [1985], *Operations en K-théorie algébrique*, Can. J. Math. **37** No. 3, 488-550.
[25] Soulé, C. [1995], *Letter to S. Bloch.*
[26] Thomason, R. W. and Trobaugh, R. [1990], *Higher algebraic K-theory of schemes and of derived categories*, in The Grothendieck Festschrift vol. III, Birkhäuser, Boston.

Fields Institute Communications
Volume **16**, 1997

Quasi-Motives of Curves

Stephen Lichtenbaum
Department of Mathematics
Brown University
Providence, RI USA 02912-2591
Stephen_Lichtenbaum@brown.edu

Abstract. We introduce the Borel-Moore and cohomology with compact supports 1-motives of a curve C, and extend our earlier work relating the homological 1-motive of C to the Suslin complex of C to include these new 1-motives as well as Deligne's original (cohomological) 1-motive.

Introduction

In this paper we discuss the homology, Borel-Moore homology, cohomology, and cohomology with compact supports of a curve C over an algebraically closed field k, from a motivic point of view. We begin by defining the Deligne 1-motive of C corresponding to each of the four theories, the cohomology 1-motive being the one defined by Deligne in [2]. If T is any of the four above theories, call the corresponding 1-motive $\mathcal{H}_T(C)$.

We next define for each theory T, two quasi-isomorphic complexes $Z_T(C)$ and $G_T(C)$, where $Z_T(C)$ is an infinite bounded-above cycle complex and $G_T(C)$ is a two-term Gersten-type complex. If T denotes homology then $Z_T(C)$ is the Suslin complex of C (see [5] and [9]) and when T is Borel-Moore homology, $Z_T(C)$ is Bloch's higher Chow group complex [1]. If T is Borel-Moore homology, $G_T(C)$ is the Gersten complex of C, called G'_m by Deninger ([3]) and the quasi-isomorphism of $G_T(C)$ and $Z_T(C)$ is a special case of a theorem of Nart [6].

We then show that in the sense of derived categories, the complexes $G_T(C)(k)$ (or equivalently, $Z_T(C)(k)$) have filtrations whose graded pieces are $\mathcal{H}_T(C)(k)$ (corresponding to the first homology or cohomology of C) and two other terms which may naturally be interpreted as the zeroth and second homology or cohomology of C.

We also note that complexes similar to our $Z_T(C)$ but defined for any variety V have been defined by Friedlander and Voevodsky [4].

1991 *Mathematics Subject Classification.* Primary 14F99; Secondary 14H99.
The author has been partially supported by N.S.F. grant DMS-9307671 and would like to thank the I. H. E. S. for its generous hospitality.

These theorems are all "shadow versions" of what should be true; namely that there exist a category $\mathcal{M}$ of motives and four objects $\tilde{Z}_T(C)$ in $\mathcal{M}$, the homology, Borel-Moore homology, cohomology, and cohomology with compact supports motives of C. The category $\mathcal{M}$ should contain the category of 1-motives as a full subcategory, and the motives $\tilde{Z}_T(C)$ should be filtered by three objects, one being $\mathcal{H}_T(C)$ and the others corresponding in a natural way to the zeroth and second cohomology motives of C. We hope to soon return to this point of view in an upcoming paper with Tom Goodwillie.

1 The 1-Motive of a Curve

We first recall from [2] the definition of a 1-motive.

Definition 1.1 *A one-motive M consists of a connected commutative algebraic group G over k which is an extension of an abelian variety A by a torus T, together with a finitely generated free abelian group X and a map u from X to $G(k)$. We say $M = (G, A, T, X, u)$.*

Let E be a smooth curve over k, and let $\mathcal{P}$ be a finite collection of disjoint finite sets $\mathcal{P}_i$ of closed points P_{ij} of E. Then there exists a unique curve $E^* = E/\mathcal{P}$ and a map π from E to E^* such that if D is any curve over k, and ϕ is a map from E to D, then ϕ factors through π if and only if $\phi(P_{ij}) = \phi(P_{ik})$ for all i, j, k. E^* is a curve with only nodal singularities. A curve of this form is said to be seminormal. It is immediate that if C is any curve, then there exists a seminormal curve C' and a map π from C' to C such that π is a bijection, and the pair (C', π) is unique up to canonical isomorphism. C' is called the *seminormalization* of C.

Now let C be any seminormal curve. Let $\bar{C}$ be the normalization of C, $\hat{C}$ the smooth completion of C, and $\hat{\bar{C}}$ the smooth completion of $\bar{C}$ (the normalization of $\hat{C}$). C is described uniquely by the data consisting of the projective smooth curve $\hat{\bar{C}}$, a finite set of disjoint finite sets of points P_{ij} on $\hat{\bar{C}}$, and a finite set S of points on $\hat{\bar{C}}$ disjoint from the union of the P_{ij}'s. C is given by $(\hat{\bar{C}} - S)/[P_{ij} \to Q_i]$, that is, by taking out the set S and identifying all the points P_{ij} for fixed i to one point Q_i of C. We write $C = (\hat{\bar{C}} - S)/[P \to Q]$.

Let I be the set of connected components C_i of $\hat{\bar{C}}$. Let ρ be the map from $\mathbf{Z}[S]$ to $\mathbf{Z}[I]$ defined by sending each point to the connected component containing it. Let $\mathbf{Z}_0[S]$ be the kernel of ρ, and let u' be the map from $\mathbf{Z}[S]$ to $Pic(\hat{\bar{C}})$ sending each point to its divisor class.

Definition 1.2 *Following Deligne [2], we define the* cohomological 1-motive *(which Deligne just calls the 1-motive) $\mathcal{H}^1(C)$ to be $\mathbf{Z}_0[S] \xrightarrow{u} Pic^0(\hat{\bar{C}}/[P \to Q])$, where the map u is induced by u'.*

(Recall that Pic^0 of a curve is the group of all line bundles whose degree is zero on each irreducible component.)

Definition 1.3 *(See [5]) We define the* homological 1-motive *$\mathcal{H}_1(C)$ to be $\mathbf{Z}_0[P \to Q] \xrightarrow{v} Pic^0(\hat{\bar{C}}/S)$, where $\mathbf{Z}_0[P \to Q]$ is the subset of $\mathbf{Z}[P_{ij}]$ which is mapped to zero in both $\mathbf{Z}[Q_i]$ and $\mathbf{Z}[I]$.*

Definition 1.4 *We define the* Borel-Moore homological 1-motive $\mathcal{H}_1^{BM}(C)$ *to be* $\mathbf{Z}_0[P \to Q, S] \to Pic^0(\hat{\hat{C}})$ *where* $\mathbf{Z}_0[P \to Q, S]$ *is the subset of the group* $\mathbf{Z}[P, S]$ *which is mapped to zero in both* $\mathbf{Z}[Q, S]$ *and* $\mathbf{Z}[I]$.

Definition 1.5 *We define the* cohomology with compact support 1-motive $\mathcal{H}_c^1(C)$ *to be* $Pic^0(\hat{\hat{C}}/S, [P \to Q])$. *(We are looking here at the curve obtained from* $\hat{\hat{C}}$ *by identifying the points of* S *to one point and then identifying the* P_{ij}*'s for fixed* i.*).*

If C is an arbitrary curve, we define each of the above four 1-motives of C to be the corresponding 1-motive of its seminormalization.

In [2], Deligne associates with every 1-motive M over $\mathbf{C}$ a mixed Hodge structure $T(M)$ and shows that T embeds the category of 1-motives as a certain full subcategory of the category of mixed Hodge structures. He also defines a mixed Hodge structure $U^i(V)$ associated with any cohomology group $H^i(V)$ of an arbitrary variety V over $\mathbf{C}$. Deligne shows that if C is any curve the mixed Hodge structure $T(\mathcal{H}^1(C))$ may be naturally identified with $U^1(C)$.

We may also associate a mixed Hodge structure $U_i(V)$ with any homology group $H_i(V)$ simply by taking the mixed Hodge structure dual to $U^i(V)$. Deligne also introduces in [2] the notion of duality of 1-motives, and shows that the functor T is compatible with this duality.

Since Ramachandran has shown in [7] that the homological 1-motive $\mathcal{H}_1(C)$ is dual to the cohomological 1-motive $\mathcal{H}^1(C)$, it follows that $T(\mathcal{H}_1(C))$ may be identified with $U_1(C)$.

It should not be too difficult to show that the analogous statements hold for Borel-Moore homology and cohomology with compact supports, i.e., that $T(\mathcal{H}_1^{BM}(C)) = U_1^{BM}(C)$ and $T(\mathcal{H}_c^1(C)) = U_c^1(C)$. Also, of course, $\mathcal{H}_1^{BM}(C)$ and $\mathcal{H}_c^1(C)$ should be dual 1-motives for C a curve over an arbitrary algebraically closed field k.

2 The Acyclicity Theorem

Let A_n be the polynomial ring $k[t_0, \ldots, t_n]/(t_0 + \cdots + t_n - 1)$. If ϕ is an order-preserving map from $[0, n]$ to $[0, m]$, ϕ induces a map ϕ^* from A_m to A_n by

$$\phi^*(t_i) = \sum_{\phi(j)=i} t_j$$

where we put $\phi^*(t_i) = 0$ if $\phi^{-1}(i)$ is empty. This is immediately seen to give $A_\bullet$ the structure of a simplicial ring. If we let $\Delta^n = Spec(A_n)$, $\Delta^\bullet$ becomes a cosimplicial k-scheme.

Let d_j be the map from $[0, n-1]$ to $[0, n]$ such that $d_j(i) = i$ for $i < j$ and $d_j(i) = i+1$ for $i \geq j$. Let s_i be the map from $[0, n]$ to $[0, n-1]$ such that $s_i(j) = j$ for $j \leq i$ and $s_i(j) = j - 1$ for $j > i$. We have

1) $d_j^*(t_i) = t_i$ for $i \leq j - 1$
2) $d_j^*(t_j) = 0$
3) $d_j^*(t_i) = t_{i-1}$ for $i \geq j + 1$

Let C be a normal curve over k, and let S be a finite set of closed points of C. Let K_n (resp. K'_n) be the multiplicative group of functions f which are non-zero on every component of $\Delta^n \times C$ such that f satisfies a) (resp. a') and b):

a) Supp $(f) \subseteq \Delta^n \times C$ is quasi-finite over Δ^n. a') Supp (f) has proper intersection with all faces of $\Delta^n \times C$. b) f is identically 1 on $\Delta^n \times S$.

Observe that if ϕ is any cosimplicial map from Δ^m to Δ^n, $\phi^*(f) = f \circ (\phi \times 1)$ is in K_m (resp. K'_m) if f is in K_n (resp. in K'_n). So $K_\bullet$ (resp. $K'_\bullet$) has a natural structure of simplicial abelian group, induced by $f \mapsto \phi^*(F)$.

Theorem 2.1 *$K_\bullet$ and $K'_\bullet$ are acyclic.*

Proof (We give the proof for $K_\bullet$ but the same proof (without the necessity for the perturbation) works for $K'\bullet$). We begin by introducing a modified version of a homotopy defined by Suslin and Voevodsky [9]. Let f in K_n be such that $d_j^* f = 1$ for all j. Let $g_i = (t_{i+1} + \cdots + t_{n+1}) + (t_0 + \cdots + t_i)s_i^*(f)$. We observe that

a) $j < i \Rightarrow d_j^*(g_i) = 1$ since $d_j^* s_i^* = s_{i-1}^* d_j^*$.

b) $j > (i+1) \Rightarrow d_j^*(g_i) = 1$ similarly.

c) $j = i \Rightarrow d_i^*(g_i) = (t_0 + \cdots + t_{i-1})f + (t_i + \cdots + t_n)$.

d) $j = i+1 \Rightarrow d_{i+1}^*(g_i) = (t_0 + \cdots + t_i)f + (t_{i+1} + \cdots + t_n)$. In particular, $d_0^*(g_0) = 1$ and $d_{n+1}^*(g_n) = f$.

Now let $g = g_n - g_{n-1} + g_{n-2} - \cdots - g_0 + 1$ if n is odd, and let $g = g_n - g_{n-1} + g_{n-2} - \cdots + g_0$ if n is even. We see that $d_{n+1}^*(g) = f$, $d_i^*(g) = 1$ for $i \leq n$. Formally, f is a boundary. However, there is no reason for g to be in K_{n+1}, so we have to perturb it. We first need:

Lemma 2.2 *The maximal order of the poles of g on the curve $P \times C$ for P a closed point of Δ^{n+1} is bounded by M independent of P.*

Proof We may clearly assume C is projective and connected. Let P' be a closed point of Δ^n. Since $(f) = \sum n_i D_i$ is quasi-finite over Δ^n, no $D_i \supseteq P' \times C$. So $v_D(f) = 0$ for all $D \supseteq P' \times C$, which implies that f is in $O_{P' \times C}$ (the local ring of the generic point of $P' \times C$) since $O_{P' \times C} = \bigcap O_D$, $D \supseteq P' \times C$.

Let P be a closed point of Δ^{n+1}. It follows that $s_i^*(f)$ is in $O_{P \times C}$, and hence each g_i and then g is in $O_{P \times C}$. Let $C' = Spec(R')$ be an affine open subset of C, and let E be the field of functions of R'. Note $E = k(C)$. Let $R = A_{n+1} \otimes R'$, so $Spec(R) = \Delta^{n+1} \times C'$. We may write $O_{P \times C} = R_{\mathcal{P}}$, where $\mathcal{P}$ is the prime ideal of R corresponding to $P \times C'$.

Since g is in $R_{\mathcal{P}}$, we may write $g = f_P / h_P$, with f_P in R, h_P not in $\mathcal{P}$. If u is in R, recall that $V(u)$ is the closed set of prime ideals $\mathcal{Q}$ of R such that u is in $\mathcal{Q}$. Let $U_P = Spec(R) - V(h_P)$. U_P is open in $Spec(R)$, and contains $\mathcal{P}$. $U = \bigcup_P U_P$ is an open subset of $Spec(R)$ containing all points P where P is the generic point of $P \times C'$ for P closed in Δ^{n+1}. Since R is Noetherian, every open subset is quasicompact so there exist finitely many P_i such that $U = \bigcup_{i=1}^n U_{P_i}$.

Consequently, there exist finitely many elements β_i, γ_i, such that $g = \beta_i / \gamma_i$ and for each $\mathcal{P}$ there exists an i such that $\gamma_i \notin \mathcal{P}$. Let $\bar{\gamma_i}$ be the image of γ_i in the residue field $k(P \times C)$ of $O_{P \times C}$.

Let $\beta_i = \sum_j a_{ij} \otimes e_{ij}$, $\gamma_i = \sum_k a'_{ik} \otimes e'_{ik}$, with $a_{ij}, a'_{ik} \in A_{n+1}, e_{ij}, e'_{ik} \in R' \subseteq E$.

The maximal order of any pole of β_i is bounded by the maximal order of pole of any e_{ij} on C. When $\gamma_i \notin \mathcal{P}$, $\bar{\gamma_i}$ is not identically zero on $P \times C$, so any zero of $\bar{\gamma_i}$ has order $\leq$ the sum of the orders of the zeroes of $\bar{\gamma_i}$ on $P \times C$ which is equal to the sum of the orders of the poles of $\bar{\gamma}_i$ on $P \times C$. But if T is any point of C $\bar{\gamma}_i$ can only have a pole at $P \times T$ if some e'_{ik} has a pole at T and its order is bounded by the maximal order of any such pole of e'_{ik}. So the sum of the orders of the poles of $\bar{\gamma}_i$ on $P \times C$ is bounded independent of P, and we have shown that the order of all poles of g on $P \times C$ is also bounded independent of P.

Now choose a point Q in $C - S$ and choose $\alpha \in k(C)$ such that $\alpha = 0$ on S and α has a pole at Q of order greater than M. Let $g' = g + t_0 \dots t_{n+1}\alpha$. Clearly g' still satisfies $d^*_{n+1}(g') = f$ and $d^*_i(g') = 1$ for $i \leq n$. We claim that g' is identically 1 on $\Delta^{n+1} \times S$, and that $Supp(g')$ is quasi-finite over $\Delta^{n+1} -$ (faces of Δ^{n+1}).

Since f is identically 1 on $\Delta^n \times S$, first g_i, then g, then g' are identically 1 on $\Delta^{n+1} \times S$. It is clear that no pole of g_i contains $P \times C$ for any P, so the same is true for g and hence g'. But by construction g' has a pole at $P \times Q$ for all P, so cannot be identically zero on $P \times C$ either, as long as $t_0, \dots t_{n+1} \neq 0$ at P, i.e., as long as P is not in any face of Δ^{n+1}.

Now let P be in the j-th face Δ^{n+1}_j of Δ^{n+1}. We know that g (and hence g') has no polar divisor containing $P \times C$. But $d^*_j(g)$ is either 1 or f, so by hypothesis has no component of its divisor containing $P \times C \subseteq \Delta^{n+1}_j$. It follows that g (and hence g') has no component of its divisor containing $P \times C$, i.e., that $Supp(g) \cap P \times C$ is quasi-finite, so $g' \in K_{n+1}$.

3 The Quasi-Motives of a Curve

Let C be a curve with completion $\hat{C}$ and normalization $\bar{C}$. Let $e(C)$ (resp. $e_0(C)$) be the number of connected components of C (resp. the number of complete connected components of C). Let $f(C)$ (resp. $f_0(C)$) be the number of irreducible components of C (resp. the number of complete irreducible components of C). Note that $e(C) = e(\hat{C})$, $f(C) = f(\bar{C}) = f(\hat{C}) = f(\hat{\hat{C}})$, and $f_0(C) = f_0(\bar{C})$. Let $\tilde{C}$ be the seminormalization of C, so that $\tilde{C} = (\hat{\hat{C}} - S)/[P \to Q]$.

Let $Div(C) = \mathbf{Z}[C]$ be the free abelian group on the closed points of C. (Note that $Div(C) = Div(\tilde{C})$). If S is a finite subset of C, let $\mathbf{Z}[S] \subseteq Div(C)$ be the free abelian group generated by S. If $\pi : C \to D$ is a surjective map of curves, let $\mathbf{Z}[\pi]$ be the kernel of the natural map from $\mathbf{Z}[C]$ to $\mathbf{Z}[D]$, and let $\mathbf{Z}[S, \pi]$ be the subgroup of $Div(C)$ generated by $\mathbf{Z}[S]$ and $\mathbf{Z}[\pi]$.

Let $I(C)$ be the set of connected components of C. There is a natural map ρ from $Div(C)$ to $\mathbf{Z}[I(C)]$ given by taking a point to the connected component containing it. Let $\mathbf{Z}_0[P \to Q] = \mathbf{Z}[P \to Q] \cap Ker\rho$.

Lemma 3.1 *Let $C = (\hat{\hat{C}} - S)/[P \to Q]$ as usual. Let π be the natural map from $\hat{\hat{C}}$ to $\hat{C}$. Then the rank of the image J of $\mathbf{Z}[S, \pi]$ in $\mathbf{Z}[I(C)]$ is given by $f(C) - e_0(C)$.*

Proof J is generated by connected components containing points of S and differences of connected components lying over the same connected component of C, hence $\mathbf{Z}[I(\hat{\hat{C}})]/J$ is naturally isomorphic to the free abelian group on the complete connected components of C, which has rank $e_0(C)$.

Let K^*_S be the group of all functions f on $\hat{\hat{C}}$ not identically zero on any component of $\hat{\hat{C}}$ such that f is identically 1 on S. Let $G_*(C)$ denote the complex

$K_S^* \to Div(C)$, where the map is given by first taking the divisor of f in $Div(\bar{C})$ and then projecting to $Div(C)$. Note that $G_*(C) = G_*(\tilde{C})$. We define the *homology quasi-motive* to be the class in the derived category of $G_*(C)$.

We first observe that the exactness of

$$0 \to \mathbf{Z}[P \to Q] \to Div(\bar{C}) \to Div(C) \to 0$$

implies that the natural map from $(K_S^* \oplus \mathbf{Z}[P \to Q]) \to Div(\bar{C}))$ to $K_S^* \to Div(C)$ is a quasi-isomorphism.

We have the commutative diagram of exact sequences:

$$\begin{array}{ccc}
0 & & 0 \\
\downarrow & & \downarrow \\
K_S^* & \longrightarrow & K_S^*/(k^*)^f \cap K_S^* \\
\downarrow & & \downarrow \\
K_S^* \oplus \mathbf{Z}[P \to Q] & \longrightarrow & Div(\bar{C}) \\
\downarrow & & \downarrow \\
\mathbf{Z}[P \to Q] & \longrightarrow & Pic(\hat{\bar{C}}/S) \\
\downarrow & & \downarrow \\
0 & & 0
\end{array}$$

Note that the complex $K_S^* \to K_S^*/(k^*)^f \cap K_S^*$ is quasi-isomorphic to $k^*[-1]^{f_0}$. We have the further commutative diagram of exact sequences:

$$\begin{array}{ccc}
0 & & 0 \\
\downarrow & & \downarrow \\
\mathbf{Z}_0[P \to Q] & \longrightarrow & Pic^0(\hat{\bar{C}}/S) \\
\downarrow & & \downarrow \\
\mathbf{Z}[P \to Q] & \longrightarrow & Pic(\hat{\bar{C}}/S) \\
\downarrow & & \downarrow \\
\mathbf{Z}^{f(C)-e(C)} & \longrightarrow & \mathbf{Z}^{f(C)} \\
\downarrow & & \downarrow \\
0 & & 0
\end{array}$$

The left-hand column is exact by Lemma 3.1, with $S = \phi$. Finally note that the last complex $\mathbf{Z}^{f(C)-e(C)} \to \mathbf{Z}^{f(C)}$ is quasi-isomorphic to $\mathbf{Z}^{e(C)}$. We have proved

Proposition 3.2 *The homology quasi-motive $G_*(C)$ has a filtration whose graded pieces are quasi-isomorphic to $\mathbf{Z}^{e(C)}$, $\mathcal{H}_1(C)(k)$, and $k^*[-1]^{f_0(C)}$,which correspond naturally to the zeroth, first, and second cohomology groups of C.*

Definition 3.3 *We define $Sus^\bullet(C)$ by putting $Sus^n(C)$ = cycles on $\Delta^n \times C$ which are finite and surjective over Δ^n. We define the Suslin complex $Z_*(C)$ to be the non-normalized complex of abelian groups associated with the simplicial abelian group $Sus^\bullet(C)$.*

(Recall that if A^n is a simplicial abelian group with face maps d_i the non-normalized complex associated with A^n is the complex of groups A^n and maps given by the alternating sum of the d_i's.) Clearly $Z_*(C)$ is naturally isomorphic to $Z_*(\tilde{C})$.

Theorem 3.4 *$G_*(C)$ is quasi-isomorphic to $Z_*(C)$.*

Proof First let $C = \hat{C} - S$ be a normal curve with normal completion $\hat{C}$. Let $s = \#(S)$ and let K_n be the group of functions f on $\Delta^n \times \hat{C}$ such that (f) is finite over Δ^n and f is identically 1 on $\Delta^n \times S$. Let $L_n \supseteq K_n$ be the group of functions f such that $(f) \cap (\Delta^n \times S) = \phi$. Let θ_n and ψ_n be the maps from K_n and L_n to $Sus^n(C)$ defined by taking a function to its divisor. (Note that $(f) \subseteq C \times \Delta^n$ is automatically finite over Δ^n.)

The exact sequence $0 \to K_n \to L_n \to (k^*)^s \to 0$ gives rise to the exact sequence

$$0 \to Ker(\theta_n) \to Ker(\psi_n) \to (k^*)^s \to Cok(\theta_n) \to Cok(\psi_n) \to 0$$

The exact sequence

$$0 \to (k^*)^{e(\hat{C})} \to L_n \xrightarrow{\psi_n} Sus^n(C) \to Pic(C \times \Delta^n) = Pic(C) \to 0$$

shows that $Ker(\psi_n)$ and $Cok(\psi_n)$ are independent of n and hence so are $Ker(\theta_n)$ and $Cok(\theta_n)$.

Now consider the commutative diagram

$$\begin{array}{ccccccccc}
 & & \downarrow & & \downarrow & & \downarrow & & \downarrow & & \\
0 & \longrightarrow & Ker(\theta_2) & \longrightarrow & K_2 & \xrightarrow{\theta_2} & Z_2(C) & \longrightarrow & Cok(\theta_2) & \longrightarrow & 0 \\
 & & \downarrow & & \downarrow & & \downarrow & & \downarrow & & \\
0 & \longrightarrow & Ker(\theta_1) & \longrightarrow & K_1 & \xrightarrow{\theta_1} & Z_1(C) & \longrightarrow & Cok(\theta_1) & \longrightarrow & 0 \\
 & & \downarrow & & \downarrow & & \downarrow & & \downarrow & & \\
 & & 0 & \longrightarrow & K_0 & \xrightarrow{\theta_0} & Z_0(C) & \longrightarrow & 0 & & \\
 & & & & \downarrow & & \downarrow & & & & \\
 & & & & 0 & & 0 & & & &
\end{array}$$

where the columns are all the non-normalized complexes associated with the indicated simplicial abelian groups.

We have seen that the first and last columns come from constant simplicial abelian groups truncated at degree one, so are exact. The second column is acyclic by Theorem 2.1 (applied to $\hat{C}$). The rows from the first on are all visibly exact. So the third column and the bottom row are both quasi-isomorphic to the single complex associated with the total complex, hence with each other. But the third column is $Z_*(C)$, and the bottom row is $G_*(C)$.

Now we consider the case of general C. We may clearly assume that C is seminormal, so write C as $\bar{C}/[P \to Q]$. We have the exact sequence

$$0 \to \mathbf{Z}[P \to Q] \to Sus^n(\bar{C}) \to Sus^n(C) \to 0.$$

Since $\mathbf{Z}[P \to Q]$ is independent of n if we replace $Sus^n(\bar{C})$ by $Sus^n(C)$ in the above diagram, everything is still independent of n, and so the argument still works.

We now replay the above argument for the case of Borel-Moore homology. Let C be an arbitrary curve over k. Let C^i be the set of codimension-i points x of C, and let $\kappa(x)$ be the residue field of x. We define the *Borel-Moore homology quasi-motive* to be the class in the derived category of the Gersten complex

$$G_*^{BM}(C) = \coprod_{x \in C^0} \kappa(x)^* \to \coprod_{x \in C^1} \mathbf{Z}.$$

Writing this complex economically as $A^1(C) \to A^0(C)$, we remark that $A^0(C)$ is just $Div(C)$, that $A^1(C) = A^1(\hat{\tilde{C}})$ and that $A^0(C) = A^0(\tilde{C})$. In particular the quasi-motive of a curve and its seminormalization are the same. Observe that $G_*^{BM}(C)$ is just the complex called $G'_m(C)$ by Deninger in [3].

Consider the commutative diagram

$$\begin{array}{ccccccccc}
 & & & & 0 & & 0 & & \\
 & & & & \downarrow & & \downarrow & & \\
 & & & & \mathbf{Z}[S, P \to Q] & \longrightarrow & Pic(\hat{\tilde{C}}) & & \\
 & & & & \downarrow & & \| & & \\
0 \longrightarrow & (k^*)^f & \longrightarrow & A^1(\hat{\tilde{C}}) & \longrightarrow & Div(\hat{\tilde{C}}) & \longrightarrow & Pic(\hat{\tilde{C}}) & \longrightarrow 0 \\
 & & & \| & & \downarrow & & & \\
 & & & A^1(C) & \longrightarrow & Div(C) & & & \\
 & & & & & \downarrow & & & \\
 & & & & & 0 & & &
\end{array}$$

where $A^1(C)$ is put in degree -1 and $Div(C)$ in degree 0.

This implies that there exists a triangle in the derived category:

$$(k^*)^f[-1] \to [A^1(C) \to Div(C)] \to [\mathbf{Z}[S, P \to Q] \to Pic(\hat{\tilde{C}})] \to (k^*)^f$$

and Lemma 3.1 implies that the complex $\mathbf{Z}[S, P \to Q] \to Pic(\hat{\tilde{C}})$ is filtered by the complexes $\mathbf{Z}_0[S, P \to Q] \to Pic_0(\hat{\tilde{C}})$ and $\mathbf{Z}^{f(C)-e_0(C)} \to \mathbf{Z}^{f(C)}$. Since the latter complex is quasi-isomorphic to $\mathbf{Z}^{e_0(C)}$, we have proved

Proposition 3.5 *The Borel-Moore quasi-motive of C has a filtration whose graded pieces are quasi-isomorphic to $\mathbf{Z}^{e_0(C)}$, $\mathcal{H}_1^{BM}(C)(k)$, and $(k^*)[-1]^{f(C)}$ which correspond naturally to the zeroth, first and second Borel-Moore homology groups of C.*

Definition 3.6 *Let $Div_g(\Delta^n \times C)$ be the group of divisors on $\Delta^n \times C$ which intersect all faces $\Delta^i \times C$ properly. (Observe that $Div_g(\Delta^\bullet \times C)$ is the codimension-1 higher Chow group complex of C.) Let $Z_*^{BM}(C)$ be the non-normalized complex of abelian groups corresponding to the simplicial abelian group $Div_g(\Delta^\bullet \times C)$.*

Theorem 3.7 *$G_*^{BM}(C)$ is quasi-isomorphic to $Z_*^{BM}(C)$.*

Proof Since $Z_*^{BM}(C)$ is isomorphic to $Z_*^{BM}(\tilde{C})$, we may assume that C is semi-normal. Assume first that C is normal, and let $\tilde{K_n}$ be the group of functions whose divisors are in $Div_g(\Delta^n \times C)$. Theorem 2.1 again implies that the complex $\tilde{K}_\bullet$ is acyclic. Let $U(\Delta^n \times C)$ be the subgroup of $\tilde{K_n}$ consisting of functions constant on each component.

We now consider the commutative diagram

$$\begin{array}{ccccccccc}
 & & \downarrow & & \downarrow & & \downarrow & & \downarrow & & \\
0 & \longrightarrow & U(\Delta^2 \times C) & \longrightarrow & \tilde{K}_2 & \longrightarrow & Z_2^{BM}(C) & \longrightarrow & Pic(\Delta^2 \times C) & \longrightarrow & 0 \\
 & & \downarrow & & \downarrow & & \downarrow & & \downarrow & & \\
0 & \longrightarrow & U(\Delta^1 \times C) & \longrightarrow & \tilde{K}_1 & \longrightarrow & Z_1^{BM}(C) & \longrightarrow & Pic(\Delta^1 \times C) & \longrightarrow & 0 \\
 & & \downarrow & & \downarrow & & \downarrow & & \downarrow & & \\
 & & 0 & \longrightarrow & \tilde{K}_0 & \longrightarrow & Z_0^{BM}(C) & \longrightarrow & 0 & & \\
 & & & & \downarrow & & \downarrow & & & & \\
 & & & & 0 & \longrightarrow & 0 & & & &
\end{array}$$

As in the case of homology, all the rows but the last and all the columns but the third are exact (recall that $U(\Delta^n \times C) = U(C)$ and $Pic(\Delta^n \times C) = Pic(C)([\mathrm{T}]))$, so the higher Chow group complex $Z_*^{BM}(C) = Div_g(\Delta^\bullet \times C)$ is quasi-isomorphic to $G_*^{BM}(C) = K^* \to Div(C)$.

Again as in the case of homology, we use the exactness of the sequence

$$0 \to \mathbf{Z}[P \to Q] \to Div_g(\Delta^n \times \bar{C}) \to Div_g(\Delta^n \times C) \to 0$$

to prove the theorem in the general case.

Note that we can replace the higher Chow group complex $Div_g(\Delta^\bullet \times C)$ by the complex $Div_{qf}(\Delta^\bullet \times C)$ where $Div_{qf}(\Delta^n \times C)$ is the group of divisors on $\Delta^n \times C$ which are quasi-finite over Δ^n. The proof of the quasi-isomorphism with the "Gersten complex" $G_*^{BM}(C)$ goes over without change, except that we now use the original version of Theorem 2.1 (with $S = \phi$). We thus give a new proof in the case of curves of the theorem of Suslin ([8], Theorem 2.1) except that Suslin's theorem only applies to affine varieties.

We now come to the cohomology quasi-motive of a curve. We recall that the group $CDiv(C)$ of *Cartier divisors* on a curve C is defined to be $H^0(C, \underline{K}^*/\underline{O}^*)$, where $\underline{K}^*$ is the sheaf of total quotient rings on C, and $\underline{O}^*$ is the sheaf of units on C. Let $G^*(C)$ be the complex $H^0(C, \underline{K}^*) \to CDiv(C)$ (where the map is the obvious one). We define the *cohomology quasi-motive* of C to be the class in the derived category of $G^*(C)$. Now assume that C is reduced. We next observe that if $\hat{C}$ is the normal completion of C and $C = \hat{C} - S$, then the natural map from $[H^0(C, \underline{K}^*) \oplus$

$\mathbf{Z}[S] \to CDiv(\hat{C})]$ to $[H^0(C, \underline{K}^*) \to CDiv(C)]$ is a quasi-isomorphism. Let $U = \Gamma(\hat{C}, \underline{O}^*)$.

We now look at the commutative diagram

$$\begin{array}{ccc} 0 & & 0 \\ \downarrow & & \downarrow \\ H^0(C, \underline{K}^*) & \longrightarrow & H^0(C, \underline{K}^*)/U \\ \downarrow & & \downarrow \\ H^0(C, \underline{K}^*) \oplus \mathbf{Z}[S] & \longrightarrow & CDiv(\hat{C}) \\ \downarrow & & \downarrow \\ \mathbf{Z}[S] & \longrightarrow & Pic(\hat{C}) \\ \downarrow & & \downarrow \\ 0 & & 0 \end{array}$$

and observe that the complex $H^0(C, \underline{K}^*) \to H^0(C, \underline{K}^*)/U$ is quasi-isomorphic to $U[-1]$. So the cohomology quasi-motive of C is filtered by $\mathbf{Z}[S] \to Pic(\hat{C})$ and $U[-1]$. U is isomorphic to $(k^*)^{e(C)}$. Now consider the commutative diagram

$$\begin{array}{ccc} 0 & & 0 \\ \downarrow & & \downarrow \\ \mathbf{Z}_0[S] & \longrightarrow & Pic^0(\hat{C}) \\ \downarrow & & \downarrow \\ \mathbf{Z}[S] & \longrightarrow & Pic(\hat{C}) \\ \downarrow & & \downarrow \\ \mathbf{Z}^{f(C)-f_0(C)} & \longrightarrow & \mathbf{Z}^{f(C)} \\ \downarrow & & \downarrow \\ 0 & & 0 \end{array}$$

and observe that the bottom complex is quasi-isomorphic to $\mathbf{Z}^{f_0(C)}$. We have proved

Proposition 3.8 *Let C be a seminormal curve. The cohomological quasi-motive of C is filtered by $(k^*)^{e(C)}[-1]$, corresponding to $H^0(C)$, $\mathbf{Z}_0[S] \to Pic^0(\hat{C})$, which is equal to $\mathcal{H}^1(C)(k)$, and $\mathbf{Z}^{f_0(C)}$, corresponding to $H^2(C)$.*

Definition 3.9 *Let $Z^*(C)$ be the non-normalized complex associated with the simplicial abelian group $CDiv_{qf}(\Delta^\bullet)$ defined by letting $CDiv_{qf}(\Delta^n \times C)$ be the group of Cartier Divisors on $\Delta^n \times C$ whose suppport is quasi-finite over Δ^n. Let $(K_n)_{qf}(C) \subseteq H^0(C \times \Delta^n, \underline{K}^*)$ be the group of functions on $\Delta^n \times C$ whose divisors are quasi-finite over Δ^n. Observe that $(K_n)_{qf}(C) = (K_n)_{qf}(\bar{C})$.*

Theorem 3.10 *Let C be a seminormal curve. Then $G^*(C)$ is quasi-isomorphic to $Z^*(C)$.*

Proof We consider the commutative diagram

$$
\begin{array}{ccccccccc}
 & & \downarrow & & \downarrow & & \downarrow & & \downarrow & & \\
0 & \longrightarrow & U(\Delta^2 \times C) & \longrightarrow & (K_2)_{qf} & \longrightarrow & Z^2(C) & \longrightarrow & Pic(\Delta^2 \times C) & \longrightarrow & 0 \\
 & & \downarrow & & \downarrow & & \downarrow & & \downarrow & & \\
0 & \longrightarrow & U(\Delta^1 \times C) & \longrightarrow & (K_1)_{qf} & \longrightarrow & Z^1(C) & \longrightarrow & Pic(\Delta^1 \times C) & \longrightarrow & 0 \\
 & & & & \downarrow & & \downarrow & & & & \\
 & & & & (K_0)_{qf} & \longrightarrow & Z^0(C) & & & & \\
 & & & & \downarrow & & \downarrow & & & & \\
 & & & & 0 & & 0 & & & &
\end{array}
$$

Since C is seminormal, $U(\Delta^n \times C) = U(C)$ and $Pic(\Delta^n \times C) = Pic(C)$. Now Theorem 2.1 (applied to $\bar{C}$) implies that all the columns except the third are acyclic. Also it is immediate that all the rows except the last are acyclic, which as before implies that $G^*(C)$ is quasi-isomorphic to $Z^*(C)$.

We now consider cohomology with compact support. Let $C = (\hat{\bar{C}} - S)/[P \to Q]$ be a seminormal curve. We identify S with its image in C. Let K_S^* be the group of functions on C which are 1 on S. We may identify K_S^* with the group of functions on $\hat{\bar{C}}$ which are 1 on S. Define the *cohomology with compact support quasi-motive* to be the class in the derived category of the complex

$$G_c^*(C) = K_S^* \to CDiv(C).$$

Let $U_S(\hat{C})$ be the group of locally constant functions on $\hat{C}$ which are equal to 1 at every point of S. Note that $U_S(\hat{C})$ is isomorphic to $(k^*)^{e_0(C)}$.

Lemma 3.11. *There is an exact sequence*

$$0 \to U_S(\hat{C}) \to K_S^* \to CDiv(C) \to Pic(\hat{C}/S) \to 0.$$

Proof Since Pic of a semi-local ring is zero, every divisor class on C may be represented by a Cartier divisor with support not meeting S. The rest is straightforward.

This yields a filtration of $G_c^*(C)$ by $(k^*)^{e_0(C)}[-1]$ and $Pic(\hat{C}/S)$. In turn $Pic(\hat{C}/S)$ is filtered by $Pic^0(\hat{C}/S)$ and $\mathbf{Z}^{f(C)}$. Since $Pic^0(\hat{C}/S)$ is equal to $Pic^0(\hat{\bar{C}}/[P \to Q], S) = \mathcal{H}_c^1(C)(k)$, we have proved

Proposition 3.12 *Let C be a seminormal curve. The cohomology with compact support quasi-motive of C has a filtration whose graded pieces are quasi-isomorphic to $(k^*)^{e_0(C)}[-1]$ corresponding to $H^0(C)$, $\mathcal{H}_c^1(C)(k)$, and $\mathbf{Z}^{f(C)}$ corresponding to $H^2(C)$.*

Definition 3.13 *Let $Z_c^*(C)$be the non-normalized complex of abelian groups associated with the simplicial abelian group $CDiv_f(\Delta^\bullet \times C)$ defined by letting $CDiv_f(\Delta^n \times C)$ be the group of Cartier divisors on $\Delta^n \times C$ whose support is finite over Δ^n.*

Theorem 3.14 *If C is seminormal the complex $G_c^*(C)$ is quasi-isomorphic to $Z_c^*(C)$.*

Proof Let K_n be the group of functions f on $\Delta^n \times C$ such that (f) is finite over Δ^n and f is identically 1 on $\Delta^n \times S$. We introduce the following commutative diagram:

$$
\begin{array}{ccccccccccc}
 & & \downarrow & & \downarrow & & \downarrow & & \downarrow & & \\
0 & \longrightarrow & U'(\Delta^2 \times C) & \longrightarrow & K_2 & \longrightarrow & Z_c^2(C) & \longrightarrow & Pic(\Delta^2 \times C) & \longrightarrow & 0 \\
 & & \downarrow & & \downarrow & & \downarrow & & \downarrow & & \\
0 & \longrightarrow & U'(\Delta^1 \times C) & \longrightarrow & K_1 & \longrightarrow & Z_c^1(C) & \longrightarrow & Pic(\Delta^1 \times C) & \longrightarrow & 0 \\
 & & & & \downarrow & & \downarrow & & & & \\
 & & & & K_0 & \longrightarrow & Z_c^0(C) & & & & \\
 & & & & \downarrow & & \downarrow & & & & \\
 & & & & 0 & & 0 & & & &
\end{array}
$$

A by now familiar argument shows again that $G_c^*(C)$ is quasi-isomorphic to $Z_c^*(C)$.

4 Variants and Questions

For any variety V we can define the *homological quasi-motive* (resp. the *Borel-Moore homological quasi-motive* of V to be the class in the derived category of the non-normalized complex associated with the simplicial abelian group $Sus^n(V) =$ the group of cycles on $\Delta^n \times V$ which are finite (resp. quasi-finite) over Δ^n, the definitions given earlier being the special cases when V is a curve. The theorem of Suslin quoted earlier implies that, if V is an affine variety of pure dimension d, the Borel-Moore homological quasi-motive is quasi-isomorphic to Bloch's codimension d higher Chow -group complex of V. Suslin says in [8] "this result should hold for arbitrary varieties but this seems to be a rather delicate problem".

There should also be a Bloch-type version of the Suslin complex of V. Define $\widetilde{Sus}^n(V)$ to be the group of cycles in $\Delta^n \times V$ which are proper over Δ^n, have dimension equal to the dimension of V, and which intersect all faces properly. Presumably Suslin's methods would also show that the natural inclusion of $Sus^\bullet(V)$ in $\widetilde{Sus}^\bullet(V)$ is a quasi-isomorphism.

The situation for cohomology is less clear. Note first that all the cohomology theorems have been stated only for seminormal curves. We could of course define the cohomological quasi-motives of a curve to be those of its seminormalization, and thereby have all the theorems hold in general. Since Deligne's definition in [2] of the cohomological 1-motive of a curve (and all the usual cohomology theories) depend only on the seminormalization, this is not so unreasonable. However, the motivic cohomology of a curve and its seminormalization definitely do not agree, so perhaps some caution should be exercised.

We leave to the reader the job of constructing the higher-Chow-group versions of the cohomology quasi-motives of curves.

Since we do not know which, if any, of the various higher codimension analogues of Cartier divisors give the best properties, we do not try to give a general definition of cohomological quasi-motives. However, in [4], Friedlander and Voevodsky do give definitions in the general case of all four different quasi-motives, and at least in the case of homology their definition gives Suslin's "algebraic singular homology" just as ours does.

Finally, we have used the terms " homology quasi-motive" and "cohomology quasi-motive", rather than "motivic homology" and "motivic cohomology". Although of course the motivic homology or cohomology complexes, etc. should be very closely related to the homology or cohomology quasi-motives, they may not be identical. We hope to return to this point soon.

References

[1] Bloch, S. [1986] *Algebraic cycles and higher K-theory* Adv. Math. **61**, 267-304.

[2] Deligne, P. [1974] *Théorie de Hodge. III* Inst. Hautes Études Sci. Publ. Math. **44**, 5-77.

[3] Deninger, C. [1987] *Duality in the étale cohomology of one-dimensional proper schemes and generalizations* Math. Ann. **277**, 529-541.

[4] Friedlander, E. and Voevodsky, V. *Bivariant cycle cohomology*, preprint.

[5] Lichtenbaum, S. [1993] *Suslin homology and Deligne 1-motives*, Algebraic K-Theory and Algebraic Topology, Kluwer Academic Publishers, the Netherlands, pp. 189-196.

[6] Nart, E. [1989] *The Bloch complex in codimension one and arithmetic duality* J. Number Theory **32**, 321-331.

[7] Ramachandran, N. [1996] *Mixed motives of schemes*, Brown Ph. D. thesis.

[8] Suslin, A. *Higher Chow groups of affine varieties and étale cohomology*, preprint.

[9] Suslin, A. and Voevodsky, V. [1996] *Singular homology of abstract algebraic varieties* Invent. math. **123**, 61-94.

[10] Traverso, C. [1970] *Seminormality and Picard group* Ann.Scuola Norm. Sup. Pisa (3) **24**, 585-595.

Fields Institute Communications
Volume **16**, 1997

A Chain Complex for the Spectrum Homology of the Algebraic K-Theory of an Exact Category

Randy McCarthy
Department of Mathematics
University of Illinois
Urbana, IL USA 61801
randy@math.uiuc.edu

Abstract. Using a modified version of Mac Lane's Q–construction for computing the stable homology of groups [7], we obtain a simple and explicit chain complex which computes the spectrum homology for the algebraic K-theory spectrum of any exact category. Using this, we can obtain a strictly associative differential graded algebra whose homology is multiplicatively that of the rationalized algebraic K-theory of a commutative ring. After further modifications, these constructions also provide a new model for the algebraic K-theory spectrum with a natural Atiyah-Hirzebruch spectral sequence whose E^1–term is given by Mac Lane's Q-construction (for exact categories).

Introduction

For a ring R, the algebraic K-theory groups of R are often described as the homotopy groups of a space, $K_n(R) = \pi_n(K(R))$, whose homology in non-zero degrees is equivalent to the homology of $BGL(R)$. However, $K(R)$ is the associated space of a connective spectrum $K(R) = \Omega^\infty \mathbf{K}(R)$. For a spectrum, $\mathbf{X}$, the spectrum homology groups of $\mathbf{X}$ are the homotopy groups of $\mathbf{X} \wedge \mathbf{H}A$ where $\mathbf{H}A$ is the Eilenberg-Mac Lane spectrum of the abelian group A. There is a Hurewicz map from a spectrum to its homology spectrum and for spectra of the kind we will be using, a Whitehead theorem applies, which states that a map of two spectra is an equivalence (isomorphism on all homotopy groups) if and only if it is an equivalence on integral homology. Also, one has that the Hurewicz map is rationally an isomorphism (see for example [1], page 203). More specifically, it produces an isomorphism

$$\pi_*(\mathbf{X}) \otimes_{\mathbf{Z}} \mathbf{Q} \xrightarrow{\cong} H_*(\mathbf{X}; \mathbf{Q}) \cong \operatorname{Prim} H_*(\Omega^\infty \mathbf{X}; \mathbf{Q}).$$

For these reasons, it seems reasonable to attempt to examine the stable homology of the algebraic K-theory spectrum $\mathbf{K}(R)$ and more generally of $\mathbf{K}(\mathcal{A})$ for an exact category $\mathcal{A}$.

1991 *Mathematics Subject Classification.* Primary 18A40, 19D55; Secondary 55N20, 55P42.
This work was supported by National Science Foundation grant # 1-5-30943.

Our approach is to generalize a construction by Mac Lane in [7]. Here, he provides us with an explicit, functorial chain construction for any abelian group A, called $Q(A)$, whose homology is the spectrum homology of the Eilenberg-Mac Lane spectrum of A. Furthermore, if A is a commutative ring, $Q(A)$ becomes a differential graded algebra whose multiplicative structure is that of the ringed spectrum on homology. Our construction is given in (2.1). In Section 1 we present the notion of *additive in a stable range* which we use in Section 2 to prove our construction provides the correct homology groups. In Section 3 we construct explicit chain maps from the homology chain complexes of $S.$ and $G.$ to our construction and show that on homology these provide the standard maps from $H(\Omega^\infty \mathbf{X}; A)$ to $H(\mathbf{X}; A)$. In Section 4 we examine the multiplicative properties of our complex and show that they agree up to homology with those from algebraic K–theory. In Section 5 we extend our techniques so that they can similarly be applied to generalized homology theories. This provides a new model for the K-theory spectrum of an exact category when one uses stable homotopy as coefficients. We also use the techniques in Section 5 to demonstrate how the chain complexes in Section 2 arise in the Atiyah-Hirzebruch spectral sequence and similarly generalize it for additive functors from exact categories to spectra in general.

The author would like to thank Alexander Nenashev and Mark Walker for their interest and many helpful insights while this paper was being prepared.

Preliminaries

In this section we recall the definition of the S construction from [11] and establish some notation.

For $q \in \mathbf{N}$, let $[q]$ denote the poset $\{0 < \cdots < q\}$ which we will often view as a category. For $\mathcal{C}$ and $\mathcal{D}$ categories, let $\mathrm{Fun}(\mathcal{C}, \mathcal{D})$ denote the category of functors from $\mathcal{C}$ to $\mathcal{D}$ and morphisms the natural transformations of these. For $\mathcal{D}$ a category, let the *arrow category*, $\mathrm{Ar}\mathcal{D}$, be $\mathrm{Fun}([1], \mathcal{D})$.

Let $\mathcal{E}$ be an exact category considered as a category with cofibrations ([11], first few lines) by setting the subcategory of cofibrations to be the admissible monomorphisms. We let $\mathrm{Ex}(\mathrm{Ar}[q], \mathcal{E})$ be the full subcategory of $\mathrm{Fun}(\mathrm{Ar}[q], \mathcal{E})$ whose objects are the functors F such that $F(j \to j) = *$ and for every triple $i \le j \le k$ in $[q]$,

$$F(i \to j) \to F(i \to k) \to F(j \to k)$$

is a short exact sequence. Setting $S_{[q]}\mathcal{E} = \mathrm{Ex}(\mathrm{Ar}[q], \mathcal{E})$ we obtain a simplicial exact category and we write $S\mathcal{E}$ for both this simplicial exact category and the associated simplicial set we obtain by taking the set of objects degreewise.

For posets $[q_1], \ldots, [q_n]$ we write $\mathrm{Ex}(\mathrm{Ar}[q_1] \times \cdots \times \mathrm{Ar}[q_n]; \mathcal{E})$ for the set of *multi-exact* functors (exact in each position separately). In particular, there is a natural isomorphism of categories

$$\mathrm{Ex}(\mathrm{Ar}[m] \times \mathrm{Ar}[n]; \mathcal{E}) \cong \mathrm{Ex}(\mathrm{Ar}[m]; \mathrm{Ex}(\mathrm{Ar}[n]; \mathcal{E})).$$

We can iterate the S construction and there is a natural isomorphism of multi-simplicial exact categories

$$S^{(n)}_{[q_1] \times \cdots \times [q_n]}\mathcal{E} \cong \mathrm{Ex}(\mathrm{Ar}[q_1] \times \cdots \times \mathrm{Ar}[q_n], \mathcal{E}).$$

For $\mathcal{E}$ an exact category, we write $\mathbf{K}\mathcal{E}$ for the connective spectrum $\{\Omega|S^{(n+1)}\mathcal{E}|\}_{n \ge 0}$ with structure maps arising from the realization of the simplicial maps

$$\Delta^1/\partial\Delta^1 \wedge S^{(n-1)}\mathcal{E} \to S^{(n)}\mathcal{E}$$

obtained by the identification $\mathcal{E} \cong S_1\mathcal{E}$.

1 Making a Functor Additive by Stabilization

In this section we assemble some basics about very general functors. Our methods are direct extensions of those by Dold and Puppe in [3]. We will make no notational distinction between a (multi-dimensional) simplicial abelian group and its associated (multi-dimensional) chain complex. By a *chain complex* we will always mean a complex which is bounded below and homologically trivial in negative dimensions (i.e., connective). Given a multi-dimensional chain complex, we will consider it as a chain complex by taking Tot (using products). What follows has various standard generalizations to various categories with cofibrations but as these extensions are "straightforward" for the expert and do little more than cloud the essential ideas we will keep our attention to exact categories. Also, we will be assuming our functors are defined for all exact categories but it is clear that the constructions work equally well for functors defined for any category $\mathcal{E}$ of exact categories such that S restricts to a functor from $\mathcal{E}$ to simplicial $\mathcal{E}$ objects. For example, the category of semi-simple exact categories (every short exact sequence splits) is such a class.

Let F be any functor from (small) linear categories (with a distinguished zero object 0) to chain complexes. It will be convenient for us to assume further that F is *reduced*—$F(0) = 0$. We will often consider a group as a chain complex concentrated in dimension 0.

We note that we are not assuming that F takes naturally equivalent linear categories to homotopic chain complexes. If F does this, we will say that F is an *equivalence functor.*

If $G.$ is any functor from some category $\mathcal{C}$ of categories to (small) simplicial linear categories, then we can of course compose functors to obtain a new functor $FG.$ from $\mathcal{C}$ to simplicial chain complexes which we once again consider a functor to chain complexes by taking Tot. By definition, if $\mathcal{A}_*$ is a simplicial (small) linear category, then $F\mathcal{A}_*$ is the simplicial chain complex obtained by applying F degreewise.

Lemma 1.1 *The composite functor FS is an equivalence functor.*

Proof (see also [11]): Let $\sigma : f \xrightarrow{\cong} g$ be a natural isomorphism between two exact functors from $\mathcal{A}$ to $\mathcal{B}$. That is, f and g are functors from $\mathcal{A}$ to $\mathcal{B}$ which take those short exact sequences which are defined in $\mathcal{A}$ to short exact sequences defined in $\mathcal{B}$. We define a simplicial homotopy h (for the definition, see for Example [8], Section 5) by (exact) functors from $S.f$ to $S.g$. Since F is functorial, $F(h)$ will be a simplicial chain homotopy and hence $Tot(F(h))$ a chain homotopy from $FS(f)$ to $FS(g)$. We will only define the $\{h_i\}$ and leave the details of checking the simplicial relations to the interested reader. For $0 \leq i \leq n$ we define $h_i : S_n\mathcal{A} \to S_{n+1}\mathcal{B}$ to be the exact functor determined by (the quotients are clear)

$$(A_1 \rightarrowtail \cdots \rightarrowtail A_n) \mapsto (F(A_1) \rightarrowtail \cdot \rightarrowtail F(A_i) \xrightarrow{\sigma A_i} G(A_i) \rightarrowtail \cdots \rightarrowtail G(A_n)).$$

□

Next we want to examine how F behaves with respect to the product of two linear categories. We will say that F is *product preserving* if for any two (small) exact categories $\mathcal{A}$ and $\mathcal{B}$, the natural projection map ρ of simplicial abelian groups $F(\mathcal{A} \times \mathcal{B})$ to $F(\mathcal{A}) \times F(\mathcal{B})$ is a homotopy equivalence. Not all the functors we will want to consider will be product preserving and we define the *second cross effect* of F to measure this failure to preserve products.

Definition 1.2 Let $\mathcal{A}$ and $\mathcal{B}$ be exact categories. We define cr_2F to be any of the naturally equivalent bi-functors from exact categories to simplicial chain complexes determined by:

$$cr_2F(\mathcal{A},\mathcal{B}) = ker[F(\mathcal{A}\times\mathcal{B}) \xrightarrow{\rho_*} F(\mathcal{A})\times F(\mathcal{B})].$$

Thus, cr_2F is a symmetric bi-functor such that $cr_2F(0,\mathcal{B}) = cr_2F(\mathcal{A},0) = 0$.

We will say that F is a p-product functor if F preserves products in a range $0 \leq i \leq p$ (that is, $\pi_i(\rho)$ is an isomorphism for all $0 \leq i \leq p-1$ and π_p is surjective). Thus, F is a p-product functor if and only if cr_2F takes values in p–connected simplicial chain complexes.

For an m-dimensional multi-simplicial linear category $\mathcal{A}_*$, we call its *total* degree Tot_q to be *p–reduced* if $\mathcal{A}_{q_1,\dots,q_m} = 0$ for all $0 \leq \Sigma_{i=1}^m q_i \leq p-1$. For example, $S^{(n)}\mathcal{C}$ is $(n-1)$-reduced for all exact categories $\mathcal{C}$.

Lemma 1.3 *If $F.$ is a p-product functor and $\mathcal{A}_*$ and $\mathcal{B}_*$ are m and n reduced simplicial exact categories, then*

$$F.(\mathcal{A}_*\times\mathcal{B}_*) \xrightarrow{\rho} F.(\mathcal{A}_*)\times F.(\mathcal{B}_*)$$

is a $(p+m+n+1)$–connected map of simplicial chain complexes.

Proof We want to show that $cr_2F(\mathcal{A}_*,\mathcal{B}_*)$ is an $(p+m+n)$–connected simplicial chain complex. By the Eilenberg-Zilber theorem, for each q, $Tot(cr_2F_q(\mathcal{A}_*,\mathcal{B}_*))$ is naturally chain homotopy equivalent to an $(m+n+1)$–reduced chain complex. Thus, $cr_2F.(A_*,B_*)$ is naturally equivalent to a simplicial object of $(m+n+1)$–reduced chain complexes which (because F is a p–product functor) is p–connected in each fixed chain dimension. □

Corollary 1.4 *If F is a p-product functor then $FS^{(n)}$ is a $p+2n-1$-product functor.*

Now we want to establish the notion of "additive in a stable range." By the proof of additivity found in [9], for any F

$$FS.S_2\mathcal{C} \overset{d_0\times d_2}{\longrightarrow} FS.(\mathcal{C}\times\mathcal{C})$$

is an equivalence. It follows that if F is a p–product functor, then the natural map $F(SS_2) \overset{d_0\times d_2}{\longrightarrow} FS\times FS$ is an equivalence in a p–range. If $FS_2 \longrightarrow F\times F$ is an equivalence in a p range then we say that F is *additive* in a p range.

Corollary 1.5 *For any $n \geq 1$, the functor $FS^{(n)}$ is a reduced equivalence functor which is a $2n-2$ product functor and additive in a $2n-2$ range.*

For X a chain complex, we let $X[z]$ be the new chain complex with $X[z]_n = X_{n-z}$ and $\partial[z]_n = \partial_{n-z}$.

Definition 1.6 For any closed category, we define

$$F_*^{st}(\mathcal{A}) = \lim_{n\to\infty} FS^{(n)}\mathcal{A}[-n]$$

which is a natural additive equivalence functor. We let $\alpha : F \to F^{st}$ be the natural transformation obtained by the structure maps for the limit system.

Lemma 1.7 *If F is a product preserving equivalence functor then $\Omega FS. \longrightarrow F^{st}_*$ is a homotopy equivalence for all exact categories. The argument is exactly like that of 1.5.5 [11] and follows from 2.2 of [4]. In general, F is an additive functor (additive in an ∞-range) if and only if $F \to F^{st}$ is an equivalence for all exact categories.*

□

Let $B.$ be the bar construction for abelian groups. For F a (reduced) functor from exact categories to abelian groups, we can define a map of simplicial abelian groups

$$\beta : B.F \to FS.$$

$$\beta_{[n]} = F(\hat{\sigma}_0) + \cdots + F(\hat{\sigma}_{n-1}) : F^{\oplus n} \to F(S_n)$$

where $\hat{\sigma}_i : \mathcal{C} \cong S_1\mathcal{C} \to S_n\mathcal{C}$ is

$$\hat{\sigma}_i = s_{n-1} \circ \cdots \circ s_{i+1} \circ s_{i-1} \circ \cdots \circ s_0.$$

Proposition 1.8 *A functor F is p–additive if and only if the map $B.F \xrightarrow{\beta} FS.$ is p–connected.*

Proof By induction, if F is p–additive, then $\beta_{[n]}$ is p–connected for all n and hence β is p–connected by a standard spectral sequence argument. Conversely, if β is p–connected, then $\beta^k = \overbrace{\beta \circ \cdots \circ \beta}^{k \text{ times}}$ from B^kF to $F(S^{(k)}\ )$ is $p+k$ connected for all k. Hence, the map α from F to F^{st} is p connected but F^{st} is additive and hence F is p–additive. □

Examples

1) Let $\mathbf{Z}$ be the functor which takes a (small) exact category $\mathcal{C}$ to the reduced free abelian group generated by the set of objects of $\mathcal{C}$. That is,

$$\mathbf{Z}(\mathcal{C}) = cokernel[\mathbf{Z}[0] \to \mathbf{Z}[Obj(\mathcal{C})]].$$

Then $\mathbf{Z}^{st}$ is the stable homology functor and

$$H_*(\mathbf{Z}^{st}(\mathcal{C})) = H_*(\mathbf{K}(\mathcal{C})) = \pi_*(\mathbf{K}(\mathcal{C}) \wedge \mathbf{HZ}).$$

2) Let $\mathcal{A}$ be an exact category. Let **Hom** be the functor defined by

$$\mathbf{Hom}(\mathcal{A}) = \bigoplus_{A \in \mathcal{A}} Hom_{\mathcal{A}}(A, A).$$

Then, by Section 2 of [4],

$$\mathbf{Hom}^{st}(\mathcal{A}) = THH(\mathcal{A})$$

where THH is the topological Hochschild homology of $\mathcal{A}$.

3) Let R be a ring and let $M_1, \ldots, M_n$ be R–bimodules. Let $\mathcal{P}$ be the category of finitely generated projective (left) R–modules and let $Q_*(R)$ be Mac Lane's Q–construction. Let G be the functor from exact right R–module categories to abelian groups defined by

$$G(\mathcal{A}) = \bigoplus_{A \in \mathcal{A}} \bigotimes_{i=1}^{n} Hom_{\mathcal{A}}(A, A) \otimes_R M_i$$

with Σ_n–action obtained by permuting the tensor factors. By Theorem 4.1 of [10]

$$G^{st}(\mathcal{P}) \sim_{\Sigma_n} \mathbf{Z}[\Sigma_n] \otimes_{\mathbf{Z}[C_n]} HH(Q_*(R)^{\otimes n}; M_1 \otimes_{\mathbf{Z}} \cdots \otimes_{\mathbf{Z}} M_n)$$

where HH is the Hochschild homology complex for $Q_*(R)^{\otimes n}$ acting on the bimodule by

$$(m_1 \otimes \cdots \otimes m_n) * (q_1 \otimes \cdots \otimes q_n) = (m_1 q_1 \otimes m_2 q_2 \otimes \cdots \otimes m_n q_n)$$

$$(q_1 \otimes \cdots \otimes q_n) * (m_1 \otimes \cdots \otimes m_n) = (q_n m_1 \otimes q_1 m_2 \otimes \cdots \otimes q_{n-1} m_n)$$

and C_n is the cyclic group of n–elements which acts by cyclic permutations—the equivalence is weakly Σ_n–equivariant.

2 Mac Lane's Q'-Construction for Exact Categories

Now we generalize the Q'–construction of Mac Lane in [7] to exact categories. Our guide is to use the S construction for exact categories as one uses the bar construction B for abelian groups. Let $\mathcal{A}$ be an exact category. Consider the three exact functors

$$d_0, d_1, d_2 : S_2\mathcal{A} \longrightarrow \mathcal{A}$$

defined by

$$\begin{aligned} d_0(A \to B \to C) &= \ C \\ d_1(A \to B \to C) &= \ B \\ d_2(A \to B \to C) &= \ A. \end{aligned}$$

We set $S_2^{(0)}\mathcal{A} = \mathcal{A}$ and for $n \geq 1$ we set

$$S_2^{(n)}\mathcal{A} = \overbrace{S_2 \cdots S_2}^{n \text{ times}}\mathcal{A}.$$

For $1 \leq i \leq n$ and $0 \leq k \leq 2$ we set

$$d_k(i) = S_2^{(n-i)}\ d_k\ S_2^{(i-1)} : S_2^{(n)}\mathcal{A} \to S_2^{(n-1)}\mathcal{A}$$

which are again exact functors.

For F a functor from exact categories to chain complexes, we define a new functor $Q'_*(F;\ )$ from exact categories to chain complexes by:

$$Q'_n(F;\mathcal{A}) = F(S_2^{(n)}\mathcal{A})$$

$$\partial_n : Q'_n(F;\mathcal{A}) \to Q'_{n-1}(F;\mathcal{A})$$

$$\partial_n = \Sigma_{i=1}^{n}(-1)^i[F(d_1(i)) - F(d_2(i)) - F(d_0(i))].$$

If F is a functor from exact categories to simplicial chain complexes, we let $Q'_*(F;\ )$ be the functor from exact categories to simplicial chain complexes obtained by applying Q'_* degreewise to F.

We now want to "prune" the complex Q' to obtain a new complex L. This should be thought of being analogous to modifying a cubical construction to obtain the correct chain complex. We call this new complex L for "linear." There is another pruning procedure one can use that is more analogous to moding out by all degenerate simplices which we discuss below. If one further modes out all "slabs" then what one obtains should be more properly called "Q" from [7] and is defined explicitly in 5.3.

Consider the two exact functors

$$s_0, s_1 : \mathcal{A} \to S_2\mathcal{A}$$

given by

$$\begin{aligned} s_0(A) &= (A \xrightarrow{=} A \to 0) \\ s_1(A) &= (0 \to A \xrightarrow{=} A). \end{aligned}$$

We define a chain map $\hat{s}_0$ from $\Sigma Q'(F;\mathcal{A})$ to $Q'(F;\mathcal{A})$ by

$$F(S_2^{(n-1)}s_0) : F(S_2^{(n-1)}\mathcal{A}) \to F(S_2^n\mathcal{A}).$$

This is seen to be a chain map by using the identities

$$d_k(i)\ s_l = \begin{cases} s_l d_k(i-1) & \text{if } i>1 \\ id & \text{if } l=0,\ k=0,1 \\ id & \text{if } l=1,\ k=1,2 \\ 0 & \text{otherwise} \end{cases}$$

and the fact that the differential of $\Sigma Q'$ is $(-1)^{n-1}\partial_{n-1}$ in dimension n. We define a chain map $\hat{s}_1$ similarly using s_1 instead of s_0.

Definition 2.1 For F a (reduced) functor from exact categories to chain complexes, we let $L(F)$ be the functor from exact categories to chain complexes defined by

$$L(F;\mathcal{A}) = \text{ cokernel } [\Sigma Q'(F;\mathcal{A}) \oplus \Sigma Q'(F;\mathcal{A}) \stackrel{\hat{s}_0+\hat{s}_1}{\longrightarrow} Q'(F;\mathcal{A})].$$

If F is a (reduced) functor from exact categories to simplicial chain complexes, we let $L(F)$ be the functor from exact categories to simplicial chain complexes obtained by applying L degreewise. We let

$$F \stackrel{\eta}{\longrightarrow} L(F)$$

be the natural transformation obtained by using $L_0(F) = F$.

Note For a fixed dimension n, $(\hat{s}_0 + \hat{s}_1)_n$ has a retract $(F(S_2^{(n-1)}d_0), F(S_2^{(n-1)}d_2))$. This splitting is levelwise natural but not a chain map.

We now define an analogous construction L^{red} which we will show in 2.5 is naturally quasi-isomorphic to L. For $1 \le i \le n-1$ and $0 \le k \le 1$ we set

$$s_k(i) = S_2^{(n-i-1)}\ s_k\ S_2^{(i-1)} : S_2^{(n-1)}\mathcal{A} \to S_2^{(n)}\mathcal{A}$$

which are again exact functors. We define

$$L^{red}(F;\mathcal{A})_n = \text{cokernel}\left[\bigoplus_{i=1}^{n-1}\bigoplus_{k=0}^{1} F(S_2^{(n-1)}\mathcal{A}) \xrightarrow{F(s_k(i))} F(S_2^{(n)}\mathcal{A})\right].$$

As before, it is straightforward to check that the quotient map $Q'_n \to L_n^{red}$ becomes a chain map and that we have a sequence of natural chain maps

$$Q'(F;\mathcal{A}) \to L(F;\mathcal{A}) \to L^{red}(F;\mathcal{A}).$$

Lemma 2.2 *For any reduced F, the functors $Q'(F)$, $L(F)$ and $L^{red}(F)$ are reduced, equivalence and product preserving linear functors for all exact categories.*

Proof The fact that $Q'(F)$, $L(F)$ and $L^{red}(F)$ are reduced is clear. In order to establish the other three properties we will apply the following general scheme. First note that it suffices to establish the result for Q' by chain homotopies which pass to L or L^{red}. Next, suppose we are given an exact functor h from $\mathcal{A}$ to $S_2\mathcal{B}$. Define $h_n : Q'(F;\mathcal{A})_n \to Q'(F;\mathcal{B})_{n+1}$ by $h_n = F(S_2^{(n)}\ h)$. A straightforward calculation yields the identity:

$$\partial_{n+1}h_n + h_{n-1}\partial_n = [d_1(1) - d_0(1) - d_2(1)]h_n$$

and hence

$$Q'(F;d_1h) \simeq Q'(F;d_0h) + Q'(F;d_2h).$$

Since the homotopy was given by an exact functor, it will naturally pass to the quotient complexes L and L^{red}.

To show that $Q'(F)$ is an equivalence preserving functor, it suffices to show that if f and g are isomorphic exact functors from $\mathcal{A}$ to $\mathcal{B}$ then $Q'(F;f) \simeq Q'(F;g)$. Let σ be a natural isomorphism from f to g. We define h to be the exact functor from $\mathcal{A}$ to $S_2\mathcal{B}$ determined by

$$h(A) = f(A) \xrightarrow{\sigma_A} g(A) \to 0$$

and so $Q'(F;g) \simeq Q'(F;f) + 0$.

Let $\mathcal{A}$ and $\mathcal{B}$ be two exact categories. Let ρ_A and ρ_B be the projections of $\mathcal{A} \times \mathcal{B}$ to $\mathcal{A}$ and $\mathcal{B}$. Let i_A be the exact functor from $\mathcal{A}$ to $\mathcal{A} \times \mathcal{B}$ defined by $A \mapsto A \times 0_B$. Define i_B similarly. Thus, $(Q'(F;\rho_A), Q'(F;\rho_B))$ from $Q'(F;\mathcal{A}\times\mathcal{B})$ to $Q'(F;\mathcal{A}) \times Q'(F;\mathcal{B})$ has a section given by $Q'(F;i_A) + Q'(F;i_B)$ and we show that the non-identity composite is homotopic to the identity. We define the exact functor h from $\mathcal{A} \times \mathcal{B}$ to $S_2(\mathcal{A} \times \mathcal{B})$ by

$$h\begin{pmatrix} A \\ \times \\ B \end{pmatrix} = \begin{pmatrix} A \xrightarrow{=} A \longrightarrow 0 \\ \times \\ 0 \longrightarrow B \xrightarrow{=} B \end{pmatrix}$$

Thus, $id_{Q'(F;\mathcal{A}\times\mathcal{B})} \simeq Q'(F;d_0h) + Q'(F;d_2h)$ and the result follows.

Let $\mathcal{A}$ be an exact category. The map $(Q'(F;d_0), Q'(F;d_2))$ from $Q'(F;S_2\mathcal{A})$ to $Q'(F;\mathcal{A}) \oplus Q'(F;\mathcal{A})$ has a section given by $Q'(F;s_0) + Q'(F;s_1)$. We show that the nonidentity composite is homotopic to the identity. Define the exact functor h from $S_2\mathcal{A}$ to $S_2^{(2)}\mathcal{A}$ by

$$\begin{pmatrix} A'' \\ \downarrow \\ A \\ \downarrow \\ A' \end{pmatrix} \xrightarrow{h} \begin{pmatrix} A'' & \xrightarrow{=} & A'' & \longrightarrow & 0 \\ \downarrow{=} & & \downarrow & & \downarrow \\ A'' & \longrightarrow & A'' & \longrightarrow & A' \\ \downarrow & & \downarrow & & \downarrow{=} \\ 0 & \longrightarrow & A' & \xrightarrow{=} & A' \end{pmatrix}$$

Thus, $id_{Q'(F;S_2\mathcal{A})} \simeq Q'(F;d_0h) + Q'(F;d_2h)$ and the result follows. □

Lemma 2.3 *If F is additive, then $\eta : F \to L(F)$ and $\eta^{red} : F \to L^{red}(F)$ are equivalences. If F is additive in a p–range, then η is p–connected.*

Proof If F is an additive functor to chain complexes, then

$$F(S_2^{(n)}\mathcal{A}) \xrightarrow{\cong} F(S_2^{(n-1)}\mathcal{A}) \oplus F(S_2^{(n-1)}\mathcal{A})$$

by $(F(d_0(1)), F(d_2(1)))$. Since $(\hat{s}_0 + \hat{s}_1)_n$ is a section to this map, we see that $(\hat{s}_0 + \hat{s}_1)_n$ is an isomorphism for all $n \geq 1$ and hence $L(F;\mathcal{A})$ is concentrated in degree 0. Thus, η is an isomorphism. A similar argument applies if we use L^{red} instead of L.

More generally, if F is additive in a p–range, the map $(F(d_0(1)), F(d_2(1)))$ is p–connected and hence so is $(\hat{s}_0 + \hat{s}_1)_n$ for all $n \geq 1$. Considering the spectral sequences for $Q'(F;\mathcal{A})$, $\Sigma Q'(F;\mathcal{A}) \oplus \Sigma Q'(F;\mathcal{A})$ and $L(F;\mathcal{A})$ obtained by taking the homology in the direction along F first, we see that η is at least p–connected. □

Lemma 2.4 *There is a natural isomorphism $L(FS;\mathcal{A}) \cong L(F;S\mathcal{A})$ such that the following diagram commutes:*

$$\begin{array}{ccc} L(\mathcal{A}) & \longrightarrow & L(FS;\mathcal{A})[-1] \\ \| & & \downarrow\cong \\ L(\mathcal{A}) & \longrightarrow & L(F;S\mathcal{A})[-1]. \end{array}$$

The same isomorphism restricts to give a similar result if we replace L by L^{red}.

Proof For $n, q \in \mathbf{N}^{\times 2}$, let $\tau_n(q)$ be the isomorphism of categories

$$\overbrace{Ar[2] \times \cdots \times \mathrm{Ar}[2]}^{n \text{ times}} \times \mathrm{Ar}[q] \xrightarrow{\tau_n(q)} Ar[q] \times \overbrace{Ar[2] \times \cdots \times \mathrm{Ar}[2]}^{n \text{ times}}$$

determined by $a_1 \times \cdots \times a_n \times q \mapsto q \times a_1 \times \cdots \times a_n$. We obtain natural isomorphisms

$$\begin{array}{c} Q'(F;S_q\mathcal{A})_n = \mathrm{Ex}(\mathrm{Ar}[q] \times \overbrace{Ar[2] \times \cdots \times \mathrm{Ar}[2]}^{n \text{ times}};\mathcal{A}) \\ \downarrow \tau_n(q)^* \\ Q'(FS_q;\mathcal{A})_n = \mathrm{Ex}(\overbrace{Ar[2] \times \cdots \times \mathrm{Ar}[2]}^{n \text{ times}} \times \mathrm{Ar}[q];\mathcal{A}) \end{array}$$

which assemble compatibly to produce an isomorphism of simplicial chain complexes $Q'(F;S\mathcal{A}) \cong Q'(FS;\mathcal{A})$ which restricts to give an appropriate isomorphism $L(F;S\mathcal{A}) \cong L(FS;\mathcal{A})$. By naturality and the fact that this isomorphism is defined by exact functors, this also restricts to L^{red}. □

Theorem 2.5 *For any reduced F, the functors F^{st}, $L(F)$ and $L^{red}(F)$ are naturally equivalent. That is, there is a natural diagram of equivalences which connects them. In particular, the map $L(\alpha) : L(F) \to L(F^{st})$ is a quasi-isomorphism for all functors F.*

Proof By Lemma 2.4 we obtain a commuting diagram:

$$\begin{array}{ccc} F & \xrightarrow{\alpha} & F^{st} \\ \downarrow\eta & & \downarrow\eta \\ L(F) & \xrightarrow{L(\alpha)} & L(F^{st}) \\ \downarrow = & & \downarrow\cong \\ L(F) & \xrightarrow{\alpha} & L(F)^{st}. \end{array}$$

The map $L(F) \xrightarrow{\alpha} L(F)^{st}$ is an equivalence by Lemmas 2.2 and 1.7. The map $F^{st} \xrightarrow{\eta} L(F^{st})$ is an equivalence by 1.7 and 2.3. The same argument applies if we use L^{red} in place of L. □

Example Let $\mathbf{Z}$ be the functor of Example 1.8. Then, for any closed category $\mathcal{C}$,

$$H_*L(\mathbf{Z};\mathcal{C}) \cong H_*(\mathbf{Z}^{st}(\mathcal{C})) = H_*(\mathbf{K}(\mathcal{C})).$$

Recall that if $\mathbf{X}$ is a connective CW-spectrum, then

$$\pi_*(\mathbf{X}) \otimes_{\mathbf{Z}} \mathbf{Q} \cong H_*(\mathbf{X};\mathbf{Q})$$

by the Hurewicz map (see for example [1], page 203). Thus, for $\mathcal{A}$ an exact category, there is a natural isomorphism

$$K_n(\mathcal{A}) \otimes_{\mathbf{Z}} \mathbf{Q} \cong H_*(L(\mathbf{Q};\mathcal{A})).$$

3 A Natural Map from $F(S.\ )$ to $L(F;\ )[1]$

Recall that for a chain complex X, we let $X[1]$ be the new chain complex obtained by setting $X[1]_n = X_{n-1}$ and $\partial_{X[1]} = \partial_X$. In this section, we define a map γ from FS to $L(F)$ which can be used to produce the natural transformation $H_*(K(\mathcal{C});A) \to H_*(\mathbf{K}(\mathcal{C});A)$.

For $n \geq 2$, let $t(n) : S_n \to S_2 S_{n-1}$ be defined by

$$A_1 \rightarrowtail \cdots \rightarrowtail A_n \in S_n\mathcal{A} \xrightarrow{t(n)} \begin{pmatrix} A_1 & \rightarrowtail & \cdots & \rightarrowtail & A_{n-2} & \rightarrowtail & A_{n-1} \\ \| & & & & \| & & \downarrow \\ A_1 & \rightarrowtail & \cdots & \rightarrowtail & A_{n-2} & \rightarrowtail & A_n \\ \downarrow & & & & \downarrow & & \downarrow \\ 0 & = & \cdots & = & 0 & \rightarrowtail & A_n/A_{n-1} \end{pmatrix}$$

(with the obvious choices for quotients). Thus, $t(n)$ is a natural exact functor and we define

$$t_n : S_n \longrightarrow S_2^{(n-1)} \quad \text{by} \quad t_n = S_2^{(n-2)}t(2) \circ \cdots \circ S_2 t(n-1) \circ t(n).$$

We extend this definition by setting $t_1 : S_1 \to S_2^{(0)} = \mathrm{Obj}$ and $t_0 : S_0 \to S_2^{(-1)} = 0$ to be the natural isomorphisms. For F a functor from exact categories to (simplicial) abelian groups, we will slightly abuse notation and let $F(t_n)$ represent the composite natural transformation

$$F(S_n\ ) \longrightarrow F(S_2^{(n-1)}\ ) \longrightarrow L_{n-1}(F).$$

Lemma 3.1 *The natural transformations $F(t_n)$ assemble to produce a natural chain map*

$$\gamma : FS \longrightarrow L(F)[1].$$

Proof We first note the two relations

$$\partial_{L(F)} \circ F(t_1) = F(t_0) \circ \partial_{F(S)} \quad \text{and} \quad \partial_{L(F)} \circ F(t_2) = F(t_1) \circ \partial_{F(S)}$$

which follow immediately from the definitions.

It is straightforward to establish the following identities for $n \geq 2$:

$$\begin{aligned} d_0(j)t_{n-1} &= t_{n-2}\,\overbrace{s_0 \cdots s_0}^{j-1 \text{ times}}\,\overbrace{d_0 \cdots d_0}^{j \text{ times}} \\ d_1(j)t_{n-1} &= t_{n-2}d_j \\ d_2(j)t_{n-1} &= t_{n-2}s_{n-1}s_{n-2}\cdots s_{j+1}d_{j+1}d_{j+2}\cdots d_n \end{aligned}$$

which one can use to establish

$$\left[\sum_{j=1}^{n-1}(-1)^j[F(d_1(j)) - F(d_0(j)) - F(d_2(j))]\right] \circ F(t_{n-1})$$

$$\|$$

$$\sum_{j=1}^{n-1} F(t_{n-2})(-1)^j$$

$$\left[F(d_j) - F(\overbrace{s_0 \cdots s_0}^{j-1 \text{ times}}\,\overbrace{d_0 \cdots d_0}^{j \text{ times}}) - F(s_{n-1}s_{n-2}\cdots s_{j+1}d_{j+1}d_{j+2}\cdots d_n)\right]$$

$$\|$$

$$F(t_{n-2})\left[F(d_0)+\sum_{j=2}^{n-1}(-1)^jF(d_j)+(-1)^nF(d_n)\right]$$

$$-F(t_{n-2})\left[\sum_{j=1}^{n-2}(-1)^jF(\overbrace{s_0\cdots s_0}^{j-1\text{ times}}\overbrace{d_0\cdots d_0}^{j\text{ times}})\right]$$

$$-F(t_{n-2})\left[\sum_{j=1}^{n-2}(-1)^jF(s_{n-1}s_{n-2}\cdots s_{j+1}d_{j+1}d_{j+2}\cdots d_n)\right].$$

This last term is $F(t_{n-2})\circ\partial_{F(S)}$ up to error terms which are seen to be zero in $L(F)$ (but not in $Q'(F)$) by using the identities

$$\begin{aligned} t_{n-1}s_0 &= s_0t_{n-2}\\ t_{n-1}s_{n-1} &= s_1t_{n-2}. \end{aligned}$$

□

Proposition 3.2 *The following are equivalent for a reduced functor F:*

(i) *F is p–additive*
(ii) *$\alpha : F\to F^{st}$ is p–connected*
(iii) *$\eta : F\to L(F)$ is p–connected*

if these conditions hold then:

(*) *$\gamma : FS\to L(F)[1]$ is $p+2$–connected.*

Proof The equivalence of (i) and (ii) is Lemma 1.8. By Lemma 2.2, $L(F)$ is additive and hence (i) and (iii) are equivalent by 2.3. Let f be the natural transformation of chain complexes from $F[1]$ to FS obtained by using the identification $\mathcal{C}\cong S_1\mathcal{C}$. The composite $\gamma\circ f : F[1]\to L(F)$ is simply the inclusion of F into $L_0(F)$ shifted by chain dimension 1 and hence is $\eta[1]$ where η is the natural transformation used in Lemma 2.3. Consider the commuting diagram:

$$\begin{array}{ccccc} F[1] & \xrightarrow{f} & FS & \xrightarrow{\gamma} & L(F)[1]\\ \downarrow{\scriptstyle \alpha[1]} & & \downarrow{\scriptstyle \alpha} & & \downarrow{\scriptstyle L(\alpha)[1]}\\ F^{st}[1] & \xrightarrow{f} & F^{st}S & \xrightarrow{\gamma} & L(F^{st})[1]. \end{array}$$

By Lemma 2.3, the composite along the bottom is an equivalence. Since F^{st} is additive, the bottom f map is an equivalence (by 1.8) and hence the lower γ is an equivalence. If F is p–additive, then FS is $p+2$ additive and α is $p+2$–connective by Lemma 1.8–hence the upper γ is $p+2$–connective. □

Corollary 3.3 *Using the functor* $\mathbf{Z}$ *of Example 1.8.1,*

$$\gamma : \mathbf{Z}(S.\mathcal{C})\to L(\mathbf{Z};\mathcal{C})[1]$$

realizes the standard map from the homology of the associated space to the homology of the spectrum for the spectrum $\{S^{(l+1)}\mathcal{C}|l\geq 0\}$.

Proof This follows from considering the commutative diagram:

$$\begin{array}{ccccc} \mathbf{Z}(S.\mathcal{C}) & \xrightarrow{\alpha} & \lim \mathbf{Z}(S^{(n+1)}\mathcal{C})[-n] & = & \mathbf{Z}^{st}(S.\mathcal{C}) \\ \downarrow \gamma & & & & \downarrow \gamma \\ L(\mathbf{Z};\mathcal{C})[1] & \xrightarrow{L(\alpha)[1]} & & & L(\mathbf{Z}^{st};\mathcal{C})[1] \end{array}$$

where the right vertical map is an equivalence by Proposition 3.2 and the lower map is an equivalence by Theorem 2.5. □

Let $G.$ be the model from [5] for algebraic K-theory. Thus, $G.\mathcal{C}$ is the simplicial pullback of

$$\begin{array}{ccc} & & PS.\mathcal{C} \\ & & \downarrow d_0 \\ PS.\mathcal{C} & \xrightarrow{d_0} & S.\mathcal{C} \end{array}$$

where PS is the simplicial path space of S. We will represent the elements by

$$G_n\mathcal{C} = \left\{ \frac{A_0 \rightarrowtail \cdots \rightarrowtail A_n}{B_0 \rightarrowtail \cdots \rightarrowtail B_n} \ ; \ \text{(plus choices)} \right\}.$$

Let π_t and π_b be the two projection maps from G to PS determined by

$$\pi_t \left(\frac{A_0 \rightarrowtail \cdots \rightarrowtail A_n}{B_0 \rightarrowtail \cdots \rightarrowtail B_n} \right) = (A_0 \rightarrowtail \cdots \rightarrowtail A_n)$$

$$\pi_b \left(\frac{A_0 \rightarrowtail \cdots \rightarrowtail A_n}{B_0 \rightarrowtail \cdots \rightarrowtail B_n} \right) = (B_0 \rightarrowtail \cdots \rightarrowtail B_n).$$

We define a chain map λ from $F(G.\ )$ to $F(S.\ )[-1]$ by

$$\lambda = F(\pi_t) - F(\pi_b).$$

Corollary 3.4 *For $\mathcal{C}$ an* exact *category, the composite map*

$$\gamma \circ \lambda : \mathbf{Z}(G.\mathcal{C}) \to L(\mathbf{Z};\mathcal{C})$$

realizes the standard map from the homology of the associated space to the homology of the spectrum for the K-theory spectrum of $\mathcal{C}$: $\{\Omega S^{(l+1)}\mathcal{C} | l \geq 0\}$.

Proof By [5], $|G.\mathcal{C}| \simeq \Omega|S.\mathcal{C}|$ and hence $G.\mathcal{C}$ can be used as a model for the associated infinite loop space of $\mathbf{K}(\mathcal{C})$. Since Ω of a connective spectrum simply shifts the spectrum homology down by one dimension, it suffices to observe that the map γ realizes the standard map of the delooped spectrum

$$H_*(|G.\mathcal{C}|) \cong H_*(\Omega|S.\mathcal{C}|) \xrightarrow{f} H_{*+1}(|S.\mathcal{C}|)$$

where f is the composite map

$$H_*(\Omega|S.\mathcal{C}|) \cong H_{*+1}(\Sigma\Omega|S.\mathcal{C}|) \xrightarrow{H_*(coadj)} .H_{*+1}(|S.\mathcal{C}|)$$

(where *coadj* is the coadjoint map). □

4 Product Structures

Let T be a bi-exact functor from $\mathcal{A} \times \mathcal{B}$ to $\mathcal{C}$. Recall that this simply means that for all $A \in \mathcal{A}$ and $B \in \mathcal{B}$, the functors $T(A;\ )$ and $T(\ ;B)$ are exact. For all $m, n \geq 0$, we can produce new bi-exact functors

$$T(m,n) : S_m\mathcal{A} \times S_n\mathcal{B} \to S_{m+n}\mathcal{C}$$

$$\begin{array}{c}(A_1 \rightarrowtail \cdots \rightarrowtail A_m)\\ \times \\ (B_1 \rightarrowtail \cdots \rightarrowtail B_n)\end{array} \quad \mapsto \quad \left(\begin{array}{ccc} T(A_1;B_1) & \rightarrowtail \cdots \rightarrowtail & T(A_m;B_1) \\ \downarrow & & \downarrow \\ \vdots & \ddots & \vdots \\ \downarrow & & \downarrow \\ T(A_1;B_m) & \rightarrowtail \cdots \rightarrowtail & T(A_m;B_n) \end{array}\right)$$

with appropriate choices (see [11], page 342). Hence we can iterate this to bi-exact functors like $T(m,n)(p,q)$ from $S_mS_n\mathcal{A} \times S_pS_q\mathcal{B} \to S_mS_nS_pS_q\mathcal{C}$.

We define $T^{(m,n)}$ to be the induced bi-exact functors defined for all $m, n \geq 0$ by:

$$S_2^{(m)}\mathcal{A} \times S_2^{(n)}\mathcal{B} \xrightarrow{T^{(m,n)}} S_2^{(m+n)}\mathcal{C}$$

$$T^{(m,n)} = T(i_k, j_k) \cdots (i_2, j_2)(i_1, j_1) \qquad k = \max\{m, n\} \qquad \begin{array}{l} i_t = \begin{cases} 2 & \text{if } t \leq m \\ 1 & \text{if } t > m \end{cases} \\ \\ j_t = \begin{cases} 2 & \text{if } t \leq n \\ 1 & \text{if } t > n \end{cases} \end{array}$$

Let F be a (reduced) functor from exact categories to chain complexes of R-modules, where R is a ring. Let $\mathbf{R}$ be the functor of Example 1.8.1. Following the manner in which products are defined using Mac Lane's Q—construction ([7]) we define a pairing:

$$Q'(\mathbf{R};\mathcal{A}) \otimes_{\mathbf{Z}} Q'(F;\mathcal{B}) \xrightarrow{\mu(T)} Q'(F;\mathcal{C})$$

as follows. Given $\vec{A} \in S_2^{(m)}\mathcal{A}$, we obtain an R-module map

$$F(S^{(n)}\mathcal{B}) \xrightarrow{F(T^{(m,n)}(\vec{A};))} F(S^{(m+n)}\mathcal{C}).$$

Extending by R-linearity and letting m and n vary, we obtain a pairing for all $m, n \geq 0$

$$\mathbf{R}(S_2^{(m)}\mathcal{A}) \otimes_{\mathbf{Z}} F(S_2^{(n)}\mathcal{B}) \xrightarrow{\mu(T)^{(m,n)}} F(S_2^{(m+n)}\mathcal{C}).$$

Lemma 4.1 *If we write $t \star u$ for $\mu(T)(t \otimes u)$ then*

$$\partial\ (t \star u) = \partial(t) \star u + (-1)^{deg(t)} t \star \partial(u).$$

Proof We first observe that for $0 \leq k \leq 2$ we have the identities for $t \in Q'_m(\mathcal{A})$ and $u \in Q'_n(\mathcal{B})$:

$$F(d_k(i))(t \star u) = \begin{cases} F(d_k(i))(t) \star u & \text{if } 1 \leq i \leq m \\ t \star F(d_k(i-m)(u)) & \text{if } m+1 \leq i \leq m+n. \end{cases}$$

Letting $F_k(i) = F(d_k(i))$ we compute:

$$\partial(t \star u) = \sum_{i=1}^{m+n} (-1)^i [F_1(i) - F_0(i) - F_2(i)](t \star u)$$

$$= \sum_{i=1}^{m} (-1)^i [F_1(i) - F_0(i) - F_2(i)](t \star u)$$

$$+ \sum_{i=m+1}^{m+n} (-1)^i [F_1(i) - F_0(i) - F_2(i)](t \star u)$$

$$= \sum_{i=1}^{m} (-1)^i [F_1(i)(t) - F_0(i)(t) - F_2(i)(t)] \star u$$

$$+(-1)^m t \star [\sum_{j=1}^{n} (-1)^j [F_1(i)(u) - F_0(i)(u) - F_2(i)](u)]$$

$$= \partial(t) \star u + (-1)^{deg(t)} t \star \partial(u).$$

□

Corollary 4.2 *Given a bi-exact functor* $T : \mathcal{A} \times \mathcal{B} \to \mathcal{C}$*, R a ring and F a (reduced) functor from exact categories to chain complexes of R-modules, there is a well-defined pairing*

$$L(\mathbf{R}; \mathcal{A}) \otimes_{\mathbf{Z}} L(F; \mathcal{B}) \xrightarrow{\mu(T)} L(F; \mathcal{C})$$

such that $\partial(t \star u) = \partial(t) \star u + (-1)^{deg(t)} t \star \partial(u)$*. This follows by noting that* $\mu(T)$ *is defined by exact functors and hence passes to the quotient of* Q'*.*

□

Definition 4.3 The following is a list of definitions/lemmas which are now immediate to verify with the constructions above.

(a) If T is *strictly associative*—$T(T; id) = T(id; T)$—from $\mathcal{A} \times \mathcal{A} \times \mathcal{A}$ to $\mathcal{A}$ then $L(\mathbf{R}; \mathcal{A})$ becomes a strictly associative differential graded algebra using $\mu(T)$. If T has a *strict unit*—$A \in \mathcal{A}$ such that $T(A; \ ) = id_{\mathcal{A}} = T(\ ; A)$—then $L(\mathbf{R}; \mathcal{A})$ is strictly unital with unit $1 \cdot [A] \in L_0(\mathbf{R}; \mathcal{A})$.

(b) If T is strictly associative and M is a *strict left T-module*—a bifunctor from $\mathcal{B} \times \mathcal{A}$ to $\mathcal{B}$ such that $M(M; id) = M(id; T)$—and F takes values in chain complexes of R-modules then $L(F; \mathcal{B})$ is a differential graded left $L(\mathbf{R}; \mathcal{A})$—module. If M is a *strictly unital* left T-module— T is unital with unit $A \in \mathcal{A}$ and $M(\ ; A) = id_{\mathcal{B}}$—then $L(F; \mathcal{B})$ is a unital differential graded left $L(\mathbf{R}; \mathcal{A})$—module.

Recall that a bi-exact functor T produces a pairing of K-theory spectra $\mathbf{K}\mathcal{A} \wedge \mathbf{K}\mathcal{B} \to \mathbf{K}\mathcal{C}$ by realizing the bi-simplicial maps

$$S^{(m)}\mathcal{A} \times S^{(n)}\mathcal{B} \xrightarrow{T} S^{(m+n)}\mathcal{C}$$

(which passes to the smash product because $T(0;\ ) = 0 = T(\ ;0)$). Thus, if one takes the spectrum homology with coefficients in a ring R, one obtains a pairing of spectra which can be realized using the bi-simplicial maps

$$\mathbf{R}(S^{(m)}\mathcal{A}) \otimes_{\mathbf{Z}} \mathbf{R}(S^{(n)}\mathcal{B}) \xrightarrow{T} \mathbf{R}(S^{(m+n)}\mathcal{C}). \qquad *$$

Proposition 4.4 *Given a bi-exact functor $T : \mathcal{A} \times \mathcal{B} \to \mathcal{C}$, the pairing determined by Corollary 4.2 agrees up to homology with the pairing of spectra determined in (*) above.*

Proof Once again let $B.$ denote the bar construction for abelian groups and Tot the total complex of a multi-dimensional chain complex. We will not distinguish notationally between a multi-simplicial abelian group and its associated multi-dimensional chain complex. Since $Tot_k[B^{(n)}G]$ is 0 if $k \leq n-1$ and G if $k = n$, there is a natural transformation from $G[n]$ to $Tot[B^{(n)}G]$ which is seen to be an equivalence by considering the normalized complex of $B^{(n)}G$. Since $L(\mathbf{R};\ )$ is additive, we obtain a natural equivalence

$$L(\mathbf{R};\ )[n] \to Tot[B^{(n)}L(\mathbf{R};\ )] \xrightarrow{\beta} Tot[L(\mathbf{R}; S^{(n)}\ )] \qquad **$$

(where β is as in 1.8).

The multi-simplicial map η from $\mathbf{R}(S^{(n)}\ )$ to $L(\mathbf{R}; S^{(n)}\ )$ is $2n-2$– connected by 1.5 and 2.3 and since Tot preserves connectivity, $Tot(\eta)$ is also $2n-2$–connected.

Using the multi-simplicial bi-exact functor produced by T from $S^{(m)}\mathcal{A} \times S^{(n)}\mathcal{B}$ to $S^{(m+n)}\mathcal{C}$ and the naturallity of the construction of the pairing in 4.1, we obtain a multi-simplicial pairing

$$L(\mathbf{R}; S^{(m)}\mathcal{A}) \otimes_{\mathbf{Z}} L(\mathbf{R}; S^{(n)}\mathcal{B}) \xrightarrow{T} L(\mathbf{R}; S^{(m+n)}\mathcal{C})$$

which is compatible with the pairing (*) via the map η. Taking Tot's, this pairing is still compatible with $Tot(\eta)$ since Tot is natural and tensor preserving (there are only finitely many terms in each product, hence actually a direct sum). The natural equivalence (**) also preserves the pairings and hence the result is true in a stable range. Taking limits with respect to all m and n's we see the result is stably true on homology. □

Corollary 4.5 *The natural chain map $\lambda \circ \gamma$ of 3.4 is multiplicative on homology.*

Proof This follows from the fact that the natural map from $H_*(\Omega^\infty \mathbf{X})$ to $H_*(\mathbf{X})$ preserves products and 3.4. □

Example Let A be a commutative ring and let $\mathcal{F}_A$ be a category equivalent to the category of finitely generated free modules of A but having *exactly one* object $[n]$ for each rank $n \geq 0$ with a chosen basis $\{e_1, \dots, e_n\}$. We define T to be the bi-exact functor from $\mathcal{F}_A \times \mathcal{F}_A$ to $\mathcal{F}_A$ defined by the dictionary ordering [essentially $T(e(m)_i \times e(n)_j) = e(mn)_l$ where $l = i + (j-1)m$]:

$$T([m] \times [n]) = [mn]$$

$$Hom_A([m],[n]) \times Hom_A([m'],[n']) \mapsto Hom_A([mm'],[nn'])$$

$$M \times M' \mapsto T(M, M')_{p,q} = M_{a,b} \cdot M'_{x,y}$$

$$\text{where} \qquad p = a + (x-1)m \text{ and } q = b + (y-1)m.$$

Then T is strictly associative and unital (as in 4.3) and hence for any ring R, $L(\mathbf{R};\mathcal{F}_A)$ is a unital differential graded algebra with unit $1_R[1] \in L_0(\mathbf{R};\mathcal{F}_A)$ which on homology realizes the product of the spectrum homology for $\mathbf{K}(\mathcal{F}_A)$ with coefficients in R. If R is commutative this is commutative on homology.

5 Using Cotriples, Mac Lane's Q-Construction and the Atiyah-Hirzebruch Spectral Sequence

We recall that a *cotriple* $(\perp,\ \epsilon,\ \delta)$ in a category $\mathcal{A}$ is a functor $\perp\colon \mathcal{A} \to \mathcal{A}$ together with natural transformations $\epsilon : \perp \Rightarrow \mathrm{id}_{\mathcal{A}}$ and $\delta : \perp \Rightarrow \perp\perp$ such that the following diagrams commute:

$$\begin{array}{ccc} \perp & \xrightarrow{\delta} & \perp(\perp) \\ \big\downarrow{\scriptstyle \delta} & & \big\downarrow{\scriptstyle \delta_\perp} \\ \perp(\perp) & \xrightarrow{\perp\delta} & =\perp\perp(\perp) \end{array} \qquad \begin{array}{ccccc} & =\swarrow & \perp & \searrow= & \\ & & \big\downarrow{\scriptstyle \delta} & & \\ \perp & \xleftarrow{\perp\epsilon} & \perp(\perp) & \xrightarrow{\epsilon_\perp} & \perp \end{array}$$

We recall that the most common manner in which one encounters cotriples is as follows: Given a pair of adjoint functors (F, U) then $(\perp = FU, \epsilon, F(\eta))$ is a cotriple when ϵ is a counit and η is a unit of the adjoint pair.

Given a cotriple $\perp$ on $\mathcal{A}$ and an object $A \in \mathcal{A}$, we let $\perp_*\ A$ be the following simplicial object in $\mathcal{A}$:

$$[n] \mapsto \perp_n A = \perp^{(n+1)} A = \overbrace{\perp \cdots \perp}^{n+1 \text{ times}} A$$

$$d_i = \perp^{(i)} \epsilon \perp^{(n-i)} : \perp^{(n+1)} A \to \perp^{(n)} A$$

$$s_i = \perp^{(i)} \delta \perp^{(n-i)} : \perp^{(n+1)} A \to \perp^{(n+2)} A.$$

We observe that $\perp_*$ is augmented over $\mathrm{id}_{\mathcal{A}}$ by ϵ. That is, if we consider $(\mathrm{id}_{\mathcal{A}}, \mathrm{id}, \mathrm{id})$ as the trivial cotriple, then ϵ gives a natural simplicial map from $\perp_*$ to id_* and id_* is the trivial simplicial $\mathcal{A}$–object.

Example 5.1 Mac Lane's Q–construction

Let $\mathcal{A}b$ be the category of abelian groups. Let $\mathcal{A}b^{\mathcal{A}b}$ be the category of endfunctors of $\mathcal{A}b$ such that $F(0) = 0$ and let $\mathcal{B}$ be the category of functors from $\mathcal{A}b \times \mathcal{A}b$ to $\mathcal{A}b$ such that $F(0, A) = 0 = F(A, 0)$ for all $A \in \mathcal{A}b$. Let $\Delta : \mathcal{B} \to \mathcal{A}b^{\mathcal{A}b}$ be the diagonal functor: $\Delta(F)(A) = F(A, A)$. We can obtain a right adjoint to Δ by choosing a model for the *second cross effect* of a functor. That is, define $cr_2F \in \mathcal{B}$ to be any functor determined by the identity:

$$cr_2F(A, B) \oplus F(A) \oplus F(B) \cong F(A \oplus B).$$

Let $\perp = \Delta(cr_2F)$ then $\perp_*\ F(A)$ is a simplicial abelian group augmented over $F(A)$. Taking the normalized chain complex of $\perp_*\ F(A)$ and then the mapping cone of the chain map to $F(A)$ we obtain Mac Lane's *Q–construction* (see [7]).

Example 5.2 L^{red} once again

Let $\mathcal{E}x$ be a category of (small) exact categories closed under taking S_2 (the exact category of short exact sequences). Let $\mathcal{A}b^{\mathcal{E}x}$ be the category of functors from $\mathcal{E}x$ to $\mathcal{A}b$ such that $F(*) = 0$. We define a cotriple $\perp$ on $\mathcal{A}b^{\mathcal{E}x}$ similar to the one in example 1 but which does not appear to arise from a pair of adjoint functors. Let $d_0, d_1, d_2 : S_2 \Rightarrow \text{id}$ and $s_0, s_1 : \text{id} \Rightarrow S_2$ be the natural transformations:

$$d_i(A \to B \to C) = \begin{cases} C & \text{if } i = 0 \\ B & \text{if } i = 1 \\ A & \text{if } i = 2 \end{cases} \qquad s_i(D) = \begin{cases} 0 = 0 \to D & \text{if } i = 0 \\ 0 \to D = D & \text{if } i = 1. \end{cases}$$

Set

$$\perp F = \text{cokernel } [F \oplus F \overset{F(s_0)+F(s_1)}{\longrightarrow} F(S_2)]$$

and let $\epsilon : \perp F \Rightarrow \text{id}$ be the unique map on the cokernel produced by $F(d_0) - F(d_1) + F(d_2) : F(S_2) \to F$. In order to define $\eta : \perp \Rightarrow \perp\perp$ we first note that $\perp\perp F$ is naturally the total cokernel (cokernel of rows of the cokernel of columns (or visa versa) or simply the zero homology of the total complex) of the cube with $\sigma = F(s_0) + F(s_1)$:

$$\begin{array}{ccc} F \oplus F & \overset{\sigma}{\longrightarrow} & F(S_2) \\ \oplus \quad \oplus & \oplus & \oplus \\ F \oplus F & \overset{\sigma}{\longrightarrow} & F(S_2) \\ \downarrow{\scriptstyle \sigma \oplus \sigma} & & \downarrow{\scriptstyle \sigma \oplus \sigma} \\ F(S_2) \oplus F(S_2) & \overset{\sigma \oplus \sigma}{\longrightarrow} & F(S_2^{(2)}). \end{array}$$

Let γ be the natural transformation from S_2 to $S_2^{(2)}$ defined by:

$$(A \to B \to C) \overset{\gamma}{\mapsto} \left(\begin{array}{ccccc} A & \overset{=}{\longrightarrow} & A & \longrightarrow & 0 \\ \downarrow{\scriptstyle =} & & \downarrow & & \downarrow \\ A & \longrightarrow & B & \longrightarrow & C \\ \downarrow & & \downarrow & & \downarrow{\scriptstyle =} \\ 0 & \longrightarrow & C & \overset{=}{\longrightarrow} & C \end{array} \right)$$

then we define η to be the unique map obtained from $F(\gamma) : F(S_2) \to F(S_2^{(2)})$ on cokernels. It is straightforward to check that this is indeed a cotriple and that if we take the augmented chain complex of $\perp_* F(X) \to F(X)$ we recover $L^{red}(X; F)$ of Section 2.

Definition 5.3 For $F \in \mathcal{A}b^{\mathcal{E}x}$ we define the *Q–construction* of F to be the chain complex obtained by taking the mapping cone of $N(\perp_* F) \to F$ (where N is the normalization of the simplicial abelian object). We note that there is a natural transformation $L^{red}(F) \to Q(F)$ which is a natural chain homotopy equivalence.

Conventions For us a "space" will mean a compactly generated topological space with a nondegenerate basepoint. A "spectrum" will simply mean a sequence of spaces $\mathbf{E} = \{E_n\}_{n \in \mathbf{N}}$ with structure maps $E_n \to \Omega E_{n+1}$ such that $\pi_i(\mathbf{E}) = \lim \pi_{i+n}(E_n)$ is 0 if $i < 0$ and a morphism of spectra will be a sequence of continuous pointed maps which strictly commute with the structure maps.

Let $\mathcal{E}x$ be a category of (small) exact categories closed under S_2 and let $\mathcal{S}pec^{\mathcal{E}x}$ be the category of functors from $\mathcal{E}x$ to spectra such that $F(*)$ is the trivial spectrum consisting of only the basepoint and identity map. For each $i \in \mathbf{Z}$ we let $\pi_i F$ be the functor from $\mathcal{E}x$ to abelian groups obtained by composing with the i-th homotopy group. We recall that $f \in \mathcal{A}b^{\mathcal{E}x}$ is *additive* if and only if

$$f(S_2 X) \cong f(X) \oplus f(X)$$

by $f(d_0), f(d_2)$ (with inverse necessarily given by $f(s_0)+f(s_1)$). We call a functor $F \in \mathcal{S}pec^{\mathcal{E}x}$ *linear* if for all $X \in \mathcal{E}x$,

$$\begin{array}{ccc} F(S_2X) & \stackrel{F(d_0)}{\longrightarrow} & F(X) \\ \Big\downarrow {\scriptstyle F(d_2)} & & \Big\downarrow \\ F(X) & \longrightarrow & F(*) = * \end{array} \qquad (lin)$$

is cartesian (a homotopy pull-back diagram).

Lemma 5.4 *A functor F is linear if and only if $\pi_i F$ is linear for all i.*

Proof Since $F(s_0)$ is a section of $F(d_0)$ and $F(s_1)$ is a section to $F(d_2)$ and these are as spectra, we see that $F(S_2) \simeq F \times F \times cr(F)$ where $cr(F) \in \mathcal{S}pec^{\mathcal{E}x}$. Thus,

$$\pi_i F(S_2X) \cong \pi_i(F(X)) \oplus \pi_i(F(X)) \oplus \pi_i(cr(F)(X))$$

with

$$\pi_i(cr(F)(X)) \cong cr_2(\pi_i F)(X)$$

and the result follows.

What we want to construct is a functor $D : \mathcal{S}pec^{\mathcal{E}x} \to \mathcal{S}pec^{\mathcal{E}x}$ with a natural transformation p : id $\Rightarrow D$ such that $D(F)$ is always linear and p is universal with respect to linearization up to natural equivalence. We will construct D by using a cotriple. To define the cotriple we first choose some model for the iterated homotopy fiber of a cubical diagram of spectra which is functorial with respect to maps of cubical diagrams and such that the fiber of a (n+1)–cube obtained from an n–cube by mapping into only the trivial spectrum is once again the fiber of the n–cube and not only homotopic to it. This can be done, for example, by using the methods of ([2] XI, Section 3–5) or equivalently Definition 5.1 of [6].

Definition 5.5 We define the cotriple $(\perp, \epsilon, \delta)$ on $\mathcal{S}pec^{\mathcal{E}x}$ by (notation as in Example 2):

$$\perp F = \text{hfib} \left(\begin{array}{ccc} F(S_2) & \stackrel{F(d_0)}{\longrightarrow} & F \\ \Big\downarrow {\scriptstyle F(d_2)} & & \Big\downarrow \\ F & \longrightarrow & F(*) = * \end{array} \right)$$

We define $\epsilon : \perp F \Rightarrow$ id to be the natural transformation obtained by taking the homotopy fiber of the following natural map of diagrams:

$$\left(\begin{array}{ccc} F(S_2) & \stackrel{F(d_0)}{\longrightarrow} & F \\ \Big\downarrow {\scriptstyle F(d_2)} & & \Big\downarrow \\ F & \longrightarrow & F(*) = * \end{array} \right) \left(\begin{array}{cc} F(d_1) & * \\ * & * \end{array} \right) \atop \longrightarrow \left(\begin{array}{ccc} F & \longrightarrow & * \\ \Big\downarrow & & \Big\downarrow \\ * & \longrightarrow & * \end{array} \right)$$

We define $\eta : \perp F \Rightarrow \perp\perp F$ to be the natural transformation obtained by taking the homotopy fiber of the following natural map of diagram of spaces (γ is defined in 5.2):

$$
\begin{array}{ccccc}
\begin{array}{ccc} F(S_2) & \xrightarrow{F(d_0)} & F \\ \big\downarrow{\scriptstyle F(d_2)} & & \big\downarrow \\ F & \longrightarrow & * \end{array}
& &
\begin{pmatrix} F(d_0) & * \\ \mathrm{id} & * \end{pmatrix} \atop \longrightarrow
& &
\begin{array}{ccc} F & \xrightarrow{=} & F \\ \big\downarrow & & \big\downarrow \\ * & \longrightarrow & * \end{array}
\\
\begin{pmatrix} F(d_2) & \mathrm{id} \\ * & * \end{pmatrix}\Big\downarrow & & & & \\
\begin{array}{ccc} F & \longrightarrow & * \\ =\big\downarrow & & \big\downarrow \\ F & \longrightarrow & * \end{array}
& & \longrightarrow & &
\begin{array}{ccc} * & \longrightarrow & * \\ \big\downarrow & & \big\downarrow \\ * & \longrightarrow & * \end{array}
\end{array}
$$

$$
\begin{pmatrix} F(\gamma) & F(s_0) \\ F(s_1) & * \end{pmatrix}\Big\downarrow
$$

$$
\begin{array}{ccccc}
\begin{array}{ccc} F(S_2^{(2)}) & \xrightarrow{F(S_2 d_0)} & F(S_2) \\ \big\downarrow{\scriptstyle F(S_2 d_2)} & & \big\downarrow \\ F(S_2) & \longrightarrow & * \end{array}
& &
\begin{pmatrix} F(d_0 S_2) & F(d_0) \\ F(d_0) & * \end{pmatrix} \atop \longrightarrow
& &
\begin{array}{ccc} F(S_2) & \xrightarrow{F(d_0)} & F \\ \big\downarrow{\scriptstyle F(d_1)} & & \big\downarrow \\ F & \longrightarrow & * \end{array}
\\
\begin{pmatrix} F(d_2 S_2) & F(d_2) \\ F(d_2) & * \end{pmatrix}\Big\downarrow & & & & \Big\downarrow \\
\begin{array}{ccc} F(S_2) & \xrightarrow{F(d_0)} & F \\ \big\downarrow{\scriptstyle F(d_2)} & & \big\downarrow \\ F & \longrightarrow & * \end{array}
& & \longrightarrow & &
\begin{array}{ccc} * & \longrightarrow & * \\ \big\downarrow & & \big\downarrow \\ * & \longrightarrow & * \end{array}
\end{array}
$$

It is straightforward to check that this gives a cotriple–details left to the interested reader.

Given a functor $F \in \mathcal{S}pec^{\mathcal{E}x}$, we obtain a simplicial functor $\perp_* F$ which takes values in spectra. By realizing, we obtain a functor $|\perp_* F| \in \mathcal{S}pec^{\mathcal{E}x}$ with a natural transformation to F. We now choose some functorial model for the homotopy cofiber of a map of spectra which we can do, for example, using the methods of XII. Section 2 of [2].

Definition 5.6 We define $D : \mathcal{S}pec^{\mathcal{E}x} \to \mathcal{S}pec^{\mathcal{E}x}$ to be the functor

$$D(F) = \text{hcofiber}[\perp_* F \to F]$$

and the natural transformation $p : \mathrm{id} \Rightarrow D$ obtained by: $\perp_* F \to F \xrightarrow{p} \mathrm{Cof}$.

Lemma 5.7 *For $F \in \mathcal{S}pec^{\mathcal{E}x}$, DF is linear and any natural transformation $\tau : F \Rightarrow G$ with G linear factors up to natural homotopy by:*

$$
\begin{array}{ccc}
F & \xrightarrow{p} & DF \\
\big\downarrow{\scriptstyle \tau} & & \big\downarrow{\scriptstyle D\tau} \\
G & \xrightarrow{p\ (\simeq)} & DG.
\end{array}
$$

Proof To show a functor F is linear is equivalent to showing $\perp F$ is equivalent to a point. Since we are in connective spectra (and not just spaces), finite homotopy inverse limits commute with homotopy colimits and so $\perp D(F) \simeq D(\perp F)$. But $D(\perp F) \simeq *$ as $\perp_* (\perp F)$ is the simplicial path space of $\perp_* F$ and hence the augmentation map $\perp_* (\perp F) \to (\perp F)$ is a simplicial homotopy equivalence with inverse given by the unused degeneracy. If G is linear, then $\perp G \simeq *$ and hence $\perp_* G \simeq *$ so $G \xrightarrow{p} D(G)$ is an equivalence and the rest of the proposition follows by naturallity.

Example (5.2+) Given a functor $F \in \mathcal{A}b^{\mathcal{E}x}$, we can consider it as a functor to spectra, HF, by associating the Eilenberg-Mac Lane spectra to an abelian group (functorially–say by iterated bar constructions). Using the natural isomorphism

$$\text{cokernel}[F \oplus F \xrightarrow{F(s_0)+F(s_1)} F(S_2)] \cong \text{kernel}[F(S_2) \xrightarrow{F(d_0),F(d_2)} F \oplus F]$$

we see that $\perp_* HF$ is naturally equivalent to $H(\perp_* F)$ (as defined in Example 5.2) and hence that $\pi_*(D(HF)) \cong H_*(L^{red}(F))$.

Lemma 5.8 *For* $F \in \mathcal{S}pec^{\mathcal{E}x}$,

$$DF \simeq \text{hocolim}_n\ \Omega^n F(S^{(n)}).$$

Proof Let $St(F) = \text{hocolim}_n\ \Omega^n F(S^{(n)})$ and $F \xrightarrow{i} St(F)$ the natural map obtained by the limit maps. Thus, St is also a functor of $\mathcal{S}pec^{\mathcal{E}x}$ to itself with a natural transformation from the identity. By Corollary 1.5, $\pi_i F(S^{(n)})$ is additive in a $(2n-2)$–range (that is, diagram (lin) is cartesian in a $(2n-2)$–range) and hence $\Omega^n F(S^{(n)})$ is linear in an $(n-2)$–range and $St(F)$ is linear. Thus, by Lemma 5.7 we see that $St(F) \xrightarrow{p} D(ST(F))$ is an equivalence for all F. Similarly, since $\perp$ is given by a finite homotopy inverse limit, there is a natural equivalence

$$St(\perp (F)) = \text{hocolim}_n\ \Omega^n \perp F(S^{(n)}) \xrightarrow{\simeq} \perp \text{hocolim}_n\ \Omega^n F(S^{(n)}) = \perp (St(F))$$

which factors $D(i)$ as the natural composite $D(F) \xrightarrow{i_D} St(D(F)) \xrightarrow{\simeq} D(St(F))$. By standard arguments (see for example 1.5.5 of [11] or 2.2 of [4]), since $D(F)$ is linear, $D(F) \xrightarrow{\simeq} \Omega D(F)(S^1)$ and hence i_D is an equivalence. The result now follows from the following commutative diagram:

$$\begin{array}{ccc}
F & \xrightarrow{i} & St(F) \\
\downarrow{\scriptstyle p} & & \downarrow{\scriptstyle p_{St}\ (\simeq)} \\
D(F) & \xrightarrow{D(i)} & D(St(F)) \\
 & (\simeq) \searrow i_D & \uparrow{\scriptstyle \simeq} \\
 & & St(D(F)).
\end{array}$$

Example 5.9 Let $\mathbf{E}$ be a spectrum and define $\mathbf{E}$ to be the functor

$$\mathbf{E}(X) = Obj(X) \wedge \mathbf{E} = Obj(X) \otimes \mathbf{E}.$$

Then,

$$\begin{aligned}
\pi_i(D\mathbf{E}(X)) &\cong \pi_i(\text{hocolim}\ \Omega^n S^{(n)} X \otimes \mathbf{E}) \\
&\cong \lim \pi_{i+n}(S^{(n)} X \otimes \mathbf{E}) \\
&= \lim H_{i+n}(S^{(n)} X; \mathbf{E}) \\
&= H_i(\mathbf{K}(X); \mathbf{E}).
\end{aligned}$$

In particular, if $\mathbf{S}$ is the sphere spectrum (the suspension spectrum of two points) then $\pi_i(D(\mathbf{S})(X)) \cong K_i(X)$.

Proposition 5.10 *(The Atiyah-Hirzebruch Spectral Sequence) For* $F \in \mathcal{S}pec^{\mathcal{E}x}$ *there is a natural convergent first quadrant spectral sequence with*

$$E^1_{n,q} = L^{red}(X;\pi_n F)_q \Rightarrow \pi_{n+q} DF(X)$$

or equivalently

$$E^1_{n,q} = Q(X;\pi_n F)_q \Rightarrow \pi_{n+q} DF(X).$$

When $F = - \otimes \mathbf{E}$ *for some spectrum* $\mathbf{E}$ *this recovers the Atiyah-Hirzebruch spectral sequence for computing* $H_*(\mathbf{K}(X);\mathbf{E})$ *[1, III.7].*

Proof Let I be the category with two objects 0 and 1 and one non-trivial morphism $0 \to 1$ (equivalently, I is the category associated to the po-set $0 < 1$). Thus, $D(F)(X) = \mathrm{hocolim}_I \perp F$ where $\perp F(0) = \perp_* F(X)$, $\perp F(1) = F$ and $\perp F(0 \to 1)$ is the augmentation map. Using the simplicial replacement functor of [2] (XII.5.1), we obtain $D(F)(X)$ as a natural bi-simplicial spectrum

$$D(F)(X) \simeq |[p],[q] \mapsto \bigvee_{a_0 \to \cdots \to a_p \in N_p I} \perp F(a_0)_{[q]}|.$$

By taking homotopy groups first in the spectrum direction we obtain a convergent spectral sequence whose $E^1_{n,*}$ is given by the homology of the total complex associated to the bisimplicial abelian group

$$[p],[q] \mapsto \bigoplus_{a_0 \to \cdots \to a_p \in N_p I} \pi_n[\perp F(a_0)_{[q]}].$$

Normalizing this bicomplex in the simplicial direction of the nerve of I (the first position as we've written it) we see that $E^1_{n,*}$ is naturally chain homotopic to the total complex of the simplicial chain complex

$$[q] \mapsto \left(\pi_n F(X) \leftarrow \pi_n[\perp F(X)_{[q]}]\right)$$

given by the augmentation map. Using the isomorphism $\pi_n[\perp^{q+1} F(X)] \cong \perp^{q+1} [\pi_n F(X)]$ we see that this chain complex is isomorphic to $L^{red}(X;\pi_n F)$ and hence the first result. Since L^{red} is naturally chain homotopy equivalent to Q by normalization we are free to choose our models.

References

[1] Adams, J. F. [1974], *Stable homotopy and generalised homology*, The University of Chicago Press, Chicago.

[2] Bousfield, A. K. and Kan, D. M. [1972], *Homotopy limits, completions and localizations*, Springer LNM **304**.

[3] Dold, A. and Puppe, D. [1961], *Homologie nicht-additiver Functoren. Anwendungen*, Ann. Ins. Fourier, **11**, 201–312.

[4] Dundas, B. and McCarthy, R. [1994], *Stable K-theory and topological Hochschild homology*, Annals of Math. **140**, 685–701.

[5] Gillet, H. and Grayson, D. [1987], *On the loop space of the Q-construction*, Illinois Journal of Mathematics **31**, 574–597.

[6] Goodwillie, T. [1992], *Calculus II: Analytic functors*, K-Theory **5**, 295–332.

[7] Mac Lane, S. [1956], *Homologie des anneaux et des modules*, CBRM, Colloque de topologie algebrique, Louvain, 55–80.

[8] May, P. [1967], *Simplicial objects in algebraic topology*, D. Van Nostrand Co.

[9] McCarthy, R. [1993], *On fundamental theorems of algebraic K-theory*, Topology **32**, 325–328.

[10] McCarthy, R. [1995], *On the computation of stabilized tensor functors and the relative algebraic K-theory of dual numbers*, Transactions of the AMS, to appear.

[11] Waldhausen, F. [1985], *Algebraic K-theory of spaces*, Springer Lecture Notes in Math. **1126**, 318–419.

Fields Institute Communications
Volume **16**, 1997

Hypercohomology Spectra and Thomason's Descent Theorem

Stephen A. Mitchell
Department of Mathematics
University of Washington
Seattle, WA, USA 98195
mitchell@math.washington.edu

Dedicated to the memory of Bob Thomason

Abstract. The proof of Thomason's theorem on Bott-periodic algebraic K-theory is discussed in detail. The discussion includes an exposition of the homotopy theory of presheaves of spectra on a Grothendieck site, from two points of view: Thomason's theory of hypercohomology spectra, and the closed model category structure of Jardine. Background material on schemes and sites is provided.

Introduction

The celebrated Lichtenbaum-Quillen conjectures predict that for a sufficiently nice scheme and given prime ℓ, the ℓ-adic algebraic K-groups of X are closely related to the ℓ-adic étale cohomology groups of X. More precisely, one version of the conjectures asserts that there is a descent spectral sequence of Atiyah-Hirzebruch type

$$H^p_{\acute{e}t}(X; Z_\ell(\frac{q}{2})) \Rightarrow \pi_{q-p}(KX^\wedge_\ell)$$

but with the convergence only valid in sufficiently high degrees. Here the coefficient sheaves are Tate twists of the ℓ-adic integers, and are to be interpreted as zero if q is odd. Throughout this paper, étale cohomology is continuous étale cohomology [**19**], and the indicated abutment of the spectral sequence consists of the homotopy groups of the Bousfield ℓ-adic completion of the spectrum KX, not the naive ℓ-adic completion of the K-groups.

In a remarkable paper [**42**], Thomason proved the Lichtenbaum-Quillen conjectures for a certain localized form of K-theory - so-called "Bott-periodic" K-theory. The first step was the development of an elaborate theory of *hypercohomology spectra* $\mathbb{H}^{\cdot}(X; \mathcal{E})$ associated to étale presheaves of spectra $\mathcal{E}$ - or more generally, to presheaves of spectra on a Grothendieck site. These hypercohomology spectra are

---

1991 *Mathematics Subject Classification.* Primary 19; Secondary 55.
Supported by a grant from the National Science Foundation.

by their very construction naturally equipped with a suitable descent spectral sequence, and there is a natural map $\mathcal{E}(X) \longrightarrow \mathbb{H}^{\cdot}(X;\mathcal{E})$. In particular, the K-theory of a scheme X maps to its associated hypercohomology spectrum $\mathbb{H}^{\cdot}_{\acute{e}t}(X;K)$.

Fix a prime ℓ, which we will assume is odd in order to simplify the discussion. Let $L(-)$ denote Bousfield localization with respect to complex K-theory, and let $\hat{L}(-)$ denote its ℓ-adic completion. The main theorem of [**42**] can be stated as follows.

Theorem 0.1 *Let X be a separated noetherian regular scheme of finite Krull dimension, with sufficiently nice residue fields of characteristic different from ℓ. Then the natural map $KX \longrightarrow \mathbb{H}^{\cdot}_{\acute{e}t}(X;K)$ induces a weak equivalence*

$$\hat{L}K(X) \xrightarrow{\cong} \hat{L}\mathbb{H}^{\cdot}_{\acute{e}t}(X;K)$$

and there is a convergent descent spectral sequence

$$H^p_{\acute{e}t}(X; Z_\ell(\frac{q}{2})) \Rightarrow \pi_{q-p}\hat{L}KX.$$

For a precise statement see Theorem 4.13. Several remarks are in order:

- Usually the theorem is stated in terms of Bott-periodic K-theory. There are non-nilpotent elements

$$\beta_n \in \pi_{2(\ell-1)\ell^{n-1}} K\mathbb{Z} \wedge M\mathbb{Z}/\ell^n$$

  that ultimately arise from a primitive ℓ^n-th root of unity ($M\mathbb{Z}/\ell^n$ is the mod ℓ^n Moore spectrum). Using the $K\mathbb{Z}$-algebra structure on KX, we can form the mapping telescope $\beta_n^{-1}KX \wedge M\mathbb{Z}/\ell^n$, and it is known that this mapping telescope is weak equivalent to $\hat{L}KX \wedge M\mathbb{Z}/\ell^n$. Thus one gets a spectral sequence going from mod ℓ^n cohomology to mod ℓ^n Bott-periodic K-theory. In this paper, however, the emphasis will be on Bousfield localization, with the Bott elements entering only as tools in the proofs. For more information on Bott elements, see [**31**] and the references cited there, as well as the recent paper by B. Kahn [**24**].
- The localization functor $\hat{L}$ is only the first of an infinite sequence of localization functors $\hat{L}_n$ defined on spectra. Thus from a global homotopy-theoretic perspective, the localization imposed by Thomason is *a priori* rather drastic. However the author showed in [**30**] that Theorem 0.1 remains true with $\hat{L}$ replaced by $\hat{L}_n$ for any n, including the "harmonic localization" $\hat{L}_\infty$, so this is not such a drastic localization after all. For further discussion of this point, see [**31**].
- The second statement of the theorem is not merely a corollary of the first. The difficulty is with the identification of the E_2-term; nowadays this depends on the Gabber rigidity theorem for hensel local rings and Suslin's theorem on algebraically closed fields. When Thomason first proved Theorem 0.1, however, these results were not available and he had to prove them in localized form.
- As a corollary to another major work [**44**], Thomason showed that if K-theory is replaced by Bass K-theory, the assumption that X is separated and regular can be dropped. This more general result will not be considered here.

Thomason's theorem has many interesting applications. For example, let F be a totally real number field with ring of integers $\mathcal{O}_F$. Let $LK_n(-) = \pi_n LK(-)$. Then up to powers of 2 the value of the Dedekind zeta function at the odd negative integers is given by the amazing formula

$$\zeta_F(1-2n) = \frac{\mid LK_{4n-2}\mathcal{O}_F \mid}{\mid LK_{4n-1}\mathcal{O}_F \mid}$$

The point is that the descent spectral sequence collapses, and the theorem of Wiles [**46**] proving Iwasawa's "Main Conjecture" computes the given zeta values in terms of étale cohomology.

Another application, due to Bill Dwyer and the author [**9**], uses Thomason's theorem to compute the ℓ-adic complex K-theory of the spectrum $K\mathcal{O}_F[\frac{1}{\ell}]$ in terms of the Iwasawa theory of the cyclotomic Z_ℓ-extension (F an arbitary number field). In the case of abelian fields, this leads to a computation of the complex K-theory of the space $BGL\mathcal{O}_F[\frac{1}{\ell}]^+$. This work also yields an explicit conjectural description of the Hopf algebra $H_*(BGL\mathcal{O}_F[\frac{1}{\ell}]^+; \mathbb{Z}/\ell)$, and leads to some partial computations in [**32**] and [**33**].

Further applications and motivation can be found in [**42**], [**43**], and [**31**]. The purpose of the present paper, however, is to explain the *proof* of Thomason's Theorem, and the vast array of machinery underlying it. It took the author rather a long time to progress from "thrill seeker" to "reckless cheat" to "honest man" (Thomason's amusing terminology; see [**42**], Introduction). Our hope is that this essay will smooth the way for others. At the same time, we will survey the work of Jardine ([**20**], [**21**], [**22**]) that recasts the entire theory in terms of closed model categories. In this approach, the role of the hypercohomology presheaf $\mathbb{H}^{\cdot}(-;\mathcal{E})$ is played by a *globally fibrant model* for $\mathcal{E}$. Here again, the technical prerequisites are formidable; our goal is to help make these works more accessible.

The paper is organized as follows. Each of the five sections has its own introduction. The first two sections are intended for readers who may not be familiar with schemes and sites. Section 1 is an introduction to Thomason's theorem for fields, eschewing all scheme-theoretic language to minimize the prerequisites. Section 2 provides a short introduction to sites and cohomology, and includes a discussion of the very useful Nisnevich site.

The experts can proceed directly to the third section. Here we introduce presheaves of spectra on a Grothendieck site, and the associated hypercohomology spectrum. The latter construction is a direct generalization to spectra of the Godement resolution of an abelian sheaf. Then we consider Jardine's closed model category structure for presheaves of spectra on a site.

Section 4 begins by discussing the descent theorems of Brown-Gersten and Nisnevich. Then the Nisnevich descent theorem is used to deduce Theorem 0.1 for schemes from Theorem 0.1 for fields. Section 5 sketches the ingenious proof for fields.

Notation, terminology, and a confession of sins. The mod n Moore spectrum is denoted $M\mathbb{Z}/n$. For a spectrum $\mathcal{E}$, the ℓ-adic completion (=Bousfield localization with respect to $M\mathbb{Z}/\ell$) is denoted $\mathcal{E}^\wedge$. We write $L\mathcal{E}$ for Bousfield localization with respect to complex K-theory, and abbreviate $(L\mathcal{E})^\wedge$ as $\hat{L}\mathcal{E}$. Recall also that $\hat{L}(-)$ is the same thing as Bousfield localization with respect to mod ℓ complex K-theory.

Bear in mind that L commutes with pseudo-filtered (e.g., direct) colimits, but $\hat{L}$ does not. Readers unfamiliar with Bousfield localization might consult [**31**].

If S is a topological space, possibly discrete, $Map(S, X)$ denotes the space (spectrum) of pointed maps to the pointed space (spectrum) X.

If A is a presheaf of sets or abelian groups, we usually write aA for the sheafification of A, but when $A = \pi_q\mathcal{E}$ is a presheaf of homotopy groups attached to a presheaf of spectra, we follow a common practice and write $\tilde{\pi}_q\mathcal{E}$ for $a\pi_q\mathcal{E}$.

The K-theory spectrum of the category of locally free sheaves on a scheme X is denoted KX. The K-theory spectrum attached to the coherent sheaves is denoted $K'X$, since the alternate notation GX would just about exhaust the already overworked letter G.

Now, as to our sins: Among the many technical problems that arise in the theory discussed here, there are several that we have chosen to ignore. Let us at least be explicit about what they are. First, there are the usual set-theoretic difficulties - for example, many of our categorical definitions only make sense if the category in question is small. Occasionally hypotheses of this sort are made explicitly, but more often it is left to the reader to make the necessary adjustments. Second, there is the problem that some of the "functors" lurking in the background are in fact only pseudo-functors - for example, the "functor" that assigns to each ring its category of finitely-generated projective modules. In addition, the passage from schemes to K-theory spectra is left inside a black box - particularly when it comes to questions involving the multiplicative structure. For a careful and rigorous treatment of all these matters, see [**22**].

Contents

1 Thomason's Theorem for Fields

Let F be a field, ℓ a prime not equal to the characteristic of F. One of the fundamental problems of algebraic K-theory is to compute the ℓ-adic K theory of F, where by "ℓ-adic K-theory" we mean the homotopy groups of the ℓ-adic completion $KF^{\wedge}$. The following beautiful theorem was proved by Suslin in 1984 ([**38**], [**39**]).

Theorem 1.1 *Let F be a separably closed field, ℓ a prime not equal to the characteristic of F. Then*

$$\pi_{2n}KF^{\wedge} \cong Z_{\ell}(n)$$

for $n \geq 0$, and $\pi_{odd}KF^{\wedge} = 0$. In fact there is a weak equivalence $KF^{\wedge} \cong bu^{\wedge}$, where bu is the -1-connected cover of the complex K-theory spectrum.

Here Z_{ℓ}(n) denotes the n-th Tate twist of the ℓ-adic integers. In other words,

$$Z_{\ell}(n) = \otimes^n Z_{\ell}(1)$$

where $Z_{\ell}(1)$ denotes the inverse limit of the groups $\mu_{\ell^{\nu}}(F)$.

Theorem 1.1 was originally stated for algebraically closed fields, but the extension to the separably closed case is immediate because ℓ-adic K-theory (with $\ell \neq char\, F$) is invariant under purely inseparable extension ([**35**], 7, 4.8). We also note that there is some choice involved in the weak equivalence $KF^{\wedge} \cong bu^{\wedge}$. It is defined by a diagram

$$KF^{\wedge} \longleftarrow KE^{\wedge} \longrightarrow bu^{\wedge}.$$

Here E is the algebraic closure of the prime subfield, and the first map is just the canonical inclusion. This map is a weak equivalence [**38**]. If Char F=0, the second map arises from a choice of embedding of E into the complex numbers. If Char F=p, it arises in effect from a choice of Brauer lift (or equivalently, by a choice of embedding of the Witt vectors of E into $\mathbb{C}$).

Given Suslin's theorem, it is natural to try to compute the K-theory of F by "descent" from its separable closure. In other words, one wants some kind of spectral sequence whose input is $\pi_* K\bar{F}^{\wedge}$ as Galois module and whose output is $\pi_* KF^{\wedge}$. The way to approach such a problem is to cook up an auxiliary spectrum Y which by its very construction admits such a spectral sequence, and which is equipped with a natural map $KF^{\wedge} \to Y$. Then one tries to show that this map is a weak equivalence. In fact, however, life is not so simple; it turns out that the best one can hope for is a weak equivalence on some connected cover. All this will be discussed further once we have Thomason's descent machinery in place. We will first try to motivate this machinery via the prototypical example of homotopy fixed point sets.

1.1 Homotopy fixed point sets. Let G be a discrete group, X a G-spectrum. Here we simply mean that each space X_n of the spectrum is a pointed G-space, and the structure maps $X_n \longrightarrow \Omega X_{n+1}$ are G-maps (with G acting trivially on the loop coordinate). The homotopy fixed-point spectrum is

$$X^{hG} = map^G(EG_+, X).$$

Here the n-th space of the spectrum on the right is the function space of equivariant pointed maps $Map^G(EG_+, X_n)$. The structure maps are the usual, obvious ones. Filtering EG by skeleta exhibits X^{hG} as the inverse limit of a tower of fibrations

$$\dots \longleftarrow Map^G((EG)_+^{n-1}, X) \longleftarrow Map^G((EG)_+^n, X) \longleftarrow \dots$$

in which the fibres have the form $\prod \Omega^n X$. This yields a "conditionally convergent" right half-plane cohomology spectral sequence - the "descent spectral sequence"

$$E_2^{p,-q} = H^p(G; \pi_q X) \Longrightarrow \pi_{q-p} X.$$

See Section 3.1 for a discussion of convergence. Here it will suffice to note that any of the following conditions will guarantee convergence:

- G has finite cohomological dimension - or more generally, there exists a positive integer d such that $H^p(G; \pi_q X) = 0$ for all $p > d$ and all q. In this case the spectral sequence collapses at E_{d+1}.
- The spectrum X is bounded above - that is, there exists an integer M such that $\pi_q X = 0$ for all $q > M$. In this case the spectral sequence is formally the same as a first quadrant spectral sequence.
- The groups $H^p(G; \pi_q X)$ are all finite.

There is a natural map $X^G \longrightarrow X^{hG}$ defined as the composite

$$X^G = Map^G(S^0, X) \longrightarrow Map^G(EG_+, X)$$

induced by the augmentation $EG_+ \longrightarrow S^0$. If this map is a weak equivalence, then the descent spectral sequence converges to the homotopy of the actual fixed point set.

There is another way to describe X^{hG} that will be essential later. Recall that EG_+ is the geometric realization of a certain simplicial pointed G-set $E.G_+$ whose n-th level is G_+^{n+1}. Thinking of $E.G_+$ as a simplicial pointed discrete space, we can apply the functor $Map^G(-, X)$ to obtain a cosimplicial spectrum that we will denote $S^{\cdot}(G, X)$. Then

$$X^{hG} = Tot\, S^{\cdot}(G, X)$$

where "Tot" is the Bousfield-Kan total space construction [**6**] applied levelwise. It is shown in [**6**] that this Tot is the inverse limit of a tower of fibrations - and indeed it is just the same tower constructed above. Moreover, there is a weak equivalence

$$X^{hG} \cong holim_{\Delta} S^{\cdot}(G, X)$$

and it is this holim construction that we will actually use.

Roughly speaking, we want to apply this machinery to the case $X = K\bar{F}$, where $\bar{F}$ is the separable closure of F, with $G = G_F$ the absolute Galois group. However we need to modify the homotopy fixed point construction in two ways. First, G is a profinite group and we want to take into account this profinite structure. In particular we want a spectral sequence involving Galois cohomology.[1] Secondly, we will change our point of view by replacing the G_F-spectrum $K\bar{F}$ by the "presheaf of spectra" $L \mapsto KL$, L a finite separable extension of F (for an elegant treatment of homotopy fixed point spaces for Galois groups, see [**15**]). We first digress to discuss Galois cohomology.

[1]If G_F is finite, no modifications are needed. However fields with nontrivial but finite absolute Galois group are extremely rare - the example of the real numbers notwithstanding! See ([**3**], X, Theorem 2.1) for a discussion of this claim.

1.2 Galois cohomology. Let G be a profinite group.[2] This is equivalent to saying that G is a compact and totally disconnected topological group, but in more practical terms it means that G is the inverse limit of its finite quotients G/U, with U varying over the *closed* normal subgroups of finite index (note that the closed subgroups of finite index are precisely the open subgroups). A *discrete G-module* M is a G-module such that M is the union of the various fixed submodules M^U. The cohomology groups $H^*(G;M)$ can be defined in two equivalent ways.

- "Sheaf": The category of discrete G-modules is abelian and has enough injectives. The functors $H^*(G;M)$ are the derived functors of the left exact functor $M \to M^G$.
- "Cech": $H^*(G;M) = colim_U H^*(G/U;M^U)$. Here the terms in the colimit are just ordinary group cohomology. The maps in the colimit combine the inclusions $M^U \subset M^V$ with pullback along the quotients $G/V \to G/U$ in the obvious way.

The reason for the slogans "sheaf" and "Cech" may not be apparent, but should become clear in the next section.

Here is one way to see that the two definitions agree. We include this here because it helps to illuminate some of the spectrum-level constructions that come later:

Form the simplicial G-set $E.G$ as usual. In this case it is a simplicial space with the profinite topology on the terms G^{n+1}. Apply the functor $Map(-,M)$ - continuous M-valued functions - to obtain a cosimplicial G-module $Map(E.G, M)$. Here the G-action is the usual left action whose fixed points yield the group of equivariant maps. The natural augmentation $M \to Map(G, M)$ makes the associated cochain complex into a resolution of M by acyclic G-modules. Applying the G-invariants functor then yields a cosimplicial abelian group $T^{\cdot}(G, M)$. Of course, this is nothing but the usual bar-type complex (a.k.a. the Godement complex, cf. 2.6). The associated cohomology groups are then precisely the groups $H^*(G;M)$ according to the "sheaf" definition.

On the other hand, continuity in this context just means that

$$Map(G^n, M) = colim_U Map((G/U)^n, M).$$

It follows easily that

$$T^{\cdot}(G, M) = colim_U T^{\cdot}(G/U, M^U)$$

so this is the same as "Cech" cohomology.

There is yet another description that will be useful (compare Section 3.1).

Proposition 1.2 *For a discrete G-module M,* $H^p(G, M) = lim^p_{\Delta} T^{\cdot}(G, M)$.

We will also need to consider cohomology with coefficients in non-discrete modules such as $Z_\ell(n)$. The category of inverse systems $\{M_n\}$ of discrete G-modules is also abelian with enough injectives, and the composite functor $\{M_n\} \mapsto lim^0_n M_n^G$ is left exact. Its derived functors will be denoted $H^*(G; \{M_n\})$. This is a special case of "continuous" cohomology in the sense of Janssen **[19]** and will arise in its more general form in 2.7. Often it is written with a subscript viz. H^*_{cont}, but we

[2] Logically, the cohomology to be discussed here should be called "profinite group cohomology". However it turns out that every profinite group is the Galois group of some Galois extension **[45]**, so the theories of Galois extensions and profinite groups are abstractly equivalent.

can dispense with this because as far as we are concerned continuous cohomology is the only cohomology. Note that there are short exact sequences

$$0 \longrightarrow lim^1_n H^{p-1}(G; M_n) \longrightarrow H^p(G; \{M_n\}) \longrightarrow lim^0_n H^p(G; M_n) \longrightarrow 0.$$

If M is given as an inverse limit of discrete modules - $M = lim^0_n M_n$ - we will write simply $H^*(G; M)$ for $H^*(G; \{M_n\})$. The main case of interest is $Z_\ell(n) = lim^0_m(\mathbb{Z}/\ell^m(n))$.

If L/F is a Galois extension, we will usually write $H^*(L/F; M)$ for $H^*(G(L/F); M)$ and simply $H^*(F; M)$ for $H^*(\bar{F}/F; M)$.

We say that $cd_\ell G = d$ if d is maximal such that there exists a discrete ℓ-torsion module M with $H^d(G; M) \neq 0$. An argument using "Shapiro's lemma" shows that if H is any closed subgroup of G, $cd_\ell H \leq cd_\ell G$. Translating into field notation, this implies that if E is any separable algebraic extension of F, $cd_\ell E \leq cd_\ell F$.

Some examples:

- If F is separably closed, obviously cd_ℓF=0 for all ℓ. In fact it is not hard to show that cd_ℓF=0 if and only if F has no finite separable ℓ-extensions.
- If F is finite, cd_ℓF=1.
- Tsen's theorem: If F is a function field of dimension one over a separably closed field k (i.e., F is a finite separable extension of $k(T)$), $cd_\ell F \leq 1$.
- If F is a number field and either ℓ is odd or F is pure imaginary, cd_ℓF=2. Otherwise $cd_2 F = \infty$.

Proposition 1.3 *If F has transcendence degree n over a subfield F_0, then $cd_\ell F \leq cd_\ell F_0 + n$. (Usually equality holds; see ([**3**], X, 2.1.))*

The proof involves inducting over a "Tate-Tsen" filtration; compare the proof of Theorem 5.1 below.

One similarly defines cohomological dimension for inverse systems of ℓ-torsion modules.

1.3 Hypercohomology spectra for fields. A *presheaf of spectra* on F is a covariant functor $\mathcal{E}$ from the category of finite separable extensions L of F (which we assume to be subfields of a fixed separable closure $\bar{F}$). Later this definition will be modified and put in its proper context of "presheaves of spectra on the étale site of F", but for now we want to keep things as simple as possible. The main example we have in mind if of course the presheaf $L \mapsto KL$.

The *stalk* of a presheaf $\mathcal{E}$ is the colimit over finite separable extensions L of the spectra $\mathcal{E}$(L). In practice, the functor $\mathcal{E}$ will be defined on infinite extensions as well and will commute with colimits (up to weak equivalence), so we will write $\mathcal{E}_{\bar{F}}$ for the stalk of $\mathcal{E}$. The stalk provides the connection between the presheaf point of view and the G_F-spectrum point of view: each presheaf has its stalk, and conversely given a G_F -spectrum X we get a presheaf $\mathcal{E}$ with $\mathcal{E}(L) = X^{G_L}$.

We will give two equivalent versions of the hypercohomology spectrum $\mathbb{H}^{\cdot}$(F,$\mathcal{E}$). The first is to be taken as the definition. For any profinite group G and spectrum X, $Map(G, X)$ will by definition mean $colim_U Map(G/U, X)$ (where as usual U ranges over open normal subgroups). Passing to equivariant maps, this means that

$$Map^G(G, X) = colim_U Map^{G/U}(G/U, X^U)$$

- "Sheaf": Apply the functor $Map^{G_F}(-, \mathcal{E}_{\bar{F}})$ (continuous maps, as always) to $E.G$, obtaining a cosimplicial spectrum $T^{\cdot}(F, \mathcal{E})$. Define
$$\mathbb{H}^{\cdot}(F, \mathcal{E}) = holim_{\Delta} T^{\cdot}(F, \mathcal{E})$$
- "Cech": Let $S^{\cdot}(F, \mathcal{E})$ denote the cosimplicial spectrum $colim_L S^{\cdot}(G(L/F), \mathcal{E}(L))$. Define
$$\check{\mathbb{H}}^{\cdot}(F, \mathcal{E}) = holim_{\Delta} S^{\cdot}(F, \mathcal{E}).$$

In fact these two spectra are naturally weak equivalent. The point is that the obvious map of cosimplicial spectra

$$S^{\cdot}(F, \mathcal{E}) \longrightarrow T^{\cdot}(F, \mathcal{E})$$

is a weak equivalence at each cosimplicial level and hence induces a weak equivalence on $holim_{\Delta}$.

It is important to note that the two limit operations in the definition of the Cech hypercohomology spectrum do not, in general, commute up to weak equivalence (but see Example 3.22 below).

Theorem 1.4 *Let F be any field, $\mathcal{E}$ a presheaf of spectra on F. There is a conditionally convergent right half-plane cohomology spectral sequence*

$$E_2^{p,-q} = H^p(F, \pi_q \mathcal{E}_{\bar{F}}) \Rightarrow \pi_{q-p} \mathbb{H}^{\cdot}(F, \mathcal{E}).$$

Our main example is the case $\mathcal{E}^n = \mathcal{K}(-) \wedge M\mathbb{Z}/\ell^n$, where $\mathcal{K}$ is the K-theory presheaf. Then Suslin's theorem yields

Corollary 1.5 *For any field F and prime ℓ, there is a conditionally convergent fourth quadrant cohomology spectral sequence*

$$E_2^{p,-q} = H^p(F; Z_\ell(\frac{q}{2})) \Rightarrow \pi_{q-p} \mathbb{H}^{\cdot}(F; \mathcal{K})^{\wedge}$$

($E_2^{p,-q}$ is zero for q odd or $q < 0$).

Note that Suslin's theorem is needed to identify the E_2-term. Also, we have implicitly used the fact that smashing with a finite spectrum commutes with hypercohomology, in order to make the identification

$$holim_n \mathbb{H}^{\cdot}(F, \mathcal{K} \wedge M\mathbb{Z}/\ell^n) = holim_n \mathbb{H}^{\cdot}(F, \mathcal{K}) \wedge M\mathbb{Z}/\ell^n.$$

This will be justified in a more general context in Section 3. An explanation of why we get continuous cohomology can be found in Section 4.11.

Note that the spectral sequence of the corollary - the descent spectral sequence for K-theory - will converge if, for example, $cd_\ell F < \infty$. When $F = \mathbb{R}$ and ℓ=2, we have $cd_2 \mathbb{R} = \infty$ but one can show the spectral sequence converges nevertheless.

Note also that when $cd_\ell F = d < \infty$, we conclude that $\mathbb{H}^{\cdot}(F, \mathcal{K})^{\wedge}$ is (-d-1)-connected, and

$$\pi_{-d} \mathbb{H}^{\cdot}(F, \mathcal{K})^{\wedge} = H^d(F; Z_\ell(0).$$

This group can easily be nonzero - for example, if F is any finite field, ℓ any prime, we have $\pi_{-1} = Z_\ell$.

For any presheaf $\mathcal{E}$, there is a natural augmentation map $\mathcal{E}(F) \rightarrow \mathbb{H}^{\cdot}(F; \mathcal{E})$ (analogous to $X^G \rightarrow X^{hG}$). In this context, the Lichtenbaum-Quillen conjecture asserts:

Conjecture 1.6 *If $cd_\ell F < \infty$, the augmentation $KF^{\wedge} \longrightarrow \mathbb{H}^{\cdot}(F, \mathcal{K})^{\wedge}$ is a weak equivalence on some connected cover.*

There are some explicit conjectures as to what the isomorphism range is. Clearly one must at least pass to the (-1)-connected cover of the hypercohomology spectrum - but this is not enough. Suppose for example that $cd_\ell F \leq 2$. Then the spectral sequence collapses and

$$\pi_0 \mathbb{H}^{\cdot}(F, \mathcal{K})^{\wedge} \cong Z_\ell \oplus H^2(F; Z_\ell(1)).$$

The term $H^2(F; Z_\ell(1))$ is closely related to the Brauer group and is typically nonzero. Hence in this case, there is at best an isomorphism above dimension zero.

1.4 Statement of Thomason's Descent Theorem for Fields.

Definition 1.7 *A field F is ℓ-good if char $F \neq \ell$ and F has finite transcendence degree over a subfield E satisfying $cd_\ell E_\infty \leq 1$, where E_∞ is obtained from E by adjoining all the ℓ-power roots of unity. If $\ell = 2$ we also require $\sqrt{-1} \in E$.*

The condition $\sqrt{-1} \in F$ when $\ell = 2$ is needed for two reasons. First of all, we will need to know that $cd_\ell F < \infty$. For that we could get by with the weaker assumption $cd_\ell(E_\infty/E) \leq 1$. However one also needs $\sqrt{-1} \in F$ for the construction of Bott elements. Examples of fields E as above include separably closed fields, finite fields, local fields, and number fields (with $\sqrt{-1} \in F$ if $\ell = 2$). Note that an ℓ-good field has finite cd_ℓ.

Theorem 1.8 *Let F be an ℓ-good field. Then the natural augmentation $KF \longrightarrow \mathbb{H}^{\cdot}(F, \mathcal{K})$ induces a weak equivalence*

$$\hat{L}KF \cong \hat{L}\mathbb{H}^{\cdot}(F, \mathcal{K}),$$

and there is a convergent spectral sequence,

$$E_2^{p,-q} = H^p(F; Z_\ell(\frac{q}{2})) \Rightarrow \pi_{q-p}\hat{L}KF.$$

One can show that the functor L commutes with hypercohomology (see Corollary 3.12), so $\hat{L}\mathbb{H}^{\cdot}(F, \mathcal{K}) = (\mathbb{H}^{\cdot}(F, L\mathcal{K}))^{\wedge}$. Then the spectral sequence is obtained as before.

The theorem can be reformulated in terms of "Bott-periodic" K-theory. Let $kF = KF \wedge M\mathbb{Z}/\ell$. Then the theorem is equivalent to the assertion that $LkF \longrightarrow \mathbb{H}^{\cdot}(F; Lk)$ is a weak equivalence, and moreover an easy transfer argument reduces to the case when F contains ℓ-th roots of unity. In that case we fix once and for all a Bott element $\beta \in \pi_2 kF$. Then by a theorem of Snaith and Dwyer, $LkF \cong \beta^{-1}kF$. Theorem 1.8 is then equivalent to:

Theorem 1.9 *Let F be an ℓ-good field containing ℓ-th roots of unity. Then the natural augmentation $kF \longrightarrow \mathbb{H}^{\cdot}(F, k)$ induces a weak equivalence*

$$\beta^{-1}kF \cong \beta^{-1}\mathbb{H}^{\cdot}(F; k).$$

The proof of this theorem will be sketched in the final section.

2 Schemes, Sites and Cohomology

This section is directed strictly from one non-expert to another. Its purpose is to provide, in condensed form, the technical background on sites and cohomology necessary to penetrate [**42**] and [**22**].

Just keeping track of the scheme-theoretic hypotheses in [**42**] is already a non-trivial task. The most basic assumption will usually be that "X is a separated noetherian regular scheme, of finite Krull dimension", so we briefly review these notions in the first subsection. In the next two subsections we recall, for the reader's convenience, the definitions of étale morphisms and hensel local rings.

In the fourth subsection we introduce Grothendieck sites. A *site* is a beautiful and far-reaching generalization of a topological space. It is an extremely general object, which at the same time is far simpler than any of its special cases. The examples of most importance for this paper are the étale, Nisnevich and Zariski sites of a scheme. The Nisnevich site lies between the étale and Zariski sites, and serves to bridge the gap between fields and general schemes. Next we introduce presheaves and sheaves, and discuss the various direct and inverse image functors associated to a site morphism.

The *topos* associated to a site is its category of sheaves of sets. According to [**3**] (IV, introduction) or [**23**] (introduction), it is the topos that is of paramount importance - different sites can yield the same topos, and a choice of site is just a particular way to construct the topos. Be that as it may, it is not the point of view that will be taken here. Our ultimate goal is to study the homotopy theory of *pre*sheaves of spectra on a site, and it would seem to be highly desirable - even essential - to work on an actual site rather than the disembodied topos. Thus topos theory *per se* will make only a minor appearance.

One place where the topos does enter is in the next section, on points and stalks. There is a beautiful theory ([**3**], IV.6) extending the concept of stalk at a point, familiar from classical sheaf theory, to general sites. The idea is that it is the inverse image or stalk functor on sheaves - associated to inclusion of a point in the usual sense - which carries all the essential information. Formulated in this way, the classical theory of points and stalks carries over to general sites to a remarkable degree. The most noteworthy difference is that a given site may or may not have "enough points", meaning that stalks detect isomorphisms in the usual way. The étale, Nisnevich and Zariski sites have enough points.

The next two subsections discuss sheaf cohomology, including some results on étale and Nisnevich cohomological dimension that are crucial for the applications to spectra. Then we introduce Čech cohomology and compare it with sheaf cohomology.

In the final section on noetherian sites we discuss two crucial results on commuting sheaf cohomology with filtered colimits. The first of these involves colimits in the sheaf variable, and is straightforward. The second result - the so-called "Continuity Theorem" has to do with co-filtered inverse limits in the scheme variable, with the étale topology (or more generally inverse limits of sites, but discretion seems the better part of valor here). This theorem is somewhat more difficult to state, and much more difficult to prove. In the affine case it says roughly that étale cohomology commutes with filtered colimits of rings.

For a quick introduction to sites and étale cohomology, see [**41**] or [**1**]. The definitive treatment is the original source [**3**]. For the Nisnevich site see [**34**], [**25**], Appendix E in [**44**] and Chapter 7 of [**22**].

2.1 Schemes. We recall that a scheme X is *separated* if the diagonal Δ_X is closed in $X \times X$. For example any affine scheme or algebraic variety is separated, whereas e.g. affine n-space "with a point doubled" is not. More generally a morphism $X \longrightarrow Y$ is separated if Δ_X is closed in $X \times_Y X$. Separation hypotheses enter in various ways. For example, in general the intersection of two affine open subsets need not be affine (consider the example above with $n \geq 2$), but this is easily checked to be true for separated schemes. Another point is that valuations behave well on separated schemes ([**18**], II.6, remark at the bottom of p. 130). General considerations like these lie behind the separated hypothesis in the resolution theorem below.

A scheme X is *noetherian* if it has a finite open affine cover

$$X = \cup Spec\, A_i$$

with A_i a noetherian ring. This implies that the underlying space of X is noetherian and in particular quasi-compact ([**18**], II, Ex. 2.13). The noetherian hypothesis is convenient for mostly obvious reasons. In some cases it could be replaced by "locally noetherian" - or even omitted entirely, provided that "finite-type" assumptions are replaced by their "finitely-presented" analogues. We will usually assume that schemes are at least locally noetherian. For a general discussion of finiteness hypotheses, beyond the noetherian condition, see ([**16**], 6).

The *Krull dimension* of X is the maximal length n of a descending chain

$$X_0 \supset X_1 \supset \ldots \supset X_n$$

of irreducible closed subsets. It's worth pointing out here that although *local* noetherian rings always have finite Krull dimension, this isn't true in general (see [**12**], Exercise 9.6). Hence our typical hypothesis "let X be a separated noetherian scheme of finite Krull dimension" is not redundant.

X is *regular* if all of its local rings $\mathcal{O}_x$ are regular. Here a local ring R with maximal ideal m and residue field $k = R/m$ is regular if it is noetherian and $dim_k(m/m^2)$ =Krull dim X. For example, Dedekind domains are regular, and an algebraic variety over a field is regular if and only if it is smooth. The key property of regular schemes that we need is (see for example [**18**], III, Exercise 6.9).

Theorem 2.1 *Let X be a separated noetherian regular scheme. Then every coherent sheaf on X admits a finite resolution by locally free sheaves.*

We remind the reader that coherent (resp. locally free) sheaves correspond to finitely-generated (resp. finitely-generated projective) modules in the affine case. Theorem 2.1 is the ultimate source of the separated and regular hypothesis that occurs in several of the main theorems below.

2.2 Étale morphisms. A morphism of locally noetherian schemes $X \xrightarrow{f} Y$ is *étale* if it is

- *flat*: for all $x \in X$ the induced homomorphism $\mathcal{O}_{f(x)} \longrightarrow \mathcal{O}_x$ of local rings is flat
- *unramified*: the fibre of f over $Spec\, k = y \in Y$ is a finite disjoint union of finite separable extensions $Spec\, k_i$ of the residue field of y. In other words, there is a pullback diagram

$$\begin{array}{ccc} \coprod_i Spec\, k_i & \longrightarrow & X \\ \downarrow & & \downarrow \\ Spec\, k & \longrightarrow & Y \end{array}$$

- *locally of finite type*: There is an affine open cover U_i of Y and for each i an affine open cover V_{ij} of $f^{-1}U_i$ such that the ring homomorphism corresponding to each $V_{ij} \longrightarrow U_i$ has finite type. (For general schemes one requires that f is locally of finite presentation; see e.g. [**41**].)

The "unramified" condition has an alternative formulation. Let $R \rightarrow S$ be a map of rings, $\Omega_{S/R}$ the "module of relative Kahler differentials" (i.e., the R-module on symbols ds, $s \in S$ modulo the Leibniz rule - see ([**18**], II.8)). Then $Spec\, S \longrightarrow Spec\, R$ is unramified if and only if $\Omega_{S/R} = 0$, with a similar result holding for arbitrary schemes (cf. [**29**], I, 3.5).

We will invariably assume that our étale maps actually have *finite* type; this just means that in the third condition above we have for each fixed i a finite such cover V_{ij} (and since our schemes will usually be noetherian, the cover U_i will also be finite).

The basic examples of an étale morphism that the reader should keep in mind are the following:

- A smooth unramified covering map between complex algebraic varieties.
- The morphism $Spec\, L \longrightarrow Spec\, F$ corresponding to finite separable field extension L/F.
- The morphism $Spec\, S \longrightarrow Spec\, R$ corresponding to an unramified finite extension of Dedekind domains.

In the first and third examples, one should picture more generally a map which is not onto but rather has image some open subset of the target. For example in the case of an arbitrary finite extension of Dedekind domains, the associated map of schemes is étale over the complement of the finite set of ramified primes.

We note that étale maps are stable under base change and composition, any open immersion is étale, and any étale map is an open map.

2.3 Hensel local rings. Let R be a local ring with maximal ideal m and residue field k. Then R is a *hensel local ring*, or is *henselian*, if every finite R-algebra S splits as a finite product of local R-algebras. This is equivalent to R satisfying "Hensel's lemma" on lifting factorizations of polynomials from k to R. Any local ring R admits a *henselization* $R \longrightarrow R^h$ which is initial among local homomorphisms to henselian rings. One can think of henselization as a kind of algebraic approximation to the analytic process of completion. Every complete local ring is henselian, and in general R^h sits between R and the completion of R.

If S is a finite R-algebra and is étale over R, then S splits as a product of local R-algebras which are themselves henselian.

Theorem 2.2 *Let R be a hensel local ring. Then the functor* $S \mapsto S \otimes_R k$ *is an equivalence between the category of finite étale R-algebras and the category of finite étale k-algebras.*

A *strict hensel* (or *strictly henselian*, or "strictly local") ring) is a hensel local ring R with separably closed residue field. Note that in this case, the preceding theorem says that every finite étale R-algebra splits as a product of copies of R.

There is also a *strict henselization* functor $R \longrightarrow R^{sh}$. For further discussion see Section 2.6 below.

2.4 Sites. A *site* $\Phi = (\mathcal{C}, \mathcal{T})$ consists of (1) a category $\mathcal{C}$ with pullbacks; and (2) a Grothendieck topology $\mathcal{T}$ on $\mathcal{C}$ - that is, for each object U of $\mathcal{C}$ a collection of families of morphisms $\{U_i \longrightarrow U\}$ called covering families, such that every cover of a cover is a cover, every pullback of a cover is a cover, and every family consisting of a single isomorphism is a cover. All sites considered in this paper will also have terminal objects and hence have all finite limits.

The reader can easily guess the precise statements needed in (2). See e.g. [**1**], [**41**], or the original source [**3**]. We remark that the definition of site given here is less general than the usual one.

There are several examples that will be important for us. Let X be a locally noetherian scheme.

- *The Zariski site* X_{Zar}. The underlying category $\mathcal{C}$ is the category of open subsets of X (for the Zariski topology), with morphisms the inclusions. A family $\{U_i \longrightarrow U\}$ is a covering family if the U_i cover U in the usual sense. This site makes sense for any topological space, and is the motivating example.
- *The étale site* $X_{\acute{e}t}$. $\mathcal{C}$ is the category of schemes étale and of finite type over X. A family $\{U_i \longrightarrow U\}$ is a covering family if the images of the U_i cover U in the usual sense.
- *The restricted étale site* $X_{r\acute{e}t}$ of a separated noetherian scheme. This is defined as in the previous example, except that we require the scheme over X, $U \longrightarrow X$ to be separated over X, and the covering families are to have only finitely many elements $U_i \longrightarrow U$. (It follows automatically that U_i is separated over U as well.) One advantage of using the restricted site is that a covering family $\{U_i \longrightarrow U\}$ can be regarded as a single surjective étale map $U' \longrightarrow U$, where $U' = \coprod U_i$. For the restricted étale site of a general scheme, see ([**41**], II, 1.5).
- *The Nisnevich site* X_{Nis} [**34**]. The underlying category is the same as for the étale site. To describe the covering families, we borrow a term from number theory. Given an étale $h : U' \longrightarrow U$, we say that $x \in U$ is *completely decomposed* in U' if there is a point $y \in h^{-1}x$ such that the induced map on residue fields $k(x) \longrightarrow k(y)$ is an isomorphism. A covering family for the Nisnevich topology is a covering family $\{U_i \longrightarrow U\}$ for the étale topology such that for all $x \in U$, x is completely decomposed in at least one U_i. This site will be discussed further in Section 2.6.
- *The site* G_{fsets} *attached to a profinite group* G. A G-set is discrete if all its isotropy groups in G are open. We take for $\mathcal{C}$ the discrete finite G-sets. Covering families are just the surjective families.
- *"Big" sites.* The étale site discussed above is analogous to a single topological space. The étale morphisms $U \rightarrow X$ play the role of the open subsets. But we could also consider some much larger category of schemes - even all schemes - and still get a Grothendieck site, using the same covering families. For example we could take all schemes X locally of finite type over a fixed base scheme S, and equip each X with the étale topology. This kind of a site (or its Zariski or Nisnevich analogue) is roughly analogous to the category of all topological spaces, and will be informally referred to as a "big", or "sufficiently large" site.

Two further examples can be found in the next section.

A morphism of sites $f : \Phi \longrightarrow \Phi'$ consists of a functor $f^{-1} : \mathcal{C}' \longrightarrow \mathcal{C}$ that preserves pullbacks and preserves covering families in the obvious sense.[3] If the sites have terminal objects, and f^{-1} preserves them, then f^{-1} preserves all finite limits. This will be the case in the examples considered in this paper, with one exception discussed in Section 2.5 below. An equivalence of sites is a morphism such that f^{-1} is an equivalence of categories and any inverse functor is also a site morphism. Exercise: show that if F is a field, the site $(G_F)_{fsets}$ is equivalent to the étale site of F.

Our convention that the site morphism goes in the opposite direction is in agreement with some sources but not with others (e.g. [**1**] or [**41**], in which our morphism f would be written as a "morphism of topologies" $\mathcal{T}' \longrightarrow \mathcal{T}$). The reason for the convention is just that a morphism of schemes $X \longrightarrow Y$ then induces site morphisms in the same direction $X_{Zar} \longrightarrow Y_{Zar}$, $X_{ét} \longrightarrow Y_{ét}$, etc., and a continuous homomorphism $G \longrightarrow H$ of profinite groups induces a morphism $G_{fsets} \longrightarrow H_{fsets}$. On the other hand it must be admitted that there are some other cases where our convention looks slightly odd. For example, it is clear that X_{Zar} and X_{Nis} are "sub-sites" of $X_{ét}$, etc., but in our notation the evident site morphisms go the other way:

$$X_{ét} \longrightarrow X_{Nis} \longrightarrow X_{Zar}.$$

However, even in this case, our convention helps the wandering topologist to keep track of things such as the Leray spectral sequence described below. Site morphisms associated to inclusion of a sub-site as above will be called *restriction* morphisms.

2.5 Presheaves and sheaves. A presheaf of sets on a site $\Phi = (\mathcal{C}, \mathcal{T})$ is just a contravariant functor from $\mathcal{C}$ to sets. An abelian presheaf is a presheaf of abelian groups. Similarly one can define presheaves with values in any category (simplicial sets, spectra, etc.), but for now we will focus on presheaves of sets or abelian groups. A *sheaf* is a presheaf $\mathcal{E}$ that satisfies the usual type of exactness condition with respect to covering families.

There is an exact sheafification functor a, left adjoint to the inclusion functor i of presheaves in sheaves.[4] A presheaf of sets $\mathcal{E}$ is *representable* if $\mathcal{E}(U) = Hom\,(U, W)$ for some object W. If $\mathcal{C}$ is an arbitary category with pullbacks, the finest topology such that every representable presheaf is a sheaf is called the *canonical* topology - see e.g. [**41**]. On most of the sites encountered in practice, every representable presheaf is a sheaf - in other words, the topology in question is no finer than the canonical topology. In the general case the phrase "sheaf represented by W" secretly means the sheafification of $Hom\,(-, W)$. The category $\mathcal{C}$ also admits a minimal topology, called the "chaotic" topology, in which the only covering families are the isomorphisms. In this topology, *every* presheaf is a sheaf.

Given a site morphism $f : \Phi \longrightarrow \Phi'$, one can define various direct and inverse image functors on sheaves and presheaves. Since these functors are extremely important, and keeping them all straight gets a bit confusing, we will make a few remarks that may be of help.

---

[3]The notation f^{-1} for the functor is meant to be suggestive - perhaps dangerously so, but we're going to take the risk.

[4]Later, when we consider presheaves of homotopy groups $\pi_q\mathcal{E}$, the associated sheaf will be denoted $\tilde{\pi}_q\mathcal{E}$.

Consider a functor $g : I \longrightarrow J$ between small categories. Let $\mathcal{D}$ be another category, which has all limits and colimits, and consider the restriction functor $\mathcal{D}^g : \mathcal{D}^J \longrightarrow \mathcal{D}^I$ on functor categories (or "diagrams"). This functor has both a left adjoint G_L and a right adjoint G_R, also know as Kan extensions ([**26**], X). Bearing in mind that presheaves are contravariant functors, we can apply this to our site morphism f, with $g = f^{-1}$. We write $f_\#$ (direct image) for the restriction functor, $f^\#$ (inverse image) for the left adjoint and $f^!$ for the right adjoint. Thus if $\mathcal{E}$ is a presheaf on Φ, $f_\#\mathcal{E}(U) = \mathcal{E}(f^{-1}U)$. And if $\mathcal{F}$ is a presheaf on Φ', $f^\#\mathcal{F}(W) = colim_V \mathcal{F}(V)$, where the colimit is taken over the opposite of the over-category $(f^{-1} \downarrow U)$. If the sites have terminal objects and f^{-1} preserves them, $(f^{-1} \downarrow U)^{op}$ is filtered. Otherwise it is only pseudofiltered.

One would like to promote these functors to the sheaf level. There is no problem with the direct image functor because it preserves sheaves and so immediately yields a direct image functor on sheaves, denoted f_*. On the other hand the symmetry between left and right adjoints breaks down. The reason is clear - the sheafification functor a is a *left* adjoint. Hence, even though $f^\#$ may fail to preserve sheaves, we get a left adjoint to f_* by setting $f^* = af^\# i$. The analogous attempt to extend $f^!$ fails simply because a appears on the wrong side. One nevertheless gets a right adjoint $f^!$ on the sheaf level in certain situations - e.g., inclusion of a closed subscheme in the Zariski or étale topologies.

To confuse matters still further, there are situations where we have site morphisms going in both directions. Specifically, consider a fixed object U of a site that has a terminal object X. Define the local site Φ/U to have underlying category the category of objects over U and covering families the same as those in Φ. The simple example to keep in mind is an open subscheme in the Zariski topology. There is an obvious site morphism $j : \Phi/U \longrightarrow \Phi$ with $j^{-1}V = U \times_X V$. But there is another obvious site morphism $k : \Phi \longrightarrow \Phi/U$ with $k^{-1}(V \rightarrow U) = V$. (Note that k^{-1} does not preserve terminal objects or products.) Inspection shows that $j^\# = k_\#$. Hence $j^\#$ preserves sheaves and has a *left* adjoint $j_!$ $(= k^\#)$, called "extension by zero" because in the Zariski example mentioned above we have $j_!\mathcal{E}(V) = \mathcal{E}(V)$ if $V \subset U$ and $j_!\mathcal{E}(V) = 0$ otherwise. In general we have

$$j_!\mathcal{E}(V) = \coprod_{Hom(V,U)} \mathcal{E}(V)$$

The category of presheaves on the site Φ is sometimes denoted $\Phi^\wedge$. The category of sheaves (the "topos") is sometimes denoted $\Phi^\sim$.

2.6 Points and stalks. In this section presheaves and sheaves will be set-valued unless otherwise mentioned.

In classical sheaf theory, on the Zariski site of a scheme X, the notion of "stalk" at a point $x \in X$ is extremely useful. For example, stalks detect isomorphisms in the sense that a map of sheaves is an isomorphism if and only if it induces an isomorphism on all stalks. It isn't obvious how to define stalks on an arbitrary site, because it isn't even clear what a "point" should be. The key is to observe that the category of sheaves on the Zariski site of $x = Spec\, k$ is the category of sets, and that the corresponding stalk functor $X^\sim_{Zar} \longrightarrow Sets$ is a left adjoint that commutes with finite limits. Thus on a site Φ we *define* a point p as a pair of adjoint functors

$$p^* : \Phi^\sim \longrightarrow Sets, \quad p_* : Sets \longrightarrow \Phi^\sim$$

where the left adjoint p^* is required to commute with finite limits. The *stalk* of a sheaf F is then defined by $F_p = p^*F$. In fact the point p is determined by a functor p^* that commutes with finite limits and all colimits, because the right adjoint exists automatically. Indeed one has the explicit formula for a set S

$$p_*S(U) = Hom_{sets}(p^*\tilde{U}, S)$$

which is easily verified using the colimit formula below for p^* ($\tilde{U}$ is the sheaf represented by the object U).

A *neighborhood* of a point p is a pair (U, a) with U an object of the site and $a \in p^*\tilde{U}$. Equivalently, this can be thought of as a lift of the point to the local site Φ/U. The neighborhoods form a category $\mathcal{N}_p$ in an obvious way, and this category is co-filtered. For any sheaf F we have

$$F_p = colim_{\mathcal{N}_p} F(U)$$

or more generally for a presheaf F

$$aF_p = colim_{\mathcal{N}_p} F(U).$$

A set of points of a site is said to be *sufficient* if it detects isomorphisms of sheaves in the sense described above. A site has *enough points* if it has a sufficient set of points. Not all sites have enough points.[5] However there is the following elegant and useful criterion ([**3**], IV, 6.5). We say that a set of points of Φ *detects covering families* provided that whenever $\{U_i \longrightarrow U\}$ is a family of morphisms and $\{p^*U_i \longrightarrow p^*U\}$ is a surjective family of sets for all points p in the set, then $\{U_i \longrightarrow U\}$ is a covering family. Then the criterion states that a set of points of Φ is sufficient if and only if it detects covering families.

In the case of the Zariski site, one can show that the new definitions agree with the old - although it isn't obvious that every site-theoretic point comes from a classical point (see [**23**], 7.24, for a quick proof). Note that in this case the stalk of the structure sheaf is the Zariski local ring $\mathcal{O}_x$.

Now consider the étale site of a scheme X. Given an ordinary point $x \longrightarrow X$, $x = Spec\, k$, choose a separable closure $\bar{x} = Spec\, \bar{k}$. Then the category of sheaves on $\bar{x}_{ét}$ is the category of sets, and the site morphism $p : \bar{x}_{ét} \longrightarrow X$ induces a functor p^* on sheaves that commutes with finite limits and all colimits. Thus each such "geometric point" $\bar{x} \longrightarrow X$ defines a site-theoretic point of $X_{ét}$. One can easily check that a neighborhood of $\bar{x} \longrightarrow X$ is the same thing as an étale $U \longrightarrow X$ together with a lift

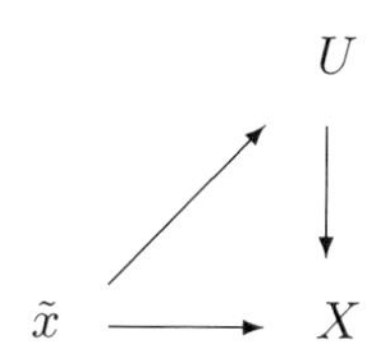

It follows that the stalk at $\bar{x}$ of the structure sheaf $U \longrightarrow \mathcal{O}_U$ is the strict henselization $\mathcal{O}^{sh}_{\bar{x}}$ of $\mathcal{O}_x$. Thus the rings $\mathcal{O}^{sh}_{\bar{x}}$ are the local rings for the étale topology. Finally, it is easy to see that geometric points detect covering families, so the étale site has enough points.

[5] See ([**3**], IV, 7.4) for an interesting example -involving Lebesgue measure on the unit interval - of a site with no points.

Contemplating the vast gap between the Zariski local ring $\mathcal{O}_x$ and its strict henselization $\mathcal{O}_x^{sh}$, one longs for an intermediate world in which the local rings are the ordinary henselizations $\mathcal{O}_x^h$. The Nisnevich site is precisely such a world.[6]

Consider first the Nisnevich site of $x = Spec\,k$, k a field. Call a presheaf F *additive* if $F(U\coprod V) = F(U) \times F(V)$. Then it is easily seen that the sheaves on x_{Nis} are precisely the additive presheaves (in effect, any covering family takes the form $U\coprod W \longrightarrow U$). Hence sheaves are the same thing as presheaves on the full subcategory of connected étale x-schemes. In particular, for any such scheme x' (i.e. $x' = Spec\,k'$, with k' a finite separable extension of k), the sections functor $F \longrightarrow F(x')$ is exact.

For a general scheme X, consider an ordinary point $i : x \longrightarrow X$, and $x' \longrightarrow x$ as above. Define $p^* : X_{Nis}^{\sim} \longrightarrow Sets$ by $p^*F = j^*_{Nis}F(x')$, where $j : x' \longrightarrow X$. Then, thanks to the exactness of the sections functor, p^* defines a point of X_{Nis}. Informally, then, a point of the Nisnevich site is denoted $x' \longrightarrow x$ or $x' \longrightarrow X$. A neighborhood of such a point is the same thing as a commutative diagram

$$\begin{array}{ccc} x' & \longrightarrow & U \\ \downarrow & & \downarrow \\ x & \longrightarrow & X \end{array}$$

Note that x' need not be a residue field of U - in general, it is only a finite extension of such a field. However the neighborhoods for which x' *is* a residue field of U form a cofinal subcategory of $\mathcal{N}_p$.

It follows that the Nisnevich stalk at $x \in X$ of the structure sheaf is $\mathcal{O}_x^h$, the henselization of $\mathcal{O}_x$. More generally the stalk at $x' \longrightarrow x$ is $\mathcal{O}_{x'}^h$, the hensel extension of $\mathcal{O}_x^h$ corresponding to $x' \longrightarrow x$ under the equivalence of Theorem 2.2. We will call this ring the "henselization of $\mathcal{O}_x$ at $x' \longrightarrow x$". Thus the local rings for the Nisnevich topology are the henselizations $\mathcal{O}_{x'}^h$.

Finally, it is easy to check that the points described above detect covering families, and hence the Nisnevich site has enough points.

2.7 Sheaf Cohomology. Let A be an abelian sheaf, U an object of $\mathcal{C}$. The category of sheaves is abelian with enough injectives, and the sheaf cohomology groups $H^p(U; A)$ are the p-th derived functors of the left exact "sections on U" functor $U \longrightarrow A(U)$. For the sites $X_{Zar}, X_{\acute{e}t}$ and so on we will write this as $H^p_{Zar}(U; A)$, $H^p_{\acute{e}t}(U; A)$ etc.

When the site has enough points, there is a functorial *Godement resolution* defined as follows. For each point p, the adjoint functors p^*, p_* defined a monad ([**26**], VI), and any monad defines a cosimplicial object in a standard way. Here we will take the product over a sufficient set of points, obtaining a monad on sheaves given by

$$T(A) = \prod_p p_* p^* A$$

The resulting cosimplicial sheaf $T^{\cdot}(A)$ provides a resolution of A by sheaves which, although not necessarily injective, are "flasque" and in particular acyclic

[6] As will be seen, however, the Nisnevich topology is much closer to the Zariski topology than it is to the étale topology.

for cohomology. Hence, taking sections over U of the Godement resolution yields a functorial complex for computing cohomology.

One of the main tools for computing sheaf cohomology is the fantastically general "Leray" or "change-of-site" spectral sequence:

Theorem 2.3 *Let $f : \Phi \longrightarrow \Phi'$ be a site morphism, A a sheaf on Φ. Then for any object U in $\mathcal{C}$ there is a first quadrant cohomology spectral sequence*

$$E_2^{p,q} = H^p(U; R^q f_* A) \Rightarrow H^{p+q}(f^{-1}U; A)$$

Here the $R^q f_* A$ are the "higher direct image sheaves" which are the derived functors of f_*. They can be described as the sheafifications of the presheaves $U \longrightarrow H^q(f^{-1}U; A)$. In the case of a morphism of schemes $X \longrightarrow Y$, we would typically take Y=U, $X = f^{-1}U$ and get the classical Leray spectral sequence of $X_{Zar} \longrightarrow Y_{Zar}$, its counterpart for the étale site, etc. For the site morphism associated to a surjective homomorphism of profinite groups, we get the Hochshild-Serre spectral sequence.

Now consider the case of a "restriction" morphism such as $X_{\acute{e}t} \longrightarrow X_{Zar}$. In this case the "direct image" functor is just the obvious restriction of sheaves to the smaller category. Nevertheless one gets a very interesting spectral sequence. For example in the étale-to-Zariski case one easily derives the generalized "Hilbert's Theorem 90" asserting that,

$$H^1_{\acute{e}t}(X, G_m) = H^1_{Zar}(X, G_m) = Pic\, X.$$

One can also use the spectral sequence to show that a given sub-site has the same cohomology as the ambient site. We'll call the following result the "weak comparison lemma" ([**41**], 3.9.3).

Proposition 2.4 *Let $f : \Phi \longrightarrow \Phi'$ be a restriction morphism of sites. Suppose that for every object U in $\mathcal{C}'$ and every covering family $\{V_i \longrightarrow U\}$ in the ambient site Φ, there are covering families $\{U_{ij} \longrightarrow V_i\}$ with U_{ij} in $\mathcal{C}'$ and $\{U_{ij} \longrightarrow U\}$ in $\mathcal{T}'$. Then for any sheaf A on Φ and object U in Φ',*

$$H^*_{\Phi'}(U, f_* A) \cong H^*_{\Phi}(U, A).$$

This applies for example to the restriction from the étale site of a separated noetherian scheme X to its restricted site as above. In fact in this case a stronger condition holds and one can apply the usual "comparison lemma" ([**41**], 3.9.1) to conclude the actual sheaf categories are equivalent. Another illustration is provided by the following important theorem:

Theorem 2.5 *Let R be a Hensel local ring with maximal ideal m and residue field $R/m = k$. Then for any sheaf A on $(Spec\, R)_{\acute{e}t}$, the morphism $j : Spec\, k \longrightarrow Spec\, R$ induces an isomorphism*

$$j^* : H^*_{\acute{e}t}(Spec\, R; A) \cong H^*_{\acute{e}t}(Spec\, k, j^* A).$$

The proof of this fact doesn't use the Leray spectral sequence of j. Instead one constructs a site morphism going the other way, as follows. Let us abbreviate $X = (Spec\, R)$ and $x = (Spec\, k)$. Let Φ denote the sub-site of $X_{\acute{e}t}$ in which both the objects and covering families consist of a single map $Y \rightarrow X$ corresponding to a finite étale R-algebra S. Then we have site morphisms $X_{\acute{e}t} \longrightarrow \Phi \longrightarrow x_{\acute{e}t}$ with the first a cohomology isomorphism by the weak comparison lemma and the second an equivalence of sites. Letting f denote the composite morphism, one has $f_* = j^*$ and the theorem follows easily.

Finally, the notion of continuous Galois cohomology discussed in Section 1 extends immediately to an arbitrary Grothendieck site. The category of inverse systems of sheaves has enough injectives, and following [**19**] we define continuous cohomology as the derived functors of the composite functor $lim \circ$ (global sections).

2.8 Cohomological dimension. We will frequently need to know that a scheme X has finite cohomological dimension for sheaves (or ℓ-torsion sheaves) on the étale, Zariski or Nisnevich sites. In general, for an object X of a site Φ, the cohomological dimension of X is the maximal n (possibly infinite) such that there exists a sheaf A with $H^n_\Phi(X; A) \neq 0$. The mod ℓ cohomological dimension is defined similarly, with the requirement that A is an ℓ-torsion sheaf. In the case of a scheme X, and one of the sites mentioned above, we will write this as $cd^{Zar}(X)$, $cd^{ét}_\ell X$, etc.

The following theorem of Grothendieck applies to any noetherian topological space X (see [**18**], III, 2.7).

Theorem 2.6 *Let X be a noetherian scheme of Krull dimension n. Then* $cd^{Zar}(X) \leq n$.

The next result, due to Kato and Saito (see [**34**]), is another indication that the Nisnevich topology is not so far from the Zariski topology.

Theorem 2.7 *Let X be a noetherian scheme of Krull dimension n. Then* $cd^{Nis}(X) \leq n$.

In the étale case we need to pay attention to the cohomological dimension of the residue fields of X. We say that X is *uniformly ℓ-bounded*, with *bound d*, if for all residue fields $k(x)$ we have $cd_\ell k(x) \leq d$. Here cd_ℓ refers to Galois cohomological dimension, or equivalently $cd^{ét}_\ell Spec\, k(x)$.

Theorem 2.8 *Let X be a noetherian scheme of Krull dimension n, and suppose X is uniformly ℓ-bounded with bound d. Then $cd^{ét}_\ell X \leq n + d$. In particular, $cd^{ét}_\ell X$ is finite.*

There is an elegant proof of this fact based on the Nisnevich site and the Kato-Seito theorem. We sketch it here as an illustration of how the Nisnevich site functions as an intermediary between fields and schemes.

Let A be an ℓ-torsion sheaf on $X_{ét}$, and consider the Leray spectral sequence of the site morphism $f : X_{ét} \longrightarrow X_{Nis}$. We have

$$E_2^{p,q} = H^p_{Nis}(X; R^q f_* A)$$

and hence $E_2^{p,q} = 0$ for $p > n$ by the Kato-Saito theorem. On the other hand we claim that the sheaves $R^q f_* A$ on X_{Nis} are identically zero for $q > d$, whence the result. It is equivalent to show that the stalks are zero. In the notation of Section 2.6, the stalk at a point x is the sheaf on x_{Nis} that assigns to each $x' \rightarrow x$ the group $H^q_{ét}(\mathcal{O}^h_{x'}; g^* A)$, where $\mathcal{O}^h_{x'}$ is the hensel local ring attached to x' and g is the natural map to X (here we have made use of the "continuity theorem" (Theorem 2.11 below)). But this is isomorphic to $H^q_{ét}(x'; A)$, by 2.5. Since k'/k is a finite separable extension, $cd_\ell k' \leq cd_\ell k$, and hence $H^q_{ét}(x'; A) = 0$ for $q > d$.

For example, suppose X is a variety of dimension n over a field k with $cd_\ell k < \infty$. Then the residue fields of X have transcendence degree at most n over k. Hence by

1.3, we get $cd_\ell^{ét} X \leq 2n + cd_\ell k$. Or if X is the ring of integers in a number field F, with either ℓ odd or F pure imaginary, then $cd_\ell^{ét} X \leq 3$.

Finally, we remark that the "uniformly ℓ-bounded" hypothesis in the above theorem is inherited by U of $X_{ét}$ (with the same bound), because the residue fields of U are finite separable extensions of the residue fields of X. Hence if in addition U has finite type over X, all the hypotheses of the above theorem are inherited by X, and $cd_\ell^{ét} U \leq n + d$.

2.9 Čech cohomology. The usual definition of Čech cohomology for spaces goes through essentially verbatim for an arbitrary Grothendieck site. Let X be an object of the site Φ, F an abelian presheaf. Given a covering family $\mathcal{U} = \{U_i \longrightarrow U\}$, one forms the associated augmented Čech complex

$$0 \longrightarrow F(U) \longrightarrow \prod_i F(U_i) \longrightarrow \prod_{i,j} F(U_i \times_U U_j) \longrightarrow \dots.$$

The cohomology of this complex (after removing F(U)) is denoted $\check{H}^*(\mathcal{U}/X; F)$ and called Čech cohomology with respect to the cover $\mathcal{U}$. Čech cohomology proper is defined by passing to a filtered colimit over covers (with the caveat that the covers themselves do not form a co-filtered category under refinement - see for example [**41**]), and denoted $\check{H}(X; F)$.

It is not hard to see that the groups $\check{H}^*(\mathcal{U}/X; F)$ are the derived functors of $\check{H}^0(\mathcal{U}/X; F)$. Since $\check{H}^0(\mathcal{U}/X; F) = F(X)$ if F is a sheaf, the global sections functor for sheaves factors as inclusion into the presheaves followed by $\check{H}^0$. This leads to a spectral sequence we will call "Čech descent for sheaf cohomology":

$$E_2^{p,q} = \check{H}^p(U/X; H^q(-, A)) \Rightarrow H^{p+q}(X; A).$$

By passing to the limit in the spectral sequence above we get another spectral sequence

$$E_2^{p,q} = \check{H}^p(X; H^q(-, A)) \Rightarrow H^{p+q}(X; A).$$

Exercise: Study the case when $X = Spec\, F$, F a field. Show that the Čech complex obtained from the nerve of a finite Galois extension L/F is just the usual bar complex, the Čech cohomology groups are exactly the groups labelled "Čech" in Section 1.2, and the Čech descent spectral sequence associated to L/F is just the Hochshild-Serre spectral sequence of the group extension $G_L \rightarrow G_F \rightarrow G(L/F)$.

Now consider the question of whether Čech and sheaf cohomology agree. This certainly isn't true in general, even on the site of a topological space. However from the Čech descent spectral sequence we see that it would be enough to show that for a *sheaf* A, $\check{H}^p(X; H^q(-, A)) = 0$ for $q > 0$. The crudest way to guarantee these groups vanish would be to have some cofinal system of covers $\mathcal{U} \longrightarrow X$ such that $H^q(\mathcal{U}_X^n; A) = 0$ for all U, n, and $q > 0$. For example, this is how one shows that on the Zariski site of a separated noetherian scheme, sheaf and Čech cohomology agree for *coherent* sheaves ([**18**], III, 4.5). In this case one uses coverings by affine schemes, which are acyclic for coherent sheaf cohomology by a classical theorem of Serre.

This is far too much to expect in general, however. As a next attempt, one might start from the observation that on an arbitrary site, the sheafification of the presheaf $H^q(-,A)$ is zero when $q>0$. This amounts to the fact that for any W and $\alpha \in H^q(W;A)$, there is a covering $\mathcal{V} \longrightarrow W$ such that α restricts to zero on $\mathcal{V}$. So what one needs to know is:

(*) For any covering $\mathcal{U} \longrightarrow X$ with Čech nerve $\mathcal{U}_X^{\cdot}$ and any covering $\mathcal{V} \longrightarrow \mathcal{U}_X^n$, there is a covering $\mathcal{U}' \longrightarrow \mathcal{U}$ such that $(\mathcal{U}')_X^n \longrightarrow \mathcal{U}_X^n$ factors through $\mathcal{V}$.

Then we can conclude that every Čech cochain on $\mathcal{U}/X$ with coefficients in $H^q(-,A)$ vanishes when we pass to the limit in $\mathcal{U}$. This is exactly what Artin [**2**] shows to get the following theorem (see also [**42**], 1.53).

Theorem 2.9 *Let X be any quasi-compact scheme such that every finite subset of X lies in an open affine subscheme - for example, suppose X is quasi-projective over a noetherian ring. Then for any additive presheaf A on $X_{ét}$,*

$$H^*_{ét}(X;aA) \cong \check{H}^*_{ét}(X;A).$$

This result is a special result about the étale site, and leaves open the question of whether there isn't a way to compute cohomology from covers more generally. An elegant answer to this question was provided by Verdier ([**3**], V, 7). The idea is to replace the Čech nerves by a more general simplicial covering object called a *hypercover*. The resulting cohomology theory, which we will call *Verdier cohomology*, agrees with sheaf cohomology on an arbitrary site - essentially because the appropriate analogue of (*) holds in general.

2.10 Noetherian sites. A site Φ is said to be *noetherian* if for every object U and every covering family $\{U_i \longrightarrow U\}$, there is a finite subset $\{U_{i_1},...,U_{i_n}\}$ which is also a covering family.[7] For example, if X is a noetherian scheme then the étale, Nisnevich and Zariski sites of X are noetherian (although in the case of the Nisnevich site this isn't obvious - see [**44**], Appendix E, 6a). In fact it is enough to assume only that X is quasicompact, provided that the definition of étale morphism is modified as alluded to earlier (see [**41**], II, 1.5.1).

Proposition 2.10 *Let Φ be a noetherian site, A_i a filtered (or even pseudo-filtered) system of sheaves on Φ. Then for all objects U, there is a natural isomorphism*

$$colim_i\, H^*_\Phi(U;A_i) \xrightarrow{\cong} H^*_\Phi(U;colim_i A_i).$$

To see why this is true, consider for the moment an arbitrary site, a presheaf F and a covering family $U_i \longrightarrow U$. Consider also the associated Čech complex

$$0 \longrightarrow F(U) \longrightarrow \prod_i F(U_i) \longrightarrow \prod_{i,j} F(U_i \times_U U_j) \longrightarrow \dots.$$

Then by definition F is a sheaf if and only if for all such covering families of all objects U, this complex is exact through the first three arrows, and is a *flasque* presheaf if and only if all such sequences are exact from the next term on (i.e., all the higher Čech cohomology groups vanish). We recall that injective sheaves are flasque, and that sheaf cohomology can be computed from flasque resolutions (see for example [**41**], I, 3.5). Thus the proof of the theorem boils down to showing that

[7] The reader might be wondering why we don't call this a quasicompact site. Note, however, that the definition generalizes the notion of a topological space X with the property that *every* open subset of X is quasicompact - that is, a noetherian topological space.

pseudo-filtered colimits take sheaves to sheaves, and flasque presheaves to flasque presheaves. This is clear provided that pseudo-filtered colimits commute with the products in the Čech complex above. But if we now assume the site is noetherian, then we need only consider finite coverings, in which case all of these products are finite and the theorem follows.

The second important result of this section concerns the étale site of an inverse limit of schemes. Suppose R_i is a filtered system of rings, and $R = colim_i\, R_i$. Then $Spec\, R$ is the (inverse) limit of the R_i in the category of schemes. More generally, suppose $\mathcal{A}_i$ is a filtered system of quasi-coherent $\mathcal{O}_Y$-algebras over a fixed base scheme Y. Then one can define associated schemes $Spec\, \mathcal{A}_i$ (see [**18**], II, Exercise 5.17), affine[8] over Y, and $Spec\,(colim_i \mathcal{A}_i)$ is again the limit of the $Spec\, \mathcal{A}_i$ in the category of schemes. Using this construction, it is not hard to show that if X_i is any co-filtered system of schemes with affine transition maps $X_i \rightarrow X_j$, then the limit of the X_i exists. However the only case we're really going to use is the affine case $X_i = Spec\, R_i$, so the reader can feel free to picture this case, if desired.

Noetherian hypotheses are a bit awkward in this setting, because obviously a filtered colimit of noetherian rings need not be noetherian. For that reason we will work a bit more generally and assume we are given the following data:

- a cofiltered system of quasi-compact and quasi-separated[9] schemes X_i, with affine transition maps. Such a system will be called admissible.
- an abelian presheaf F on a sufficiently large étale site (i.e., that includes all X_i as well as X, etc.). The presheaf F is assumed to be continuous in the sense that for any admissible system U_i in the site, with limit U, the natural map $colim_i\, F(U_i) \longrightarrow F(U)$ is an isomorphism.

For example, any filtered system of rings, or of separated noetherian schemes with affine transition maps, defines an admissible system. There is no need to fuss over the precise meaning of "sufficiently large" in the second condition above, because in practice F will be a presheaf defined on *all* schemes, e.g., the K-theory presheaf, which is continuous as discussed in the next section. The following theorem is the *Continuity Theorem.*

Theorem 2.11 *Let X_i be an admissible system of schemes with limit X, and let F be a continous presheaf on a sufficiently large étale site. Then there is a natural isomorphism*

$$colim_i\, H^*_{\acute{e}t}(X_i; aF_i) \xrightarrow{\cong} H^*_{\acute{e}t}(X; aF).$$

The proof of this important theorem is quite pretty, and as we will need one of the details, we will give a very brief sketch (see [**1**]). The beautiful fact is that essentially any morphism of schemes $f : U \rightarrow V$ over X is pulled back from a morphism $f_i : U_i \rightarrow V_i$ over some X_i, and moreover any reasonable property of f can be arranged to hold for f_i. Given a scheme U over X, we call U_i over X_i a *source* of U if U is isomorphic as X-scheme to $U_i \times_{X_i} X$, and similarly for maps of X-schemes.

[8]A morphism $f : X \longrightarrow Y$ is *affine* if Y has an open affine cover U_i such that each $f^{-1}U_i$ is affine.

[9]A scheme is *quasi-separated* if the unique map to $Spec\, \mathbb{Z}$ is quasicompact (i.e., the inverse image of every quasicompact open set is quasicompact). Every separated scheme is quasi-separated; see ([**16**], I.6) for this and related issues.

Lemma 2.12 *Any étale* $f : U \longrightarrow X$ *has an étale source* $U_i \longrightarrow X_i$, *which can be taken surjective if* f *is surjective. More generally, any morphism* $U \rightarrow V$ *of schemes étale over* X *has a source* $U_i \longrightarrow V_i$ *with the same properties. Sources are essentially unique, in the sense that any two sources of* $U \rightarrow V$ *agree after base change to some higher* X_j.

In fact by theorems of Grothendieck ([**17**], IV.8 and IV.11), the lemma holds with "étale" replaced by virtually any property P stable under base change - separated, finite, affine, projective, etc. The hardest step is obtaining a flat source.

Now let G be a presheaf on our big étale site. Ignoring the fact that G is already defined on X, we define a presheaf eG on X by

$$eG = colim_i \pi_i^{\#} G_i$$

where $\pi_i^{\#}$ is the presheaf inverse image functor and G_i is the restriction of G to X_i. Recall that the inverse image functor itself involves a gargantuan filtered colimit, so this definition appears to be a bit complicated. However it is easy to see on inspection:

Lemma 2.13 *Let* U *be étale over* X, *and let* U_i *be any source of* U. *Then there is a natural isomorphism*

$$colim_j G_j(U_j) \xrightarrow{\cong} (eG)(U)$$

where the colimit is over all $i \rightarrow j$ *and* U_j *is the base change of* U_i *to* X_j.

Note that if G is continuous, this shows that eG is isomorphic to the restriction of G to $X_{\acute{e}t}$. Using this last lemma, and the fact that $X_{\acute{e}t}$ is a noetherian site, it is easy to show that the functor e is exact, takes sheaves to sheaves, and takes flasque presheaves to flasque presheaves. The argument is just as in Theorem 2.10, and the proof of Theorem 2.11 is then completed the same way.

Here is the sort of application we have in mind. Let X be a scheme, F a sheaf on $X_{\acute{e}t}$.

Corollary 2.14 *Consider the presheaf* $U \mapsto H^*_{\acute{e}t}(U; F)$ *on the Zariski, Nisnevich and étale sites of* X. *Then*

(i) *The Zariski stalk at* $x \in X$ *is* $H^*_{\acute{e}t}(Spec\, \mathcal{O}_x; eF)$.
(ii) *The Nisnevich stalk at* $x' \rightarrow x$ *is* $H^*_{\acute{e}t}(Spec\, \mathcal{O}^h_{x'}; eF)$.
(iii) *The étale stalk at* $\bar{x} \rightarrow x$ *is* $H^*_{\acute{e}t}(Spec\, \mathcal{O}^{sh}_{\bar{x}}; eF)$ *(=0 for* $* > 0$*).*

Note we are not assuming F is continuous, or even defined on the various local rings, in the above corollary. If it happens to be continuous, that's fine, but the corollary as stated is needed for example in Theorem 2.8. Its proof is easy from the continuity theorem and the definition of stalks; recall that in all three cases these can be computed from a cofiltered limit of affine schemes. Of course the fact that the étale stalk is zero follows from more elementary considerations.

3 Hypercohomology Spectra on Schemes

In this section we introduce and compare the cohomological descent machines of Thomason ([**42**], see especially Sections 1 and 5) and Jardine ([**20**], [**21**], [**22**]).

We begin with some homotopy-theoretic preliminaries, including a few words on closed model categories and our conventions on spectral sequences. It is a basic fact of life that homotopy limits (and various other constructions) do not commute with filtered colimits. Nevertheless, there are many situations where a filtered colimit *can* be "commuted" with whatever other construction is at hand. The "colimit lemma" (3.3) is an explicit and practical criterion for deciding the issue; it will be used frequently in the sequel.

Next we introduce Thomason's spectrum-level Godement construction. This is the analogue, for presheaves of spectra, of the hypercohomology of a chain complex of sheaves. (The reader is urged to consult Thomason's "Scholium of Great Enlightenment" at this point[10] ([**42**], 5.32).) It can also be regarded as a generalization of the homotopy fixed point set discussed in Section 1. Virtually any of the standard theorems on abelian sheaf cohomology can be extended to hypercohomology spectra - although typically some boundedness or finite cohomological dimension hypotheses must be imposed. Two important examples are given, including the spectrum-level continuity theorem. Further examples can be found in ([**42**], 1). We also give conditions for commuting smash products and Bousfield localization with hypercohomology.

It should be noted that the Thomason-Godement construction, as given here, only applies to sites with enough points, although according to Thomason ([**42**], 1.34), there is a modified version that works in the general case.

The next section surveys the elegant work of Jardine ([**20**], [**21**]) that establishes a closed model category structure on presheaves of spectra. In this approach, Thomason's hypercohomology spectrum $\mathbb{H}^{\cdot}(-;\mathcal{E})$ is replaced by a globally fibrant model for $\mathcal{E}$. It is not as obvious in this setting that there is a descent spectral sequence, so we explain this point in more detail. Finally, we compare the globally fibrant model with $\mathbb{H}^{\cdot}(-;\mathcal{E})$ (on a site with enough points). The conclusion is that, subject to a finite cohomological dimension assumption, the two constructions agree.

In the last section we give a brief introduction to Čech hypercohomology spectra ([**42**], 1). Once again, results from abelian sheaf theory generalize to the spectrum setting without much difficulty - for example, "Čech descent for sheaf cohomology", and Artin's theorem [**2**] relating Čech and sheaf cohomology on the étale site.

Throughout this work, the adjective *pointwise* always indicates that a property of presheaves or morphisms of presheaves holds for the sections on any object U of the site. For example a map $\mathcal{E}\longrightarrow\mathcal{F}$ of presheaves of spectra is a pointwise weak equivalence if each map $\mathcal{E}(U)\longrightarrow\mathcal{F}(U)$ is a weak equivalence, and so on.

3.1 Preliminaries on homotopy theory.

3.1.1 *Spectral sequences.* In this section we explain our conventions on spectral sequences. For a very complete analysis of convergence see [**4**], where the term "conditionally convergent" is coined.

Suppose given a tower of fibrations of spectra

$$* \xleftarrow{f_0} X_0 \xleftarrow{f_1} X_1 \xleftarrow{f_2} \ldots$$

and let W_n denote the homotopy-fibre of f_n. Applying π_* yields an exact couple and a spectral sequence which we index by setting

$$E_1^{p,-q} = \pi_{q-p} W_p$$

This is a right half-plane cohomology spectral sequence, with the "Atiyah-Hirzebruch" style of indexing (motivated by the example in which X_n is the function

[10]The referee points out that the "scholium" is not as enlightening as it could be, because the assertions found there are neither proved nor obvious. For proofs, see Chapter 4 of [**22**].

spectrum $F(A_n, E)$ for some spectrum E and CW-complex A with skeletal filtration A_n).

Let $F^p = F^p\pi_*X$ denote the kernel of the projection map $\pi_*X \longrightarrow \pi_*X_{p-1}$. The spectral sequence *converges* if (i) the natural inclusion $F^p/F^{p+1} \subset E_\infty^{p,*}$ is an isomorphism for all p; and (ii) $lim_n^1 \pi_* X_n = 0$. In fact the spectral sequence is always *conditionally convergent* - that is, it converges if and only if $lim_r^1 E_r^{p,q} = 0$. Thus the spectral sequence automatically converges if the E_2-term is bounded on the right, bounded below, or is finite in each bidegree.

For example, suppose I is a small category and $i \longrightarrow X_i$ a functor to spectra. Then the methods of ([**6**], XI), lead to a tower of fibrations and a spectral sequence

$$E_2^{p,-q} = lim_i^p \pi_q X_i \Rightarrow \pi_{q-p} holim_i X_i$$

where lim^p is the p-th derived functor of the inverse limit. If $I = \Delta$, so that $X^{\cdot}$ is a cosimplicial spectrum, we have $E_2^{p,-q} = \pi^p(\pi_q X^{\cdot})$, where π^p denotes the cohomotopy of a cosimplicial abelian group - that is, the cohomology of the associated chain complex ([**6**], XI, 7.3).

Finally, we remark that we could also consider an extended tower

$$\ldots \longleftarrow X_{-1} \longleftarrow X_0 \longleftarrow X_1 \longleftarrow \ldots$$

such as the Postnikov tower of a non-connective spectrum. This leads to a similar spectral sequence, but now the spectral sequence is no longer confined to a half-plane and the convergence question becomes more delicate. Nevertheless, in some cases it is still conditionally convergent. Suppose, for example, that $\pi_k X_n = 0$ for $k > n$. Then it is even possible to re-index the spectral sequence so that it becomes a right half-plane spectral sequence - compare ([**42**], p. 542). See Theorem 3.19 below for an example of this situation.

3.1.2 *Closed model categories.* A *closed model category* is a category $\mathcal{C}$ equipped with three distinguished classes of morphisms, called weak equivalences, fibrations and cofibrations, that are closed under composition and contain all identities. These classes are subject to several axioms that codify the abstract properties of a "homotopy theory".

(CM1): $\mathcal{C}$ has all finite limits and colimits.

(CM2): If $f = gh$ is a composite morphism and any two of f, g, h are weak equivalences, then so is the third.

(CM3): All three classes of morphisms are closed under retracts.

(CM4): Given a commutative square

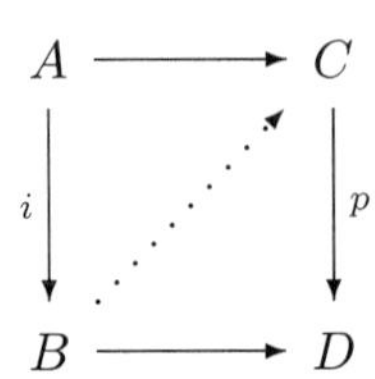

the dotted arrow exists whenever (i) i is an acyclic cofibration and p is a fibration, or (ii) i is a cofibration and p is an acyclic fibration ("acyclic" is shorthand for weak equivalence).

(CM5): Any map $X \to Y$ can be factored in two ways: (i) as $X \xrightarrow{i} W \xrightarrow{p} Y$, where i is a cofibration and p is an acyclic fibration; and (ii) as $X \xrightarrow{j} Z \xrightarrow{q} Y$, where j is an acylic cofibration and p is a fibration.

Let ϕ and $*$ denote respectively "the" initial and terminal objects of the category (usually, for us, $\phi = *$). An object X is *cofibrant* if $\phi \to X$ is a cofibration and *fibrant* if $X \to *$ is a fibration. Part of the general philosophy is that homotopy theory works best with a cofibrant source and a fibrant target (see [**11**], 3.18 and 3.22). In practice one of these two conditions holds trivially for all - or almost all - objects of the category, while the other imposes possibly stringent conditions on an object X. For example, in the category of presheaves of spectra discussed below, cofibrancy is a harmless condition to impose (although not all objects are cofibrant), whereas fibrancy or the lack of it is essentially the descent problem.

The most basic example for present purposes is the category of pointed simplicial sets. We say that a map of pointed simplicial sets is a weak equivalence if the induced map on realizations is a weak equivalence in the usual topological sense, is a cofibration if it is injective, and is a fibration if it is a Kan fibration. With these definitions, pointed simplicial sets form a closed model category [**36**]; of course this works in the unpointed category also. Every object is cofibrant, and the fibrant objects are the Kan complexes.

This closed model category structure was extended to spectra in [**5**]. Here a spectrum is defined to be a sequence of pointed simplicial sets X_n, $n \geq 0$, equipped with structure maps $S^1 \wedge X_n \to X_{n+1}$. The homotopy groups of a spectrum X are defined as usual by

$$\pi_n X = colim_k\, \pi_{n+k} \mid X_{n+k} \mid$$

with the colimit taken in the evident way using the structure maps. We say that a map of spectra $f : X \to Y$ is a weak equivalence if it induces an isomorphism on homotopy groups, and is a cofibration if for all $n \geq 0$ the natural map from the pushout P_n in the diagram

$$\begin{array}{ccc} S^1 \wedge X_{n-1} & \longrightarrow & S^1 \wedge Y_{n-1} \\ \downarrow & & \downarrow \\ X_n & \longrightarrow & P_n \end{array}$$

to Y_n is a cofibration of simplicial sets (in particular f is a spacewise cofibration). Finally, we say that f is a fibration if it has the right lifting property as in CM4(ii) with respect to all acyclic cofibrations. With these definitions, the category of spectra is a closed model category [**5**]. The fibrations are in particular spacewise fibrations, and can be described explicitly (see [**5**]). It turns out that the fibrant spectra are precisely the Ω-spectra (in the classical sense) whose constituent spaces are Kan complexes. The cofibrant spectra are just the spectra with injective structure maps.

3.1.3 *Filtered colimits of spectra.* Consider the closed model category of pointed simplicial sets, as defined above. Then:

Proposition 3.1 *Filtered colimits preserve cofibrations and fibrations (in particular, a filtered colimit of Kan complexes is a Kan complex). Filtered colimits of Kan complexes preserve weak equivalences.*

For cofibrations, this is obvious, since a cofibration is just an injective map. The proof for fibrations may be sketched as follows. Call a simplicial set K *small* if it has only finitely many nondegenerate simplices. Then it is easy to see that the functor $Hom(K, -)$ commutes with filtered colimits (even when the maps in the colimit are not injective!). But Kan fibrations are defined in terms of the right lifting property with respect to certain maps of small simplicial sets (inclusion of the boundary minus one face of Δ^n into Δ^n), and the result follows. A similar argument applies to weak equivalences, provided that the topological homotopy groups occurring in the definition of weak equivalence can be replaced by simplicial homotopy groups. This is possible since we have assumed the spaces in question are Kan complexes.

Similar arguments apply to the Bousfield-Friedlander closed model category structure on spectra, although the situation is complicated slightly by the different notion of fibration. Call a spectrum X a Kan spectrum if each space X_n is a Kan complex. The result is:

Proposition 3.2 *Filtered colimits of spectra preserve cofibrations and fibrations, and filtered colimits of Kan spectra preserve weak equivalences.*

In fact the assertions for cofibrations and weak equivalences hold for pseudofiltered colimits. We note that up to functorial weak equivalence, any spectrum can be replaced by a Kan spectrum; this will be done tacitly, when necessary, in the sequel.

We now turn to the (critical!) question of when homotopy limits and related constructions commute with filtered colimits. Our goal is to give a simple criterion that is easily applied in practice - although it will take longer to state this criterion than to prove it. Throughout this section "spectral sequence" means right half-plane cohomology spectral sequence.

We assume given the following data:

- A spectrum X and a filtered system of spectra X_i, $i \in I$.
- A compatible family of maps $f_i : X_i \longrightarrow X$, so there is an induced map $colim\, X_i \longrightarrow X$.
- Spectral sequences $E_* X_i$ and $E_* X$ converging conditionally to $\pi_* X_i$ and $\pi_* X$ respectively, and that are natural with respect to the maps in the filtered system and with respect to the f_i. (In particular, the associated filtrations on $\pi_* X_i$ and $\pi_* X$ are assumed natural in this sense.)

We say that a spectral sequence is *bounded on the right* if there exists d such that $E_2^{p,*} = 0$ for $p > d$, and is *bounded below* if there exists m such that $E_2 *, q = 0$ for $q < m$. Note that either condition implies convergence. If we are given a family $E_*(i)$ of spectral sequences, we say that $E_*(i)$ is *uniformly* bounded on the right if there is a fixed d that works for all i, etc. The next proposition will be called the "colimit lemma".

Proposition 3.3 *Suppose that (i) either the E_2-terms $E_2(X_i)$ are uniformly bounded on the right, or else they are uniformly bounded below; and (ii) the map $f_* : colim_i\, E_2(X_i) \longrightarrow E_2(X)$ is an isomorphism. Then the map $f : colim_i X_i \longrightarrow X$ is a weak equivalence.*

Proof Since the index category I is filtered, the colimit of the spectral sequences $E_*(i)$ is still a spectral sequence. However it need not, in general, converge to the homotopy of the colimit - the trouble is that nontrivial elements of $\pi_* colim_i X_i = colim_i \pi_* X_i$ can have infinite filtration. But hypothesis (i) rules out this possibility; the induced filtration on $colim_i \pi_q X_i$ is finite for each q, and the colimit of the spectral sequences converges to the homotopy of the colimit. Then condition (ii) implies that the target spectral sequence is also convergent, the induced map on E_∞-terms is an isomorphism, and f is a weak equivalence.

As an illustration, suppose that I and J are small categories with I filtered and J arbitrary. Let $X_{i,j}$ be an $(I \times J)$-diagram of spectra, and consider the natural map

$$f : colim_i\, holim_j\, X_{i,j} \longrightarrow holim_j\, colim_i X_{i,j}.$$

Here "holim" means the homotopy limit as in [**6**], extended to spectra as in [**42**], Section 5. For present purposes, all the reader needs to know is that (1) holim is a functor from J-diagrams Y_j of spectra to spectra; and (2) there is a natural conditionally convergent spectral sequence,

$$E_2^{p,-q} = lim_j^p(\pi_q Y_j) \Rightarrow \pi_{q-p} holim\, Y_j$$

where lim^p denotes the p-th derived functor of the inverse limit. Recall that a spectrum X is *bounded above* if its homotopy groups vanish above some dimension. Then we obtain as an immediate corollary:

Proposition 3.4 *Suppose that (i) either there exists d such that for all q, i the groups $lim_j^p(\pi_q X_{i,j})$ vanish for $p > d$, or else the $X_{i,j}$ are uniformly bounded above; and (ii) the natural map*

$$colim_i\, lim_j^p \pi_* X_{i,j} \longrightarrow lim_j^p\, colim_i \pi_* X_{i,j}$$

is an isomorphism. Then the map

$$f : colim_i\, holim_j\, X_{i,j} \longrightarrow holim_j\, colim_i\, X_{i,j}$$

is a weak equivalence.

For example if $J = \Delta$, so we have a filtered system of cosimplicial spectra, condition (ii) holds automatically but condition (i) is a nontrivial assumption about finite cohomological dimension. On the other hand if J is the category $0 \leftarrow 1 \leftarrow 2 \leftarrow ...$, so a J-diagram is a tower, then condition (i) holds with $d = 1$ but condition (ii) can easily fail. Let's call a tower of spectra Y_j *stable* if for each fixed q there is an m_q such that $\pi_q Y_j \rightarrow \pi_q Y_{j-1}$ is an isomorphism for all $n > m_q$ (for example, a Postnikov tower). We call an I-diagram of towers X_{ij} as above *uniformly stable* if each tower X_{ij} (i fixed) is stable and moreover each number m_q depends only on q and not on i. In this case $\pi_q lim_j\, X_{ij} = \pi_q X_{i,n}$ with $n >> 0$ depending only on q. Then (ii) obviously holds and we get the easy but important corollary:

Corollary 3.5 *Let X_{ij} be a filtered system (in i) of towers (in j), and suppose the system is uniformly stable. Then*

$$colim_i\, lim_j X_{i,j} \xrightarrow{\cong} lim_j\, colim_i X_{i,j}.$$

Remark 3.6 *The assumption on E_2-terms in the colimit lemma can, of course, be replaced by the same assumption on some other fixed E_r-term. An interesting example is discussed in the introduction to Section 4.*

3.2 The Godement construction for spectra. Let Φ be a site with enough points - for example, the Zariski, étale, or Nisnevich site of a scheme X. Let $\mathcal{E}$ be a presheaf of spectra on Φ. Then the construction of the Godement resolution for ordinary abelian sheaves carries over almost verbatim to presheaves of spectra. In fact let $\mathcal{A}$ be any category with filtered colimits, products and coproducts, and consider an $\mathcal{A}$-valued presheaf F. Then we can *define* the stalk $F_p = p^*F$ at a point p as a filtered colimit over neighborhoods of p as in 2.6. The formula given there for the right adjoint p_* works fine too. The Godement construction then produces a cosimplicial $\mathcal{A}$-valued presheaf $T^{\cdot}F$, whose sections on X give a cosimplicial object in $\mathcal{A}$. In particular we get a Godement functor $\mathcal{E} \longrightarrow T^{\cdot}\mathcal{E}$ from presheaves of spectra to cosimplicial presheaves of spectra, and we define the *hypercohomology of* X *with coefficients in* $\mathcal{E}$ by

$$\mathbb{H}^{\cdot}_{\Phi}(X; \mathcal{E}) = holim_{\Delta} T^{\cdot}\mathcal{E}(X).$$

In fact $\mathbb{H}^{\cdot}_{\Phi}(-; \mathcal{E})$ is itself a presheaf on the site. Note that $\pi_* T^{\cdot}\mathcal{E} = T^{\cdot}\pi_*\mathcal{E}$. Hence the Bousfield-Kan machinery discussed above yields a conditionally convergent spectral sequence

$$E_2^{p,-q} = H^p_{\Phi}(X; \tilde{\pi}_q\mathcal{E}) \Rightarrow \pi_{q-p}\mathbb{H}^{\cdot}_{\Phi}(X; \mathcal{E}).$$

Many basic results about cohomology can now be extended rather easily to hypercohomology spectra. We illustrate this with some examples that will be important later. Given a site Φ, a presheaf of spectra $\mathcal{E}$ and an object X of the site, we say that $cd^{\Phi}(X; \mathcal{E})$ is bounded, or $cd\,\Phi < \infty$, if for some d $H^p(X; \tilde{\pi}_q\mathcal{E}) = 0$ for all $p > d$ and all q. If the presheaf and/or the object are varying over a family, the phrase "uniformly bounded" means that a fixed d works for all presheaves and/or objects in the family. A condition is said to hold *stalkwise* if it holds for the stalks of the presheaf in question. Often this makes sense even without reference to points of the site - for example, a presheaf of spectra is *stalkwise bounded above* if for some m the *sheaves* $\tilde{\pi}_q$ vanish for all $q > m$.

Theorem 3.7 *Suppose that Φ is a noetherian site, $\mathcal{E}_i$ a pseudofiltered system of presheaves of spectra on Φ. Fix an object X of Φ and suppose that either (a) $cd^{\Phi}(X; \mathcal{E}_i)$ is uniformly bounded, or (b) the $\mathcal{E}_i$ are stalkwise uniformly bounded above. Then the natural map*

$$colim_i \mathbb{H}^{\cdot}{}_{\Phi}(X; \mathcal{E}_i) \longrightarrow \mathbb{H}^{\cdot}{}_{\Phi}(X; colim_i \mathcal{E}_i)$$

is a weak equivalence.

Proof One simply checks the conditions of the colimit Lemma 3.3. Condition (i) of that lemma holds by assumption, and condition (ii) is immediate from the algebraic version 2.10 .

This result is important because it will allow us to commute smash products and Bousfield localizations with hypercohomology (see the end of this section).

Similarly, we can promote the continuity Theorem 2.11 to the spectrum level. Thus suppose X_i is an admissible system of schemes (i.e., a cofiltered system of quasicompact and quasiseparated schemes, with affine transition maps), with $X = lim_i X_i$. Recall that the main case to keep in mind is that of a filtered system of rings, such as the system used to define henselizations. Let $\mathcal{E}$ be a presheaf of ℓ-torsion spectra on a sufficiently large étale site, and suppose that $\mathcal{E}$ is *continuous* in the sense that for any admissible system U_i in the site, the natural map $colim_i\, \mathcal{E}(U_i) \longrightarrow \mathcal{E}(U)$ is a weak equivalence.

Theorem 3.8 *Suppose that X_i is an admissible system of schemes, $\mathcal{E}$ a continuous presheaf of ℓ-torsion spectra as above. Suppose also that either (a) $cd^{\Phi}(X_i;\mathcal{E})$ is uniformly bounded, or (b) $\mathcal{E}$ is stalkwise bounded above. Then the natural map*

$$colim_i\, \mathbb{H}^{\cdot}_{\acute{e}t}(X_i;\mathcal{E}) \longrightarrow \mathbb{H}^{\cdot}_{\acute{e}t}(X;\mathcal{E})$$

is a weak equivalence.

Again, the proof is immediate from the colimit lemma together with the continuity Theorem 2.11. The one point to note is that the presheaves $\pi_q\mathcal{E}$ are continuous by assumption, and hence so are the associated sheaves $\tilde{\pi}_q\mathcal{E}$.

The main example of a continuous presheaf of spectra is the K-theory presheaf. It is continuous by a theorem of Quillen ([**35**], 7, 2.2). The point is that the category of locally free sheaves on the limit scheme X is the "colimit" of the corresponding categories on the X_i.

Our main application of continuity is to the determination of the stalks of hypercohomology presheaves.

Theorem 3.9 *Let X be a separated noetherian scheme of finite Krull dimension. Suppose X is uniformly ℓ-bounded. Let $\mathcal{E}$ be a presheaf of ℓ-torsion spectra on $X_{\acute{e}t}$ and regard $\mathcal{F} = \mathbb{H}^{\cdot}_{\acute{e}t}(-;\mathcal{E})$ as a presheaf on X_{Zar}, X_{Nis}, or $X_{\acute{e}t}$. Then*

(i) *The Zariski stalk of $\mathcal{F}$ at x is $\mathbb{H}^{\cdot}_{\acute{e}t}(\mathcal{O}_x;\mathcal{E})$.*
(ii) *The Nisnevich stalk of $\mathcal{F}$ at $x' \to x$ is $\mathbb{H}^{\cdot}_{\acute{e}t}(\mathcal{O}^h_{x'};\mathcal{E})$.*
(iii) *The étale stalk of $\mathcal{F}$ at $\bar{x} \to x$ is $\mathbb{H}^{\cdot}_{\acute{e}t}(\mathcal{O}^{sh}_{\bar{x}};\mathcal{E})$.*

Proof Consider for example case (iii), where we can work on the restricted étale site. If $U = Spec\, \mathcal{O}^{sh}_{\bar{x}}$, then $U = lim_i U_i$, where U_i ranges over the cofiltered system of étale neighborhoods of $\bar{x}$. The U_i can be taken affine and hence form an admissible system. Moreover the assumptions on X are inherited by the U_i, so that $cd^{\acute{e}t}_{\ell}(U_i)$ is bounded uniformly in i. Hence Theorem 3.8 applies, completing the proof.

Now suppose W is a fixed spectrum. Then for any presheaf of spectra $\mathcal{E}$ there is a natural map

$$W \wedge \mathbb{H}^{\cdot}_{\Phi}(X;\mathcal{E}) \longrightarrow \mathbb{H}^{\cdot}_{\Phi}(X; W \wedge \mathcal{E})$$

which in general need not be a weak equivalence. However smashing with a finite spectrum commutes with arbitrary homotopy limits and products, hence:

Proposition 3.10 *If W is a finite spectrum, then for any presheaf of spectra $\mathcal{E}$ and object X the map*

$$W \wedge \mathbb{H}^{\cdot}_{\Phi}(X;\mathcal{E}) \longrightarrow \mathbb{H}^{\cdot}_{\Phi}(X; W \wedge \mathcal{E})$$

is a weak equivalence.

If W is not finite then some condition has to be imposed on the site, the object, the presheaf or all three. The next result is not the most general one can imagine, but takes the form that we will typically need.

Proposition 3.11 *Suppose $cd_\ell^\Phi(X) < \infty$, and that either W is an ℓ-torsion spectrum or $\mathcal{E}$ is a presheaf of ℓ-torsion spectra. Then the map*

$$W \wedge \mathbb{H}^{\cdot}_{\Phi}(X;\mathcal{E}) \longrightarrow \mathbb{H}^{\cdot}_{\Phi}(X; W \wedge \mathcal{E})$$

is a weak equivalence.

Proof W is a directed colimit of finite spectra W_i, which can be taken ℓ-torsion if W is. Then

$$W \wedge \mathbb{H}^{\cdot}_{\Phi}(X;\mathcal{E}) = colim_i\, W_i \wedge \mathbb{H}^{\cdot}_{\Phi}(X;\mathcal{E}) = colim_i\, \mathbb{H}^{\cdot}_{\Phi}(X; W_i \wedge \mathcal{E})$$

by the preceding proposition. Now each presheaf $W_i \wedge \mathcal{E}$ is a presheaf of ℓ-torsion spectra, so Theorem 3.7 applies.

Finally, recall that a spectrum E is *smashing* if $L_E Z = Z \wedge L_E S^0$ for all Z, or equivalently L_E commutes with directed colimits. For example the complex K-theory spectrum is smashing, but the mod ℓ Moore spectrum is not (completion doesn't commute with direct limits).

Corollary 3.12 *Suppose $cd_\ell^\Phi(X) < \infty$, E is smashing, and that either $L_E S^0$ is an ℓ-torsion spectrum or $\mathcal{E}$ is a presheaf of ℓ-torsion spectra. Then the natural map*

$$L_E \mathbb{H}^{\cdot}_{\Phi}(X;\mathcal{E}) \longrightarrow \mathbb{H}^{\cdot}_{\Phi}(X; L_E\mathcal{E})$$

is a weak equivalence.

3.3 Jardine's closed model category. Let Φ be a site and consider the category $\mathcal{P}$ of presheaves of spectra on Φ. We would like to define a closed model category structure on $\mathcal{P}$ in a way that depends on the Grothendieck topology, not just the underlying category. In particular the weak equivalences should depend on the topology, and the fibrant objects should be the objects satisfying descent.

The most natural definition of weak equivalence is stalkwise weak equivalence. We say that a map of presheaves $\mathcal{E} \longrightarrow \mathcal{F}$ is a *stalkwise weak equivalence* if the induced map of associated abelian *sheaves* $\tilde{\pi}_*\mathcal{E} \longrightarrow \tilde{\pi}_*\mathcal{F}$ is an isomorphism. Note this definition makes sense even when Φ doesn't have enough points. If Φ does have enough points, as is the case in all the examples we care about, then a map is a stalkwise weak equivalence if and only if the induced maps on stalks are weak equivalences of spectra. However even in this case it will be convenient to have available the definition as stated. Note that every pointwise weak equivalence is a stalkwise weak equivalence, since pointwise weak equivalence just means that the induced map on homotopy presheaves is already an isomorphism. The converse is obviously false.

Remark Any stalkwise weak equivalence is also a stalkwise homology isomorphism. This is clear for a site with enough points, but not at all obvious in the general case. See ([**22**], 2.9).

A map of presheaves is a *cofibration* if it is a pointwise cofibration, and is a *fibration* if it has the right lifting property with respect to cofibrations that are also stalkwise weak equivalences. The following theorem is due to Jardine [**21**]. Earlier results of this type, but for *sheaves* of simplicial sets, were proved by Brown-Gersten [**7**] in the Zariski case and Joyal in the general case.

Theorem 3.13 *Let Φ be an arbitrary site. Then with the above definitions, the category $\mathcal{P}$ of presheaves of spectra on Φ is a closed model category.*

In fact Jardine shows that $\mathcal{P}$ is a proper simplicial model category. The proof of Theorem 3.13 is discussed at the end of this section.

In order to avoid possible confusion with either pointwise fibrations or the local fibrations of [**20**], the fibrations in this closed model category structure are called *global* fibrations. Similarly a presheaf is *globally fibrant* if it is fibrant in the closed model category structure of Theorem 3.13. Note that the cofibrant objects are just the presheaves which are pointwise cofibrant in the sense of Bousfield-Friedlander. In particular any presheaf is pointwise weak equivalent to a cofibrant presheaf, so there is no harm in assuming that our presheaves are cofibrant.

Now by CM5, given any presheaf $\mathcal{E}$ we can find a globally fibrant presheaf $G\mathcal{E}$ and a map $\mathcal{E}\longrightarrow G\mathcal{E}$ which is both a cofibration and a stalkwise weak equivalence. It follows from the proof of Theorem 3.13 that $G\mathcal{E}$ can be chosen functorially. It is convenient, however, not to be tied down to any particular choice; any such stalkwise weak equivalence (whether or not it is a cofibration) will be called a *globally fibrant model* for $\mathcal{E}$. This globally fibrant model (which we will see shortly is well-defined up to pointwise weak equivalence) will serve as an alternative to, and generalization of, Thomason's hypercohomology spectrum. Before justifying this assertion, we summarize a few properties of Jardine's closed model category structure, properties which begin to illustrate the elegance of this approach. We first consider the behaviour of the three classes of morphisms under direct image, inverse image, and extension by zero.

Proposition 3.14 *Let $f:\Phi\longrightarrow\Phi'$ be a site morphism. Then the presheaf inverse image functor $f^{\#}$ preserves cofibrations and stalkwise weak equivalences, and the direct image functor $f_{\#}$ preserves cofibrations and global fibrations.*

The proof of the first statement is easy, using the fact that $f^{\#}$ is defined in terms of pseudofiltered colimits of spectra. Obviously $f_{\#}$ preserves cofibrations. That $f_{\#}$ preserves global fibrations is a formal consequence of the fact that it has a left adjoint, namely $f^{\#}$, that preserves cofibrations and weak equivalences. It is clear that $f^{\#}$ does not generally preserve global fibrations, and $f_{\#}$ does not generally preserve stalkwise weak equivalences (consider for example a scheme with the Zariski topology and a map to Spec of a field).

Here is a simple example that will be used later. Consider a scheme X and a presheaf $\mathcal{E}$ which is globally fibrant on the étale site. Applying the proposition to the restriction morphisms $X_{\acute{e}t}\longrightarrow X_{Nis}\longrightarrow X_{Zar}$ shows that $\mathcal{E}$ is globally fibrant on the Nisnevich and Zariski sites as well.

Corollary 3.15 *Let $j:\Phi/U\longrightarrow\Phi$ denote inclusion of a local site as in Section 2.5. Then the extension by zero functor $j_!$ preserves cofibrations and stalkwise weak equivalences, and $j^{\#}$ preserves global fibrations.*

This follows immediately because $j^{\#}=k_{\#}$ and $j_!=k^{\#}$, where k is as in Section 2.5. We also deduce:

Corollary 3.16 *Every global fibration is a pointwise fibration.*

For suppose given a global fibration $\pi:\mathcal{E}\longrightarrow\mathcal{F}$. We wish to show that for every object U of the site, $\mathcal{E}(U)\longrightarrow\mathcal{F}(U)$ is a fibration of spectra. But the "sections on U" functor has a left-adjoint that preserves cofibrations and stalkwise weak equivalences - namely, the functor $E \mapsto j_! c_E$, where c_E is constant presheaf on Φ/U associated to the spectrum E.

As noted above, a stalkwise weak equivalence need not be a pointwise weak equivalence. There is one case, however, where we do get pointwise weak equivalence. I thank Paul Goerss and the referee for providing the proof of the following important proposition.

Proposition 3.17 *Suppose $f : \mathcal{E} \longrightarrow \mathcal{F}$ is a stalkwise weak equivalence of globally fibrant presheaves. Then f is a pointwise weak equivalence.*

Using CM5 and CM2, we reduce immediately to two cases: either f is a global fibration, or f is a cofibration. If f is a global fibration, the adjoint functor argument used above shows that f is a pointwise weak equivalence. If f is a cofibration, then f is a *homotopy equivalence* - in fact, $\mathcal{E}$ is a deformation retract of $\mathcal{F}$. To explain this assertion, we need a tiny bit of the "simplicial" model category structure on presheaves of spectra. If $\mathcal{A}$ is any presheaf of spectra, let $\mathcal{A}^{\Delta^1}$ denote the path object of simplicial maps $\Delta^1 \longrightarrow \mathcal{A}$ (defined pointwise and levelwise). Homotopies can then be defined using this path object, in the usual way. Now since $\mathcal{E}$ is globally fibrant, there is a lift r in the diagram

$$\begin{array}{ccc} \mathcal{E} & \xrightarrow{=} & \mathcal{E} \\ {\scriptstyle f}\downarrow & \overset{r}{\nearrow} & \downarrow \\ \mathcal{F} & \longrightarrow & * \end{array}$$

so that $rf = 1_{\mathcal{E}}$. On the other hand, since $\mathcal{F}$ is globally fibrant, the obvious map $\mathcal{F}^{\Delta^1} \longrightarrow \mathcal{F} \times \mathcal{F}$ is a global fibration (this is a simple special case of Quillen's axiom "SM7" for simplicial model categories). Hence there is a lift in the diagram

$$\begin{array}{ccc} \mathcal{E} & \longrightarrow & \mathcal{F}^{\Delta^1} \\ {\scriptstyle f}\downarrow & \nearrow & \downarrow \\ \mathcal{F} & \xrightarrow[(1,fr)]{} & \mathcal{F} \end{array}$$

where the top map is the constant homotopy of f. This lift is a homotopy $1_{\mathcal{F}} \cong fr$. Thus f is a homotopy equivalence as claimed. It follows immediately that for any object U, $\mathcal{E}(U) \longrightarrow \mathcal{F}(U)$ is a homotopy equivalence of fibrant spectra and hence a weak equivalence.

In particular this shows that globally fibrant models are well-defined up to pointwise weak equivalence.

Elegant though this machinery may be, it is still in need of justification. The crucial test is that there should be a descent spectral sequence for the pointwise homotopy of a globally fibrant presheaf. The key step in setting up such a spectral sequence is an analysis of globally fibrant models for Eilenberg-Maclane spectra. So let A be a presheaf of abelian goups. An Eilenberg-Maclane presheaf of type (A, n) is a presheaf of spectra $\mathcal{A}$ such that $\pi_n \mathcal{A}$ is isomorphic to A and $\pi_k \mathcal{A}$ is identically zero for $k \neq n$.

Proposition 3.18 *Let $\mathcal{A}$ be an Eilenberg-Maclane presheaf of spectra of type (A, n), with globally fibrant model $G\mathcal{A}$. Then for all $k \in \mathbb{Z}$ and all objects U of the site,*

$$\pi_k(G\mathcal{A})(U) \cong H^{n-k}(U; aA).$$

We sketch a proof (see [**22**], Ch.4 for a proof that is more in the spirit of Thomason's "Scholium of Great Enlightenment"). It is convenient to work with spaces (=simplicial sets), so we first reduce to that case and use the closed model category structure for presheaves of spaces (see the end of this section). Since formal suspension and desuspension preserve globally fibrant models, we can assume $k \geq 0$. Since $\mathcal{A}$ can be assumed pointwise fibrant, and the zero-th space presheaf of $G\mathcal{A}$ is a globally fibrant presheaf of spaces, we reduce to the analogous statement for Eilenberg-Maclane presheaves of spaces. We will continue to use the notation $\mathcal{A}$ for such a presheaf.

Now let $\mathcal{P}_{ss}, \mathcal{P}_{sag}, \mathcal{P}_{ch}$ denote respectively the categories of presheaves of simplicial sets, simplicial abelian groups, and non-negatively graded chain complexes. There is an evident pair of adjoint functors

$$\mathbb{Z} : \mathcal{P}_{ss} \longrightarrow \mathcal{P}_{sag},\ i : \mathcal{P}_{sag} \longrightarrow \mathcal{P}_{ss},$$

as well as a pair of mutually inverse equivalences of categories

$$N : \mathcal{P}_{sag} \longrightarrow \mathcal{P}_{ch},\ T : \mathcal{P}_{ch} \longrightarrow \mathcal{P}_{sag}.$$

(See [**27**]; N is the normalized chain complex functor.) Replacing $\mathcal{A}$ by the n-th Postnikov section of $\mathbb{Z}\mathcal{A}$, we can assume $\mathcal{A}$ is a simplicial abelian group. In fact up to natural pointwise weak equivalence we can even assume $\mathcal{A} = T(C_A)$, where C_A is the chain complex consisting of the presheaf A in degree n and zero elsewhere. Now let $aA \longrightarrow I.$ be an injective resolution of the sheaf aA, and form the chain complex J_A which is zero above degree n, is equal to I_{n-k} for $0 < k < n$ and with $I_0 = Ker\,(I_n \rightarrow I_{n-1})$. Then the natural map $C_A \longrightarrow J_A$ is a stalkwise homology isomorphism, and the pointwise homology of J_A is given by $H_k J_A = H^{n-k}(U; aA)$.

We claim that iTJ_A is a globally fibrant model for $\mathcal{A}$. Since homology in $\mathcal{P}_{ch}$ corresponds to homotopy in $\mathcal{P}_{sag}$, we have a stalkwise weak equivalence $\mathcal{A} \longrightarrow T(J_A)$. Finally, it is not hard to show that iTJ_A is globally fibrant - for example, by using the argument for the analogous theorem in [**15**].

To construct the descent spectral sequence, we start from a Postnikov tower $\{P^n\mathcal{E}\}$ of the presheaf $\mathcal{E}$. Let F_n denote the fibre of $P^n\mathcal{E} \longrightarrow P^{n-1}\mathcal{E}$. Easy formal arguments lead to a commutative ladder

$$\begin{array}{ccccccc}
\cdots & \longleftarrow & P^{n-1}\mathcal{E} & \longleftarrow & P^n\mathcal{E} & \longleftarrow & \cdots \\
 & & \big\downarrow & & \big\downarrow & & \\
\cdots & \longleftarrow & GP^{n-1}\mathcal{E} & \longleftarrow & GP^n\mathcal{E} & \longleftarrow & \cdots
\end{array}$$

in which (i) the bottom row consist of global fibrations and globally fibrant objects; (ii) $lim_n GP^n\mathcal{E}$ is globally fibrant; and (iii) the fibre of $GP^n\mathcal{E} \longrightarrow GP^{n-1}\mathcal{E}$ is a globally fibrant model GF_n of the Eilenberg-Maclane presheaf F_n. Hence we have

$$\pi_k(GF_n)(U) = H^{n-k}(U; \tilde{\pi}_n\mathcal{E}).$$

If we assume the cohomology groups on the right vanish for $n-k >> 0$, then both towers are uniformly stable (see 3.1.3) and hence

$$\mathcal{E} = lim_n P^n\mathcal{E} \longrightarrow lim_n GP^n\mathcal{E}$$

is a stalkwise weak equivalence. In other words, $lim_n GP^n\mathcal{E}$ is a globally fibrant model for $\mathcal{E}$. The tower of fibrations $GP^n\mathcal{E}$ then yields a spectral sequence with

$$E_1^{p,-q} = H^{2p-q}(X; \tilde{\pi}_p\mathcal{E}).$$

The indexing may look bizarre to some readers, but we can formally reindex the spectral sequence to get the usual form of the E_2-term (see [**42**], p. 542; note the same peculiar indexing shows up if one constructs the Atiyah-Hirzebruch spectral sequence by a Postnikov resolution of the target instead of a skeletal filtration of the source). The result is:

Theorem 3.19 *Let $\mathcal{E}$ be a presheaf of spectra on a site Φ. Let X be an object of the site and suppose $cd^\Phi(X;\mathcal{E}) = d < \infty$. Then there is a convergent right half-plane cohomology spectral sequence*

$$E_2^{p,-q} = H^p_\Phi(X; \tilde{\pi}_q\mathcal{E}) \Rightarrow \pi_{q-p}(G\mathcal{E})(X).$$

From this spectral sequence it is morally clear that the constructions of Jardine and Thomason must agree on a site with enough points, at least under finite cohomological dimension hypotheses. To prove this, we need to check two things.

Proposition 3.20 *Let Φ be a site with enough points, $\mathcal{E}$ a presheaf of spectra on Φ. Then the hypercohomology presheaf $\mathbb{H}^{\cdot}(-;\mathcal{E})$ is globally fibrant. If $cd^\Phi\mathcal{E} < \infty$, then the natural map $\mathcal{E} \longrightarrow \mathbb{H}^{\cdot}(-;\mathcal{E})$ is a stalkwise weak equivalence.*

Thus the hypercohomology presheaf of $\mathcal{E}$ is a globally fibrant model of $\mathcal{E}$. To get the stalkwise weak equivalence in the proposition, consider the map on stalks at p

$$f : colim\, \mathcal{E}(U) \longrightarrow colim\, \mathbb{H}^{\cdot}(U;\mathcal{E})$$

where the colimits are over the filtered category $\mathcal{N}_p^{op}$. The finite cohomological dimension hypothesis ensures that the colimit of the descent spectral sequences for $\mathbb{H}^{\cdot}(U;\mathcal{E})$ converges to the homotopy of the colimit (Proposition 3.3). This colimit spectral sequence collapses to its vertical edge, and $\pi_* f$ can be identified with the natural isomorphism $(\pi_*\mathcal{E})_p \cong (\tilde{\pi}_*\mathcal{E})_p$.

The adjoint functor trick used in (3.14) shows that each cosimplicial term of the cosimplicial presheaf $T^{\cdot}\mathcal{E}$ is a globally fibrant presheaf. One then has to show that the homotopy limit remains globally fibrant, using the techniques of [**6**]; see ([**20**], 3.3).

We conclude this section with a discussion of the proof of Theorem 3.13. The theorem was proved first for presheaves of simplicial sets in [**20**], and then extended to spectra in [**21**], using the techniques of [**5**]. The key point is to get a closed model category structure on the category $\mathcal{P}_{ssets}$ of presheaves of simplicial sets.

In $\mathcal{P}_{ssets}$ the definition of stalkwise weak equivalence is not as simple as it is for spectra, because one cannot even define presheaves of homotopy groups without a compatible family of basepoints. Suppose that E is a presheaf of simplicial sets and U is an object of the site. Each choice of basepoint $e_0 \in E(U)_0$ defines a compatible family of basepoints on the local site Φ/U, because U is a terminal object of the latter. Hence for $n > 0$ we get a presheaf of homotopy groups $\pi_n(E\,|_U; e_0)$ on Φ/U. If $n = 0$ we have a presheaf of sets $\pi_0 E$ on Φ. We then say that a map $f : E \longrightarrow F$ is a stalkwise weak equivalence if (i) $\tilde{\pi}_0(f)$ is an isomorphism of sheaves; and (ii) for all objects U, basepoints $e_0 \in E(U)_0$, and $n > 0$, the induced map $\tilde{\pi}_n(E\,|_U; e_0) \longrightarrow \tilde{\pi}_n(F\,|_U; f(e_0))$ is an isomorphism of sheaves on Φ/U.

It is not hard to check that if the site has enough points, then f is a stalkwise weak equivalence if and only if $p^*E \longrightarrow p^*F$ is a weak equivalence for a sufficient set of points p. In this case several steps of the proof of Theorem 3.13 become much easier.

As in the case of spectra, the cofibrations are the pointwise cofibrations and the global fibrations are defined by the right lifting property with respect to acyclic cofibrations. Axioms CM1 and CM3 (as well as the mostly redundant requirement that the three classes are closed under composition and contain all identities) are then easily verified. If the site has enough points, CM2 is obvious (CM2 is also obvious for spectra). In the general case, however, one of the three implications of CM2 runs into a problem with basepoints, as the reader will quickly discover on inspection (see [**20**], 1.11 and 2.1). Half of CM4 is true by definition. It turns out that the other half follows formally from CM5 by a trick due to Joyal, using nothing more than the fact that cofibrations are stable under cobase change ([**20**], p. 64).

The crux of the matter, then, is CM5. This axiom is usually proved using Quillen's "small object argument". Consider first the problem of factoring a map $f : E \longrightarrow F$ as an acylic cofibration followed by a global fibration. To make the small object argument work, two ingredients are needed. First, we need to know that acyclic cofibrations are stable under (i) cobase change, and (ii) pseudo-filtered colimits. If the site has enough points, (i) is obvious ((i) is also obvious for spectra, because of the Mayer-Vietoris sequence in homotopy associated to a pushout square). The general case of (i) is much harder ([**20**], 2.2). Condition (ii) is clear in all cases.

Second, we need to know that there is a set S of acyclic cofibrations such that (i) S detects global fibrations (in the obvious sense); and (ii) S is "small", in a sense to be indicated. Here we fix an infinite cardinal number α greater than the cardinality of the set of morphisms of the underlying category of the site. Then for S we take a set of representatives of the isomorphism classes of acyclic cofibrations $A \longrightarrow B$ such that the cardinality of $B(U)_n$ is at most α for all objects U and all $n \geq 0$. The key fact is that this S detects global fibrations; see ([**20**], Lemma 2.4), for the interesting proof.

Returning to an arbitrary morphism $f : E \longrightarrow F$, the "small object argument" (or "infinite glueing construction" in [**11**]) begins by forming the pushout

$$\begin{array}{ccc} \coprod A_i & \longrightarrow & E \\ \big\downarrow & & \big\downarrow{\scriptstyle j} \\ \coprod B_i & \longrightarrow & E_1 \end{array}$$

in which $A_i \longrightarrow B_i$ is ranging over all maps of elements of S into f. This yields a factorization $E \longrightarrow E_1 \longrightarrow F$ in which the first map is an acyclic cofibration. The second map is not a global fibration, but at least all the lifts associated to the given

$A_i \longrightarrow B_i$ have been "filled in". Iterating this construction some transfinite number of times (taking colimits at the limit ordinals) yields a factorization

$$E \xrightarrow{i} E_\infty \xrightarrow{p} F$$

in which E_∞ is a certain colimit over some large ordinal. The point now is that if we choose this ordinal to be sufficiently large (bigger than the cardinality of the power set of the set of morphisms will do), then the cardinality bound defining the elements of S forces every map of an element $A \longrightarrow B$ of S into f to factor through some lower stage of the colimit. Then the desired lift exists by construction.

The second part of CM5 can also be achieved by a small object argument, or by the following cheap trick. In the closed model category structure on simplicial sets, the factorizations of CM5 can be chosen in a functorial way. Hence any map of presheaves $f : E \longrightarrow F$ can be factored as

$$\begin{array}{ccccc} E & \xrightarrow{j} & D & \xrightarrow{q} & F \\ & & {\scriptstyle k}\downarrow & \nearrow{\scriptstyle r} & \\ & & C & & \end{array}$$

where j is a cofibration, q is a pointwise weak equivalence, and q has been factored as an acylic cofibration k followed by a global fibration r, using the part of CM5 already proved. Then kj is a cofibration and r is an acyclic global fibration (using CM2). See also the more interesting argument in [**20**], which shows that the acyclic fibration can be taken to be a *pointwise* weak equivalence.

3.4 Čech hypercohomology spectra. Our discussion of Čech hypercohomology will be brief and somewhat imprecise. See ([**42**], 1), for details. Fix a site Φ and an object X of the site. Let $\mathcal{U} = \{U_i \longrightarrow X\}$ be a covering family. As a cosimplicial object, the complex defining ordinary Čech cohomology (2.9) makes perfectly good sense for presheaves with values in any category with products. In particular, given a presheaf of spectra $\mathcal{E}$, we obtain in this way a cosimplicial spectrum $\check{C}^{\cdot}(\mathcal{U}/X;\mathcal{E})$. Taking an appropriate filtered colimit over a cofinal family of covers leads to the absolute Čech complex $\check{C}^{\cdot}(X;\mathcal{E})$. The corresponding Čech hypercohomology spectra are then defined by

$$\check{\mathbb{H}}^{\cdot}(\mathcal{U}/X;\mathcal{E}) = holim_{\Delta}\check{C}^{\cdot}(\mathcal{U}/X;\mathcal{E})$$

and

$$\check{\mathbb{H}}^{\cdot}(X;\mathcal{E}) = holim_{\Delta} colim_{\mathcal{U}}\check{C}^{\cdot}(\mathcal{U}/X;\mathcal{E}).$$

As with the Godement complex, each of these formulae lead to a descent spectral sequence with E_2-term given by the appropriate Čech cohomology groups. Of course one cannot, in general, reverse the order of the holim and the colim in the second formula above. Using (3.3), however, we obtain:

Proposition 3.21 *If $\mathcal{E}$ is pointwise bounded above, then the natural map*

$$colim_{\mathcal{U}}\check{\mathbb{H}}^{\cdot}(\mathcal{U}/X;\mathcal{E}) \longrightarrow \check{\mathbb{H}}^{\cdot}(X;\mathcal{E})$$

is a weak equivalence.

(Note that if $\pi_q\mathcal{E} = 0$ for $q > m$, then for all $\mathcal{U}$ we have $\pi_q\check{\mathbb{H}}^{\cdot}(\mathcal{U}/X;\mathcal{E}) = 0$ for $q > m$.)

Example 3.22 *Consider the étale site of a field k. We can take the finite Galois extensions of k as cofinal system of covers. Then for $\mathcal{E}$ pointwise bounded above, we have (reverting to the notation of Section 1)*

$$colim_{\alpha}\mathbb{H}^{\cdot}(G_{\alpha};\mathcal{E}) \cong \mathbb{H}^{\cdot}(k;\mathcal{E})$$

where G_α ranges over the Galois groups of such extensions. More generally this works with any Galois extension M of k in place of the separable closure.

"Čech descent for sheaf cohomology" takes the following form:

Theorem 3.23 *Suppose that either $\mathcal{E}$ is stalkwise bounded above, or that there is a uniform bound on $cd(U;\mathcal{E})$ (uniform in U, as U ranges over the individual objects of all covering families in sight). Then there is a natural weak equivalence*

$$\mathbb{H}^{\cdot}(X;\mathcal{E}) \cong \check{\mathbb{H}}^{\cdot}(\mathcal{U}/X;\mathbb{H}^{\cdot}(-;\mathcal{E})).$$

Example 3.24 *In the notation of the preceding example, suppose that either $\mathcal{E}$ is stalkwise bounded above or that $\mathcal{E}$ is an ℓ-torsion spectrum and $cd_{\ell}k = d < \infty$. Let k'/k be a Galois extension with group G. Then*

$$\mathbb{H}^{\cdot}(k;\mathcal{E}) \cong \mathbb{H}^{\cdot}(G;\mathbb{H}^{\cdot}(k';\mathcal{E})).$$

Note $cd_{\ell}k' \leq d$.

As a final and very pretty example we show how Artin's Theorem 2.9 extends to hypercohomology spectra ([**42**], Proposition 1.54).

Theorem 3.25 *Suppose X is quasi-projective over a noetherian ring of finite Krull dimension, and $\mathcal{E}$ is an additive presheaf of spectra on $X_{\acute{e}t}$. Then there is a natural chain of weak equivalences between $\mathbb{H}^{\cdot}_{\acute{e}t}(X;\mathcal{E})$ and $\check{\mathbb{H}}^{\cdot}_{\acute{e}t}(X;\mathcal{E})$.*

To prove this one first reduces to the case where $\mathcal{E}$ is pointwise bounded above, by using a Postnikov resolution. Then consider the natural maps

$$\check{\mathbb{H}}^{\cdot}_{\acute{e}t}(X;\mathcal{E}) \xrightarrow{\alpha} \check{\mathbb{H}}^{\cdot}_{\acute{e}t}(X;\mathbb{H}^{\cdot}_{\acute{e}t}(-;\mathcal{E})) \xleftarrow{\beta} \mathbb{H}^{\cdot}_{\acute{e}t}(X;\mathcal{E}).$$

Note that β is a weak equivalence by Theorem 3.23. As for α, note first that the Čech hypercohomology spectral sequences converge, so it is enough to show that α is an isomorphism on E_2-terms - i.e., the map of coefficient presheaves $\pi_q\mathcal{E}\longrightarrow\pi_q\mathbb{H}^{\cdot}_{\acute{e}t}(-;\mathcal{E})$ induces an isomorphism on Čech cohomology. But this map is an isomorphism on stalks by the second part of Proposition 3.20, and hence induces an isomorphism on sheaf cohomology of the associated sheaves. Then Artin's Theorem 2.9 completes the proof.

In conclusion, we note that it might be interesting to study Verdier hypercohomology spectra (see Section 2.9).

4 Descent Theorems in K-Theory

Let $\mathcal{E}$ be a presheaf of spectra on a site Φ, X an object of the site. If the site has enough points, so that Thomason's hypercohomology spectrum $\mathbb{H}^{\cdot}{}_{\Phi}(X;\mathcal{E})$ is defined, then the problem of *descent* is to determine whether or not the natural map $\eta : \mathcal{E}(X)\longrightarrow\mathbb{H}^{\cdot}{}_{\Phi}(X;\mathcal{E})$ is a weak equivalence. If it is a weak equivalence, then we get a descent spectral sequence as in Section 3, but now converging to $\pi_*\mathcal{E}(X)$. Typically X is a terminal object of the site, and in the cases of interest any proof of descent for X will prove descent for an arbitrary object of the site as well. So we may drop the assumption of enough points, and (bearing in mind Proposition 3.20) rephrase the descent problem in terms of Jardine's closed model category structure.

Descent Problem: Let $\eta : \mathcal{E}\longrightarrow G\mathcal{E}$ be a globally fibrant model for the presheaf $\mathcal{E}$. Is η a pointwise weak equivalence?

Recall that the map η is always a *stalkwise* weak equivalence, by definition.

Now let X be a separated noetherian regular scheme of finite Krull dimension, and let $\mathcal{E} = K$ denote the K-theory presheaf on one of the sites $X_{Zar}, X_{Nis}, X_{\acute{e}t}$. Then K satisfies descent on X_{Zar} by the Brown-Gersten theorem, and satisfies descent on X_{Nis} by a theorem of Nisnevich; these results are discussed in the first two sections below. Descent on the étale site is precisely the subject of the Lichtenbaum-Quillen conjectures. In this case, however, there are two complicating factors. The first is that, for technical reasons, we need to work at a fixed prime ℓ with the ℓ-adic completions. The second and more serious complication is that it simply isn't true that K-theory satisfies ℓ-adic descent. In fact, in the form stated above descent fails even for finite fields F, because the ℓ-adic hypercohomology spectrum has a nonvanishing π_{-1}, isomorphic to $H^1_{\acute{e}t}(Spec\, F; Z_\ell(0)) \cong Z_\ell$. A natural reaction to this annoyance is to replace GK (or equivalently, $\mathbb{H}^{\cdot}$) by its (-1)-connected cover. But then, for example when F is a number field, we find that $\pi_0(GK)^\wedge$ contains a contribution from the Brauer group that is simply not present in K_0F. It seems that the best one can hope for is an equivalence above some dimension depending on the ℓ-adic étale cohomological dimension d of the scheme X.

ℓ-adic Étale Descent Problem for K-theory: Is the globally fibrant model $\eta : K\longrightarrow G_{\acute{e}t}K$ a pointwise weak equivalence after ℓ-adic completion and passing to some connected cover?

One guess is that $\pi_k\eta^\wedge$ is an isomorphism for $k > 2d$; for another see ([**22**], Conjectures 7.35 and 7.36). The Beilenson-Lichtenbaum conjectures suggest yet another guess; see for example [**40**].

Thomason's descent theorem [**42**], which forms the main subject of this paper, says that η induces an isomorphism on "Bott-periodic" homotopy mod powers of ℓ. Equivalently, $\hat{L}(\eta)$ is a weak equivalence, where $\hat{L}$ denotes Bousfield localization with respect to mod ℓ topological complex K-theory. Spectra of the form $\hat{L}$(-) are always non-connective (or trivial!), so connectivity obstructions of the sort discussed above do not arise. The proof we will sketch has two main steps - (i) the proof for fields, all discussion of which is postponed to the final Section 5; and (ii) the deduction of the general case from (i). With the Thomason-Jardine machinery of Section 3 and the Nisnevich descent theorem in hand, the proof of step (ii) is remarkably simple.

For practical applications, the descent problem as formulated above is still missing two ingredients. First, the main point of establishing a pointwise weak equivalence $\mathcal{E} \longrightarrow G\mathcal{E}$ is not merely to get a descent spectral sequence, but to get one whose E_2-term is actually computable. This means that we need to at least identify the sheaves associated to the presheaves $\pi_q\mathcal{E}$. We state this as:

Descent Problem, Addendum: "Determine" the associated sheaves $\tilde{\pi}_*\mathcal{E}$.

In practice this amounts to identifying the stalks, and by the continuity Theorem 2.11 this in turn reduces to computing $\pi_*\mathcal{E}$ on the appropriate local rings. Thus in the K-theory setting, we need to know the K-theory of the Zariski, hensel, or strict hensel local rings, as the case may be. In the strict hensel case the answer is completely known, thanks to the Gillet-Thomason-Suslin-Gabber rigidity theorem and Suslin's theorem on algebraically closed fields. This is discussed further in Section 4.11 below; in particular we show how Thomason's descent theorem yields a spectral sequence with the desired E_2-term.

Second, one needs to know that the descent spectral sequence converges. The cheapest way to ensure convergence is to assume that the presheaf (or even the site) in question has finite cohomological dimension. We generally do so in order to simplify the exposition. It is worth noting, however, that there are interesting cases where the cohomological dimension is infinite but one has convergence nevertheless. For example, suppose $\ell = 2$ and $X = Spec\, R$ with R either a number field admitting a real embedding or a ring of S-integers in such a field. Then $cd_2^{\acute{e}t} X = \infty$. Nevertheless, the 2-adic descent spectral sequence for $\mathbb{H}^{\cdot}_{\acute{e}t}(X; K)$ converges - in fact, the filtration has length at most 3 and the spectral sequence collapses at E_4. This follows from the analogous statement for $\mathbb{R}$, together with classical calculations of Tate. As a result, many of the general results of the last section still apply (see Remark 3.6).

Finally, some comments may be in order concerning our scheme-theoretic hypotheses. The three K-theory descent theorems stated here all assume that X is separated and regular. The sole reason for this assumption is to ensure that $KX \cong K'X$, where $K'X$ is the K-theory spectrum associated to the exact category of coherent sheaves on X. This is an instance of "reduction by resolution" [**35**], which in turn depends on Theorem 2.1. Thus the "separated and regular" hypothesis can be dropped if we are only interested in $K'X$. It can also be dropped if we replace K-theory with Bass K-theory, using the deep theorem of Thomason [**44**].

For Thomason's étale descent theorem, we also assume that X is uniformly ℓ-bounded (i.e., there is a uniform bound on $cd_\ell k$ as k ranges over the residue fields of X). One reason for this assumption (which is satisfied in most cases of interest) is to ensure that $cd_\ell X < \infty$, so that the descent spectral sequence converges. Moreover it is a property inherited by every object of the (restricted) étale site of X. Hence we are free to pass back and forth between hypercohomology presheaves *à* la Thomason and globally fibrant models *à* la Jardine, using Proposition 3.20.

Some further (but mild) restriction on the residue fields is necessary. Our assumption will be that the residue fields are "ℓ-good" as defined in Section 4.4. This hypothesis, which is taken from [**22**], could undoubtedly be weakened somewhat; it has the advantage that it is easily remembered and applied.

4.1 Brown-Gersten descent. Let X be a scheme, $\mathcal{E}$ a presheaf of spectra on the Zariski site of X. We say that $\mathcal{E}$ is a Mayer-Vietoris presheaf if for every pair of open sets $U, V \subset X$, the diagram

$$\begin{array}{ccc} \mathcal{E}(U \cup V) & \longrightarrow & \mathcal{E}(U) \\ \downarrow & & \downarrow \\ \mathcal{E}(V) & \longrightarrow & \mathcal{E}(U \cap V) \end{array}$$

is a homotopy fibre square. Alternatively, we can express this as an excision condition for closed pairs. Given closed subschemes Y, Z with $Z \subset Y$, let $\Gamma_{Y,Z}\mathcal{E}$ denote the presheaf assigning to each open U the homotopy fibre of $\mathcal{E}(U-Z) \longrightarrow \mathcal{E}(U-Y)$. We write Γ_Y for $\Gamma_{Y,\phi}$. Then $\mathcal{E}$ *satisfies excision* if $\Gamma_{Y,Z}\mathcal{E}$ depends only on $Y - Z$ - that is, if $(Y,Z) \subset (Y', Z')$ then the natural map $\Gamma_{Y,Z}\mathcal{E} \longrightarrow \Gamma_{Y',Z'}\mathcal{E}$ is a pointwise weak equivalence. It is easy to check that $\mathcal{E}$ is Mayer-Vietoris if and only if $\mathcal{E}$ satisfies excision.

Remark Quillen's localization sequence for K'-theory [**35**] shows that K' satisfies excision and hence defines a Mayer-Vietoris presheaf. If X is separated and regular, then the K-theory presheaf is also a Mayer-Vietoris presheaf, since then $K \cong K'$.

The next theorem is due to Brown and Gersten [**7**].

Theorem 4.1 *Let X be a noetherian scheme of finite Krull dimension, and suppose that $\mathcal{E}$ is a Mayer-Vietoris presheaf of spectra on the Zariski site of X. Then $\mathcal{E}$ satisfies Zariski descent: any globally fibrant model*

$$\mathcal{E} \longrightarrow G_{Zar}\mathcal{E} = \mathbb{H}^{\cdot}{}_{Zar}(-;\mathcal{E})$$

is a pointwise weak equivalence.

Corollary 4.2 *If X is a separated noetherian regular scheme the K-theory presheaf satisfies Zariski descent. In particular there is a convergent spectral sequence*

$$E_2^{p,-q} = H^p_{Zar}(X; \tilde{\pi}_q K) \Rightarrow \pi_{q-p}K(X).$$

Here is a quick proof of the theorem, paraphrasing Thomason's Exercise 2.5 [**42**]. Call a presheaf of spectra *quasi-fibrant* if it is pointwise weak equivalent to a globally fibrant presheaf. We will use the following easy facts: (a) if any two terms of a pointwise fibre sequence are quasi-fibrant, so is the third; and (b) a pseudo-filtered colimit of quasi-fibrant presheaves is quasi-fibrant (this follows from Proposition 3.7, since X_{Zar} is a noetherian site of finite cohomological dimension).

Let Y be a closed subscheme of X. For each closed $Z \subset Y$ of codimension at least one, the sequence

$$\Gamma_Z\mathcal{E} \longrightarrow \Gamma_Y\mathcal{E} \longrightarrow \Gamma_{Y,Z}\mathcal{E}$$

is a pointwise fibre sequence by the "Quetzlcoatl lemma" (Thomason's picturesque term for the homotopy-theoretic analogue of the snake lemma). Passing to the colimit over the directed set of all such Z yields a pointwise fibre sequence

$$colim_Z \Gamma_Z\mathcal{E} \longrightarrow \Gamma_Y\mathcal{E} \longrightarrow C_Y\mathcal{E}$$

where $C_Y\mathcal{E} = colim_Z\Gamma_{Y,Z}\mathcal{E}$. If Y is irreducible with generic point x, we also write $\Gamma_x\mathcal{E} = C_Y\mathcal{E}$ (the reason for this apparently superfluous notation will appear when we consider Nisnevich descent). Using excision, Quetzlcoatl, and induction, one obtains:

Lemma 4.3 *Suppose Y has irreducible components $Y_1, ..., Y_m$. Then there is a pointwise weak equivalence*

$$\bigvee C_{Y_i}\mathcal{E} \xrightarrow{\cong} C_Y\mathcal{E}.$$

Now let

$$S^p\mathcal{E} = colim_{codim\, Y \geq p}\Gamma_Y\mathcal{E}.$$

We will show by descending induction on p that $S^p\mathcal{E}$ is quasi-fibrant. This will prove the theorem, since $S^0\mathcal{E} = \mathcal{E}$. To start the induction, note that if X has Krull dimension n, $S^{n+1}\mathcal{E} = *$. At the inductive step, define $C^p\mathcal{E}$ by the pointwise cofibre sequence

$$S^{p+1}\mathcal{E} \longrightarrow S^p\mathcal{E} \longrightarrow C^p\mathcal{E}.$$

It is enough to show that $C^p\mathcal{E}$ is quasi-fibrant.

Lemma 4.4 *There is a pointwise weak equivalence*

$$\bigvee_{codim\, x=p} \Gamma_x\mathcal{E} \longrightarrow C^p\mathcal{E}.$$

The proof is easy from Lemma 4.3. Thus we have reduced to showing that $\Gamma_x\mathcal{E}$ is quasi-fibrant for all $x \in X$.

Lemma 4.5 *Let $i : x \rightarrow X$ denote the inclusion. Then the natural map $\Gamma_x\mathcal{E} \longrightarrow i_\# i^\# \Gamma_x\mathcal{E}$ is a pointwise weak equivalence.*

To prove this lemma, it is enough to check (i) $(colim_Z\Gamma_{Y,Z}\mathcal{E})(U)$ is weakly contractible if x is not in U; and (ii) $(colim_Z\Gamma_{Y,Z}\mathcal{E})(U) \longrightarrow (colim_Z\Gamma_{Y,Z}\mathcal{E})(W)$ is a weak equivalence whenever $x \in W \subset U$. Condition (i) is obvious. For condition (ii), consider the diagram

$$\begin{array}{ccccc}
\Gamma_{Y,Z}\mathcal{E}(U) & \longrightarrow & \mathcal{E}(U-Z) & \longrightarrow & \mathcal{E}(U-Y) \\
\downarrow j & & \downarrow & & \downarrow \\
\Gamma_{Y,Z}\mathcal{E}(W) & \longrightarrow & \mathcal{E}(W-Z) & \longrightarrow & \mathcal{E}(W-Y)
\end{array}$$

Note that $(W-Z) \cap (U-Y) = W-Y$. If we have $Y-Z \subset W$, then $U-Z = (U-Y) \cup (W-Z)$ and hence j is a weak equivalence by a direct application of the Mayer-Vietoris condition. But the set of Z satisfying $Y-Z \subset W$ is easily seen to be cofinal, so j is a weak equivalence after passage to the colimit, proving (ii).

A presheaf on the Zariski site of a field is quasi-fibrant if and only if its value on the empty set is weakly contractible. It follows that $i^\#\Gamma_x\mathcal{E}$ is quasi-fibrant, and since direct image functors preserve quasi-fibrant presheaves, this completes the proof of Theorem 4.1.

4.2 Nisnevich descent. Let X be a noetherian scheme of finite Krull dimension. By *excision data* we mean a morphism $f : U \longrightarrow V$ between objects of the Nisnevich site (recall that the underlying category is the same as that of the étale site), together with a closed subscheme $Z \subset V$ such that $U \times_V Z \longrightarrow Z$ is an isomorphism. A presheaf of spectra $\mathcal{E}$ on X_{Nis} satisfies *Nisnevich excision* if for all such excision data, the square

$$\begin{array}{ccc} \mathcal{E}(V) & \longrightarrow & \mathcal{E}(V-Z) \\ \downarrow & & \downarrow \\ \mathcal{E}(U) & \longrightarrow & \mathcal{E}(f^{-1}(V-Z) \end{array}$$

is a homotopy fibre square, and $\mathcal{E}(\phi) \cong *$. It is easy to check that then $\mathcal{E}$ is *additive* - that is, $\mathcal{E}(U \coprod V) \cong \mathcal{E}(U) \times \mathcal{E}(V)$ - and that the restriction of $\mathcal{E}$ to the Zariski site of any object U satisfies Zariski excision (i.e. the Mayer-Vietoris-condition of the preceding section). It is also clear that K'-theory satisfies Nisnevich excision, and hence so does K-theory if X is separated and regular.

The following theorem is the Nisnevich descent theorem [**34**].

Theorem 4.6 *Let X be a noetherian scheme of finite Krull dimension, $\mathcal{E}$ a presheaf of spectra on the Nisnevich site of X. If $\mathcal{E}$ satisfies Nisnevich excision, then $\mathcal{E}$ satisfies Nisnevich descent - that is, the globally fibrant model*

$$\mathcal{E} \longrightarrow G_{Nis}\mathcal{E} = \mathbb{H}^{\cdot}{}_{Nis}(-;\mathcal{E})$$

is a pointwise weak equivalence.

Corollary 4.7 *If X is a separated noetherian regular scheme of finite Krull dimension, the K-theory presheaf satisfies Nisnevich descent.*

The proof that we will sketch here follows [**22**], and very closely parallels the proof of the Brown-Gersten theorem. For each object $U \longrightarrow X$ of X_{Nis} and each closed subscheme $Y \subset U$, we can form the spectra $\Gamma_Y \mathcal{E}(U)$ and $C_Y \mathcal{E}(U)$ as before. Note that these are not presheaves on X_{Nis}. We can, however, define for each $x \in X$ a presheaf $\Gamma_x \mathcal{E}$ by

$$\Gamma_x \mathcal{E}(\pi : U \longrightarrow X) = C_{\overline{\pi^{-1}x}} \mathcal{E}(U).$$

The spectrum $S^p\mathcal{E}(U)$ we defined in the Zariski case is functorial in U, so we also get a presheaf $S^p\mathcal{E}$ on X_{Nis}. Let us again define $C^p\mathcal{E}$ by the pointwise cofibre sequence

$$S^{p+1}\mathcal{E} \longrightarrow S^p\mathcal{E} \longrightarrow C^p\mathcal{E}.$$

Lemma 4.8 *There is a pointwise weak equivalence*

$$\bigvee_{codim\, x=p} \Gamma_x \mathcal{E} \longrightarrow C^p \mathcal{E}.$$

The proof is easy from Lemma 4.3. Thus we have again reduced to showing that $\Gamma_x \mathcal{E}$ is quasi-fibrant.

Lemma 4.9 *Let $i : x \to X$ denote the inclusion. Then the natural map $\Gamma_x\mathcal{E} \longrightarrow i_\# i^\# \Gamma_x \mathcal{E}$ is a pointwise weak equivalence.*

Assuming this for the moment, the proof of Theorem 4.6 is completed as follows. A presheaf of spectra on the Nisnevich site of a field is quasi-fibrant if and only if it is additive (the proof is straightforward - this is the only place where the Nisnevich *topology* is used!). One easily checks that $\Gamma_x\mathcal{E}$ is additive and hence also $i^\#\Gamma_x\mathcal{E}$ is additive. Since direct images preserve quasi-fibrant presheaves, this completes the proof.

We briefly indicate the proof of Lemma 4.9, so the reader can at least see where the Nisnevich excision condition comes in. By unravelling the definitions of $i_\#$ and $i^\#$, one reduces to checking the following:

Lemma 4.10 *Suppose $f : U \longrightarrow V$ is a morphism in X_{Nis} and $z \in V$. Suppose that $f^{-1}z$ consists of a single point y, and that $y \to z$ is an isomorphism of schemes (i.e., an isomorphism on residue fields). Then the induced map $C_{\bar{z}}\mathcal{E}(V) \longrightarrow C_{\bar{y}}\mathcal{E}(U)$ is a weak equivalence.*

If $\bar{y} \longrightarrow \bar{z}$ is an isomorphism of schemes, this follows by a direct application of Nisnevich excision. But we can always reduce to this case because $\bar{y} \longrightarrow \bar{z}$ is an isomorphism on generic points by assumption, and hence restricts to an isomorphism between suitable open subschemes of $\bar{y}, \bar{z}$. See ([**22**], Lemma 7.21), for further details.

4.3 Gabber rigidity and the K-theory sheaf. On the étale site, the ℓ-adic K-theory *sheaf* is explicitly known. Let

$$R_n = \mathbb{Z}[t]/t^{\ell^n} - 1$$

and let μ_{ℓ^n} denote the sheaf on the big étale site (all schemes) represented by $Spec\, R_n$. In other words, $\mu_{\ell^n}(U)$ is the group of units of order dividing ℓ^n in the ring of regular functions on U. Next, let $\beta_n \in \pi_2(B\mathbb{Z}/\ell^n; \mathbb{Z}/\ell^n)$ correspond to the standard generator under the Bockstein isomorphism

$$\pi_2(B\mathbb{Z}/\ell^n; \mathbb{Z}/\ell^n) \cong \pi_1 B\mathbb{Z}/\ell^n = \mathbb{Z}/\ell^n.$$

The natural map

$$\Sigma^\infty B\mathbb{Z}/\ell^n_+ \longrightarrow KR_n$$

then yields a "Bott element" in $\pi_2(KR_n; \mathbb{Z}/\ell^n)$, also denoted β_n.

Now define a map of presheaves $e : \mu_{\ell^n} \longrightarrow \pi_2(K; \mathbb{Z}/\ell^n)$ by

$$e(f : U \longrightarrow Spec\, R_n) = f^*\beta_n.$$

If U is affine - say $U = Spec\, S$ - then $e(f)$ is just the image of β_n under the composite

$$B\mathbb{Z}/\ell^n \longrightarrow BGL_1 R_n \longrightarrow BGL_1 S \longrightarrow BGLS^+.$$

From this description it is clear that e is a group homomorphism. Using the ring spectrum structure on $K(-; \mathbb{Z}/\ell^n)$, we obtain more generally maps of presheaves

$$e_i : \mathbb{Z}/\ell^n(i) \equiv \otimes^i \mu_{\ell^n} \longrightarrow \pi_{2i} K(-; \mathbb{Z}/\ell^n).$$

Proposition 4.11 *Let X be a scheme over $\mathbb{Z}[1/\ell]$. Then the maps e_i induce isomorphisms of sheaves*

$$\mathbb{Z}/\ell^n(\frac{q}{2}) \xrightarrow{\cong} \tilde{\pi}_q K(-;\mathbb{Z}/\ell^n).$$

The group on the left is to be interpreted as zero if q is odd. The proof of the proposition depends on the *Gabber Rigidity Theorem* ([**13**]; earlier versions are due to Gillet-Thomason [**14**] and Suslin ([**39**])):

Theorem 4.12 *Let R be a hensel local ring with residue field k of characteristic different from ℓ. Then the reduction map $KR \longrightarrow Kk$ is a weak equivalence on ℓ-adic completions.*

Combining Gabber rigidity and Suslin's Theorem 1.1 shows that for any strictly henselian ring S over $\mathbb{Z}[1/\ell]$, $\mathbb{Z}/\ell^n(q/2) \cong \pi_q(KS;\mathbb{Z}/\ell^n)$. Hence by the continuity Theorem 2.11, e_i induces an isomorphism on stalks. This proves the proposition.

In other words, we have settled the question raised in the "Addendum" to the descent problem mentioned in the introduction. As a consequence the ℓ-adic descent spectral sequence for hypercohomology with coefficients in K-theory takes the form

$$E_2^{p,-q} = H^p_{\acute{e}t}(X; Z_\ell(\frac{q}{2})) \Rightarrow \pi_{q-p}(\mathbb{H}^{\cdot}_{\acute{e}t}(X;K)^\wedge)$$

where as always étale cohomology means continuous cohomology in the sense of Jannsen [**19**] and the groups on the right are the homotopy groups of the completion, not the completion of the homotopy groups. To see why we get continous cohomology, note first that

$$(\mathbb{H}^{\cdot}_{\acute{e}t}(X;K))^\wedge = holim_n \mathbb{H}^{\cdot}_{\acute{e}t}(X;K) \wedge M\mathbb{Z}/\ell^n = holim_n \mathbb{H}^{\cdot}_{\acute{e}t}(X; K \wedge M\mathbb{Z}/\ell^n)$$

by (3.11), and this in turn is equivalent to

$$holim_n holim_\Delta T^{\cdot}(K(-;\mathbb{Z}/\ell^n))(X) = holim_\Delta holim_n T^{\cdot}(K(-;\mathbb{Z}/\ell^n))(X).$$

Therefore we get a spectral sequence whose E_2-term is the cohomotopy of the cosimplicial abelian group $\pi_* holim_n T^{\cdot}(K(-;\mathbb{Z}/\ell^n))(X)$. Now

$$\pi_q T^{\cdot}(K(-;\mathbb{Z}/\ell^n)) = T^{\cdot}\pi_q(K(-;\mathbb{Z}/\ell^n)) = T^{\cdot}\mu_{\ell^n}\left(\frac{q}{2}\right).$$

Since the reduction maps $\mu_{\ell^{n+1}} \rightarrow \mu_{\ell^n}$ are epimorphisms of sheaves, and the composite functor $\Gamma(X,-)\circ T^{\cdot}$ is exact, it follows that the Milnor lim^1 vanishes and

$$\pi_* holim_n T^{\cdot}(K(-;\mathbb{Z}/\ell^n))(X) = lim_n \pi_q T^{\cdot}(K(-;\mathbb{Z}/\ell^n)).$$

Since $T^{\cdot}$ takes values in acyclic sheaves, it also follows that $T^{\cdot}\pi_q(K(-;\mathbb{Z}/\ell^n))$ is a resolution of $\mu_{\ell^n}(\frac{q}{2})$ by $lim^0 \circ \Gamma(X,-)$-acyclic objects. Hence the cohomotopy yields continous cohomology as claimed.

4.4 Thomason's étale descent theorem. We recall the following definition from Section 1: a field F is ℓ-good if *char* $F \neq \ell$ and F has finite transcendence degree over a subfield E satisfying $cd_\ell E_\infty \leq 1$, where E_∞ is obtained from E by adjoining all the ℓ-power roots of unity; and with $\sqrt{-1} \in E$ if $\ell = 2$. A scheme X is then said to be *ℓ-good* if all of its residue fields are ℓ-good.

Recall also (see Theorem 2.8) that a scheme X is *uniformly ℓ-bounded* if there is a uniform bound $d < \infty$ on the étale cohomological dimension of the residue fields of X. For example, if X is "uniformly ℓ-good" in the sense that the transcendence degrees occurring in the definition of ℓ-good scheme are uniformly bounded (e.g. any variety over an ℓ-good field), then X is uniformly ℓ- bounded.

We can now state the descent theorem of Thomason [**42**]:

Theorem 4.13 *Let X be a separated noetherian regular scheme of finite Krull dimension. Assume that X is ℓ-good and uniformly ℓ-bounded. Then the natural map $KX \longrightarrow \mathbb{H}^{\cdot}_{\acute{e}t}(X;K)$ induces a weak equivalence*

$$\hat{L}K(X) \xrightarrow{\cong} \hat{L}\mathbb{H}^{\cdot}_{\acute{e}t}(X;K).$$

Equivalently, in the terminology of Jardine, the localization of globally fibrant models

$$\hat{L}K \longrightarrow \hat{L}(G_{\acute{e}t}K)$$

is a pointwise weak equivalence on $X_{\acute{e}t}$.

Corollary 4.14 *There is a convergent descent spectral sequence*

$$E_2^{p,-q} = H^p_{\acute{e}t}(X; Z_\ell(\frac{q}{2})) \Rightarrow \pi_{q-p}\hat{L}KX.$$

For the corollary, note that $\hat{L}\mathbb{H}^{\cdot}_{\acute{e}t}(X;K) \cong (\mathbb{H}^{\cdot}(X;LK))^{\wedge}$ by Corollary 3.12. The indicated form of the E_2-term then follows from Gabber rigidity as in Section (4.11).

The two forms of Theorem 4.13 are equivalent by Proposition 3.20, and because all of the hypotheses on X are inherited by any object $U \rightarrow X$ of the (restricted) étale site of X.

Now, **assuming Theorem 4.13 has been proved for ℓ-good fields**, the general case is a rather easy consequence of the machinery thus far assembled. It will be convenient to use the abbreviation

$$\mathcal{L} = LK \wedge M\mathbb{Z}\ell = K \wedge LM\mathbb{Z}/\ell = K \wedge LS^0 \wedge M\mathbb{Z}/\ell.$$

Then it is enough to show that the globally fibrant model $\eta : \mathcal{L} \longrightarrow G_{\acute{e}t}\mathcal{L}$, which is by definition a stalkwise weak equivalence for the étale topology, is a pointwise weak equivalence (that is, a stalkwise weak equivalence for the chaotic topology!). Let us first show:

Lemma 4.15 *$\eta : \mathcal{L} \longrightarrow G_{\acute{e}t}\mathcal{L}$ is a stalkwise weak equivalence for the Nisnevich topology.*

Proof By continuity, this reduces to showing that when $X = Spec\, R$, R a henselian ring, the map $\mathcal{L}X \longrightarrow (G_{\acute{e}t}\mathcal{L})(X) = \mathbb{H}^{\cdot}_{\acute{e}t}(X;\mathcal{L})$ is a weak equivalence. Let k as usual denote the residue field of the maximal ideal and let $x = Spec\, k$. Consider the commutative diagram

$$\begin{array}{ccc} \mathcal{L}X & \longrightarrow & \mathcal{L}x \\ {\scriptstyle\eta}\downarrow & & \downarrow \\ \mathbb{H}^{\cdot}_{\acute{e}t}(X;\mathcal{L}) & \underset{h}{\longrightarrow} & \mathbb{H}^{\cdot}_{\acute{e}t}(x;\mathcal{L}) \end{array}$$

The top map is a weak equivalence by Gabber rigidity, and the righthand map is a weak equivalence by our assumption that the theorem has been proved for fields. The bottom map is also a weak equivalence. To see this, note that $\mathcal{L}$ is a presheaf on the big étale site, and h factors as

$$\mathbb{H}^{\cdot}_{\acute{e}t}(X;\mathcal{L}_X) \xrightarrow{a} \mathbb{H}^{\cdot}_{\acute{e}t}(x;j^{\#}(\mathcal{L}_X)) \xrightarrow{b} \mathbb{H}^{\cdot}_{\acute{e}t}(x;\mathcal{L}_x)$$

Here the notations $\mathcal{L}_X, \mathcal{L}_x$ indicate restriction from the big étale site to the appropriate small site, and $j : x \rightarrow X$ is the inclusion. The first map is a weak equivalence by a spectral sequence comparison argument using Theorem 2.5. The second map is induced by the map of presheaves $j^{\#}(\mathcal{L}_X) \longrightarrow \mathcal{L}_x$ (which exists because $\mathcal{L}$ is a presheaf on the big site). This map is a stalkwise weak equivalence by Gabber rigidity, so b is a weak equivalence as claimed. Hence η is a weak equivalence, completing the proof of the lemma.

Now consider the commutative diagram of presheaves

$$\begin{array}{ccc} \mathcal{L} & \xrightarrow{\eta} & G_{\acute{e}t}\mathcal{L} \\ \downarrow & & \downarrow \\ G_{Nis}\mathcal{L} & \underset{G_{Nis}\eta}{\longrightarrow} & G_{Nis}(G_{\acute{e}t}\mathcal{L}) \end{array}$$

We wish to show that η is a pointwise weak equivalence. The left-hand map is a pointwise weak equivalence by the Nisnevich descent theorem. The right-hand map is a pointwise weak equivalence since $G_{\acute{e}t}\mathcal{L}$ is globally fibrant on the Nisnevich site (because the étale topology is finer than the Nisnevich topology; see Proposition 3.14). Since η is a Nisnevich stalkwise weak equivalence by the lemma, the induced map $G_{Nis}\eta$ on globally fibrant models is a pointwise weak equivalence (Proposition 3.17). This completes the proof of Theorem 4.13 - assuming that it has been verified for fields, as will be done in the next section.

In conclusion, we remark that Thomason's original proof used the Brown-Gersten theorem, as Nisnevich descent was not available at the time. This approach does not work quite so neatly, because of the lack of an analogue of Gabber rigidity. See [**42**] or [**22**] for the Zariski descent proof. See also [**44**].

5 The Proof of Thomason's Theorem for Fields

In this final section we sketch the proof of Theorem 4.13 for fields, following Chapter 7 of [**22**]. Since we no longer need to consider general schemes, we write in terms of rings rather than schemes and use the following notation: F is a field, and

L is a finite separable extension of F - usually Galois with group G. X is a spectrum, a G-spectrum or a presheaf of spectra on the étale site of F. Let $k(-)$ denote the mod ℓ K-theory presheaf $K \wedge M\mathbb{Z}/\ell$. In order to avoid distracting technicalities, we assume $\ell > 3$.

Thus our goal is to show that the natural map $kF \longrightarrow \mathbb{H}^{\cdot}(F;k)$ induces a weak equivalence $LkF \longrightarrow L\mathbb{H}^{\cdot}(F;k)$, or equivalently

$$A^{-1}kF \xrightarrow{\cong} A^{-1}\mathbb{H}^{\cdot}(F;k) \cong \mathbb{H}^{\cdot}(F;A^{-1}k).$$

Here A is the Adams self-map of the Moore spectrum, and since $cd_{\ell}F < \infty$ we are free to commute A^{-1} and $\mathbb{H}^{\cdot}$ by Theorem 3.7. A theorem of Snaith and Dwyer says that inverting the Adams map is equivalent to inverting the Bott element $\beta_{\ell-1} \in k_{2\ell-2}F$. By a standard (and trivial) transfer argument we can assume F contains the ℓ-th roots of unity. Thus we have a Bott element $\beta \in \pi_2 kF$, and Theorem 4.13 for fields is equivalent to:

Theorem 5.1 *Let F be an ℓ-good field containing the ℓ-th roots of unity. Then $kF \longrightarrow \mathbb{H}^{\cdot}(F;k)$ induces a weak equivalence on Bott-periodic spectra*

$$\beta^{-1}kF \longrightarrow \beta^{-1}\mathbb{H}^{\cdot}(F;k) \cong \mathbb{H}^{\cdot}(F;\beta^{-1}k).$$

The mapping telescopes in this theorem are to be interpreted in terms of the evident kF-module spectrum structures. In particular $\mathbb{H}^{\cdot}(F;k)$ is a kF-module spectrum with multiplication

$$kF \wedge \mathbb{H}^{\cdot}(F;k) \longrightarrow \mathbb{H}^{\cdot}(F;kF \wedge k) \longrightarrow \mathbb{H}^{\cdot}(F;k).$$

Before we can even outline the proof, we need some preliminaries on the hypernorm and hypertransfer (subsections 1-4). The fifth subsection outlines the proof modulo two key lemmas, whose proofs are sketched in the sixth and last subsection.

5.1 Hyperhomology spectra. Suppose G is a discrete group and X is a G-spectrum. The *hyperhomology spectrum* $\mathbb{H}.(G;X)$ is the homotopy colimit of the G-action, which we can take to mean the spectrum $EG_+ \wedge_G X$. Filtering EG by skeleta, one easily obtains a first quadrant homology spectral sequence

$$E^2_{p,q} = H_p(G;\pi_q X) \Rightarrow \pi_{p+q}\mathbb{H}.(G;X).$$

Let $\mathcal{F}(X,Y)$ denote the function spectrum for spectra X, Y. Then if Y has trivial G-action, there is a more or less obvious universal coefficient formula

$$\mathbb{H}^{\cdot}(G;\mathcal{F}(X,Y)) \cong \mathcal{F}(\mathbb{H}.(G;X),Y).$$

5.2 The hypernorm. Let G be a finite group, X a G-spectrum. There is a "norm" map $N: X \rightarrow X$, in the stable category that is just the sum of the actions of the various elements of G. An explicit, functorial strict representative of this map is given by the composite

$$X \xrightarrow{\Delta_G} Map(G_+,X) \xleftarrow{\phi} G_+ \wedge X \xrightarrow{\nabla} X$$

where $\Delta_G(x)(g) = gx$, ∇ is the ordinary fold map, and $\phi(g \wedge x)(h) = x$ if $h = g$ and $\phi(g \wedge x)(h) = *$ otherwise. Note that $Map(G_+,X)$ and $G_+ \wedge X$ are just the product and wedge respectively of $\mid G \mid$ copies of X. Hence ϕ is a weak equivalence and the diagram represents a map in the stable category.

Now observe that there are $G\times G$-actions on all four terms of the above diagram, such that the three maps are $G \times G$-equivariant - although the actions on the two instances of X are not the same! Explicitly, for $(g,h) \in G \times G$ we have

$Map\,(G,X):\ \ [(g,h)\psi](a) = h\psi(h^{-1}ag)$
$G_+ \wedge X:\ \ (g,h)(a \wedge x) = hag^{-1} \wedge hx$
X (in source): $(g,h)x = gx$
X (in target): $(g,h)x = hx$.

Thus the norm N is a $G \times G$-equivariant map.

There is a functor from $G\times G$-spectra to spectra that takes the homotopy limit with respect to the second copy of G and then the homotopy colimit with respect to the first copy of G. Equivalently, we can think of this functor as

$$EG^1_+ \wedge_{G^1} (Map_{G^2}(EG^2_+, Y))$$

for a $G \times G$-spectrum Y. Here the superscripts serve to indicate which copy of G is acting. Applying this to the norm map and making use of the augmentation $BG_+ \longrightarrow S^0$ yields

$$EG_+ \wedge_G X \longrightarrow EG_+ \wedge_G (Map\,(BG_+, X) \longrightarrow$$
$$BG_+ \wedge Map^G(EG_+, X) \longrightarrow Map^G(EG_+, X).$$

This composite is the *hypernorm* which we write as

$$N_h : \mathbb{H}_{\cdot}(G;X) \longrightarrow \mathbb{H}^{\cdot}(G;X).$$

It has the property that the composite map

$$X \longrightarrow \mathbb{H}_{\cdot}(G;X) \longrightarrow \mathbb{H}^{\cdot}(G;X) \longrightarrow X$$

is the ordinary norm N. As a consequence we have:

Lemma 5.2 *Suppose the Tate cohomology groups $\hat{H}^p(G;\pi_q X)$ vanish for all p,q. Then the hypernorm is a weak equivalence:*

$$N_h : \mathbb{H}_{\cdot}(G;X) \xrightarrow{\cong} \mathbb{H}^{\cdot}(G;X).$$

For both spectral sequences collapse to an edge, and the induced map on homotopy can be identified with the algebraic norm map

$$n : (\pi_q X)_G \longrightarrow (\pi_q X)^G.$$

But $Ker\, n = \hat{H}^{-1}$ and $Coker\, n = \hat{H}^0$, whence the lemma.

5.3 An analogue for spectra of a theorem of Tate. Let G be a finite group. Recall that Tate cohomology is defined in the following way. Given a G-module M, let $P_{\cdot}$ and $I^{\cdot}$ denote the standard projective and injective resolutions (respectively) of M. The Tate complex $\tau^{\cdot} M$ is obtained by splicing together $(P_{\cdot})_G$ and $(I^{\cdot})^G$ via the norm map

$$\sigma : (P_0)_G \longrightarrow M_G \xrightarrow{n} M^G \longrightarrow (I^0)^G.$$

In other words, $\tau^{\cdot} M$ is the mapping cone of σ. Tate cohomology $\hat{H}^*(G;M)$ is then defined as the cohomology of this complex.

The same construction works for hypercohomology of cochain complexes - given a cochain complex $C^{\cdot}$ (which we picture running up the q-axis, in the upper half of the (p,q)-plane) we get a second quadrant double complex $(P^{\cdot\cdot})_G(C^{\cdot})$ and a first quadrant double complex $(I_{\cdot\cdot})^G(C^{\cdot})$ which can be spliced to produce an upper half-plane double complex $\tau^{\cdot\cdot}(G;C^{\cdot})$ of cohomological type.

Now suppose G is a profinite group and M is a discrete G-module. We say that $cd\, M \leq d$ if for all open subgroups V of G, $H^n(V;M) = 0$ for $n > d$. Then there is the following lemma of Tate ([**37**], Annexe 1 to Chapitre I):

Lemma 5.3 *The discrete G-module M satisfies* $cd\, M = 0$ *if and only if for all open normal subgroups U of G,* $\hat{H}^*(G/U;M^U) \equiv 0$.

Now suppose $cd\, M \leq d$. Then one easily constructs a finite resolution

$$0 \longrightarrow M \longrightarrow N^0 \longrightarrow N^1 \longrightarrow \ldots \longrightarrow N^d \longrightarrow 0$$

with $cd\, N^i = 0$ for all i (Tate, *loc. cit.*). Fix an open normal subgroup U and form the associated Tate double complex $T^{\cdot\cdot}(G/U;(N^{\cdot})^U)$ as above. Since the horizontal cohomology of this complex vanishes by Tate's lemma, the cohomology of the total complex is also trivial. We conclude:

Corollary 5.4 *The norm* $(P^{\cdot\cdot})_{G/U}(N^{\cdot U}) \longrightarrow (I^{\cdot\cdot})^{G/U}(N^{\cdot U})$ *induces an isomorphism on total cohomology.*

Since the total cohomology of the target is $H^*(G;M)$, it follows that there is a spectral sequence $E_2^{p,q} = H_{-p}(G/U;H^q(U;M)) \Rightarrow H^*(G;M)$. Re-indexing by setting $s = -p$ and $t = d - q$ yields a first quadrant homology spectral sequence - the Tate spectral sequence:

$$E^2_{s,t} = H_s(G/U;H^{d-t}(U;M)) \Rightarrow H^{d-s-t}(G;M).$$

We will not make any use of this spectral sequence, however. We have stated Corollary 5.4 only to motivate its analogue for spectra. This analogue could of course be stated purely in terms of profinite groups, but for future reference we record it in terms of Galois extensions.

Theorem 5.5 *Suppose E is a Galois extension of F and X a presheaf of spectra on* (E/F) *with* $cd\,(E/F;X) < \infty$. *Then for any finite Galois subextension* L/F *with group G, the hypernorm*

$$N_h : \mathbb{H}_{\cdot}(G;\mathbb{H}^{\cdot}(E/L;X)) \longrightarrow \mathbb{H}^{\cdot}(G;\mathbb{H}^{\cdot}(E/L;X))$$

is a weak equivalence.

Here $cd\,(E/F;X) \leq d$ if $cd\,(G(E/F);\tilde{\pi}_q X) \leq d$ for all q, in the sense described above.

Thomason's original proof is an induction on $cd\,(E/F;X)$ that proceeds as follows. If this dimension is zero, then $\pi_q(\mathbb{H}^{\cdot}(E/L;X) = H^0(E/L;\pi_q X_E)$. Hence N_h is a weak equivalence by Tate's lemma combined with Lemma 5.2.

At the inductive step, form the cofibre sequence $X \rightarrow TX \rightarrow CX$, where TX is the Thomason-Godement construction as usual and CX is the cofibre. Then $cd\, TX = 0$ and $cd\, CX < cd\, X$; the result follows easily from this. See ([**22**], 7.1), for a proof using Postnikov towers.

5.4 Transfer and hypertransfer. Let L be a finite field extension of F. In addition to the map $i : KF \longrightarrow KL$ arising from extension of scalars, there is a *transfer* map $t : KL \longrightarrow KF$ arising from restriction of scalars. We recall that the transfer has the following basic properties:

- **Index formula**. The composite $KF \xrightarrow{i} KL \xrightarrow{t} KF$ is multiplication by the index $[L : F]$.
- **Invariance under base change.** The transfer commutes with base change - that is, given an arbitrary field extension F'/F there is a commutative diagram in the stable category

$$\begin{array}{ccc} KL & \longrightarrow & KL' \\ \downarrow & & \downarrow \\ KF & \longrightarrow & KF' \end{array}$$

  where $L' = L \otimes_F F'$ and the vertical maps are transfers (of course L' may not be a field, but the transfer is defined more generally).
- **Norm formula**. If L is Galois over F with group G, then the composite $KL \xrightarrow{t} KF \xrightarrow{i} KL$ is the norm.
- **Linearity**. The transfer is a map of KF-module spectra.

We note that if M/F is an arbitrary Galois extension with trivial pro-ℓ-Sylow subgroup, it follows from the index and norm formulas that

$$(KF)_{(\ell)} \xrightarrow{\cong} ((KM)_{(\ell)})^{G(M/F)}$$

and therefore the K-theory presheaf, localized at ℓ, satisfies descent on M/F. In particular this suffices to prove the Lichtenbaum-Quillen conjectures for fields F with $cd_\ell F = 0$, since a profinite group G satisfies $cd_\ell G = 0$ if and only if the pro-ℓ-Sylow subgroup of G is trivial.

Now consider the transfer $t : KL \longrightarrow KF$ in the case of a finite Galois extension. Then t is not G-equivariant with respect to the trivial action on KF, because the forgetful functor $L - mod \rightarrow F - mod$ is only G-equivariant up to natural isomorphism. Here we are using the standard G-action in which $g \in G$ acts by tensoring up along $L \xrightarrow{g} L$. Thus the transfer is only equivariant up to homotopy. In fact, however, t is equivariant up to an infinity of higher homotopies, in the sense that t extends canonically to a map

$$t_h : \mathbb{H}.(G; KL) \longrightarrow KF$$

called the *hypertransfer*. The easiest way to think of this is to replace the G-action described above by the obvious equivalent action in which G acts by leaving the underlying F-vector space strictly alone and twisting the scalar multiplication. Then the transfer is strictly equivariant and we can define t_h as the composite

$$\mathbb{H}.(G; KL) = EG_+ \wedge_G KL \longrightarrow KL/G \longrightarrow KF.$$

At the level of plus constructions, the hypertransfer boils down to inclusion of the semi-direct product of $GL_n L$ and G into $GL_{dn} F$, where $d = [L : F]$. For a rigorous account of all this, see ([**42**], Example 2.30), and [**22**]. We note that a dual discussion applies to extension of scalars $i : KF \longrightarrow KL$, so that the augmentation $KF \longrightarrow \mathbb{H}^{\cdot}(G; KL)$ could just as well be called "hyperextension" i_h.

The hypertransfer has the following basic properties, each of which will be used at a crucial point in the proof of the main theorem below.

- **Invariance under base change.** If F' is an arbitrary field extension of F and $L' = L \otimes_F F'$, then the following diagram commutes in the stable category:

$$\begin{array}{ccc} \mathbb{H}.(G; KL) & \longrightarrow & \mathbb{H}.(G; KL') \\ \downarrow & & \downarrow \\ KF & \longrightarrow & KF' \end{array}$$

  Here the vertical maps are the hypertransfers; in the application L' will in fact be a field, but the hypertransfer could have been defined more generally.
- **Norm formula**. The composite

$$\mathbb{H}.(G; KL) \xrightarrow{t_h} KF \xrightarrow{i_h} \mathbb{H}^{\cdot}(G; KL)$$

  is the hypernorm.
- **Linearity**. The hypertransfer is a map of KF-module spectra (with respect to the obvious module structure on $\mathbb{H}.(G; KL)$).

5.5 Outline of the proof. The proof is based on the two key Lemmas 5.5 and 5.6, whose proofs are discussed in the final subsection. A *Bott inductor* for a finite Galois extension L/E is an element $\tilde{\beta} \in \pi_2 \mathbb{H}.(G; kL)$ such that $t_h(\tilde{\beta}) = \beta$. An arbitrary Galois extension M/E *admits Bott inductors* if for every finite Galois subextension a Bott inductor exists.

Lemma 5.6 *Let M/E be a Galois extension such that either (a)* $M = E_{sep}$ *and* $cd_\ell E \leq 1$*; or (b)* $M = E_\infty$ *and* $cd_\ell E_\infty \leq 1$*. Then M/E admits Bott inductors.*

Lemma 5.7 *Let M/E be a Galois extension such that* $cd_\ell(M/E) < \infty$ *and M/E admits Bott inductors. Then* $\beta^{-1}k$ *satisfies descent on M/E:*

$$\beta^{-1}kE \xrightarrow{\cong} \beta^{-1}\mathbb{H}^{\cdot}(M/E; k).$$

We also need one more general fact about hypercohomology. Consider a sequence of Galois extensions $F \subset M \subset N$, and a presheaf of spectra X on N/F. Then there is a weak equivalence

$$\mathbb{H}^{\cdot}(N/F; X) \xrightarrow{\cong} \mathbb{H}^{\cdot}(M/F; \mathbb{H}^{\cdot}(N/-; X)).$$

This can be seen in several ways. The simplest is to observe that the coefficient presheaf on the right is the direct image of $\mathbb{H}^{\cdot}(N/-; X)$ under the site morphism $G(N/F)_{fsets} \longrightarrow G(M/F)_{fsets}$; then one can use the fact that direct image functors preserve globally fibrant presheaves (Proposition 3.14). It is also easily proved directly in terms of homotopy fixed point sets (Section 1). Using this we obtain:

Lemma 5.8 *Suppose M/F has a filtration $F = M_0 \subset M_1 \subset ... \subset M_n$ with each layer M_{i+1}/M_i Galois and $cd_\ell M_{i+1}/M_i < \infty$, and suppose X is a presheaf of spectra on M_n/F satisfying*

$$X_{M_k} \xrightarrow{\cong} \mathbb{H}^{\cdot}(M_{k+1}/M_k; X)$$

for all k. Then $X \xrightarrow{\cong} \mathbb{H}^{\cdot}(F; X)$.

Proof We proceed by descending induction over the filtration. At the inductive step, there is a commutative diagram

$$\begin{array}{ccc} X_{M_k} & \longrightarrow & \mathbb{H}^{\cdot}(M_k; X) \\ \downarrow & & \downarrow \\ \mathbb{H}^{\cdot}(M_{k+1}/M_k; X) & \longrightarrow & \mathbb{H}^{\cdot}(M_{k+1}/M_k; \mathbb{H}^{\cdot}(M/-; X)) \end{array}$$

The left-hand map is a weak equivalence by assumption, and the right-hand map is a weak equivalence by the remarks above. The bottom map is a weak equivalence because by inductive hypothesis the map on stalks is a weak equivalence.

Now suppose F is an ℓ-good field containing ℓ-th roots of unity. We suppose that F has finite transcendence degree over a field F_0 with $cd_\ell F_0 = 1$; the proof in the other case is quite similar. Replacing F_0 by its algebraic closure in F if necessary, we can assume F is purely transcendental over F_0. Furthermore we can assume F is a separable algebraic extension of $E = F_0(t_1, ..., t_n)$ for some n (since both mod ℓ K-theory and étale hypercohomology are invariant under purely inseparable extension), and moreover that F is finite over E (by a direct limit argument). In other words, we can assume F is a "function field" of dimension n over F_0.

Now any such function field admits a "Tate-Tsen filtration" - that is, a filtration of F_{sep}/F of the form $F = M_0 \subset M_1 \subset ... \subset M_n$ in which each layer M_{i+1}/M_i is induced from an extension of the form N_{sep}/N with $cd_\ell N \leq 1$. For example, if $F = F_0(t)$ we can take the filtration $F_0(t) \subset (F_0)_{sep}(t) \subset F_{sep}$, in which the first layer is induced from $F_0 \subset (F_0)_{sep}$ and the second layer has $cd_\ell \leq 1$ by Tsen's theorem. A similar but somewhat more complicated argument constructs the filtration in the general case.

The extensions N_{sep}/N admit Bott inductors by Lemma 5.6. Since the hyper-transfer commutes with base change, each layer M_{i+1}/M_i therefore admits Bott inductors. Hence $X_{M_k} \xrightarrow{\cong} \mathbb{H}^{\cdot}(M_{k+1}/M_k; X)$ by Lemma 5.7. Then Lemma 5.8 completes the proof.

5.6 The key lemmas. Suppose $cd_\ell F \leq 2$, and consider the map $\eta : kF \longrightarrow \mathbb{H}^{\cdot}(F; k)$ in low degrees. One can show:

- There is a short exact sequence

$$0 \longrightarrow H^2(F; \mathbb{Z}/\ell) \longrightarrow \pi_0 \mathbb{H}^{\cdot}(F; k) \longrightarrow \mathbb{Z}/\ell \longrightarrow 0$$

  and $\mathbb{Z}/\ell = \pi_0 kF \longrightarrow \pi_0 \mathbb{H}^{\cdot}(F; k)$ maps onto the quotient term $\mathbb{Z}/\ell$. Since F contains the ℓ-th roots of unity, the norm residue symbol gives an isomorphism from $H^2(F; \mathbb{Z}/\ell)$ to the ℓ-torsion in the Brauer group of F.

- $\pi_1 \mathbb{H}^{\cdot}(F;k) = H^1(F;\mathbb{Z}/\ell) = F^{\times}/\ell$ by Hilbert's Theorem 90, and $\pi_1(\eta)$ is an isomorphism.
- Multiplication by β is an isomorphism $\pi_0 \mathbb{H}^{\cdot}(F;k) \xrightarrow{\cong} \pi_2 \mathbb{H}^{\cdot}(F;k)$. By the Mercurjev-Suslin theorem [**28**], $\pi_2(\eta)$ is an isomorphism.
- $\pi_n kF = 0$ for $n < 0$, but $\pi_{-1}\mathbb{H}^{\cdot}(F;k) \cong \pi_1 \mathbb{H}^{\cdot}(F;k)$ and $\pi_{-2}\mathbb{H}^{\cdot}(F;k) \cong \pi_0 \mathbb{H}^{\cdot}(F;k)$.

Now consider the commutative diagram in the stable category

$$\begin{array}{ccc}
\mathbb{H}.(G;kL) & \xrightarrow{\mathbb{H}.(G;\eta)} & \mathbb{H}.(G;\mathbb{H}^{\cdot}(L;k)) \\
\tau_h \Big\downarrow & & \Big\downarrow N_h \\
kF & \longrightarrow & \mathbb{H}^{\cdot}(G;\mathbb{H}^{\cdot}(L;k))
\end{array}$$

The bottom map is the composite

$$kF \longrightarrow \mathbb{H}^{\cdot}(G;kL) \xrightarrow{\cong} \mathbb{H}^{\cdot}(G;\mathbb{H}^{\cdot}(L;k))$$

One might hope to prove Lemma 5.6 as follows. If we pretend for a moment that $\pi_*(\eta)$ is an isomorphism through degree 2, then the same is true for $\mathbb{H}.(G;\eta)$, using the hyperhomology spectral sequence. Since N_h is a weak equivalence by the spectrum-level Tate theorem, it follows that $\pi_2(\tau_h)$ is an isomorphism. Hence we have a (unique) Bott inductor.

Unfortunately, as noted above, we do not have an isomorphism through degree 2. Even when $cd_{\ell}F \leq 1$, the group $\pi_{-1}\mathbb{H}^{\cdot}(F;k)$ gets in the way. In this case, however, the problem is easily circumvented by replacing the presheaf k by its 0-connected cover. The reader can then easily complete the proof of Lemma 5.6 when $cd_{\ell}F \leq 1$.

If $cd_{\ell}F = 2$, then even the modified argument fails. The problem is with the Brauer group term in $\pi_0\mathbb{H}^{\cdot}(F;k)$. This term can be eliminated by passing to more highly connected covers of k, but then we only succeed in eliminating the very group we are attempting to analyze! Nevertheless, there is a similar - but much more delicate - argument that carries the day ([**22**], Theorem 7.16). One has to pick out just the right connected covers and Postnikov sections to isolate the Bott element. Of course, the Mercurjev-Suslin theorem again plays a crucial role.

We now turn to the proof of Lemma 5.7. We wish to show that $\eta : kF \longrightarrow \mathbb{H}^{\cdot}(M/F;k)$ induces an isomorphism on $\beta^{-1}\pi_*$. Let us first show that it induces an injection. Let L/F be a finite Galois subextension with group G. Fix a Postnikov section $P^n k$ of the presheaf k. Note that $P^n k$ is a presheaf of kF-module spectra.

Lemma 5.9 *There is a map $\phi : \mathbb{H}^{\cdot}(G;P^n k) \longrightarrow \Omega^2 P^n kF$ such that the composite*

$$P^n kF \longrightarrow \mathbb{H}^{\cdot}(G;P^n k) \xrightarrow{\phi} \Omega^2 P^n kF$$

is homotopic to multiplication by β.

In fact the lemma holds with $P^n k$ replaced by k itself. The point of passing to a Postnikov section is that then

$$colim_L \mathbb{H}^{\cdot}(L/F; P^n k) \cong \mathbb{H}^{\cdot}(M/F); P^n k).$$

by Example 3.22. To show that $\beta^{-1}\eta$ is injective, we will show that if $a \in \pi_* k$ and $\eta_* a = 0$, then $\beta a = 0$. For this purpose we may replace k by a Postnikov section $P^n k$, $n >> 0$, and then it follows immediately from the lemma that $\beta a = 0$.

To construct ϕ, consider the composition θ of multiplication and the transfer:

$$kL \wedge P^n kL \longrightarrow P^n kL \longrightarrow P^n kF.$$

Taking adjoints yields a G-equivariant map $P^n kL \longrightarrow \mathcal{F}(kL, P^n kF)$, where $\mathcal{F}(-,-)$ denotes the function spectrum. We then define ϕ to be the composite

$$\mathbb{H}^{\cdot}(G; P^n kL) \longrightarrow \mathbb{H}^{\cdot}(G; \mathcal{F}(kL, P^n kF)) \cong \mathcal{F}(\mathbb{H}_{\cdot}(G; kL), P^n kF) \xrightarrow{\tilde{\beta}^*} \Omega^2 P^n kF$$

where the first map is induced by the coefficient map θ, the second is the universal coefficient equivalence, and the last map is pre-composition with a Bott inductor $\tilde{\beta}$. Using linearity of the hypertransfer, it is not hard to show that the composite map of the lemma is multiplication by β.

The proof of surjectivity is similar. Essentially, one shows that the composite $\mathbb{H}^{\cdot}(G; P^n k) \longrightarrow \Omega^2 P^n kF \longrightarrow \Omega^2 \mathbb{H}^{\cdot}(G; P^n k)$ is also multiplication by β; it then follows that for all $b \in \pi_* \mathbb{H}^{\cdot}(M/F; k)$, βb is in the image of η.

Acknowledgements

I would especially like to thank Rick Jardine for conversations and correspondence, particularly concerning his book [**22**], on which much of this paper is based. Thanks are also due to Paul Goerss, for sharing his expertise on closed model categories and related matters, and to Bill Dwyer for helpful conversations on the material of Section 5.

References

[1] Artin, M. [1962], *Grothendieck Topologies*, Harvard University lecture notes.

[2] Artin, M. [1971], *On the joins of Hensel rings*, Advances in Math **7**, 282-296.

[3] Artin, M., Grothendieck, A. and Verdier, J.-L. [1972/73], *Théorie des topos et cohomologie étale des schemas (SGA 4)*, Lecture Notes in Math 269, 270, 305, Springer, Berlin-Heidelberg-New York.

[4] Boardman, J. M. [1981], *Conditionally convergent spectral sequences*, unpublished manuscript.

[5] Bousfield, A. K. and Friedlander, E. [1978], *Homotopy theory of* Γ*-spaces, spectra, and bisimplicial sets*, Springer Lecture Notes in Math. **658**, 80-130.

[6] Bousfield, A. K. and Kan, D. [1987], *Homotopy limits, completions and localizations*, Springer Lecture Notes in Math., Vol. 304 (2nd corrected printing), Springer-Verlag, Berlin-Heidelberg-New York.

[7] Brown, K. S. and Gersten, S. M. [1973], *Algebraic K-theory as generalized cohomology theory*, Springer Lecture Notes in Math. **341**, 266-292.

[8] Dwyer, W. and Friedlander, E. [1985], *Algebraic and etale K-theory*, Trans. Amer. Math. Soc. **292**, 247-280.

[9] Dwyer, W. and Mitchell, S. [1993], *On the K-theory spectrum of a ring of algebraic integers*, preprint.

[10] Dwyer, W. and Mitchell, S., *On the K-theory spectrum of a smooth curve over a finite field*, Topology, to appear.

[11] Dwyer, W. and Spalinski [1995], *Homotopy theories and model categories*, in Handbook of Algebraic Topology, (I. M. James, ed.), Elsevier Science B.V., Amsterdam-New York.

[12] Eisenbud, D. [1995], *Commutative algebra with a view toward algebraic geometry*, Springer-Verlag, New York.

[13] Gabber, O. [1992], *K-theory of Henselian local rings and Henselian pairs*, Contemporary Math. **126**, 59-70.

[14] Gillet, H. and Thomason, R. [1984], *The K-theory of strict hensel local rings and a theorem of Suslin*, J. Pure and Applied Alg. **34**, 241-254.

[15] Goerss, P. [1995], *Homotopy fixed points for Galois groups*, Contemporary Math. **181**, 187-224.

[16] Grothendieck, A. and Dieudonné, J. [1971], *Eléments de Geometrie Algébrique I*, (EGA 1), Springer, Berlin-Heidelberg-New York.

[17] Grothendieck, A. [1966], *Étude local des schémas et des morphismes de schémas (troisieme partie)*, IHES Pub. Math. **28**.

[18] Hartshorne, R. [1977], *Algebraic Geometry*, Graduate Texts in Mathematics, Vol. 52, Springer-Verlag, New York-Heidelberg-Berlin.

[19] Jannsen, U. [1988], *Continuous étale cohomology*, Math. Ann. **280**, 207-245.

[20] Jardine, J. F. [1987], *Simplicial presheaves*, J. Pure Applied Algebra **47**, 35-87.

[21] Jardine, J. F. [1987], *Stable homotopy theory of simplicial presheaves*, Can. J. Math. **39**, 733-747.

[22] Jardine, J. F., *Generalized étale cohomology theories*, book manuscript, Progress in Math. series, Birkhauser, to appear.

[23] Johnstone, P. T. [1977], *Topos theory*, Academic Press, London-New York.

[24] Kahn, B., *Bott elements in algebraic K-theory*, Topology, to appear.

[25] Kato, K. and Seito, S. [1986], *Global class field theory of arithmetic schemes*, Contemporary Math. **55**, 255-332.

[26] MacLane, S. [1971], *Categories for the working mathematician*, Springer-Verlag, New York.

[27] May, J. P. [1992], *Simplicial Objects in Algebraic Topology*, University of Chicago Press, Chicago, (c1967).

[28] Mercurjev, A. S. and Suslin, A. A. [1983], *K-cohomology of Severi-Brauer varieties and the norm residue homomorphism*, Math. USSR Izvestia **21**, 307-340.

[29] Milne, J. S. [1980], *Etale Cohomology*, Princeton University Press, Princeton, New Jersey.

[30] Mitchell, S. A. [1990], *The Morava K-theory of algebraic K-theory spectra*, J. Algebraic K-Theory **3**, 607-626.

[31] Mitchell, S. A. [1994], *On the Lichtenbaum-Quillen conjectures from a stable homotopy-theoretic viewpoint*, in Algebraic Topology and Its Applications, MSRI Publications 27, Springer-Verlag, New York, pp. 163-240.

[32] Mitchell, S. A. [1996], *Class groups and general linear group cohomology for rings of algebraic integers*, Math. Zeit. **223**, 101-122.

[33] Mitchell, S. A., *Units and general linear group cohomology for rings of algebraic integers*, Math. Zeit., to appear.

[34] Nisnevich, Y. A. [1989], *The completely decomposed topology on schemes and associated descent spectral sequences in algebraic K-theory*, Algebraic K-theory: Connections with Geometry and Topology , NATO ASI Series C, vol. 279, Kluwer, Dordrecht, pp. 241-342.

[35] Quillen, D. [1973], *Higher Algebraic K-theory 1*, Springer Lecture Notes in Math. **341**, 85-147.

[36] Quillen, D. [1967], *Homotopical algebra*, Lecture Notes in Math. **43**, Springer, Berlin.

[37] Serre, J. P. [1973], *Cohomologie Galoisienne*, Lecture Notes in Math. **5**, Springer-Verlag.

[38] Suslin, A. [1983], *On the K-theory of algebraically closed fields*, Invent. Math. **73**, 241-245.

[39] Suslin, A. [1984], *On the K-theory of local fields*, J. Pure Appl. Alg. **34**, 301- 318.

[40] Suslin, A. and Voevodsky, V. [1995], *Bloch-Kato conjecture and motivic cohomology with finite coefficients*, preprint.

[41] Tamme, G. [1994], *Introduction to Étale Cohomology*, Springer-Verlag, New York-Heidelberg-Berlin.

[42] Thomason, R. [1985], *Algebraic K-theory and etale cohomology*, Ann. Scient. Ecole Norm. Sup. **13**, 437-552.

[43] Thomason, R. [1989], *Survey of algebraic vs. etale topological K-theory*, Contemporary Math. **83**, 393-444.

[44] Thomason, R. and Trobaugh, T. [1990], *Higher algebraic K-theory of schemes and of derived categories*, The Grothendieck Festschrift, Volume III, Progress in Mathematics v. 88, Birkhauser, pp. 247-436.

[45] Waterhouse, W. [1973], *Profinite groups are Galois groups*, Proc. Amer. Math. Soc. **42**, 639-640.

[46] Wiles, A. [1990], *The Iwasawa conjecture for totally real fields*, Annals of Math. **131**, 493-540.

Fields Institute Communications
Volume **16**, 1997

Kahler Differentials of Certain Cusps

Leslie G. Roberts
Department of Mathematics and Statistics
Queen's University
Kingston, Ontario, Canada K7L 3N6
roberts1@mast.queensu.ca

Abstract. Let $A = k \oplus t^N k[t]$, where k is a field of characteristics not equal to 2. Then I calculate $\Omega^m_{A/k}$ for $m \geq 3$. This, together with previous results, determines the Hilbert function of $\Omega^i_{A/k}$ for all $i \geq 0$.

Introduction

In my talk at the Algebraic K-Theory conference (Lake Louise, December 1987), I described the Hilbert function of $\Omega^1_{A/k}$ where $A = k \oplus t^N k[t]$. Part of my discussion mentioned $\Omega^i_{A/k}$ for $i > 1$. Someone in the audience asked how a higher module of differentials could be non-zero when the Krull dimension of A is one. Before I was able to answer the question, Bob Thomason excitedly explained that this is possible because A is singular. Such active participation in seminars was typical of Bob. Since then I have been able to complete the computation of $\Omega^*_{A/k}$ and would like to describe the result here.

My interest in Ω^* arose from its occurrence in K-Theory. My motivation for studying the cusps $A = k \oplus t^N k[t]$ is that after the planar cusps they are (arguably) both the simplest to describe and at the same time the most extreme. I have often come across ring extensions $B_1 \subseteq B$ where some element $b \in B$ satisfies $b^n \in B_1$ for n sufficiently large (for example in [9], and in a series of papers on subintegrality [5],[6],[7] [10],[11]). The cusps A are useful test rings when studying Ω^* in such cases (for example [8], Section 7).

Kahler Differentials of Cusps

Let k be a field of characteristic 0, N an integer ≥ 3, and $A = k \oplus t^N k[t]$. As a k-algebra, A is generated by $u_i := t^{N+i-1}$ $(1 \leq i \leq N)$. Let $R = k[U_1, ..., U_N]$, where the U_i are indeterminates. Mapping U_i to u_i we obtain that $A \cong k[U_1, ..., U_N]/I$.

1991 *Mathematics Subject Classification.* Primary 13N05; Secondary 14H45..

This work was supported by an NSERC grant. The author would like to thank Sue Geller for helpful discussions.

The ring A is graded by degree$(t) = 1$, equivalently degree$(u_i) = N + i - 1$ $(1 \leq i \leq N)$. Grade $\Omega^*_{A/k}$ by taking the degree of du_i also to be $N + i - 1$, so that d is of degree 0. (Alternatively one might take t to be of degree 2 and d of degree -1 so as to fit in better with the language of differential graded algebras, but I prefer to be consistent with the notation of [8]). Let $M = t^N k[t]$ and $V = M/M^2$. Then V is a graded k-vector space with basis u_i $(1 \leq i \leq N)$, degree$(u_i) = N + i - 1$. My main conclusion (see Theorem 1) is that

$$\Omega^m_{A/k} \cong \Lambda^m V \oplus \Lambda^{m+1} V \text{ for } m \geq 3.$$

The exterior power is of k-vector spaces so that $\Lambda^m V$ has k-basis $\{u_{i_1} \wedge u_{i_2} \wedge ... \wedge u_{i_m}\}$ where $i_1 < i_2 < ... < i_m$, and $u_{i_1} \wedge u_{i_2} \wedge ... \wedge u_{i_m} \in \Lambda^m V$ is identified with the element $u_{i_1} \wedge u_{i_2} \wedge ... \wedge u_{i_m} \in \Omega^m_{A/k}$. The basis element $du_{i_1} \wedge du_{i_2} \wedge ... \wedge du_{i_{m+1}}$ of $\Lambda^{m+1} V$ is identified with the element $u_{i_1} du_{i_2} \wedge ... \wedge du_{i_{m+1}} \in \Omega^m_{A/k}$ giving the copy of $\Lambda^{m+1} V$ that is included in $\Omega^m_{A/k}$. The A-module structure of $\Omega^m_{A/k}$ is what one would expect from this k-basis; namely, M kills $\Lambda^{m+1} V \subseteq \Omega^m_{A/k}$, and if $1 \leq a \leq N$ then

$$u_a du_{i_1} \wedge du_{i_2} \wedge ... \wedge du_{i_m} =$$

$$\begin{cases} (-1)^j u_{i_1} du_{i_2} \wedge ... \wedge du_{i_j} \wedge du_a \wedge du_{i_{j+1}} \wedge ... du_{i_m} & \text{if } a \notin \{i_1, ..., i_m\} \\ \qquad \text{and } i_1 < i_2 < ... < i_j < a < i_{j+1} < ... < i_m; & \\ 0 & \text{if } a \in \{i_1, ..., i_m\}. \end{cases}$$

Since $\Omega^1_{A/k}$ has been determined (as a graded k-vector space) in ([8], Section 8), and the deRham complex of A is exact ([8], 1.2), we also know $\Omega^2_{A/k}$ (at least as a graded k-vector space).

To simplify the notation a bit I will omit the $\wedge$'s. We do have to take care to remember that $du_i du_j = -du_j du_i$ if $i \neq j$, and that $du_i du_i = 0$. Sometimes it will be convenient to use powers of t instead of u's, although then we have to remember that t^i is not an element of A unless $i \geq N$.

The ideal I contains the elements $U_{i_1} U_{i_4} - U_{i_2} U_{i_3}$ where $1 \leq i_1 < i_2 \leq i_3 < i_4 \leq N$ and $i_1 + i_4 = i_2 + i_3$. These are not all the generators of I since I is not homogeneous if we give each U_i degree 1, but these generators are the only ones we need to know explicitly in order to compute $\Omega^m_{A/k}$ for $m \geq 3$. However, in order to regard A as a subring of $k[t]$ during the course of the computation we are implicitly using all the generators of I. If $N = 2$ there are no generators of this type.

I will now restate my result in a more precise form:

Theorem 1 *(a) Let k be a field of characteristic not equal to 2. Let V be a vector space of dimension $N \geq 1$ over k, and let $B = k \oplus V$ (with ring structure given by $(a_1 + v_1)(a_2 + v_2) = a_1 a_2 + a_1 v_2 + a_2 v_1$ for $a_i \in k, v_i \in V$). Then $\Omega^m_{B/k} \cong \Lambda^m V \oplus \Lambda^{m+1} V$ $(m \geq 0)$.*
(b) Let $N \geq 3$ and $A = k \oplus t^N k[t], M = t^N k[t], V = M/M^2$ and $B = k \oplus V$ (as in part (a)). Then the natural surjection $A \longrightarrow A/M^2 \cong B$ induces surjections $\pi : \Omega^m_{A/k} \longrightarrow \Omega^m_{B/k}$ which are isomorphisms if $m \geq 3$. Hence if $m \geq 3$ then $\Omega^m_{A/k} \cong \Lambda^m V \oplus \Lambda^{m+1} V$. Note that V is an N dimensional graded vector space with basis elements of degrees $N, N+1, ..., 2N-1$.

Proof (a) Suppose that V has basis $v_1, ..., v_N$. Then $B \cong k[U_1, U_2, \ldots, U_N]/J = R/J$ as a k-algebra, where $J = (\{U_iU_j\}_{1 \le i \le j \le N})R$ and U_i maps to v_i . By [4, 4.12] $\Omega^*_{B/k} \cong \Omega^*_{R/k}/(J, dJ) \cong (\oplus B dv_{i_1} dv_{i_2} ... dv_{i_m})/D$ where $0 \le m \le N$, $i_1 < i_2 < ... < i_m$ (if $m = 0$ take basis element 1), and D is the submodule generated by the image of dJ, namely by

(i) $\{v_i dv_j + v_j dv_i\}_{i \ne j}$

and

(ii) $\{v_i dv_i\}$.

(We need characteristic $\ne 2$ for (ii), since $U_i^2 \in J$ and $dU_i^2 = 2U_i dU_i$.) As a B-module $\Omega^*_{B/k}$ is generated by $\{dv_{i_1} dv_{i_2} ... dv_{i_m}\}$ where the i_j are distinct $(m \ge 0, 1 \le i \le N)$. From the anticommutativity property of exterior algebras we can assume that $i_1 < i_2 < ... < i_m$. If $a \in \{i_1, i_2, ..., i_m\}$ then we obtain that $v_a dv_{i_1} dv_{i_2} ... dv_{i_m} = 0$ by using a type (ii) generator of dJ. If $a \notin \{i_1, i_2, ..., i_m\}$ we can use a type (i) generator to reorder the subscripts obtaining $v_a dv_{i_1} dv_{i_2} ... dv_{i_m} = (-1)^j v_{i_1} dv_{i_2} ... dv_{i_j} dv_a dv_{i_{j+1}} ... dv_{i_m}$ where $i_1 < i_2 < ... < i_j < a < i_{j+1} < ...... < i_m$. Also we have $v_a v_b dv_{i_1} dv_{i_2} ... dv_{i_m} = 0$, $(1 \le a, b \le N)$ since $V^2 = 0$. This gives us a k-spanning set for $\Omega^m_{B/k} (0 \le m \le N)$, namely $S_{m1} := \{dv_{i_1} dv_{i_2} ... dv_{i_m}\}$ where $1 \le i_1 < i_2 < ... < i_m \le N$ and $S_{m2} := \{v_{i_1} dv_{i_2} ... dv_{i_{m+1}}\}$ where $1 \le i_1 < i_2 < ... < i_{m+1} \le N$. ($S_{01}$ is the singleton $\{1\}$ and S_{02} is the basis $\{v_1, ..., v_N\}$ of V. Also $S_{N2} = \emptyset$.) Write $S_1 = \bigcup_{m=0}^{N} S_{m1}$ and $S_2 = \bigcup_{m=0}^{N-1} S_{m2}$. Now note that $\beta := dv_1 dv_2 ... dv_N \ne 0$ since β does not occur as a non-zero term in any element of (J, dJ). All elements of $S_1 \bigcup S_2$ can be converted to $\pm\beta$ by applying the differential d (if our element lies in S_2) and then multiplying by some element of S_1. Therefore all elements of $S_1 \bigcup S_2$ are non-zero. Suppose we have a linear combination $s := \sum a_i x_i + \sum b_j y_j = 0$ where $a_i, b_j \in k, x_i \in S_1$ and $y_j \in S_2$. Applying d and then multiplying by an appropriate element of S_1 to get $\pm b_j \beta$ we conclude that all the b_j are zero. Then multiplying s by an appropriate element of S_1 to get $\pm a_i \beta$ we conclude that all the a_i are zero. Thus we conclude that $S_1 \bigcup S_2$ is a k-basis of $\Omega^*_{B/k}$. Finally $S_{m1} \bigcup S_{m2}$ is a k-basis for $\Omega^m_{B/k}$. The k-subspace spanned by S_{m1} can be identified with $\Lambda^m V$ and the k-subspace spanned by S_{m2} can be identified with $\Lambda^{m+1} V$, from which part (a) follows.

The case $m = 1$ is ([3], Corollary 3.2).

(b) It is clear that $A/M^2 \cong B$ and that the natural map $\Omega^m_{A/k} \longrightarrow \Omega^m_{B/k}$ is a surjection. I will first prove injectivity for $m = N$. Since $\Omega^N_{B/k}$ is a k-vector space of dimension 1 with basis $dv_1 dv_2 ... dv_N$ we need only prove that in $\Omega^N_{A/k}$ we have $u_a du_1 du_2 ... du_N = 0$ for any $a, 1 \le a \le N$. But from the elements $U_{i_1} U_{i_4} - U_{i_2} U_{i_3}$ of I that are described above, we clearly can choose one of the i_j to be a, say $i_1 = a$. Applying d to $U_{i_1} U_{i_4} - U_{i_2} U_{i_3}$ and multiplying by $dU_1 dU_2 ... dU_N$ (omitting dU_{i_4}) we obtain $u_a du_1 du_2 ... du_N = 0$ in $\Omega^N_{A/k}$ as we wished to show. The other possibilities for $a = i_s$ are handled similarly, except if $N = 3$ and $a = 2$ we will have to use $U_1 U_3 - U_2^2$ and will again need characteristic $\ne 2$. The rest of the proof of injectivity of π for $m \ge 3$ follows from a series of steps (1.1)-(1.4):

Let $T = \{1, 2, ..., N\}$. We now work in $\Omega^*_{A/k}$.

(1.1) Let $i, i+1, i+2$ be three consecutive elements of T. Then

$$u_i du_i du_{i+1} du_{i+2} = 0$$

$$u_{i+1}du_idu_{i+1}du_{i+2} = 0$$
$$u_{i+2}du_idu_{i+1}du_{i+2} = 0.$$

Proof of (1.1) Applying d to $u_{i+1}^2 = u_iu_{i+2}$ we obtain

$$2u_{i+1}du_{i+1} = u_idu_{i+2} + u_{i+2}du_i \qquad (*)$$

Multiplying (*) by du_idu_{i+2} we obtain $-2u_{i+1}du_idu_{i+1}du_{i+2} = 0$ and since $\mathrm{char}(k) \neq 2$ we obtain $u_{i+1}du_idu_{i+1}du_{i+2} = 0$. Multiplying (*) by du_idu_{i+1} we obtain $u_idu_idu_{i+1}du_{i+2} = 0$. We complete the proof of (1.1) by multiplying (*) by $du_{i+1}du_{i+2}$ to obtain $u_{i+2}du_idu_{i+1}du_{i+2} = 0$.

(1.2) Let $i < j < l$ be three distinct elements of T. If a is one of i, j, l then $u_adu_idu_jdu_l = 0$.

Proof of (1.2) Define the spread of $\{i, j, l\}$ to be $l - i$. The proof will be by induction on the spread, $l - i = 2$ being (1.1).

(1.2a) a is either i or l. Define j' so that $i + l = j + j'$. Note that $i < j' < l$ so $j' \in T$. From $u_iu_l = u_ju_{j'}$ we obtain

$$u_idu_l + u_ldu_i = u_jdu_{j'} + u_{j'}du_j \qquad (**)$$

Multiplying (**) by du_jdu_l we obtain $u_ldu_idu_jdu_l = u_jdu_{j'}du_jdu_l$. If $j = j'$ then $u_jdu_{j'}du_jdu_l = 0$ by properties of exterior algebras. If $j \neq j'$ then the spread of $\{j, j', l\}$ is either $l - j$ or $l - j'$, both of which are less than $l - i$ so $u_jdu_{j'}du_jdu_l = 0$ by induction on the spread. We conclude that $u_ldu_idu_jdu_l = 0$. Similarly multiplying (**) by du_idu_j we obtain $u_idu_idu_jdu_l = 0$.

(1.2b) $a = j$. Suppose that $j - i \geq l - j$. Write $2j = i' + l$. Then $i \leq i' < j$ so $i' \in T$ and $u_j^2 = u_{i'}u_l$. Applying d we obtain

$$2u_jdu_j = u_{i'}du_l + u_ldu_{i'} \qquad (***)$$

Multiplying (***) by du_idu_l we obtain $-2u_jdu_idu_jdu_l = -u_ldu_idu_{i'}du_l$. If $i = i'$ then $-u_ldu_idu_{i'}du_l = 0$ by properties of exterior algebras. If $i \neq i'$ then $-u_ldu_idu_{i'}du_l = 0$ by (1.2a). We conclude that $u_jdu_idu_jdu_l = 0$ as was to be shown. The case $j - i < l - j$ is similar, concluding the proof of (1.2).

(1.3) Suppose $i < j < l < m$ are four distinct elements of T. If σ is a permutation of $\{i, j, l, m\}$ then $u_{\sigma(i)}du_{\sigma(j)}du_{\sigma(l)}du_{\sigma(m)} = (-1)^{sgn(\sigma)}u_idu_jdu_ldu_m$.

Proof of (1.3) For a permutation involving only j, l, m this follows from properties of exterior algebras. It then suffices to prove (1.3) for any permutation that brings respectively j, l, or m to the first position. I will now work with powers of t rather than u 's. Hence let $u_i = t^{i_1}, u_j = t^{i_2}, u_l = t^{i_3}, u_m = t^{i_4}$ for $N \leq i_1 < i_2 < i_3 < i_4 \leq 2N - 1$.

(1.3a) bringing l or m to the front. We have $u_idu_jdu_ldu_m = t^{i_1}dt^{i_2}dt^{i_3}dt^{i_4} = t^{i_1}dt^{i_4}dt^{i_2}dt^{i_3} = dt^{i_1+i_4}dt^{i_2}dt^{i_3} - t^{i_4}dt^{i_1}dt^{i_2}dt^{i_3}$. Write $i_1 + i_4 = i_3 + \alpha$ for some α, $i_1 < \alpha < i_4$ (hence $\alpha \in T$). Then $dt^{i_1+i_4}dt^{i_2}dt^{i_3} = dt^{i_3+\alpha}dt^{i_2}dt^{i_3} = t^{\alpha}dt^{i_3}dt^{i_2}dt^{i_3} + t^{i_3}dt^{\alpha}dt^{i_2}dt^{i_3} = 0 + t^{i_3}dt^{\alpha}dt^{i_2}dt^{i_3} = 0$ (by (1.2) or because $\alpha = i_2$

or i_3). Hence $t^{i_1}dt^{i_2}dt^{i_3}dt^{i_4} = -t^{i_4}dt^{i_1}dt^{i_2}dt^{i_3}$ which is (1.3) in this case. Similarly $t^{i_1}dt^{i_2}dt^{i_3}dt^{i_4} = t^{i_3}dt^{i_1}dt^{i_2}dt^{i_4}$.

(1.3b) bringing j to the front. By (1.3a) we have $t^{i_1}dt^{i_2}dt^{i_3}dt^{i_4} - t^{i_4}dt^{i_2}dt^{i_1}dt^{i_3} = dt^{i_2+i_4}dt^{i_1}dt^{i_3} - t^{i_2}dt^{i_4}dt^{i_1}dt^{i_3}$. Write $i_2 + i_4 = i_3 + \gamma$. Then $i_2 < \gamma < i_4$ so $\gamma \in T$ and $dt^{i_2+i_4}dt^{i_1}dt^{i_3} = dt^{i_3+\gamma}dt^{i_1}dt^{i_3} = t^{\gamma}dt^{i_3}dt^{i_1}dt^{i_3} + t^{i_3}dt^{\gamma}dt^{i_1}dt^{i_3} = 0 + t^{i_3}dt^{\gamma}dt^{i_1}dt^{i_3} = 0$ (by (1.2) or because $\gamma = i_3$). Hence $t^{i_1}dt^{i_2}dt^{i_3}dt^{i_4} = -t^{i_2}dt^{i_4}dt^{i_1}dt^{i_3} = -t^{i_2}dt^{i_1}dt^{i_3}dt^{i_4}$ as required.

(1.4) If $i_1 < i_2 < i_3$ and $a \geq 2N$ then $t^a dt^{i_1}dt^{i_2}dt^{i_3} = 0$.

Proof of (1.4) We have $t^a dt^{i_1}dt^{i_2}dt^{i_3} = t^{a-N}t^N dt^{i_1}dt^{i_2}dt^{i_3}$. Note that $a - N \geq N$. If $N = i_1$ then $t^a dt^{i_1}dt^{i_2}dt^{i_3} = 0$. If $N < i_1$ then, by (1.3), $t^N dt^{i_1}dt^{i_2}dt^{i_3} = -t^{i_3}dt^N dt^{i_1}dt^{i_2}$ so $t^{a-N}t^N dt^{i_1}dt^{i_2}dt^{i_3} = -t^{a-N+i_3}dt^N dt^{i_1}dt^{i_2} =$
$-t^{a-N+i_3-N}t^N dt^N dt^{i_1}dt^{i_2} = 0$ by (1.2). Note that $a - N + i_3 - N \geq N$ so we stay within the ring A. Altogether we get $t^a dt^{i_1}dt^{i_2}dt^{i_3} = 0$, completing the proof of (1.4).

It follows from (1.1) to (1.4) that if $m \geq 3$ then $\{du_{i_1}du_{i_2}...du_{i_m}\}$ where $i_1 < i_2 < ... < i_m$, and $\{u_{i_1}du_{i_2}...du_{i_{m+1}}\}$ where $i_1 < i_2 < ... < i_{m+1}$ form a k-spanning set of $\Omega^m_{A/k}$ that is mapped bijectively by π onto the k-basis $S_{m1} \bigcup S_{m2}$ of $\Omega^m_{B/k}$. This implies 1(b), completing the proof of Theorem 1. □

In the course of the proof of Theorem 1 we also proved that $\Omega^m_{A/k}$ has the claimed A-module structure for $m \geq 3$ and that $\Omega^m_{B/k}$ has the analogous B-module structure for all m.

Theorem 1 holds if char$(k) \neq 2$. Now we revert back to char$(k) = 0$. Since A is graded, the deRham complex of A is exact [8, 1.2]. The map $d : \Omega^i_{A/k} \longrightarrow \Omega^{i+1}_{A/k}$ ($i \geq 3$) maps the copy of $\Lambda^{i+1}V$ in $\Omega^i_{A/k}$ isomorphically onto the copy of $\Lambda^{i+1}V$ in $\Omega^{i+1}_{A/k}$ so the exactness is clearly visible at $\Omega^i_{A/k}$ ($i \geq 4$). The lower part of the deRham complex yields an exact sequence $0 \longrightarrow \Omega^1_{A/k}/dA \longrightarrow \Omega^2_{A/k} \longrightarrow \Lambda^3 V \longrightarrow 0$ where $\Lambda^3 V \subseteq \Omega^3_{A/k}$ as in (1b). The Hilbert function of $\Omega^1_{A/k}/dA$ was found explicitly in ([8], 8.5). On the interval $2N < i < 3N$ it is the sequence 1 1 2 2 3 3... repeating each value (except possibly the last) twice. It is symmetric about $3N$, and at $3N$ has value one less than if the sequence stated above were to continue. Outside the interval $[2N+1, 4N-1]$ the Hilbert function of $\Omega^1_{A/k}/dA$ is 0. Let H be the Hilbert function of $\Lambda^m V, 1 \leq m \leq N$. Then $H(n)$ is equal to the number of ways that n can be written as the sum of m distinct integers in the sequence $\{N, N+1, ..., 2N-1\}$. Subtracting N from the first integer, $N+1$ from the second,... we see that $H(n)$ is the number of partitions of $n - f$ as the sum of at most m positive integers each at most $N - m$, where $f = \sum_{i=0}^{m-1}(N+i) = m(2N+m-1)/2$. Thus the Hilbert series $P(t) = \sum H(n)t^n$ is equal to $t^f \begin{bmatrix} N \\ m \end{bmatrix}$ where $\begin{bmatrix} N \\ m \end{bmatrix}$ is a Gaussian polynomial ([1], pp. 33,35), (replace the variable q used in [1] by t). As such, the sequence of coefficients is symmetric and unimodal. It is clear that if $m = 3$ the smallest n such that $H(n) \neq 0$ is $3N + 3$ and the largest is $6N - 6$ corresponding to the sums $N + (N+1) + (N+2)$ and $(2N-1) + (2N-2) + (2N-3)$.

A few examples are:

Example 2 (a) $N = 3$. Then the Hilbert series of $\Omega^1_{A/k}/dA$ is $t^7 + t^8 + t^9 + t^{10} + t^{11}$

and that of $\Lambda^3 V$ is t^{12} so the Hilbert series of $\Omega^2_{A/k}$ is $t^7+t^8+t^9+t^{10}+t^{11}+t^{12}$.
(b) $N=4$. Then the Hilbert series of $\Omega^1_{A/k}/dA$ is $t^9+t^{10}+2t^{11}+t^{12}+2t^{13}+t^{14}+t^{15}$ and that of $\Lambda^3 V$ is $t^{15}+t^{16}+t^{17}+t^{18}$ so the Hilbert series of $\Omega^2_{A/k}$ is $t^9+t^{10}+2t^{11}+t^{12}+2t^{13}+t^{14}+2t^{15}+t^{16}+t^{17}+t^{18}$. □

A k-basis for $\Lambda^2 V$ will form part, but not all, of a basis for $\Omega^1_{A/k}/dA$. For example, in (2a) ($N=3$) the Hilbert series for $\Lambda^2 V$ is $t^7+t^8+t^9$ leaving $t^{10}+t^{11}$ to come from elsewhere. A direct calculation using ([4], 4.12) shows that $t^3dt^3dt^4$ and $t^3dt^3dt^5$ (or $t^4dt^3dt^4$) are non-zero elements in $\Omega^2_{A/k}$. All these elements would be 0 if the analogue of (1b) held for $m=2$.

Example 3 Let C be the affine co-ordinate ring of the co-ordinate axes. That is, $C=k[X_1,...,X_N]/\mathcal{I}$ where $\mathcal{I}$ is generated by the products X_iX_j $(i<j)$. Let M be the ideal of C generated by $\{X_i\}_{1\le i\le N}$. Again M/M^2 is of dimension N over k, and by ([8], Section 3) and (1a) above we have that the surjection $C\longrightarrow C/M^2$ induces an isomorphism $\Omega^i_{C/k}\longrightarrow\Omega^i_{(C/M^2)/k}\cong\Lambda^iV\oplus\Lambda^{i+1}V$ $(i\ge 2)$. □

Example 4 Let $C=k[t^4,t^5,t^7]\cong k[X,Y,Z]/(XY^2-Z^2,X^2Z-Y^3,X^3-YZ)$ (by using CoCoA [2]), where $X\rightarrow t^4, Y\rightarrow t^5, Z\rightarrow t^7$. Using the presentation of ([4], 4.12) we find that $XdXdYdZ\neq 0$ in $\Omega^3_{C/k}$. To get $XdXdYdZ=0$ we would have to have a term X^2, XY, or XZ in some element of the ideal $(XY^2-Z^2,X^2Z-Y^3,X^3-YZ)$, which is clearly not the case. Thus the analogue of (1b) does not always hold for subrings of $k[t]$ generated by monomials in t.

References

[1] Andrews, G. E. [1976] *The Theory of Partitions*, Addison-Wesley.

[2] Armando, E., Giovini, A., Niesi, G. [1991] *CoCoA User's Manual, v1.5*, Dipartimento di Matematica, Universita di Genova.

[3] Cortinas, G., Geller, S. C. and Weibel, C. *The Artinian Berger Conjecture*, Mathematische Zeitschrift (to appear).

[4] Kunz, E. [1986] *Kahler Differentials*, Viehweg.

[5] Reid, L., Roberts, L. G. and Singh, B. [1993] *Finiteness of subintegrality*, Algebraic K-Theory and Algebraic Topology NATO ASI series, Series C (P. G. Goerss and J. F. Jardine, eds.), vol. 407, Kluwer Academic Publishers, Dordrecht, pp. 223–227.

[6] Reid, L., Roberts, L. G. and Singh, B. [1995] *The structure of generic subintegrality* Indian Academy of Sciences Proceedings (Math.Sci) **105**, 1–22.

[7] Reid, L., Roberts, L. G. and Singh, B. [1996] *On weak subintegrality* Journal of Pure and Applied Algebra **114**, 93-109.

[8] Roberts, L. G. [1989] *Kahler differentials and HC_1 of certain graded k-algebras,*, Algebraic K-Theory: Connections with Geometry and Topology, NATO ASI series, Series C (J. F. Jardine and V. P. Snaith, eds.), vol. 279, Kluwer Academic Publishers, Dordrecht, pp. 389–424.

[9] Roberts, L. G. and Singh, B. [1986] *Seminormality and cohomology of projective varieties* J.Alg **103**, 500–519.

[10] Roberts, L. G. and Singh, B. [1993] *Subintegrality, invertible modules and the Picard group* Compositio Math. **85**, 249–279.

[11] Roberts, L. G. and Singh, B. [1994] *Invertible modules and generic subintegrality* J.P.A.A. **95**, 331–351.

Fields Institute Communications
Volume **16**, 1997

Local Fundamental Classes Derived from Higher-Dimensional K-Groups

Victor P. Snaith
Department of Mathematics
McMaster University
Hamilton, Ontario, Canada L8S 4K1
snaith@mcmaster.ca

Abstract. If L/K is a Galois extension of local fields then local class field theory constructs a 2-extension of Galois modules between L^* and $\mathbb{Z}$. Here we construct a correspondingly natural 2-extension relating $K_3^{\text{ind}}(L)$ and $K_2(L)$, which are presumably manifestations of the weight two motivic complexes of [19] [20].

1 Introduction

In ([**27**] Example 1, pp.122-123 (lower numbering)) one finds the construction of a long exact K-theory sequence associated to the twisted polynomial ring in one variable, T, over $\overline{\mathbf{F}}_q$, the algebraic closure of $\mathbf{F}_q$. If one twists the multiplication by F, the Frobenius of $\overline{\mathbf{F}}_q/\mathbf{F}_q$ (i.e., $Tx = F(x)T$ for all $x \in \overline{\mathbf{F}}_q$) then the exact sequence takes the form

$$\ldots \longrightarrow K_i(\mathbf{F}_q) \longrightarrow K_i(\overline{\mathbf{F}}_q) \stackrel{1-F}{\longrightarrow} K_i(\overline{\mathbf{F}}_q) \longrightarrow \ldots .$$

Now suppose that $\mathbf{F}_q$ were replaced by a p-adic local field, L, with residue field $\overline{L} = \mathbf{F}_q$. If L_0 is the maximal unramified extension of L then the Galois group, $G(L_0/L)$, is isomorphic to $G(\overline{\mathbf{F}}_q/\mathbf{F}_q) \cong \hat{\mathbf{Z}}$, topologically generated by the Frobenius, F. Despite the fact that the proof of ([**27**] ibid) does not seem to generalise to the case of L_0/L, a related exact sequence is to be found in the literature concerning local class formations. Namely, ([**28**], p.202; see also [**30**], p.303, [**31**], p.9) there exists an exact sequence of the form

$$0 \longrightarrow L^* \longrightarrow L_0^* \stackrel{1/F}{\longrightarrow} L_0^* \longrightarrow \mathbf{Z} \longrightarrow 0$$

which resembles the 0- and 1-dimensional part of the previous sequence, since $K_1(L^*) \cong L^*$, $K_1(L_0^*) \cong L_0^*$ and $K_0(L) \cong \mathbf{Z}$.

The local field sequence has some important applications. When L/K is a totally ramified Galois extension of p-adic local fields with group, $G(L/K)$, the sequence represents the inverse of the fundamental class of local class field theory in

1991 *Mathematics Subject Classification.* Primary 11R37; Secondary 19F05.
Research partially supported by an NSERC grant.

$Ext_{\mathbf{Z}[G(L/K)]}(\mathbf{Z}, L^*) \cong H^2(G(L/K); L^*)$. In fact, this sequence may easily be extended to provide the local fundamental class in $H^2(G(L/K); L^*)$ for any Galois extension of p-adic local fields ([**5**], [**30**], Section 7.1.8, [**31**], Section 1.2).

A second application is to the construction of the second Chinburg invariant, $\Omega(L/K.2)$. Let $\mathcal{O}_K$ denote the ring of integers in a local or global field, K. Let L/K be a finite Galois extension of number fields. In [**5**] three class-group invariants, $\Omega(L/K, i) \in \mathcal{CL}(\mathbf{Z}[G(L/K)])$, were constructed. Here $\mathcal{CL}(\mathbf{Z}[G(L/K)])$ denotes the class group of finitely generated, locally free $\mathbf{Z}[G(L/K)]$-modules. For each finite prime, $P \triangleleft \mathcal{O}_K$, choose a prime lying over it, $Q \triangleleft \mathcal{O}_L$, so that the decomposition group of Q is isomorphic to $G(L_Q/K_P)$. We shall say that P is *tame* if L_Q/K_P is tamely ramified (i.e., $G_1(L_Q/K_P) = \{1\}$) and that P is *wild* otherwise. By a theorem of E. Noether [**11**], if P is tame then $\mathcal{O}_{L_Q}$ is a free $\mathcal{O}_{K_P}[G(L_Q/K_P)]$-module of rank one. Therefore we may choose an adèle

$$(a_P) = a \in \prod_P \mathcal{O}_{L_Q}$$

(the product being taken over finite primes of K with Q being the chosen prime over P) such that

(i) $a_P \in \mathcal{O}_{L_Q}$, $K_P[G(L_Q/K_P)]a_P = L_Q$ for each P, and
(ii) $\mathcal{O}_{K_P}[G(L_Q/K_P)]a_P = \mathcal{O}_{L_Q}$ for each tame P.

There are isomorphisms of the form

$$L \otimes_K K_P \cong \prod_{R|P} L_R \cong Ind_{G(L_Q/K_P)}^{G(L/K)}(L_Q)$$

and

$$\mathcal{O}_L \otimes_{\mathcal{O}_K} \mathcal{O}_{K_P} \cong \prod_{R|P} \mathcal{O}_{L_R} \cong Ind_{G(L_Q/K_P)}^{G(L/K)}(\mathcal{O}_{L_Q})$$

so that

(i) $K_P[G(L/K)]a_P \cong \prod_{R|P} L_R$ for each P, and
(ii) $\mathcal{O}_{K_P}[G(L/K)]a_P \cong \prod_{R|P} \mathcal{O}_{L_R}$ for each tame P.

We shall abbreviate $\mathcal{O}_{K_P}[G(L_Q/K_P)]a_P$ to X_Q and set $X = \mathcal{O}_K[G(L/K)]a$. This is to be interpreted as meaning that X is the intersection of L with the product of its P-completions, X_P, where both are considered as subgroups of the adèles. Hence X is a locally free $\mathcal{O}_K[G(L/K)]$-module whose P-completion is

$$X_P = \mathcal{O}_{K_P}[G(L/K)]a_P = Ind_{G(L_Q/K_P)}^{G(L/K)}(X_Q).$$

In addition we shall assume henceforth (replacing X by mX for a suitable integer, $m \in \mathbf{Z}$, if necessary) that the Q-adic exponential defines an isomorphism

$$exp : X_Q \xrightarrow{\cong} 1 + X_Q \subset \mathcal{O}_{L_Q}^*$$

for all wild L_Q/K_P. Since X is locally free, it is cohomologically trivial and so also is X_Q for each Q. Hence X defines a class, $X \in \mathcal{CL}(\mathbf{Z}[G(L/K)])$. For each L_Q/K_P we may choose a 2-extension of $\mathbf{Z}[G(L_Q/K_P)]$-modules

$$0 \longrightarrow L_Q^* \longrightarrow A \longrightarrow B \longrightarrow \mathbf{Z} \longrightarrow 0,$$

representing the local fundamental class in $H^2(G(L_Q/K_P); L_Q^*)$. In addition, A and B may be chosen to be cohomologically trivial with B finitely generated and

A finitely generated over $\mathbf{Z}_p[G(L_Q/K_P)]$, p being the rational prime below P. Therefore, in the 2-extension,

$$L_Q^*/(1+X_Q) \longrightarrow A/(1+X_Q) \longrightarrow B \longrightarrow \mathbf{Z}$$

the $\mathbf{Z}[G(L_Q/K_P)]$-modules, $A/(1+X_Q)$ and B are finitely generated and cohomolically trivial and hence define elements in the class-group. Define

$$\Omega(L_Q/K_P, 1+X_Q)) = [A/(1+X_Q)] - [B] \in \mathcal{CL}(\mathbf{Z}_p[G(L_Q/K_P)]).$$

This local Euler characteristic depends on the choices. However, if

$$Ind_{G(L_Q/K_P)}^{G(L/K)} : \mathcal{CL}(\mathbf{Z}_p[G(L_Q/K_P)]) \longrightarrow \mathcal{CL}(\mathbf{Z}_p[G(L/K)])$$

is the canonical homomorphism then the global Euler characteristic

$$\Omega(L/K, 2) = [X] + \sum_P Ind_{G(L_Q/K_P)}^{G(L/K)}(\Omega(L_Q/K_P, 1+X_Q))$$

is well-defined in $\mathcal{CL}(\mathbf{Z}[G(L/K)])$. In fact, the local terms in the sum are possibly non-zero only when P is wild and the construction is independent of the choices for X_Q ([**30**], Section 7.2.12, [**31**], Section 2.2.4).

Henceforth we shall denote $\Omega(L/K, 2)$ by $\Omega_0(L/K, 2)$ in order to distinguish it from the analogous invariant, $\Omega_1(L/K, 2)$, whose definition we are about to sketch and in order to suggest that these two invariants form the start of a sequence. In fact, such a sequence does exist for the Chinburg invariant, $\Omega(L/K, 3)$ of [**5**], which we shall denote by $\Omega_0(L/K, 3)$. In ([**6**]; see also [**31**] Chapter 7) a series of invariants, $\Omega_n(L/K, 3)$, are introduced which are related both to the algebraic K-groups of $\mathcal{O}_L$ in dimensions $2n, 2n+1$ and to L-function values at $s = -n$.

We are now in a position to state the main result of this paper, which is a consequence of Theorem 3.14, Proposition 4.2 and Theorem 4.6.

Theorem 1.1 *Let L/K be a Galois extension of p-adic local fields with group, $G(L/K)$. If $r = [G_0(L/K) : G_1(L/K)]$, let $\xi_r \in L_0$ be a primitive r-th root unity. Then, dependent upon the choice of ξ_r, there is a canonical 2-extension of $\mathbf{Z}[G(L/K)]$-modules of the form*

$$K_3^{ind}(L) \longrightarrow \tilde{\oplus}_{i=1}^{d} K_3^{ind}(L_0) \longrightarrow \tilde{\mathcal{W}} \longrightarrow Tors(K_2(L)).$$

(Note that the right-hand module is isomorphic to $\mu(L)$, by Proposition 2.1) such that

(i) *the groups $\tilde{\oplus}_{i=1}^{d} K_3^{ind}(L_0)$ and $\tilde{\mathcal{W}}$ are cohomologically trivial $\mathbf{Z}[G(L/K)]$-modules,*

(ii) *the equivalence class of the 2-extension is natural with respect to passage to subgroups, $G(L/N) \subseteq G(L/K)$, and to quotient groups, $G(N/K)$, by taking $G(L/N)$-fixed points when $G(L/N) \triangleleft G(L/K)$,*

(iii) *changing the choice of ξ_r alters the equivalence class of the 2-extension by the action of an automorphism of $K_3^{ind}(L)$.*

Corollary 1.2 *Let L/K be a Galois extension of p-adic local fields with group, $G(L/K)$. Then, for all i, cup-product with the 2-extension of Theorem 1.1 induces an isomorphism of Tate cohomology groups of the form*

$$\hat{H}^i(G(L/K); \mu(L)) \longrightarrow \hat{H}^{i+2}(G(L/K); K_3^{ind}(L)).$$

In order to apply the fundamental 2-extension of Theorem 1.1 to construct an invariant associated to a Galois extension of number fields, E/F,

$$\Omega_1(E/F, 2) \in \mathcal{CL}(\mathbf{Z}[G(E/F)])$$

one introduces the topological K-group, $K_3^{top}(L)$, of a local field, L, and a regulator homomorphism defined on it. We will sketch the construction here and furnish the missing details in Section 5.

From Theorem 1.1 we have a canonical class in $Ext^2_{\mathbf{Z}[G(L/K)]}(\mu(L), K_3^{ind}(L))$. From the isomorphism $K_3(L) \cong K_3(\mathcal{O}_L)$ we have a canonical map

$$K_3(L) \longrightarrow \lim_{\overleftarrow{s}} K_3(\mathcal{O}_L/(\pi_L^s)) = K_3^{top}(L)$$

inducing ([**15**], Lemma 7.1)

$$\phi_3 : K_3^{ind}(L) \longrightarrow K_3^{top}(L).$$

The target group, $K_3^{top}(L)$, is a finitely generated $\mathbf{Z}_p[G(L/K)]$-module and ϕ_3 induces an isomorphism in p-adic cohomology. In addition, we have

$$\rho : K_3^{ind}(L) \cong K_3^{ind}(\mathcal{O}_L) \longrightarrow K_3(\overline{L})$$

induced by reduction modulo π_L, where $\mathcal{O}_L/(\pi_L) = \overline{L}$ is the residue field of L.

The map

$$(\phi_3, \rho) : K_3^{ind}(L) \longrightarrow K_3^{top}(L) \oplus K_3(\overline{L})$$

induces isomorphisms in Tate cohomology in all dimensions. Hence, pushing out via (ϕ_3, ρ), we obtain a canonical 2-extension of the form

$$K_3^{top}(L) \oplus K_3(\overline{L}) \longrightarrow A \longrightarrow B \longrightarrow \mu(L)$$

in which A and B may be chosen to be cohomologically trivial with B finitely generated over $\mathbf{Z}[G(L/K)]$ and the p-adic part of A is finitely generated over $\mathbf{Z}_p[G(L/K)]$.

Now let E/F be a Galois extension of number fields and let $L = E_Q$. We have canonical isomorphisms, given by étale and syntomic Chern classes, of the form

$$ch_3^{\acute{e}t} : K_3^{top}(L) \xrightarrow{\cong} H^1_{\acute{e}t}(L; \mathbf{Z}_p(2))$$

and

$$K_3^{top}(L) \otimes \mathbf{Q}_p \cong H^1_{\acute{e}t}(L; \mathbf{Q}_p(2)) \xleftarrow{f_M} H^1(\mathcal{O}_L; S_\infty(2)_{\mathbf{Q}_p}) \xrightarrow{\alpha} L$$

such that, if $L/\mathbf{Q}_p$ is unramified, $K_3^{top}(L)/Tors$ is identified with $\mathcal{O}_L \subset L$ for almost all p.

Let X denote the finitely generated, locally free $\mathbf{Z}_p[G(E/F)]$-module used previously in the construction of the second Chinburg invariant, $\Omega_0(E/F, 2)$, (see also [**30**], Section 7.2). By construction, $X_Q \subseteq \mathcal{O}_{E_Q}$ with equality if E/F is tamely ramified at $Q \lhd \mathcal{O}_E$. We may assume that X_Q has been lifted to a p-adic lattice

$$X_Q \subseteq K_3^{top}(E_Q)$$

mapping isomorphically to $X_Q \subseteq \mathcal{O}_{E_Q}$ under the canonical map. Dividing out by X_Q yields a 2-extension of $\mathbf{Z}[G(E_Q/F_P)]$-modules of the form

$$(K_3^{top}(E_Q)/X_Q) \oplus K_3(\mathcal{O}_E/Q) \longrightarrow A/X_Q \longrightarrow B \longrightarrow \mu(E_Q)$$

in which the two middle modules are finitely generated and cohomologically trivial. Hence we may define an Euler characteristic

$$\Omega_1(E_Q/F_P, X_Q) = [A/X_Q] - [B] \in \mathcal{CL}(\mathbf{Z}[G(E_Q/F_P)]).$$

We define

$$\Omega_1(E/F, 2) = [X] + \sum_P Ind_{G(E_Q/F_P)}^{G(E/F)}(\Omega_1(E_Q/F_P, X_Q)) \in \mathcal{CL}(\mathbf{Z}[G(E/F)]).$$

In the sum, Q is a chosen prime over $P \triangleleft \mathcal{O}_F$. As explained in Section 5, only finitely many of the local invariants in the sum are non-zero and $\Omega_1(E/F, 2)$ is independent of all choices involved in the construction.

The existence of the canonical 2-extension has interesting implications for the Galois module structure of K_3^{top}, particularly when $p = 2, 3$. For example, suppose that $L/\mathbf{Q}_3$ is the unique, unramified Galois extension of degree $3n$. There is no 3-primary torsion in $K_2(L)$ so that $\hat{H}^i(\mathbf{Z}/3n; K_3^{top}(L)) = 0$ for all i. However, by [**3**], $K_3^{top}(L) \cong \mathbf{Z}/3 \oplus (\oplus_{j=1}^{3n} \mathbf{Z}_3)$. In fact, the direct sum of $3n - 1$ copies of $\mathbf{Z}_3$ constitute the augmentation ideal, whose cohomology is in odd dimensions. One copy of $\mathbf{Z}_3$ is fixed by the Galois action and has cohomology in even dimensions. The $\mathbf{Z}/3$ has cohomology in all dimensions and the Galois action arranges things so as to kill the cohomology of $K_3^{top}(L)$.

Notice that the module, X, is the same in the definition of $\Omega_0(E/F, 2)$ and $\Omega_1(E/F, 2)$. This means that the difference, $\Omega_0(E/F, 2) - \Omega_1(E/F, 2)$, is equal to the sum of local terms, each of which reflects the relationship between the (classical) K_0/K_1 local fundamental class and the new K_2/K_3 class. It will be interesting to evaluate the local terms, $\Omega_0(E_Q/F_P, X_Q) - \Omega_1(E_Q/F_P, X_Q)$.

The contents of the paper are arranged in the following manner. In Section 2 we collect all the cohomological facts about the K-groups of p-adic local fields, which we shall need in Section 3. In Section 3 we give the main construction of the paper, namely the fundamental 2-extension associated to a Galois extension of p-adic local fields, L/K, and the K-groups, $K_i(L)$ for $i = 2, 3$. In Section 4 we analyse the dependence of the fundamental 2-extension on the choices of elements and generators made in the course of the construction. In Section 5 we complete the promised details concerning the construction and naturality of the class-group invariant, $\Omega_1(E/F, 2)$, whose definition and properties were sketched in this Introduction.

The first form of this paper was written in 1992, using the results of ([**18**],[**21**],[**22**],[**23**],[**24**],[**25**],[**26**], [**37**],[**38**],[**40**])in proofs which resurfaced in greater generality in [**33**]. Both papers appear here by what may be loosely termed "popular request". As the reader who gets that far will see, the second of these papers is made possible by the outstanding results of V. Voevodsky, concerning the 2-adic Quillen-Lichtenbaum conjectures for the K-theory of fields. Since Voevodsky gave some of his first lectures on this material at the Great Lakes K-theory Conference at The Fields Institute in March 1995, the suggestion to include these papers in this volume seemed to me to be quite a good one.

This paper arose out of a project which I was trying to develop in preparation for inclusion in a graduate course to be given in the 1993/4 Fields Institute emphasis year entitled "Number Theory and L-functions". The project was to associate class-group invariants to higher algebraic K-groups of rings of algebraic integers, $\mathcal{O}_E$, in a manner generalising the association of Chinburg invariants, $\Omega_0(E/F, i)$ of [**5**], to $K_0(\mathcal{O}_E)$ and $K_1(\mathcal{O}_E) \cong \mathcal{O}_E^*$. At first sight $\Omega_0(E/F, 2)$ seemed to be the simplest Chinburg invariant to attempt to generalise. As a result this work began, using the results of ([**18**],[**21**],[**22**],[**23**],[**24**],[**25**],[**26**], [**37**],[**38**],[**40**]), before ([**31**], Chapter 7). The latter contains the first steps, developed independently at about the same time by George Pappas, in defining $\Omega_1(E/F, 3)$, based on the many corollaries of ([**18**],[**26**]) which are derived in [**15**]. In ([**6**],[**7**]), $\Omega_n(E/F, 3)$ was defined in general and evaluated in some cases.

Incidentally, I believe that one can compare the fundamental classes (or 2-extensions) of Theorem 1.1 with Lichtenbaum's complexes, $\Gamma(2)$ of ([**19**], [**20**]), for local fields. This suggests that the Galois naturality properties of our 2-extensions should be added to the litany of conjectured properties for the $\Gamma(r)$'s.

My subsequent return to the construction of $\Omega_1(E/F, 2)$ has been considerably facilitated by awareness of the results of [**15**]. I should point out that the methods of proof used here for K-theory in dimensions 2 and 3, unlike those of [**33**], work to prove the analogous results in characteristic p. This is done in [**32**].

In addition I am very grateful to David Burns, Ted Chinburg, David Holland, Manfred Kolster, Ib Madsen, George Pappas and Martin Taylor for helpful correspondence and discussions. In particular, David Burns patiently described to me his generalisation of the second Chinburg invariant. The Burns invariant has a cohomological construction - avoiding K-theory - similar to that of [**6**] and applies to any well-behaved motive endowed with a Galois structure coming from base-change. I believe we now know sufficient about regulators to be able to prove that these invariants agree, when their definition overlap. Incidentally, in Section 5.1 I have recalled the definitions and properties of several p-adic regulators. The reader may well agree with the referee that this is an appalling waste of space. While grateful for the other comments, I have decided to disregard this one (and fortunately the editor concurred!) because the Chinburg invariant construction requires a number of properties of the regulator maps which are used. Therefore I thought it would be beneficial to remind the reader of what was available and to encourage work of comparison of regulators of different ilks.

2 Algebraic K-Theory of Local Fields

Throughout this paper, if A is an abelian group, we shall denote by Tors(A) the torsion subgroup of A. Also we set ${}_nA = \{a \in A \mid na = 0\}$, $A/n = A/nA = A \otimes \mathbf{Z}/n$ and $K_i(X)$ will denote the i-dimensional Quillen K-group of the ring or scheme, X. In this section we collect some cohomological facts about the K-groups of local fields which we will need in Section 3.

Lemma 2.1 *Let L/K be a Galois extension of p-adic local fields with Galois group, $G(L/K)$. Then*

(i) *$G(L/K)$ acts trivially on $Br(L)$, the Brauer group of L, and*
(ii) *there is a $\mathbf{Z}[G(L/K)]$-isomorphism, given by the norm residue symbol,*

$$Tors(K_2(L)) \cong \mu(L)$$

whereas $\mu(L)$ denotes the group of roots of unity in L.

Proof The Brauer group of L is isomorphic to $\mathbf{Q}/\mathbf{Z}$ and part (i) follows from the fact that the norm is a $\mathbf{Z}[G(L/K)]$-isomorphism

$$N_{L/K} : Br(L) \cong \mathbf{Q}/\mathbf{Z} \xrightarrow{\cong} Br(K)$$

and that $G(L/K)$ acts trivially on ${}_mBr(K)$. The fact that the norm is an isomorphism on Brauer groups follows from the commutative diagram ([**28**], p. 193, Proposition 7)

$$\begin{array}{ccc} Br(K) & \xrightarrow{\text{inv}} & \mathbf{Q}/\mathbf{Z} \\ \downarrow{\scriptstyle i_{L/K}} & & \downarrow{\scriptstyle ([L:K]\cdot -)} \\ Br(L) & \xrightarrow{\text{inv}} & \mathbf{Q}/\mathbf{Z} \end{array}$$

in which *inv* is an isomorphism. On the other hand $Br(K) \cong H^2(K; (L^{sep})^*)$ so that

$$Br(K) \stackrel{i_{L/K}}{\longrightarrow} Br(L) \stackrel{N_{L/K}}{\longrightarrow} Br(K)$$

is also multiplication by $[L : K]$ ([**29**], p. 12, Lemma 2.39). Therefore, up to isomorphism, we may identify $N_{L/K}$ with a map, $N : \mathbf{Q}/\mathbf{Z} \longrightarrow \mathbf{Q}/\mathbf{Z}$, such that $N([L : K] \cdot x) = [L : K] \cdot x$ for all $x \in \mathbf{Q}/\mathbf{Z}$. The only such N is the identity map.

For part (ii), let $\mu(L) \cong \mathbf{Z}/m$ then, by [**23**], there is an isomorphism of the form

$$K_2(L) \cong \mathbf{Z}/m \oplus D_L$$

where D_L is a uniquely divisible group. Therefore, via the norm residue isomorphism ([**24**], [**25**]), we have a sequence of $\mathbf{Z}[G(L/K)]$-isomorphisms of the form

$$\mathbf{Z}/m \cong K_2(L)/m \cong H^2(L; \mu_m(2)) \cong H^2(L; \mu_m(1)) \otimes \mu_m(1) \cong \ {}_mBr(L) \otimes \mu_m(1)$$

where $\mu_m(n)$ denotes the n-fold tensor product of the module of m-th roots of unity. By part (i) the Galois action is trivial on ${}_mBr(L) \cong \mathbf{Z}/m$, which completes the proof. □

Lemma 2.2 *Let L/K be an extension of p-adic local fields, not necessarily Galois, of degree $[L : K] = d$. Suppose that $\mu(K) \cong \mathbf{Z}/m$ and $\mu(L) \cong \mathbf{Z}/mn$. Then*

(i) *the norm, $N_{L/K} : Tors(K_2(L) \longrightarrow Tors(K_2(K))$, is surjective, and*
(ii) *identifying these cyclic groups with subgroups of $\mathbf{Q}/\mathbf{Z}$, when L/K is Galois, the natural homomorphism*

$$i_{L/K} : \mathbf{Z}/m \cong Tors(K_2(K)) \longrightarrow Tors(K_2(L)) \cong \mathbf{Z}/mn$$

is given, up to automorphism, by multiplication by a where $\mathrm{Norm}_{L/K}(\Xi_{mn}) = \xi_m^a$ (i.e., $1/m$ is mapped to $a/m = an/mn$). Here ξ_s denotes a primitive s-th root of unity.

Proof We have a commutative diagram in which the middle isomorphisms are given by the norm residue homomorphism ([**24**], [**25**]).

$$\begin{array}{ccccccc} K_2(L)/mn & \longrightarrow & K_2(L)/m & \xrightarrow{\cong} & H^2(L; \mu_m(1)) & \xrightarrow{\text{inv}} & \mathbf{Q}/\mathbf{Z} \\ \downarrow{\scriptstyle N_{L/K}} & & \downarrow{\scriptstyle N_{L/K}} & & \downarrow{\scriptstyle N_{L/K}} & & \downarrow{\scriptstyle \cong} \\ K_2(K)/mn & \xrightarrow{\cong} & K_2(K)/m & \xrightarrow{\cong} & H^2(K; \mu_m(1)) & \xrightarrow{\text{inv}} & \mathbf{Q}/\mathbf{Z} \end{array}$$

Hence the norm, $N_{L/K} : Tors(K_2(L) \longrightarrow Tors(K_2(K))$ is surjective.

To establish part (ii) let $x \in Tors(K_2(K)) \cong K_2(K)/mn$ be a generator. We may choose a generator, $y \in Tors(K_2(L)) \cong K_2(L)/mn$, whose norm is equal to x. Hence $i_{L/K}(x) = i_{L/K}(N_{L/K}(y))$. On the other hand, there are isomorphisms

$$K_2(L)/mn \cong H^2(L;\mu_{mn}(2)) \cong H^2(L;\mu_{mn}(1)) \otimes \mu_{mn}(1)$$

under which y corresponds to $f \otimes \xi_{mn}$ for some generator, $f \in H^2(L;\mu_{mn}(1)) \cong {}_{mn}Br(L)$. In cohomology the composition, $i_{L/K} \cdot N_{L/K}$, is evaluated by the double coset formula ([**29**] p.13 Theorem 2.43) but in the Galois situation the formula simplifies ([**29**], p.14, Corollary 2.45) to the form

$$\begin{aligned} i_{L/K}(N_{L/K}(f \otimes \xi_{mn})) &= \sum_{g \in G(L/K)} g^*(f \otimes \xi_{mn}) \\ &= \sum_{g \in G(L/K)} f \otimes g(\xi_{mn}) \\ &= f \otimes Norm_{L/K}(\xi_{mn}) \\ &= f \otimes \xi_m^a, \end{aligned}$$

since $G(L/K)$ acts trivially on f, by Lemma 2.1(i). In $\mathbf{Q}/\mathbf{Z}$ we may suppose that $y = 1/mn$ and then $i_{L/K}(x)$ corresponds to $a/m = an/mn$, as required. □

2.3 Now let L/K be a totally ramified Galois extension of p-adic local fields with Galois group, $G(L/K)$. Let K_0/K denote the maximal unramified extension and set $L_0 = K_0L$. Let $F \in G(K_0/K) \cong \hat{\mathbf{Z}} \cong G(L_0/L)$ denote the topological generator given by the Frobenius then we have an isomorphism of the form

$$G(L_0/K) \cong G(L/K) \times G(K_0/K) \cong G(L/K) \times \hat{\mathbf{Z}}.$$

For each integer, s, let $K(s) \subset K_0$ denote the fixed field of $s\hat{\mathbf{Z}}$ and set $L(s) = K(s)L$.

Let $K_3^{ind}(L)$ denote the indecomposable K-group in dimension three (i.e., the quotient of $K_3(L)$ by the subgroup generated by triple products of elements of $K_1(L) \cong L^*$). The norm on $K_3(L)$ induces a norm homomorphism on $K_3^{ind}(L)$.

Lemma 2.4 *In the notation of 2.3 the norm*

$$N_{L(s)/L} : K_3^{ind}(L(s)) \longrightarrow K_3^{ind}(L)$$

is surjective if $HCF(p,s) = 1$ and otherwise the cokernel is a cyclic $\mathbf{Z}_p$-module.

Proof If $\mathcal{O}_L$ denotes the ring of integers in L then the localisation exact sequence of ([**27**], Section 5) implies that $K_3(\mathcal{O}_L) \cong K_3(L)$. Let $\overline{L}$ denote the residue field. By ([**37**]; see also [**15**], Section 7) the canonical map

$$K_3(\mathcal{O}_L) \longrightarrow K_3(\overline{L})$$

is a split surjection whose kernel is finitely generated over the p-adic integers (up to an unknown but uniquely divisible part on which the norm is surjective, by a transfer argument). In fact, $K_3(\overline{L})$ is a finite group of order prime to p. However, since $L(s)/L$ is unramified, the norm

$$N_{L(s)/L} : K_3(\mathcal{O}_{L(s)}) \longrightarrow K_3(\mathcal{O}_L)$$

induces the norm on K-groups of residue fields

$$N_{\overline{L(s)}/\overline{L}} : K_3(\overline{L(s)}) \longrightarrow K_3(\overline{L}).$$

Since K_2 of a finite field vanishes and $K_3(\overline{L(s)}) = K_3^{ind}(\overline{L(s)})$, ([**15**]; see Proposition 2.7(proof)) implies that $\hat{H}^i(G(L(s)/L); K_3(\overline{L(s)})) = 0$ for all i. In particular, when $i = 0$ we see that $N_{\overline{L(s)}/\overline{L}}$ is surjective.

It remains to examine the surjectivity of the norm on the p-adic part for which it suffices, by Nakayama's Lemma, to study

$$N_{L(s)/L} : K_3^{ind}(L(s))/p \longrightarrow K_3^{ind}(L)/p.$$

We may assume that L contains ξ_p, a p-th root of unity. For if not then, by a transfer argument, the norm under consideration is a summand of

$$N_{L(s)(\xi_p)/L(\xi_p)} : K_3^{ind}(L(s)(\xi_p))/p \longrightarrow K_3^{ind}(L(\xi_p))/p.$$

By ([**18**], [**26**]) we have an exact sequence of the form

$$0 \longrightarrow K_3^{ind}(L(s)(\xi_p))/p \longrightarrow H^1(L(s)(\xi_p); \mu_p(2)) \longrightarrow \ {}_pK_2(L(s)(\xi_p)) \longrightarrow 0$$

and

$$H^1(L(s)(\xi_p); \mu_p(2)) \cong H^1(L(s)(\xi_p); \mu_p(1)) \cong L(s)(\xi_p)^*/p.$$

Since $L(s)(\xi_p)/L(\xi_p)$ is unramified the p-primary roots of unity in $L(s)(\xi_p)$ all lie in $L(\xi_p)$. Hence the surjection

$$N_{L(s)(\xi_p)/L(\xi_p)} : \ {}_pK_2(L(s)(\xi_p)) \longrightarrow \ {}_pK_2(L(\xi_p))$$

is an isomorphism.

Finally, by class field theory we have an exact sequence of the form

$$L(s)(\xi_p)^* \stackrel{N_{L(s)(\xi_p)/L(\xi_p)}}{\longrightarrow} L(\xi_p)^* \longrightarrow G(L(s)(\xi_p)/L(\xi_p)) \cong \mathbf{Z}/s \longrightarrow 0$$

which means that

$$Coker(N_{L(s)(\xi_p)/L(\xi_p)} : K_3^{ind}(L(s)(\xi_p))/p \longrightarrow K_3^{ind}(L(\xi_p))/p) \cong \mathbf{Z}/s \otimes \mathbf{Z}/p.$$

Hence, if p does not divide s the K-theory norm is surjective. If p divides s then the cokernel is a $\mathbf{Z}_p$-module whose reduction modulo p is isomorphic to $\mathbf{Z}/p$. □

We shall need the following result:

Lemma 2.5 *Let p be a prime. If the direct limit*

$$\lim_{\rightarrow}(\mathbf{Z}/p^{a_1} \longrightarrow \mathbf{Z}/p^{a_2} \longrightarrow \mathbf{Z}/p^{a_3} \longrightarrow \ldots)$$

is divisible then it is either zero or $\mathbf{Q}_p/\mathbf{Z}_p$.

Proof If the limit is not zero then there exists a generator, $x \in \mathbf{Z}/p^{a_m}$ which is non-zero and infinitely p-divisible in the limit. We may as well assume that $m = 1$. Hence there exists m_1 such that $x = p^s y$ in $\mathbf{Z}/p^{a_{m_1}}$. We may as well take $m_1 = 2$. Therefore, if s is maximal then $\mathbf{Z}/p^{a_1} \longrightarrow \mathbf{Z}/p^{a_2}$ may be identified an injection $\mathbf{Z}/p^{a_1} \longrightarrow \mathbf{Z}/p^{a_1+s}$ and y is a generator of the target group. Now we may repeat the argument with x replaced by y and so, by induction, identify the limit with the union of inclusions of cyclic p-groups of monotonically increasing orders. This limit is, of course, $\mathbf{Q}_p/\mathbf{Z}_p$. □

Corollary 2.6 *In 2.4*

$$Coker(N_{L(s)/L} : K_3^{ind}(L(s)) \longrightarrow K_3^{ind}(L))$$

is a finite cyclic p-group, possibly zero. If $\kappa(s) = Ker(N_{L(s)/L})$ and

$$\mathcal{U} = \bigcup_{s=1}^{\infty} \kappa(s) \subset K_3^{ind}(L_0)$$

there is an exact sequence of $\mathbf{Z}[G(L/K)]$*-modules of the form*

$$0 \longrightarrow \mathcal{U} \longrightarrow K_3^{ind}(L_0) \longrightarrow K_3^{ind}(L) \otimes \mathbf{Q} \longrightarrow (\mathbf{Q}_p/\mathbf{Z}_p)^e \longrightarrow 0$$

where $e = 0$ *or* 1.

Proof By a transfer argument $Coker(N_{L(s)/L})$ consists of s-torsion and is therefore a finite p-group of rank one. The direct limit of the exact sequences

$$0 \longrightarrow \kappa(s) \longrightarrow K_3^{ind}(L(s)) \longrightarrow K_3^{ind}(L) \longrightarrow Coker(N_{L(s)/L}) \longrightarrow 0$$

yields the sequence involving $\mathcal{U}$.

Proposition 2.7 *Let* L/K *be a totally ramified Galois extension of* p*-adic local fields with Galois group,* $G(L/K)$ *as in Section 2.3. Then there are isomorphisms*

$$\hat{H}^s(G(L/K); K_3^{ind}(L_0)) \cong \begin{cases} \mathbf{Z}/[G(L/K) : G_1(L/K)] & \text{if } s \text{ is odd,} \\ 0 & \text{if } s \text{ is even.} \end{cases}$$

Here $\hat{H}^s$ *denotes Tate cohomology and* $G_1(L/K)$ *is the first wild ramification group of* L/K.

Proof Consider the map

$$K_2(K_0) \longrightarrow K_2(L_0).$$

Firstly we must evaluate these groups. Since $K(n)/K$ is unramified we must have $\mu(K) \cong \mathbf{Z}/p^a q_0$ and $\mu(K(n)) \cong \mathbf{Z}/p^a q_n$ where p does not divide q_i $(i \geq 0)$. Hence, by Lemma 2.2, $K_2(K(n)) \longrightarrow K_2(K(np^a))$ annihilates the p-primary torsion, since $Norm_{K(np^a)/K(n)}(\xi_{p^a}) = 1$ and so

$$K_2(K_0) \cong lim_{\overrightarrow{n}} K_2(K(n))$$

has no non-trivial p-primary torsion.

Now let l be a prime different from p. If $\overline{K(n)} = \mathbf{F}_q$ with $q = p^e$ the prime-to-p roots of unity in $K(n)$ are isomorphic to $\mathbf{F}_q^* \cong \mathbf{Z}/(q-1)$. Similarly the prime-to-p roots of unity in $K(nm)$ are isomorphic to $\mathbf{F}_{q^m}^* \cong \mathbf{Z}/(q^m - 1)$. Since $K(nm)/K(n)$ is unramified the norm, $\mathbf{F}_{q^m}^* \longrightarrow \mathbf{F}_q^*$, is surjective and Lemma 2.2(ii) implies that $K_2(K(n)) \longrightarrow K_2(K(mn))$ is injective on l-primary torsion. Since this applies to all l different from p we find that

$$Tors(K_2(K_0)) \cong (\mathbf{Q}/\mathbf{Z})[1/p].$$

Next we must evaluate the map

$$Tors(K_2(K_0)) \longrightarrow Tors(K_2(L_0)).$$

This time $\mu(K(n)) \cong \mathbf{Z}/p^a(q-1)$ and $\mu(L(n)) \cong \mathbf{Z}/p^b(q-1)$ since $L(n)/K(n)$ is totally ramified. Hence, by Lemma 2.2(ii), on prime-to-p torsion the natural map is multiplication by $[L(n) : K(n)] = [L : K]$, since $Norm_{L(n)/K(n)}(\xi_{q-1}) = \xi_{q-1}^{[L(n):K(n)]}$.

From [15] we have an exact sequence of the form

$$0 \longrightarrow H^1(G(L/K); K_3^{ind}(L_0)) \longrightarrow K_2(K_0) \longrightarrow$$

$$K_2(L_0)^{G(L/K)} \longrightarrow H^2(G(L/K); K_3^{ind}(L_0)) \longrightarrow 0.$$

The central homomorphism is an isomorphism on the uniquely divisible part and on torsion it is given by the surjection

$$([L : K] \cdot -) : (\mathbf{Q}/\mathbf{Z})[1/p] \longrightarrow (\mathbf{Q}/\mathbf{Z})[1/p].$$

Hence

$$\hat{H}^1(G(L/K); K_3^{ind}(L_0)) \cong (\mathbf{Z}/[L:K])[1/p]$$

and

$$\hat{H}^2(G(L/K); K_3^{ind}(L_0)) \cong 0.$$

When $G(L/K)$ is cyclic this completes the proof, since $\hat{H}^i(G(L/K); M) \cong \hat{H}^{i+2}(G(L/K); M)$ in that case.

In general we use the spectral sequence associated to $G(L/E) \triangleleft G(L/K)$,

$$E_2^{s,t} = \hat{H}^s(G(E/K); H^t(G(L/E); K_3^{ind}(L_0))) \Longrightarrow \hat{H}^{s+t}(G(L/K); K_3^{ind}(L_0)).$$

If $G(L/E)$ is the first wild ramification group, $G(L/E) = G_1(L/K)$ - a p-group, we may replace L/K by L/E and prove by induction that

$$\hat{H}^t(G(L/E); K_3^{ind}(L_0)) = 0$$

for all t so that

$$H^t(G(L/E); K_3^{ind}(L_0)) = \begin{cases} 0 & \text{if } i > 0, \\ K_3^{ind}(E_0) & \text{if } i = 0, \end{cases}$$

since $K_3^{ind}(L_0)^{G(L/E)} \cong K_3^{ind}(L_0^{G(L/E)}) \cong K_3^{ind}(E_0)$. When $G(L/E) = G_1(L/K)$ then the quotient, $G(E/K)$ is cyclic of order prime to p and

$$\begin{aligned} &\hat{H}^s(G(L/K); K_3^{ind}(L_0)) \\ &\cong \hat{H}^s(G(E/K); K_3^{ind}(E_0)) \\ &\cong \begin{cases} 0 & \text{if } s \text{ is even}, \\ \mathbf{Z}/[G(L/K) : G_1(L/K)] & \text{if } s \text{ is odd}, \end{cases} \end{aligned}$$

which completes the proof. □

Proposition 2.8 *Let L/K be a totally ramified Galois extension of p-adic local fields with Galois group, $G(L/K)$ as in Section 2.3. In the notation of Section 2.3 and Corollary 2.6, there is a natural exact sequence of $\mathbf{Z}[G(L/K)]$-modules of the form*

$$0 \longrightarrow K_3^{ind}(L) \longrightarrow K_3^{ind}(L_0) \xrightarrow{1-F} \mathcal{U} \longrightarrow {}_{p^\infty}K_2(L) \longrightarrow 0$$

where ${}_{p^\infty}K_2(L)$ denotes the p-primary torsion of $K_2(L)$.

Proof The proof of Proposition 2.7 shows that

$$\bigcup_n Ker(K_2(L) \longrightarrow K_2(L(n))^{G(L(n)/L)}) = {}_{p^\infty}K_2(L).$$

By [**15**]

$$Ker(K_2(L) \longrightarrow K_2(L(n))^{G(L(n)/L)}) \cong H^1(G(L(n)/L); K_3^{ind}(L(n)))$$

so that

$$\lim_{\vec{n}} H^1(G(L(n)/L); K_3^{ind}(L(n))) \cong {}_{p^\infty}K_2(L).$$

However,

$$H^1(G(L(n)/L); K_3^{ind}(L(n))) \cong \frac{Ker(N_{L(n)/L} : K_3^{ind}(L(n)) \longrightarrow K_3^{ind}(L))}{Im(1-F : K_3^{ind}(L(n)) \longrightarrow K_3^{ind}(L(n)))}$$

from which the exact sequence is obtained by taking direct limits. □

Lemma 2.9 *Let p be a prime and let $\mathbf{Z}/p$ act on $\mathbf{Q}_p/\mathbf{Z}_p$. Hence, if τ is a generator of $\mathbf{Z}/p$ then τ must act trivially when p is odd but may act by multiplication by (-1) when $p = 2$.*

The Tate cohomology groups of this module are given by

$$\hat{H}^{2s}(\mathbf{Z}/p; \mathbf{Q}_p/\mathbf{Z}_p) \cong \begin{cases} \mathbf{Z}/p & \text{if } p = 2 \text{ and } \tau(x) = -x, \\ 0 & \text{if } \tau(x) = x, \end{cases}$$

and

$$\hat{H}^{2s+1}(\mathbf{Z}/p; \mathbf{Q}_p/\mathbf{Z}_p) \cong \begin{cases} \mathbf{Z}/p & \text{if } \tau(x) = x, \\ 0 & \text{if } p = 2 \text{ and } \tau(x) = -x. \end{cases}$$

Proof In all cases the cohomology is computed from a periodic chain complex of the form

$$\ldots \xrightarrow{p} \mathbf{Q}_p/\mathbf{Z}_p \xrightarrow{0} \mathbf{Q}_p/\mathbf{Z}_p \xrightarrow{p} \mathbf{Q}_p/\mathbf{Z}_p \xrightarrow{0} \ldots$$

so that alternate dimensional cohomology groups are zero of cyclic of order p. The only difference in the cases is that the dimension of the zero homomorphism is odd when τ acts non-trivially and even otherwise. □

Theorem 2.10 *Let L/K be a totally ramified Galois extension of p-adic local fields with Galois group, $G(L/K)$, as in Section 2.3. In Corollary 2.6, if $[L : K]$ is divisible by p then $e = 0$.*

Proof If p divides the degree of the extension then we can find a cyclic subgroup, $G(L/E) \subset G(L/K)$, of order p. Therefore we may suppose that $E = K$ and that $G(L/K) \cong \mathbf{Z}/p$ generated by τ.

Suppose that $e = 1$. In this case the exact sequence of Corollary 2.6 implies, together with Proposition 2.7 and Lemma 2.9, that $\hat{H}^i(G(L/K); \mathcal{U})$ is zero in odd dimensions and equal to $\mathbf{Z}/p$ in even dimensions or vice versa.

The exact sequence of Proposition 2.8, together with Lemma 2.1(ii) and Proposition 2.7, yields an exact cohomology sequence of the form

$$\ldots \longrightarrow \hat{H}^{i+1}(G(L/K); K_3^{ind}(L)) \longrightarrow \hat{H}^i(G(L/K); \mathcal{U})$$

$$\longrightarrow \hat{H}^i(G(L/K); \mu_{p^\infty}(L)) \longrightarrow \ldots .$$

In addition, the groups $\hat{H}^i(G(L/K); \mu_{p^\infty}(L))$ will either all be zero or all be cyclic of order p since the Herbrand difference of a cyclic group acting on a finite cyclic group is zero.

Now we shall calculate $H^*(G(L/K); K_3^{ind}(L))$. Since L/K is totally ramified we must have $\mu(K) \cong \mathbf{Z}/p^a q$ and $\mu(L) \cong \mathbf{Z}/p^b q$ where p does not divide q. We shall use the exact sequence

$$0 \longrightarrow H^1(G(L/K); K_3^{ind}(L)) \longrightarrow K_2(K) \longrightarrow$$

$$K_2(L)^{G(L/K)} \longrightarrow H^2(G(L/K); K_3^{ind}(L)) \longrightarrow 0$$

to calculate $H^*(G(L/K); K_3^{ind}(L))$, using Lemma 2.2(ii) to calculate the middle map. When $a = b = 0$ the middle map is an isomorphism and all the cohomology groups are trivial. Now let us consider the cases when $a \geq 1$. In this case either $b = a$ or $b = a+1$. When $b = a$ the middle map is multiplication by p on p-primary torsion and $H^i(G(L/K); K_3^{ind}(L))$ is cyclic of order p in each dimension. When $p = 2$ and $(a, b) = (1, 2)$ the 2-primary part of the middle map is the zero map, $\mathbf{Z}/2 \longrightarrow \mathbf{Z}/2$. Similarly the 2-primary part of the middle map is zero, $\mathbf{Z}/2 \longrightarrow \mathbf{Z}/2$, when $p = 2$, $(a, b) = (1, 3)$ and $\tau(\xi_8) = \xi_8^{-1}$ but when $\tau(\xi_8) = \xi_8^3$ the middle map is an isomorphism. When $p = 2$ and $a \geq 2$ the 2-primary part of the middle map is multiplication by $1 + 2^{a-1}$, $\mathbf{Z}/2^a \longrightarrow \mathbf{Z}/2^a$. Finally, when p is odd and $a \geq 1$ the middle map is again an isomorphism, $\mathbf{Z}/p^a \longrightarrow (\mathbf{Z}/p^{a+1})^{G(L/K)}$. In all cases the finite group, $H^i(G(L/K); K_3^{ind}(L))$, is independent of i. Therefore, if $\hat{H}^i(G(L/K); \mu_{p^\infty}(L)) \cong A$ and $\hat{H}^i(G(L/K); K_3^{ind}(L)) \cong B$ for all i we obtain an exact sequence of the form

$$0 \longrightarrow A \longrightarrow B \longrightarrow \mathbf{Z}/p \longrightarrow A \longrightarrow B \longrightarrow 0.$$

Such an exact sequence is impossible, since it has a non-zero Euler characteristic. □

Lemma 2.11 *Let L/K be a totally ramified Galois extension of p-adic local fields with Galois group, $G(L/K)$, as in Section 2.3. In the situation of Section 2.3, set*

$$\mathcal{O}_{L_0} = \bigcup_n \mathcal{O}_{L(n)} \subset L_0$$

so that the natural map, $K_3^{ind}(\mathcal{O}_{L_0}) \longrightarrow K_3^{ind}(L_0)$, is an isomorphism (c.f. Lemma 2.4(proof)). Then the composition

$$K_3^{ind}(L_0) \cong K_3^{ind}(\mathcal{O}_{L_0}) \longrightarrow K_3(\overline{L_0}) \cong \mathbf{Q}/\mathbf{Z}(2)[1/p]$$

induces a cohomology isomorphism

$$\hat{H}^i(G(L/K); K_3^{ind}(L_0)) \longrightarrow \hat{H}^i(G(L/K); \mathbf{Q}/\mathbf{Z}(2)[1/p]).$$

for each i.

Proof If $G(L/E) = G_1(L/K)$, as in the proof of Proposition 2.7, then we may replace L/K by E/K. In this case we may assume that $G(L/K) \cong \mathbf{Z}/r$ with $HCF(p, r) = 1$. Then, as in the proof of Lemma 2.4, the kernel of the surjection

$$K_3^{ind}(L_0) \longrightarrow \mathbf{Q}/\mathbf{Z}(2)[1/p]$$

is r-divisible and so has trivial Tate cohomology. □

Corollary 2.12 *Let L/K be a totally ramified Galois extension of p-adic local fields with Galois group, $G(L/K)$, as in §2.3. In Proposition 2.8 the inclusion, $\mathcal{U} \subset K_3^{ind}(L_0)$, induces cohomology isomorphisms*

$$\hat{H}^i(G(L/K); \mathcal{U}) \xrightarrow{\cong} \hat{H}^i(G(L/K); K_3^{ind}(L_0))$$

for all i.

In addition,

$$(1 - F)_* : \hat{H}^i(G(L/K); K_3^{ind}(L_0)) \longrightarrow \hat{H}^i(G(L/K); \mathcal{U})$$

is the zero map.

Proof The first part follows from the exact sequence of Corollary 2.6 once we have shown that $\hat{H}^i(G(L/K); (\mathbf{Q}_p/\mathbf{Z}_p)^e) = 0$ for all i. This is clearly so if p does not divide $[L : K]$ and if p does divide $[L : K]$ then $e = 0$ by Theorem 2.10.

For the second part, as in Lemma 2.11, we may replace L/K by E/K where $G_1(L/K) = G(L/E)$. Hence we may assume that $G(L/K) \cong \mathbf{Z}/r$ with $HCF(p, r) = 1$. Also, if $|\overline{L}| = q$ then r divides $q - 1$. Now the non-zero cohomology groups are given by

$$\hat{H}^{2s+1}(G(L/K); \mathbf{Q}/\mathbf{Z}(2)[1/p]) \cong {}_r\mathbf{Q}/\mathbf{Z}(2)[1/p]$$

and on $\mathbf{Q}/\mathbf{Z}(2)[1/p]$ the Frobenius acts like $F(z) = z^{q^2}$. Therefore $F - 1$ acts like "multiplication" by $q^2 - 1 = (q-1)(q+1)$. Therefore $F - 1$ is trivial on the r-torsion of $\mathbf{Q}/\mathbf{Z}(2)[1/p]$, since r divides $q - 1$. □

3 Construction of the Canonical 2-Extension

3.1 *Galois groups in the general case*

Let L/K be a Galois extension of p-adic local fields with Galois group, $G(L/K)$. We shall relax the total ramification condition of §2.3. Set $W = L \cap K_0$ so that W/K is the maximal unramified subextension of L/K. Hence L/W is totally ramified and the sequence

$$0 \longrightarrow K_3^{ind}(L) \longrightarrow K_3^{ind}(L_0) \xrightarrow{1-F} \mathcal{U} \longrightarrow {}_{p^\infty}K_2(L) \longrightarrow 0$$

of Proposition 2.8 is a 2-extension of $\mathbf{Z}[G(L/W)]$-modules.

We wish to construct a similar 2-extension of $\mathbf{Z}[G(L/K)]$-modules. Our method will be to imitate the construction of the local fundamental class as described in ([**30**], Section 7.1; [**31**], Section 1.2).

Notice that $G(L/W) = G_0(L/K)$ and $G_1(L/W) = G_1(L/K)$, by ([**28**], p. 62) so that $r = [G(L/W) : G_1(L/W)] = [G_0(L/K) : G_1(L/K)]$. If $G(L/E) = G_1(L/K)$ then

$$G(L/K)/G(L/E) \cong G(E/K) = \{a, g \mid a^r = 1, g^d = a^c, gag^{-1} = a^v\}$$

where $< a >= G(E/W)$ and $v = |\overline{K}|$.

Now let us consider the Galois groups which will be involved in the construction. We have a map, given by the restriction of g to W and denoted by $(g|W)$, inducing a homomorphism

$$h_1 : G(K_0/K) \times G(L/K) \cong \hat{\mathbf{Z}} \times G(L/K) \longrightarrow G(W/K)$$

defined by $h_1(F^i, z) = (F^i|W) \cdot (z|W)^{-1}$. Here, as in Section 2.3, F denotes the Frobenius automorphism. To a pair, $(F^i, z) \in Ker(h_1)$, we may associate the Galois automorphism of $L_0 = LK_0$ which is equal to F^i on K_0 and to z on L. This induces an isomorphism

$$Ker(h_1) \xrightarrow{\cong} G(L_0/K).$$

If $d = [W : K]$ define $F_0 \in G(L_0/K)$ by $(F_0|K_0) = F^d$ and $(F_0|L) = 1$, the identity map.

As explained in ([**30**], pp. 303-304; [**31**], pp. 9-10), there is an isomorphism of K-algebras

$$\lambda : L \otimes_K K_0 \longrightarrow \oplus_{i=1}^d L_0$$

given by the formula $\lambda(\alpha \otimes \beta) = (F^{d-1}(\beta)\alpha, F^{d-2}(\beta)\alpha, \ldots, \beta\alpha)$ and fitting into a commutative diagram of the form

$$\begin{array}{ccc} L \otimes_K K_0 & \xrightarrow{(1 \otimes F)} & L \otimes_K K_0 \\ \downarrow \lambda & & \lambda \downarrow \\ \oplus_{i=1}^d L_0 & \xrightarrow{\hat{F}} & \oplus_{i=1}^d L_0 \end{array}$$

where $\hat{F}(x_1, \ldots, x_d) = (F_0(x_d), x_1, \ldots, x_{d-1})$. Since K-theory is additive there is an induced isomorphism

$$\lambda : K_s(L \otimes_K K_0) \longrightarrow \oplus_{i=1}^d K_s(L_0)$$

for each $s \geq 0$. This isomorphism induces a $G(L/K)$-action on the right-hand group, induced by the Galois action on L. Since the Frobenius acts on the other factor, K_0, in the tensor product we may define a $\mathbf{Z}[G(L/K)]$-homomorphism

$$\hat{F} : \oplus_{i=1}^d K_s(L_0) \longrightarrow \oplus_{i=1}^d K_s(L_0)$$

by the same formula as before. Therefore the induced map

$$\hat{F} : \oplus_{i=1}^d K_3^{ind}(L_0) \longrightarrow \oplus_{i=1}^d K_3^{ind}(L_0)$$

is also a $\mathbf{Z}[G(L/K)]$-homomorphism.

Lemma 3.2 *Suppose that $g \in G(L/K)$ satisfies $(g|W) = (F^j|W)$ for some $0 \leq j \leq d-1$. Then the action of g on $(u_1, u_2, \ldots, u_d) \in \oplus_{i=1}^d K_3^{ind}(L_0)$ is given by*

$$g(u_1, u_2, \ldots, u_d) = \hat{F}^{-j}((F^j, g)(u_1), (F^j, g)(u_2), \ldots, (F^j, g)(u_d)).$$

Proof By the functoriality of K-groups we need only verify the same formula on $(u_1, u_2, \ldots, u_d) \in \oplus_{i=1}^d L_0$. To do this we first observe that the formula is clear when $j = 0$ and, if $1 \leq j$,

$$\hat{F}^{-j}(u_1, u_2, \ldots, u_d) = (u_{j+1}, u_{j+2}, \ldots, u_d, F_0^{-1}(u_1), F_0^{-1}(u_2), \ldots, F_0^{-1}(u_j)).$$

Therefore

$$\begin{aligned} \hat{F}^{-j}((F^j, g)(\lambda(\alpha \otimes \beta))) &= \hat{F}^{-j}((F^j, g)(F^{d-1}(\beta)\alpha, F^{d-2}(\beta)\alpha, \ldots, \beta\alpha)) \\ &= \hat{F}^{-j}(F^{d+j-1}(\beta)g(\alpha), F^{d+j-2}(\beta)g(\alpha), \ldots, F^j(\beta)g(\alpha)) \\ &= (F^{d-1}(\beta)g(\alpha), F^{d-2}(\beta)g(\alpha), \ldots, \beta g(\alpha)) \\ &= \lambda(g(\alpha) \otimes \beta), \end{aligned}$$

as required. □

Definition 3.3 *Let $\mathcal{U} \subset K_3^{ind}(L_0)$ be as in Corollary 2.6. Define a subgroup, $\mathcal{W} \subset \oplus_{i=1}^d K_3^{ind}(L_0)$ by*

$$\mathcal{W} = \{(u_1, u_2, \ldots, u_d) | \sum_{i=1}^d u_i \in \mathcal{U}\}.$$

Lemma 3.4

(i) *The subgroup, $\mathcal{W} \subset \oplus_{i=1}^d K_3^{ind}(L_0)$, is a $\mathbf{Z}[G(L/K)]$-submodule, and*

(ii) *the image of* $1-\hat{F} : \oplus_{i=1}^{d} K_3^{ind}(L_0) \longrightarrow \oplus_{i=1}^{d} K_3^{ind}(L_0)$ *lies in* $\mathcal{W}$.

Proof For part (i), suppose that $(u_1, u_2, \ldots, u_d) \in \mathcal{W}$. We may suppose that each u_i lies in $K_3^{ind}(L(n))$ and that $N_{L(n)/L}(\sum_{i=1}^{d} u_i) = 0$. Since F_0 is the identity map on L the formula of Lemma 3.2 implies that the norm of the sum of the coordinates of $g(u_1, u_2, \ldots, u_d)$ is equal to

$$N_{L(n)/L}((F^j, g)(\sum_{i=1}^{d} u_i)) = g(N_{L(n)/L}(\sum_{i=1}^{d} u_i)) = 0,$$

as required.

For part (ii), observe that $(1-\hat{F})(u_1, u_2, \ldots, u_d) = (u_1 - F_0(u_d), u_2 - u_1, \ldots, u_d - u_{d-1})$ so that the sum of the coordinates is equal to $(1-F^d)(u_d)$, which lies in $\mathcal{U}$ by Proposition 2.8 applied to L/W. □

Proposition 3.5 *In the notation of Section 3.1, there is a natural exact sequence of* $\mathbf{Z}[G(L/K)]$*-modules of the form*

$$0 \longrightarrow K_3^{ind}(L) \xrightarrow{\Delta} \oplus_{i=1}^{d} K_3^{ind}(L_0) \xrightarrow{1-\hat{F}} \mathcal{W} \xrightarrow{\pi} {}_{p^\infty}K_2(L) \longrightarrow 0$$

where Δ *denotes the diagonal map and* π *is the composition of the map of Proposition 2.8 with the addition of coordinates,* $\mathcal{W} \longrightarrow \mathcal{U}$.

Proof Since the canonical map, $K_3^{ind}(L) \longrightarrow K_3^{ind}(L_0)$, is injective then so is Δ, proving exactness at $K_3^{ind}(L)$.

Clearly, $(1-\hat{F})\Delta(z) = (1-\hat{F})(z, z, \ldots, z) = (z - F_0(z), 0, \ldots, 0)$ is zero for all $z \in K_3^{ind}(L)$. On the other hand, $(z_1, z_2, \ldots, z_d) \in Ker(1-\hat{F})$ if and only if $z_1 = z_2 = \ldots = z_d$ and $F_0(z_d) = z_d$. Since F_0 is the Frobenius of $G(K_0/W)$ the exactness of the sequence of Proposition 2.8 shows that $z_d \in K_3^{ind}(L)$, proving exactness at $\oplus_{i=1}^{d} K_3^{ind}(L_0)$.

Since $\pi(1-\hat{F})(u_1, u_2, \ldots, u_d) = \pi(u_1 - F_0(u_d), u_2 - u_1, \ldots, u_d - u_{d-1})$ is equal to the image of $u_d - F_0(u_d)$ under $\mathcal{U} \longrightarrow {}_{p^\infty}K_2(L)$, the composition $\pi(1-\hat{F})$ is zero, by Proposition 2.8 applied to L/W. Also $\mathcal{W} \longrightarrow {}_{p^\infty}K_2(L)$ is surjective since $\mathcal{U} \longrightarrow {}_{p^\infty}K_2(L)$ is surjective and if $u \in \mathcal{U}$ maps to $x \in {}_{p^\infty}K_2(L)$ then so does $(u, 0, \ldots, 0) \in \mathcal{W}$.

It remains to show that $Ker(\pi) = Im(1-\hat{F})$. If $\pi(u_1, u_2, \ldots, u_d) = 0$ then, by Proposition 2.8 applied to L/W, $\sum_{i=1}^{d} u_i = (1-F_0)(z)$ for some $z \in K_3^{ind}(L_0)$. We wish to solve $(u_1, u_2, \ldots, u_d) = (z_1 - F_0(z_d), z_2 - z_1, \ldots, z_d - z_{d-1})$ for the $\{z_i\}$. Such a solution is given by setting $z_i = F_0(z) + \sum_{j=1}^{i} u_j$, which is consistent since $z_d = F_0(z) + \sum_{j=1}^{d} u_j = z$. □

Theorem 3.6 *With the notation of Section 3.1 the cohomology groups of the* $\mathbf{Z}[G(L/K)]$*-module,* $\oplus_{i=1}^{d} K_3^{ind}(L_0)$*, are given by*

$$H^i(G(L/K); \oplus_{i=1}^{d} K_3^{ind}(L_0)) \cong \begin{cases} \mathbf{Z}/r & \text{if } i \text{ is odd,} \\ 0 & \text{if } 0 < i \text{ is even,} \\ K_3^{ind}(W_0) & \text{if } i = 0. \end{cases}$$

Here $r = [G_0(L/K) : G_1(L/K)]$.

Proof One should compare the following proof with the overly cavalier treatment of the analogous reduction to the totally ramified case which I gave in ([**31**], p. 14)!

Consider the spectral sequence for ordinary group cohomology

$$E_2^{s,t} = H^s(G(W/K); H^t(G(L/W); \oplus_{i=1}^d K_3^{ind}(L_0)))$$
$$\Longrightarrow H^{s+t}(G(L/K); \oplus_{i=1}^d K_3^{ind}(L_0)).$$

As in Section 3.1, let $r = [G(L/W) : G_1(L/W)] = [G_0(L/K) : G_1(L/K)]$. By Lemma 3.2, $G(L/W)$ acts diagonally, component-by-component on $\oplus_{i=1}^d K_3^{ind}(L_0)$ so that, by Proposition 2.7 applied to L/W together with the isomorphism $(K_3^{ind}(L_0))^{G(L_0/W_0)} \cong K_3^{ind}(W_0)$,

$$H^t(G(L/W); \oplus_{i=1}^d K_3^{ind}(L_0)) \cong \begin{cases} \oplus_{i=1}^d \mathbf{Z}/r & \text{if } t \text{ is odd,} \\ 0 & \text{if } 0 < t \text{ is even,} \\ \oplus_{i=1}^d K_3^{ind}(W_0) & \text{if } t = 0. \end{cases}$$

If $g \in G(W/K) \cong \mathbf{Z}/d$ is a generator then the cohomology, $H^*(G(W/K); M)$, is computed from the cochain complex

$$\dots \longrightarrow M \xrightarrow{1-g} M \xrightarrow{N} M \xrightarrow{1-g} M \longrightarrow \dots$$

where $N = \sum_{j=0}^{d-1} g^j$. The generator, g, may be lifted to

$$(F, F) \in G(L_0/K) \subset G(K_0/K) \times G(L/K)$$

in the notation of Section 3.1. By Lemma 3.2, the action on $\oplus_{i=1}^d \mathbf{Z}/r$ is given by

$$\begin{aligned} g(a_1, \dots, a_d) &= (F(a_2), F(a_3), \dots, F(a_d), F_0^{-1}(F(a_1))) \\ &= (F(a_2), F(a_3), \dots, F(a_d), F(a_1)) \\ &= (qa_2, qa_3 \dots qa_d, qa_1) \end{aligned}$$

since F acts on $\mathbf{Z}/r = {}_r(\mathbf{Q}/\mathbf{Z})(2)[1/p]$ by "multiplication" by $q = |\overline{K}|^2$, $F_0 = F^d$ and r divides $q^d - 1$.

Therefore $Ker(1-g)$ consists of d-tuples, $(a_1, \dots, a_d)$, such that $a_i = q^{d-i}a_d$ which comprise a copy of $\mathbf{Z}/r$ generated by $(q^{d-1}, q^{d-2}, \dots, q, 1)$. On the other hand

$$\begin{aligned} N(a_1, \dots, a_d) &= (a_1, a_2, \dots, a_{d-1}, a_d) \\ &\quad + (qa_2, qa_3, \dots, qa_d, qa_1) \\ &\quad + (q^2a_3, q^2a_4, \dots, q^2a_1, q^2a_2) \\ &\quad \vdots \\ &= xq^{1-d}(q^{d-1}, q^{d-2}, \dots, q, 1) \end{aligned}$$

where $x = \sum_{i=1}^d q^{i-1}a_i \in \mathbf{Z}/r$. Therefore $H^{2s}(G(W/K); \oplus_{i=1}^d \mathbf{Z}/r)$ vanishes if $s > 0$ and is isomorphic to $\mathbf{Z}/r$ if $s = 0$.

An element, $(a_1, \dots, a_d)$, lies in $Ker(N)$ if and only if $x = \sum_{i=1}^d q^{i-1}a_i \equiv 0$ (modulo r). However, for such elements, we may solve the equation

$$(a_1, \dots, a_d) = (1-g)(b_1, \dots, b_d) = (b_1 - qb_2, b_2 - qb_3, \dots, b_d - qb_1)$$

by choosing b_1 and b_d to satisfy $b_d - qb_1 = a_d$ and then setting $b_i = q^{d-i+1}b_1 + \sum_{j=i}^d q^{j-1}a_j$ for $2 \leq i \leq d-1$. This is consistent because, in $\mathbf{Z}/r$, $a_1 + qb_2 = (a_1 + qa_2 + q^2a_3 + \ldots + q^{d-1}a_d) + q^d b_1 = b_1$. Hence $H^{2s+1}(G(W/K); \oplus_{i=1}^d \mathbf{Z}/r)$ vanishes for all $s \geq 0$.

Now let us consider $H^s(G(W/K); \oplus_{i=1}^d K_3^{ind}(W_0))$. Since $F_0 = F^d$ acts trivially on W_0 the action on $(u_1, \ldots, u_d) \in \oplus_{i=1}^d K_3^{ind}(W_0)$ is given by $g(u_1, \ldots, u_d) = (F(u_2), F(u_3), \ldots, F(u_d), F(u_1))$. Therefore $(u_1, \ldots, u_d) \in Ker(1-g)$ if and only if $u_i = F^{d-i}(u_d)$ for each $1 \leq i \leq d$. On the other hand

$$
\begin{aligned}
N(v_1, \ldots, v_d) &= (v_1, v_2, \ldots, v_{d-1}, v_d) \\
&\quad + (F(v_2), F(v_3), \ldots, F(v_d), F(v_1)) \\
&\quad + (F^2(v_3), F^2(v_4), \ldots, F^2(v_1), F^2(v_2)) \\
&\quad \vdots \\
&= (w, F^{-1}(w), F^{-2}(w), \ldots, F^{1-d}(w))
\end{aligned}
$$

where $w = \sum_{j=1}^d F^{j-1}(v_j)$. Setting $v_1 = v_2 = \ldots = v_{d-1} = 0$ and $v_d = u_d$ shows that $Ker(1-g) = Im(N)$ and $H^{2s}(G(W/K); \oplus_{i=1}^d K_3^{ind}(W_0))$ vanishes for $s > 0$ while $H^0(G(W/K); \oplus_{i=1}^d K_3^{ind}(W_0)) \cong K_3^{ind}(W_0)$.

The element, $(u_1, \ldots, u_d)$, lies in $Ker(N)$ if and only if $0 = \sum_{j=1}^d F^{j-1}(u_j) \in K_3^{ind}(W_0)$ while $(1-g)(v_1, \ldots, v_d) = (v_1 - F(v_2), v_2 - F(v_3), \ldots, v_d - F(v_1))$. In this case the equation $(1-g)(v_1, \ldots, v_d) = (u_1, \ldots, u_d)$ may be solved by choosing v_1 and v_d to satisfy $v_d - F(v_1) = u_d$ and then setting $v_i = F^{d-i+1}(v_1) + \sum_{j=i}^d F^{j-i}(u_j)$ for $2 \leq i \leq d-1$. This is consistent because $u_1 + F(v_2) = u_1 + \sum_{j=2}^d F^{j-2+1}(u_j) + F^d(v_1) = v_1$. Hence $H^{2s+1}(G(W/K); \oplus_{i=1}^d K_3^{ind}(W_0))$ vanishes for all $s \geq 0$.

Therefore $E_2^{s,t} = 0$ if $0 < s+t$ is even so that $H^{2s}(G(L/K); \oplus_{i=1}^d K_3^{ind}(L_0))$ vanishes when $s > 0$, as required. When $s+t = 2u+1$ the only contributions in the spectral sequence come from $E_2^{0,2u+1} \cong \mathbf{Z}/r$ and so $H^{2u+1}(G(L/K); \oplus_{i=1}^d K_3^{ind}(L_0)) \cong \mathbf{Z}/r$, as required. Finally, the only contribution to $H^0(G(L/K); \oplus_{i=1}^d K_3^{ind}(L_0))$ comes from $E_2^{0,0} \cong K_3^{ind}(W_0)$, which completes the proof. □

3.7 Suppose that L/K is a totally ramified Galois extension of p-adic local fields with group, $G(L/K)$ and set $r = [G(L/K) : G_1(L/K)]$. Choose a primitive r-th root of unity, ξ_r. If X is any $\mathbf{Z}[G(L/K)]$-module containing ${}_r(\mathbf{Q}/\mathbf{Z})(2)[1/p]$, depending on the choice of ξ_r, we shall define a new module $X\tilde{\oplus}\mathbf{Z}$, and an extension of the form

$$0 \longrightarrow X \longrightarrow X\tilde{\oplus}\mathbf{Z} \longrightarrow \mathbf{Z} \longrightarrow 0.$$

For example, this construction may be applied with X equal to any module in the chain

$$Tors(K_3^{ind}(L_0))[1/p] \cong (\mathbf{Q}/\mathbf{Z})(2)[1/p] \subset \mathcal{U} \subset K_3^{ind}(L_0).$$

Note that the action of $G(L/K)$ on $(\mathbf{Q}/\mathbf{Z})(2)[1/p]$ is trivial, since L/K is totally ramified, while the Frobenius, F of Section 3.1, acts like "multiplication" by t^2 where $t = |\overline{L}|$.

As an abelian group $X\tilde{\oplus}\mathbf{Z}$ is just the direct sum, $X \oplus \mathbf{Z}$. As a $\mathbf{Z}[G(L/K)]$-module the embedding of X is given by sending $x \in X$ to $(x, 0)$. If $G_1(L/K) =$

$G(L/E)$, as in Corollary 2.12 (proof), the action on $(0,1) \in X\tilde{\oplus}\mathbf{Z}$ will factor through $G(L/K)/G(L/E) \cong G(E/K)$. Furthermore, if $G(E/K) \cong \mathbf{Z}/r$ with generator, g, then $g(0,1) = (\xi_r, 1)$. This action is well-defined since $g^i(0,1) = (i\xi_r, 1)$, in additive notation.

Similarly we may define a $\mathbf{Z}[G(L/K)]$-module, $X\tilde{\oplus}\mathbf{Z}[1/p]$, and an extension of the form

$$0 \longrightarrow X \longrightarrow X\tilde{\oplus}\mathbf{Z}[1/p] \longrightarrow \mathbf{Z}[1/p] \longrightarrow 0$$

by setting

$$X\tilde{\oplus}\mathbf{Z}[1/p] = \lim_{\rightarrow}(X\tilde{\oplus}\mathbf{Z} \stackrel{(1,t)}{\longrightarrow} X\tilde{\oplus}\mathbf{Z} \stackrel{(1,t)}{\longrightarrow} X\tilde{\oplus}\mathbf{Z} \stackrel{(1,t)}{\longrightarrow} \ldots),$$

the direct limit of iterations of the map sending (z,m) to (z,mt) where $t = |\overline{L}|$. This is a $\mathbf{Z}[G(L/K)]$-module, since $t \equiv 1$ (*modulo* r).

Proposition 3.8 *In the notation of Section 3.7 let X be any of the $\mathbf{Z}[G(L/K)]$-modules in the chain*

$$Tors(K_3^{ind}(L_0))[1/p] \cong (\mathbf{Q}/\mathbf{Z})(2)[1/p] \subset \mathcal{U} \subset K_3^{ind}(L_0).$$

Then $X\tilde{\oplus}\mathbf{Z}[1/p]$ is a cohomologically trivial $\mathbf{Z}[G(L/K)]$-module.

Proof By [**37**], the composition of $\mathbf{Z}[G(L/K)]$-module homomorphisms

$$(\mathbf{Q}/\mathbf{Z})(2)[1/p] \cong Tors(K_3^{ind}(L_0))[1/p] \cong Tors(K_3^{ind}(\mathcal{O}_{L_0}))[1/p]$$

$$\longrightarrow Tors(K_3^{ind}(\overline{L_0}))[1/p] \cong (\mathbf{Q}/\mathbf{Z})(2)[1/p]$$

is an isomorphism. Therefore, by Lemma 2.11 and Corollary 2.12, it will suffice to show that

$$\hat{H}^i(G(L/K); K_3^{ind}(L_0)\tilde{\oplus}\mathbf{Z}[1/p]) \cong 0$$

for all i. If $G(L/E) = G_1(L/K)$ we have an isomorphism, by the spectral sequence used in the proof of Proposition 2.7,

$$\hat{H}^i(G(L/K); K_3^{ind}(L_0)\tilde{\oplus}\mathbf{Z}[1/p]) \cong \hat{H}^i(G(E/K); K_3^{ind}(E_0)\tilde{\oplus}\mathbf{Z}[1/p])$$

since $G(L/E)$ is a p-group acting trivially on $\mathbf{Z}[1/p]$. Now consider the exact sequence

$$0 \longrightarrow K_3^{ind}(E_0) \longrightarrow K_3^{ind}(E_0)\tilde{\oplus}\mathbf{Z}[1/p] \longrightarrow \mathbf{Z}[1/p] \longrightarrow 0.$$

It suffices to show that the resulting coboundary

$$\delta : \hat{H}^i(G(E/K); \mathbf{Z}[1/p]) \longrightarrow \hat{H}^{i+1}(G(E/K); K_3^{ind}(E_0))$$

is an isomorphism for all i. When i is odd both groups are zero but when i is even they are both isomorphic to $\mathbf{Z}/r$, by Proposition 2.7. The cohomology, $\hat{H}^i(G(E/K); M)$, is computed from the chain complex

$$\ldots \longrightarrow M \stackrel{1-g}{\longrightarrow} M \stackrel{N}{\longrightarrow} M \stackrel{1-g}{\longrightarrow} M \longrightarrow \ldots$$

where $N = \sum_{j=0}^{r-1} g^j$. Hence a generator for $\hat{H}^{2s}(G(E/K); \mathbf{Z}[1/p])$ is represented by 1. To compute $\delta[1]$ we lift 1 to $(0,1)$ and apply $1-g$ to obtain

$$(1-g)(0,1) = (0,1) - (\xi_r, 1) = (\xi_r, 0)$$

so that $\delta[1] = [\xi_r]$. However, since ξ_r generates ${}_r(\mathbf{Q}/\mathbf{Z})(2)[1/p]$, this implies that $\delta[1]$ is a generator of $\hat{H}^{2s+1}(G(E/K); K_3^{ind}(E_0))$, by Proposition 2.7. Therefore δ is an isomorphism, as required. □

3.9 In the situation of Section 3.1 we shall define a $\mathbf{Z}[G(L/K)]$-module

$$\tilde{\oplus}_{i=1}^{d} K_3^{ind}(L_0).$$

As an abelian group this module is simply the direct sum

$$(\oplus_{i=1}^{d} K_3^{ind}(L_0)) \oplus (\oplus_{i=1}^{d} \mathbf{Z}[1/p]).$$

The summand, $\oplus_{i=1}^{d} K_3^{ind}(L_0))$, has the $\mathbf{Z}[G(L/K)]$-module structure described in the proof of Theorem 3.6. Thus $g \in G(L/K)$ lifts to $(F, F) \in G(L_0/K)$ and acts like

$$g(a_1, \ldots, a_d) = (F(a_2), F(a_3), \ldots, F(a_d), F_0^{-1}(F(a_1))).$$

Any element, $h \in G(L/W)$, acts component-by-component as $h(a_1, \ldots, a_d) = (h(a_1), \ldots, h(a_d))$.

Consider the $\mathbf{Z}[G(L/K)]$-submodule, $\oplus_{i=1}^{d}(\mathbf{Q}/\mathbf{Z})(2)[1/p] \subset \oplus_{i=1}^{d} K_3^{ind}(L_0)$. Since L/W is totally ramified, $G(L/W)$ acts trivially on this submodule. Hence $G(L/K)$ acts through the quotient, $G(W/K) =< g >$, with

$$g(a_1, \ldots, a_d) = (F(a_2), F(a_3), \ldots, F(a_d), F_0^{-1}F(a_1)) = (qa_2, qa_3, \ldots, qa_d, q^{1-d}a_1)$$

where $q = v^2$ and $v = |\overline{K}|$. An arbitrary element of the $\mathbf{Z}[G(L/K)]$-module

$$Inf_{G(W/K)}^{G(L/K)} Ind_{\{1\}}^{G(W/K)}((\mathbf{Q}/\mathbf{Z})(2)[1/p])$$

has the form $z = 1 \otimes b_d + g \otimes b_{d-1} + \ldots + g^{d-1} \otimes b_1$ and $g(z) = 1 \otimes b_1 + g \otimes b_d + \ldots + g^{d-1} \otimes b_2$. Hence, if we write $z = (b_1, \ldots, b_d)$, then $g(z) = (b_2, b_3, \ldots, b_d, b_1)$.

Define an isomorphism

$$\phi : \oplus_{i=1}^{d}(\mathbf{Q}/\mathbf{Z})(2)[1/p] \xrightarrow{\cong} Inf_{G(W/K)}^{G(L/K)} Ind_{\{1\}}^{G(W/K)}((\mathbf{Q}/\mathbf{Z})(2)[1/p])$$

by $\phi(a_1, \ldots, a_d) = (q^{1-d}a_1, q^{2-d}a_2, \ldots, q^{-1}a_{d-1}, a_d)$. This is an isomorphism of $\mathbf{Z}[G(L/K)]$-modules, since

$$\begin{aligned}\phi(g(a_1, \ldots, a_d)) &= \phi(qa_2, qa_3, \ldots, qa_d, q^{1-d}a_1)\\ &= (q^{2-d}a_2, q^{3-d}a_3, \ldots, a_d, q^{1-d}a_1)\end{aligned}$$

and

$$\begin{aligned}g(\phi(a_1, \ldots, a_d)) &= g(q^{1-d}a_1, q^{2-d}a_2, \ldots, q^{-1}a_{d-1}, a_d)\\ &= (q^{2-d}a_2, q^{3-d}a_3, \ldots, a_d, q^{1-d}a_1).\end{aligned}$$

We have an isomorphism of $\mathbf{Z}[G(L/K)]$-modules

$$Inf_{G(W/K)}^{G(L/K)} Ind_{\{1\}}^{G(W/K)}((\mathbf{Q}/\mathbf{Z})(2)[1/p]) \cong Ind_{G(L/W)}^{G(L/K)}((\mathbf{Q}/\mathbf{Z})(2)[1/p])$$

where $G(L/W)$ acts trivially on $(\mathbf{Q}/\mathbf{Z})(2)[1/p]$. Hence we may form the following push-out diagram, which defines the module $\tilde{\oplus}_{i=1}^{d} K_3^{ind}(L_0)$.

$$\begin{array}{ccc}
\oplus_{i=1}^{d}(\mathbf{Q}/\mathbf{Z})(2)[1/p] \cong Ind_{G(L/W)}^{G(L/K)}((\mathbf{Q}/\mathbf{Z})(2)[1/p]) & \longrightarrow & \oplus_{i=1}^{d} K_3^{ind}(L_0) \\
\downarrow & & \downarrow \\
Ind_{G(L/W)}^{G(L/K)}((\mathbf{Q}/\mathbf{Z})(2)[1/p]\tilde{\oplus}\mathbf{Z}[1/p]) & \longrightarrow & \tilde{\oplus}_{i=1}^{d} K_3^{ind}(L_0)
\end{array}$$

Proposition 3.10 *The* $\mathbf{Z}[G(L/K)]$*-module,* $\tilde{\oplus}_{i=1}^{d} K_3^{ind}(L_0)$*, of Section 3.9 is cohomologically trivial.*

Proof We must show that $\hat{H}^i(G(L/N); \tilde{\oplus}_{i=1}^{d} K_3^{ind}(L_0)) = 0$ for all i and all $K \subseteq N \subseteq L$. If $N = K$ this follows from the long exact cohomology sequence

$$\ldots \longrightarrow \hat{H}^i(G(L/K); \tilde{\oplus}_{i=1}^{d} K_3^{ind}(L_0)) \longrightarrow$$

$$\hat{H}^{i+1}(G(L/K); Ind_{G(L/W)}^{G(L/K)}((\mathbf{Q}/\mathbf{Z})(2)[1/p])) \longrightarrow$$

$$\hat{H}^{i+1}(G(L/K); Ind_{G(L/W)}^{G(L/K)}((\mathbf{Q}/\mathbf{Z})(2)[1/p]\tilde{\oplus}\mathbf{Z}[1/p]))$$

$$\oplus \hat{H}^{i+1}(G(L/K); \oplus_{i=1}^{d} K_3^{ind}(L_0)) \longrightarrow$$

$$\hat{H}^{i+1}(G(L/K); \tilde{\oplus}_{i=1}^{d} K_3^{ind}(L_0)) \longrightarrow \ldots .$$

For $\hat{H}^i(G(L/K); Ind_{G(L/W)}^{G(L/K)}(M)) \cong \hat{H}^i(G(L/W); M)$ so that, by Theorem 3.8 and Theorem 3.6(proof), the central map may be identified with a map of the form

$$\hat{H}^{i+1}(G(L/W); (\mathbf{Q}/\mathbf{Z})(2)[1/p]) \cong \hat{H}^{i+1}(G(L/W); \oplus_{i=1}^{d}(\mathbf{Q}/\mathbf{Z})(2)[1/p])^{G(W/K)}$$

$$\longrightarrow \hat{H}^{i+1}(G(L/W); \oplus_{i=1}^{d} K_3^{ind}(L_0))^{G(W/K)}.$$

These groups are zero when $i+1$ is even and $\mathbf{Z}/r$ when $i+1$ is odd and the map is an isomorphism, by Lemma 2.11.

For a general N the induced module, $Ind_{G(L/W)}^{G(L/K)}(M))$, breaks up according to the double coset formula into summands corresponding to the $G(L/N)$-orbits of $G(L/K)/G(L/W) \cong \mathbf{Z}/d$, where the orbit of z corresponds to $Ind_{G(L/Nz(W))}^{G(L/N)}((z^{-1})^*(M))$. Of course, the modules such as $\oplus_{i=1}^{d} K_3^{ind}(L_0)$ decompose into $G(L/N)$-summands in precisely the same way and the corresponding summands are isomorphisms via the isomorphism, ϕ. Therefore the result follows by applying the above result to each summand with L/K, L/W and d replaced by L/N, $L/Nz(W)$ and $[Nz(W) : W]$, respectively. □

3.11 Now consider the $\mathbf{Z}[G(L/K)]$-submodule, $\mathcal{W} \subset \oplus_{i=1}^{d} K_3^{ind}(L_0)$, of Definition 3.3. Since $(\mathbf{Q}/\mathbf{Z})(2)[1/p] \subset \mathcal{U}$ we have $\oplus_{i=1}^{d}(\mathbf{Q}/\mathbf{Z})(2)[1/p] \subset \mathcal{W}$. Also $G(L/W)$ acts component-by-component on $\mathcal{W}$ so that $\mathcal{W} \cong (\oplus_{i=1}^{d-1} K_3^{ind}(L_0)) \oplus \mathcal{U}$ as a $\mathbf{Z}[G(L/W)]$-module. Therefore, by Corollary 2.12, the inclusion induces a cohomology isomorphism of the following form

$$H^t(G(L/W); \mathcal{W}) \xrightarrow{\cong} H^t(G(L/W); \oplus_{i=1}^d K_3^{ind}(L_0)).$$

Comparison of cohomology spectral sequences implies that the natural map

$$\hat{H}^t(G(L/K); \mathcal{W}) \longrightarrow \hat{H}^t(G(L/K); \oplus_{i=1}^d K_3^{ind}(L_0))$$

is also an isomorphism.

Define $\tilde{\mathcal{W}}$ by the pushout diagram

$$\begin{array}{ccc} \oplus_{i=1}^d (\mathbf{Q}/\mathbf{Z})(2)[1/p] \cong Ind_{G(L/W)}^{G(L/K)}((\mathbf{Q}/\mathbf{Z})(2)[1/p]) & \longrightarrow & \mathcal{W} \\ \downarrow & & \downarrow \\ Ind_{G(L/W)}^{G(L/K)}((\mathbf{Q}/\mathbf{Z})(2)[1/p]\tilde{\oplus}\mathbf{Z}[1/p]) & \longrightarrow & \tilde{\mathcal{W}} \end{array}$$

so that we immediately obtain the following result from Proposition 3.10.

Proposition 3.12 *The $\mathbf{Z}[G(L/K)]$-module, $\tilde{\mathcal{W}}$, of Section 3.11 is cohomologically trivial.*

3.13 On $(\mathbf{Q}/\mathbf{Z})(2)[1/p]$ F_0 acts by multiplication by q^d where $q = v^2$, $v = |\overline{K}|$. Hence, if $a_i \in (\mathbf{Q}/\mathbf{Z})(2)[1/p]$, then

$$(1 - \hat{F})(a_1, \ldots, a_d) = (a_1 - q^d a_d, a_2 - a_1, \ldots, a_d - a_{d-1}).$$

Now consider the isomorphism of Section 3.9

$$\phi : \oplus_{i=1}^d (\mathbf{Q}/\mathbf{Z})(2)[1/p] \xrightarrow{\cong} Ind_{G(L/W)}^{G(L/K)}((\mathbf{Q}/\mathbf{Z})(2)[1/p])$$

where, on the right, $G(L/W)$ acts trivially on $(\mathbf{Q}/\mathbf{Z})(2)[1/p]$ and

$$\phi(a_1, \ldots, a_d) = (q^{1-d}a_1, q^{2-d}a_2, \ldots, q^{-1}a_{d-1}, a_d) = \sum_{i=0}^{d-1} g^i \otimes q^{-i} a_{d-i}.$$

Hence $\hat{F}$ translates to $\phi\hat{F}\phi^{-1}$, which is given by

$$\begin{aligned}
&\phi(\hat{F}(\phi^{-1}(\sum_{i=0}^{d-1} g^i \otimes b_{d-i})))\\
&= \phi(\hat{F}(\phi^{-1}(b_1, \ldots, b_d)))\\
&= \phi(\hat{F}(q^{d-1}b_1, q^{d-2}b_2, \ldots, qb_{d-1}, b_d))\\
&= \phi(q^d b_d, q^{d-1}b_1, q^{d-2}b_2, \ldots, qb_{d-1}))\\
&= (qb_d, qb_1, \ldots, qb_{d-1})\\
&= q(1 \otimes b_{d-1} + g \otimes b_{d-2} + \ldots + g^{d-1} \otimes b_d)\\
&= qg^{d-1}(\sum_{i=0}^{d-1} g^i \otimes b_{d-i})\\
&= qg^{-1}(\sum_{i=0}^{d-1} g^i \otimes b_{d-i}).
\end{aligned}$$

Define

$$\hat{F} : Ind_{G(L/W)}^{G(L/K)}((\mathbf{Q}/\mathbf{Z})(2)[1/p]\tilde{\oplus}\mathbf{Z}[1/p]) \longrightarrow Ind_{G(L/W)}^{G(L/K)}((\mathbf{Q}/\mathbf{Z})(2)[1/p]\tilde{\oplus}\mathbf{Z}[1/p])$$

by the formula ($b \in (\mathbf{Q}/\mathbf{Z})(2)[1/p]$, $m \in \mathbf{Z}[1/p]$))

$$\hat{F}(g^i \otimes (b, m)) = g^{i-1} \otimes (qb, vm).$$

If $u \in \mathbf{Z}$ satisfies $uv \equiv 1$ (modulo r) then $g^{-i}ag^i(0,1) = a^{u^i}(0,1) = (u^i\xi_r, 1) \in (\mathbf{Q}/\mathbf{Z})(2)[1/p]\tilde{\oplus}\mathbf{Z}[1/p])$, writing the first coordinate additively, as usual. Here a and g are as in Section 3.1. Then $\hat{F}$ is a $\mathbf{Z}[G(L/K)]$-module homomorphism since

$$\begin{aligned}
\hat{F}(a(g^i \otimes (0,1))) &= \hat{F}(g^i \otimes a^{u^i}(0,1))\\
&= \hat{F}(g^i \otimes (u^i\xi_r, 1))\\
&= g^{i-1} \otimes (qu^i\xi_r, v)
\end{aligned}$$

while

$$\begin{aligned}
a(\hat{F}(g^i \otimes (0,1))) &= a(g^{i-1} \otimes (0,v))\\
&= g^{i-1} \otimes a^{u^i}(0,v))\\
&= g^{i-1} \otimes (vu^{i-1}\xi_r, v)
\end{aligned}$$

and these are equal since $vu^{i-1} \equiv quu^{i-1} \equiv qu^i$ (modulo r). Also

$$\hat{F}(g(g^i \otimes (0,1))) = \hat{F}(g^{i+1} \otimes (0,1)) = g^i \otimes (0,v)$$

while

$$g(\hat{F}(g^i \otimes (0,1))) = g(g^{i-1} \otimes (0,v)) = g^i \otimes (0,v).$$

Theorem 3.14 *In the notation of Section 3.1 and Section 3.13, there is a 2-extension of* $\mathbf{Z}[G(L/K)]$*-modules of the form*

$$K_3^{ind}(L) \xrightarrow{\Delta} \tilde{\oplus}_{i=1}^{d} K_3^{ind}(L_0) \xrightarrow{1-\hat{F}} \tilde{\mathcal{W}} \xrightarrow{(\pi,\pi_2)} {}_{p^\infty}K_2(L) \oplus \overline{L}^*.$$

Note that the right-hand module is $Tors(K_2(L)) \cong \mu(L)$, *by Proposition 2.1.*

Proof We have merely to evaluate the kernel and cokernel of the map

$$1 - \hat{F} : \tilde{\oplus}_{i=1}^{d} K_3^{ind}(L_0) \longrightarrow \tilde{\mathcal{W}}$$

where $\hat{F}$ was defined in §3.13. However, as abelian groups there are isomorphism

$$\tilde{\oplus}_{i=1}^{d} K_3^{ind}(L_0) \cong (\oplus_{i=1}^{d} K_3^{ind}(L_0)) \oplus (\oplus_{i=1}^{d} \mathbf{Z}[1/p])$$

and

$$\tilde{\mathcal{W}} \cong \mathcal{W} \oplus (\oplus_{i=1}^{d} \mathbf{Z}[1/p]) .$$

Furthermore the resulting map on the quotients of these direct sums by the submodules given by the first summands has the form

$$\lambda : \oplus_{i=1}^{d} \mathbf{Z}[1/p] \longrightarrow \oplus_{i=1}^{d} \mathbf{Z}[1/p]$$

and is given by

$$\lambda(m_1, \ldots, m_d) = (m_1 - vm_d, \ldots, m_d - vm_{d-1}).$$

If $(m_1, \ldots, m_d) \in Ker(\lambda)$ then $m_1 = vm_d, m_2 = vm_1, \ldots$ and $m_d(1 - v^d) = 0$ so that λ is injective. Also we can solve the equation $\lambda(m_1, \ldots, m_d) = (x_1, \ldots, x_d)$ by choosing $x_d, x_1, \ldots, x_{d-2}$ successively but the equations are consistent if and only if

$$m_d(1 - v^d) = v^{d-1}x_1 + v^{d-2}x_2 + \ldots + vx_{d-1} + x_d.$$

Therefore, if π_2 is defined to depend only on the coordinates in the second summand and to be given by

$$\pi_2(x_1, x_2, \ldots, x_d) = v^{d-1}x_1 + v^{d-2}x_2 + \ldots + vx_{d-1} + x_d$$

then $Ker(\pi_2) = Im(\lambda)$, which proves the exactness of the sequence.

It remains to verify that $G(L/K)$ acts on $\mathbf{Z}/(v^d - 1) \cong \overline{L}^*$ via the quotient to $G(W/K)$ whose generator acts by "multiplication" by $v = |\overline{K}|$. Certainly, $G(L/E)$ acts trivially on $\tilde{\oplus}_{i=1}^{d} K_3^{ind}(L_0)$ and $\tilde{\mathcal{W}}$. Also the action by $a \in G(E/W)$ is trivial on the second coordinates in the direct sums. Finally the generator, $g \in G(W/K)$, acts via

$$\begin{aligned} g(x_1, \ldots, x_d) &= g(\sum_{i=0}^{d-1} g^i \otimes x_{d-i}) \\ &= \sum_{i=0}^{d-1} g^{i+1} \otimes x_{d-i} \\ &= (x_2, x_3, \ldots, x_d, x_1) \end{aligned}$$

and $x_1 + vx_d + v^2x_{d-1} + \ldots + v^{d-1}x_2 \equiv v(v^{d-1}x_1 + v^{d-2}x_2 + \ldots + vx_{d-1} + x_d)$ (modulo $v^d - 1$) so that g acts on $\mathbf{Z}/(v^d - 1)$ by multiplication by v, as required. $\square$

4 Properties of the Canonical 2-Extension

4.1 Let us consider the effect of replacing ξ_r by ξ_r^t, with $HCF(r,t) = 1$, in the construction of the 2-extension of Theorem 3.14.

Let $K \subseteq W \subseteq E \subseteq L$ be p-adic local fields, as in Section 3.1. Hence W/K is the maximal unramified subextension of L/K and $[W : K] = d$. Also $G(L/E) = G_1(L/K)$, $G(E/W) \cong \mathbf{Z}/r =< a >$ and

$$G(L/K)/G(L/E) \cong G(E/K) = \{a, g \mid a^r = 1, g^d = a^c, gag^{-1} = a^v\}$$

where $v = |\overline{K}|$.

As explained earlier, there is a natural $\mathbf{Z}[G(L/K)]$-module decomposition of the form

$$K_3^{ind}(L_0) \cong (\mathbf{Q}/\mathbf{Z})(2)[1/p] \oplus \frac{K_3^{ind}(L_0)}{(Tors(K_3^{ind}(L_0)))[1/p]}.$$

By injectivity of $(\mathbf{Q}/\mathbf{Z})(2)[1/p]$, there exists an automorphism

$$\Phi_t : (\mathbf{Q}/\mathbf{Z})(2)[1/p] \xrightarrow{\cong} (\mathbf{Q}/\mathbf{Z})(2)[1/p]$$

such that $\Phi_t(\xi_r) = \xi_r^t$. Extending Φ_t by the identity on the other summand we obtain an automorphism

$$\Phi_t : K_3^{ind}(L_0) \xrightarrow{\cong} K_3^{ind}(L_0)$$

such that $\Phi_t(\xi_r) = \xi_r^t$. The subgroup, $K_3^{ind}(L)$, inherits a natural direct sum decomposition of the form

$$K_3^{ind}(L) \cong (Tors(K_3^{ind}(L)))[1/p] \oplus \frac{K_3^{ind}(L)}{(Tors(K_3^{ind}(L)))[1/p]}$$

on which Φ_t has the form $\phi_t = (\Phi_t|(Tors(K_3^{ind}(L)))[1/p]) \oplus 1$. Define an isomorphism

$$\Phi_t : (\mathbf{Q}/\mathbf{Z})(2)[1/p]\tilde{\oplus}\mathbf{Z}[1/p] \xrightarrow{\cong} (\mathbf{Q}/\mathbf{Z})(2)[1/p]\tilde{\oplus}\mathbf{Z}[1/p]$$

by $(z_1 \in (\mathbf{Q}/\mathbf{Z})(2)[1/p],\ z_2 \in \mathbf{Z}[1/p])$

$$\Phi_t(z_1, z_2) = (\Phi_t(z_1), z_2).$$

The action of $G(L/W)$ on $(\mathbf{Q}/\mathbf{Z})(2)[1/p]\tilde{\oplus}\mathbf{Z}[1/p]$ factorises through $G(E/W) \cong < a >$ and, in additive notation, $a(z_1, z_2) = (z_1 + z_2\xi_r, z_2)$. Hence $\Phi_t(a(z_1, z_2)) = (\Phi_t + z_2\xi_r^t, z_2)$ so that Φ_t is a $\mathbf{Z}[G(L/W)]$-isomorphism from $(\mathbf{Q}/\mathbf{Z})(2)[1/p]\tilde{\oplus}\mathbf{Z}[1/p]$ defined using ξ_r to $(\mathbf{Q}/\mathbf{Z})(2)[1/p]\tilde{\oplus}\mathbf{Z}[1/p]$ defined using ξ_r^t. Under the isomorphism of Section 3.9

$$\phi : \oplus_{i=1}^d (\mathbf{Q}/\mathbf{Z})(2)[1/p] \xrightarrow{\cong} Ind_{G(L/W)}^{G(L/K)}((\mathbf{Q}/\mathbf{Z})(2)[1/p])$$

$Ind_{G(L/W)}^{G(L/K)}(\Phi_t)$ is identified with $\oplus_{i=1}^d \Phi_t$, which extends to a $\mathbf{Z}[G(L/K)]$-isomorphism, $\oplus_{i=1}^d \Phi_t$, on $\oplus_{i=1}^d K_3^{ind}(L_0)$. Therefore these $\mathbf{Z}[G(L/K)]$-isomorphisms induce a $\mathbf{Z}[G(L/K)]$-isomorphism of the form

$$\tilde{\Phi}_t : \tilde{\oplus}_{i=1}^d K_3^{ind}(L_0) \xrightarrow{\cong} \tilde{\oplus}_{i=1}^d K_3^{ind}(L_0)$$

between the module defined using ξ_r and that defined using ξ_r^t. This isomorphism restricts to give $\tilde{\Phi}_t : \tilde{\mathcal{W}} \stackrel{\cong}{\longrightarrow} \tilde{\mathcal{W}}$ where the second module is defined using ξ_r^t and the first using ξ_r. In addition, in the notation of Section 3.13,

$$\begin{aligned}\tilde{\Phi}_t(\hat{F}(g^i \otimes (b,m)) &= \tilde{\Phi}_t((g^{i-1} \otimes (qb, vm)) \\ &= (g^{i-1} \otimes (q\tilde{\Phi}_t(b), vm)) \\ &= \hat{F}((g^i \otimes (\tilde{\Phi}_t(b), m)) \\ &= \hat{F}(\tilde{\Phi}_t(g^i \otimes (b,m))\end{aligned}$$

from which it is clear that $(1-\hat{F})\tilde{\Phi}_t = \tilde{\Phi}_t(1-\hat{F})$.

Therefore we have established the following result.

Proposition 4.2 *In the notation of Section 4.1 there is a commutative diagram of* $\mathbf{Z}[G(L/K)]$*-isomorphisms of the form*

$$\begin{array}{ccccccc} K_3^{ind}(L) & \stackrel{\Delta}{\longrightarrow} & \tilde{\oplus}_{i=1}^d K_3^{ind}(L_0) & \stackrel{1-\hat{F}}{\longrightarrow} & \tilde{\mathcal{W}} & \longrightarrow & Tors(K_2(L)) \\ \downarrow \phi_t & & \downarrow \tilde{\Phi}_t & & \downarrow \tilde{\Phi}_t & & \downarrow 1 \\ K_3^{ind}(L) & \stackrel{\Delta}{\longrightarrow} & \tilde{\oplus}_{i=1}^d K_3^{ind}(L_0) & \stackrel{1-\hat{F}}{\longrightarrow} & \tilde{\mathcal{W}} & \longrightarrow & Tors(K_2(L)) \end{array}$$

in which the upper sequence is that of Theorem 3.14 defined using ξ_r *while the lower sequence is that defined using* ξ_r^t.

There is a similar quasi-isomorphism of 2-extensions of Theorem 3.14 corresponding to different choices of the generators, a and g, of $G(E/K)$.

The remainder of this section will be concerned with the behaviour of the canonical 2-extension (up to quasi-isomorphism) undertaking fixed points of a normal subgroup.

Lemma 4.3 *Suppose that we are given a pushout square of* $\mathbf{Z}[G]$*-modules of the form*

$$\begin{array}{ccc} A & \stackrel{\alpha_1}{\longrightarrow} & B \\ \downarrow \beta_1 & & \downarrow \beta_2 \\ C & \stackrel{\alpha_2}{\longrightarrow} & D \end{array}$$

and that α_1 *induces isomorphisms in Tate cohomology in all dimensions and for all subgroups of G while C and D are cohomologically trivial. Let J be any subgroup of G. Then the exact sequence*

$$0 \longrightarrow A \stackrel{(\alpha_1,\beta_1)}{\longrightarrow} B \oplus C \stackrel{\alpha_2-\beta_2}{\longrightarrow} D \longrightarrow 0$$

remains exact upon taking J-fixed points.

Proof This follows from the long exact cohomology sequence

$$0 \longrightarrow A^J \longrightarrow B^J \oplus C^J \longrightarrow D^J$$

$$\longrightarrow H^1(J;A) \stackrel{(\alpha_1)_*}{\longrightarrow} H^1(J;B) \longrightarrow 0.$$

□

4.4 We shall apply this observation to the pushouts used to define the modules $\tilde{\oplus}_{i=1}^d K_3^{ind}(L_0)$ and $\tilde{\mathcal{W}}$ associated to L/K.

Suppose that G is a finite group and

$$A \longrightarrow B \longrightarrow C \longrightarrow D$$

is a 2-extension of finitely generated $\mathbf{Z}[G]$-modules in which B and C are cohomologically trivial. Let $J \triangleleft G$ be a normal subgroup and let A^J, A_J denote the J-invariants and J-coinvariants of A, respectively ([**31**] p.3). We have a commutative diagram of $\mathbf{Z}[G/J]$-modules in which the rows and columns are exact and N_J denotes the norm, $N(x) = \sum_{g \in J} gx$,

$$\begin{array}{ccccccccc}
 & & & & B_J & \longrightarrow & C_J & \longrightarrow & D_J & \longrightarrow & 0 \\
 & & & & \cong\big\downarrow N_J & & \cong\big\downarrow N_J & & & & \\
0 & \longrightarrow & A^J & \longrightarrow & B^J & \longrightarrow & C^J & & & &
\end{array}$$

resulting in the associated J-invariant/coinvariant 2-extension

$$A^J \longrightarrow B^J \longrightarrow C^J \longrightarrow D_J.$$

Let $J = G(L/M) \triangleleft G(L/K)$ and set $I = G_0(L/M)$, the inertia group. Since I is the kernel of the homomorphism $G(L/M) \longrightarrow G(\overline{L}/\overline{M}) \longrightarrow G(\overline{L}/\overline{K})$ it is a characteristic subgroup of $G(L/K)$ and hence $I \triangleleft J$. Therefore we may calculate the J-fixed points of X in two stages as $X^J = (X^I)^{J/I}$. Therefore we consider the two cases:

Case A: $I = J$, $\overline{K} = \mathbf{F}_q$, $\overline{M} = \overline{L} = \mathbf{F}_{q^d}$, and

Case B: $I = \{1\}$, $\overline{K} = \mathbf{F}_q$, $\overline{M} = \mathbf{F}_{q^a}$, $\overline{L} = \mathbf{F}_{q^d}$ and $d = as$.

In Case A, $J = G(L/M) \subseteq G(L/W)$ acts trivially on $\oplus_{i=1}^d(\mathbf{Q}/\mathbf{Z})(2)[1/p]$ and acts component by component on $\oplus_{i=1}^d K_3^{ind}(L_0)$ with fixed points equal to $\oplus_{i=1}^d K_3^{ind}(M_0)$.

Since $G(L/M) \triangleleft G(L/K)$ and $G(L/M) \subseteq G(L/W)$

$$Ind_{G(L/W)}^{G(L/K)}((\mathbf{Q}/\mathbf{Z})(2)[1/p]\tilde{\oplus}\mathbf{Z}[1/p])^{G(L/M)}$$

$$= Ind_{G(M/W)}^{G(M/K)}((((\mathbf{Q}/\mathbf{Z})(2)[1/p]\tilde{\oplus}\mathbf{Z}[1/p])^{G(L/M)}).$$

If $t = [G(L/M) : G_1(L/M)]$ then $tw = r$ and the image of $G(L/M)$ in $G(E/K)$ is the cyclic group of order t generated by a^w. Since $a^w(0,1) = (w\xi_r, 1)$ we see that

$$Ind_{G(L/W)}^{G(L/K)}((\mathbf{Q}/\mathbf{Z})(2)[1/p]\tilde{\oplus}\mathbf{Z}[1/p])^{G(L/M)}$$

$$= Ind_{G(M/W)}^{G(M/K)}((\mathbf{Q}/\mathbf{Z})(2)[1/p]\tilde{\oplus}t\mathbf{Z}[1/p]).$$

Notice that the action of $a \in G(E/K)$ on $(0,t)$ is given by $a(0,t) = (t\xi_r, t) = (\xi_w, t)$ and $w = [G_0(M/K) : G_1(M/K)]$.

Also

$$\mathcal{W}^{G(L/M)} = \bigcup_s Ker(K_3^{ind}(M(s)) \longrightarrow K_3^{ind}(M))$$

since $G(L/M) \cong G(L(s)/M(s))$ and $G(L(s)/M) \cong G(L(s)/L) \times G(L(s)/M(s))$. These observations, together with Lemma 4.3, imply that the $G(L/M)$-fixed points of the pushout squares defining $\tilde{\oplus}_{i=1}^d K_3^{ind}(L_0)$ and $\tilde{\mathcal{W}}$ for L/K are canonically isomorphic to the corresponding pushout squares for M/K. Furthermore it is clear that $\hat{F}$ for L/K induces $\hat{F}$ for M/K so that the $G(L/M)$-fixed points of the $\mathbf{Z}[G(L/K)]$-homomorphism

$$1 - \hat{F} : \tilde{\oplus}_{i=1}^d K_3^{ind}(L_0) \longrightarrow \tilde{\mathcal{W}}$$

(defined for L/K) is the $\mathbf{Z}[G(M/K)]$-homomorphism

$$1 - \hat{F} : \tilde{\oplus}_{i=1}^d K_3^{ind}(M_0) \longrightarrow \tilde{\mathcal{W}}$$

(defined for M/K).

4.5 *Case B in detail*

Now consider Case B in which $G(L/M) \cong \mathbf{Z}/s \cong G(\overline{L}/\overline{M})$ and $G(L/M) \triangleleft G(L/K)$. Since L/M is unramified and L/W is totally ramified we have

$$G(L/M) \cap G(L/W) = \{1\}.$$

Therefore we have a canonical isomorphism

$$G(L/W) \xrightarrow{\cong} G(M/W \cap M).$$

Let $b \in G(L/M)$ be a generator which maps to $F^{-a} \in G(\overline{L}/\overline{K}) \cong \mathbf{Z}/d$. In $G(M_0/M) \times G(L/M)$ we have $(F^a, 1)$ and $x = (F^{-a}, b)$ both lying in the subgroup $G(L_0/M)$. We may write $b = (F^a, 1)(F^{-a}, b) = (F^a, 1)x$. Now consider the pushout diagram defining $\tilde{\oplus}_{i=1}^d K_3^{ind}(L_0)$.

$$\begin{array}{ccc}
\oplus_{i=1}^d (\mathbf{Q}/\mathbf{Z})(2)[1/p] \cong Ind_{G(L/W)}^{G(L/K)}((\mathbf{Q}/\mathbf{Z})(2)[1/p]) & \longrightarrow & \oplus_{i=1}^d K_3^{ind}(L_0) \\
\downarrow & & \downarrow \\
Ind_{G(L/W)}^{G(L/K)}((\mathbf{Q}/\mathbf{Z})(2)[1/p]\tilde{\oplus}\mathbf{Z}[1/p]) & \longrightarrow & \tilde{\oplus}_{i=1}^d K_3^{ind}(L_0)
\end{array}$$

Upon taking $G(L/M)$-fixed points we are going to verify that the resulting pushout contains in a canonical manner the following pushout diagram, which defines

$\tilde{\oplus}_{i=1}^{a} K_3^{ind}(M_0)$, as a sub-pushout diagram.

$$\begin{array}{ccc} \oplus_{i=1}^{a}(\mathbf{Q}/\mathbf{Z})(2)[1/p] \cong Ind_{G(M/W\cap M)}^{G(M/K)}((\mathbf{Q}/\mathbf{Z})(2)[1/p]) & \longrightarrow & \oplus_{i=1}^{a} K_3^{ind}(M_0) \\ \downarrow & & \downarrow \\ Ind_{G(M/W\cap M)}^{G(M/K)}((\mathbf{Q}/\mathbf{Z})(2)[1/p]\tilde{\oplus}\mathbf{Z}[1/p]) & \longrightarrow & \tilde{\oplus}_{i=1}^{a} K_3^{ind}(M_0) \end{array}$$

The action of $b \in G(L/M)$ on $\oplus_{i=1}^{d} K_3^{ind}(L_0)$ is given by ([**30**], p.330)

$$b(z_1, \dots, z_d) = (F_0(x(z_{d-a+1})), \dots, F_0(x(z_d)), x(z_1), \dots, x(z_{d-a}))$$

where $x = (F^{-a}, b) \in G(L_0/M)$. Therefore, if $b(z_1, \dots, z_d) = (z_1, \dots, z_d)$ then $z_{d-a+1}, \dots, z_d$ determine all the $\{z_i\}$. Also, if $1 \leq j \leq a$,

$$F_0(x(z_{d-j+1})) = z_{a-j+1} = x^{-1}(z_{2a-j+1}) = \dots = x^{1-s}(z_{d-j+1})$$

but $x^{s-1}F_0x$ acts like $F^{-a(s-1)+d-a} = 1$ on K_0 and like $b^{s-1}b = 1$ on L so that the elements $\{z_{d-j+1} \in K_3^{ind}(L_0); 1 \leq j \leq a\}$ are arbitrary. In particular,

$$\oplus_{i=1}^{a} K_3^{ind}(M_0) \subset (\oplus_{i=1}^{d} K_3^{ind}(L_0))^{G(L/M)}$$

by mapping $(z_{d-a+1}, \dots, z_d)$ to $(F_0(x(z_{d-a+1})), \dots, z_{d-a+1}, \dots, z_d)$. The induced $G(M/K)$-action on the subgroup, $\oplus_{i=1}^{a} K_3^{ind}(M_0)$, is described in the following manner. If $g \in G(M/K)$ restricts to F^{-l} on $W \cap M$, where $0 \leq l \leq a-1$ since $G(W \cap M/K) \cong \mathbf{Z}/a$, then $g(z_{d-a+1}, \dots, z_d)$ corresponds to

$$\hat{F}^l((F^{-l}, g)F_0x(z_{d-a+1}), \dots, (F^{-l}, g)(z_d)).$$

If $z_{d-a+1}, \dots, z_d \in K_3^{ind}(M_0)$ and $y = (F^{-l}, g)$ then the a-tuple consisting of the last coordinates is given by

$$(y(z_{d-a-l+1}), \dots, y(z_{d-l+1}), y(z_{d-l}))$$

and $z_{d-a-l+1} = x^{-1}(z_{d-a+(a-l)+1})$ but $x^{-1} = (F^a, b^{-1})$ and b acts trivially on $K_3^{ind}(M_0)$ so that g acts via the formula

$$g(z_{d-a+1}, \dots, z_d) = (F_0y(z_{d-l+1}), \dots, y(z_{d-l}))$$

where $F_0 = F^a$. This is the correct action on

$$\oplus_{i=1}^{a} K_3^{ind}(M_0) \subset (\oplus_{i=1}^{d} K_3^{ind}(L_0))^{G(L/M)}.$$

A similar calculation applies to the $\mathbf{Z}[G(M/K)]$-module

$$\oplus_{i=1}^{a}(\mathbf{Q}/\mathbf{Z})(2)[1/p] \subset (\oplus_{i=1}^{d}(\mathbf{Q}/\mathbf{Z})(2)[1/p])^{G(L/M)}.$$

Next we must examine

$$Ind_{G(L/W)}^{G(L/K)}((\mathbf{Q}/\mathbf{Z})(2)[1/p]\tilde{\oplus}\mathbf{Z}[1/p])^{G(L/M)}.$$

We need to show that

$$Ind_{G(M/W\cap M)}^{G(M/K)}((\mathbf{Q}/\mathbf{Z})(2)[1/p]\tilde{\oplus}\mathbf{Z}[1/p])$$

embeds into these $G(L/M)$-invariants in a manner compatible with the pushout diagrams. This will imply that the pushout square defining $\tilde{\oplus}_{i=1}^{a} K_3^{ind}(M_0)$ embeds into the pushout square defining $\tilde{\oplus}_{i=1}^{d} K_3^{ind}(L_0)$.

If $g \in G(L/K)$ maps to $F \in G(\overline{L}/\overline{K})$ then coset representatives for $G(L/K)/G(L/W)$ are given by $\{b^j g^i \mid 0 \leq j \leq s-1; 0 \leq i \leq a-1\}$. If

$$z = \sum_{i,j} b^j g^i \otimes x_{aj+i} \in Ind_{G(L/W)}^{G(L/K)}((\mathbf{Q}/\mathbf{Z})(2)[1/p]\tilde{\oplus}\mathbf{Z}[1/p])^{G(L/M)}$$

then each x_{aj+i} depends only on i and

$$z = \sum_{i,j} b^j g^i \otimes x_i,$$

which is fixed by b since $b^s = 1$. A set of coset representatives for $G(M/K)/G(M/W \cap M)$ is $\{g^i \mid 0 \leq i \leq a-1\}$ and there is a $\mathbf{Z}[G(M/K)]$-module isomorphism

$$f : Ind_{G(M/W\cap M)}^{G(M/K)}((\mathbf{Q}/\mathbf{Z})(2)[1/p]\tilde{\oplus}\mathbf{Z}[1/p])$$

$$\xrightarrow{\cong} Ind_{G(L/W)}^{G(L/K)}((\mathbf{Q}/\mathbf{Z})(2)[1/p]\tilde{\oplus}\mathbf{Z}[1/p])^{G(L/M)}$$

defined by

$$f(\sum_{i=0}^{a-1} g^i \otimes x_i) = \sum_{i,j} b^j g^i \otimes x_i.$$

Next we temporarily write $\mathcal{W}_L$ and $\mathcal{W}_M$ for the module $\mathcal{W}$ associated to L/K and M/K, respectively. If $L(t)/L$ is the unramified Galois extension of degree t then, in Case B, $L(t) = M(ts)$ since L/M is unramified. From this and the isomorphism, $K_3^{ind}(M_0) \cong (K_3^{ind}(L_0))^{G(L/M)}$, one sees that $\mathcal{W}_L^{G(L/M)}$ contains $\mathcal{W}_M$, via the restriction of the injection

$$\oplus_{i=1}^{a} K_3^{ind}(M_0) \longrightarrow \oplus_{i=1}^{d} K_3^{ind}(L_0).$$

Finally we must verify that

$$1 - \hat{F} : \tilde{\oplus}_{i=1}^{d} K_3^{ind}(L_0) \longrightarrow \tilde{\mathcal{W}_L} \subset \tilde{\oplus}_{i=1}^{d} K_3^{ind}(L_0)$$

restricts to the canonical map

$$1 - \hat{F} : \tilde{\oplus}_{i=1}^{a} K_3^{ind}(M_0) \longrightarrow \tilde{\mathcal{W}_M} \subset \tilde{\oplus}_{i=1}^{d} K_3^{ind}(M_0).$$

On $Ind_{G(L/W)}^{G(L/K)}((\mathbf{Q}/\mathbf{Z})(2)[1/p]\tilde{\oplus}\mathbf{Z}[1/p])^{G(L/M)}$, by definition of $\hat{F}$ in Section 3.13,

$$\hat{F}(\sum_{i,j} b^j g^i \otimes (\beta_i, m_i)) = \sum_{i,j} b^j g^i \otimes (q\beta_i, vm_i)$$

which corresponds with $\hat{F}$ for M/K under the isomorphism, f.

On $(\oplus_{i=1}^{d} K_3^{ind}(L_0))^{G(L/M)}$ $\hat{F}$ is the Frobenius,

$$\hat{F}(z_1, \ldots, z_d) = (F_0(z_d), z_1, \ldots, z_{d-1}).$$

On $\oplus_{i=1}^{a} K_3^{ind}(M_0)$, mapped in via the "last a-tuple map" the Frobenius sends the image of $(z_{d-a+1}, \ldots, z_d)$ to $(z_{d-a}, z_{d-a+1}, \ldots, z_{d-1})$ but $z_{d-a} = x(z_{d-2a}) = \ldots =$

$F_0x^{s-1}(z_d)$ where $x = (F^{-a}, b)$. Hence F_0x^{s-1} is F^a on $K_0 = K_{nr}$ and $b^{-1} = 1$ on M. Therefore

$$(z_{d-a}, z_{d-a+1}, \ldots , z_{d-1}) = (F_0(z_d), z_1, \ldots , z_{d-1})$$

where F_0 now denotes the "Frobenius" associated with M/K in the definition of the action on $\oplus_{i=1}^a K_3^{ind}(M_0)$.

Hence we have to show that there is a commutative diagram of 2-extensions of $\mathbf{Z}[G(M/K)]$-modules of the form

$$\begin{array}{ccccccc}
K_3^{ind}(M) & \longrightarrow & \tilde{\oplus}_{i=1}^a K_3^{ind}(M_0) & \xrightarrow{1-\hat{F}} & \mathcal{W}_M & \longrightarrow & \mu(M) \\
\downarrow \cong & & \downarrow & & \downarrow & & \downarrow \cong \\
K_3^{ind}(M) & \longrightarrow & (\tilde{\oplus}_{i=1}^d K_3^{ind}(L_0))^{G(L/M)} & \xrightarrow{1-\hat{F}} & \mathcal{W}_L^{G(L/M)} & \longrightarrow & \mu(L)_{G(L/M)}
\end{array}$$

in which the lower 2-extension is the $G(L/M)$-invariant/coinvariant 2-extension derived from the fundamental 2-extension for L/K. The upper 2-extenson is the fundamental 2-extension for M/K. We have shown that the central square commutes and therefore induces a commutative diagram of this form; the issue is whether or not the induced vertical maps on the extreme left and right are isomorphisms.

The isomorphism on the left is easily seen to be the identity, since both copies of $K_3^{ind}(M)$ are mapped in via the diagonal map, Δ.

Since L/M is unramified we have

$$\mu(L) \cong \mu_{p^\infty}(L) \times \overline{L}^*$$

and therefore $G(L/M)$ acts trivially on the p-primary roots of unity and we obtain

$$\begin{aligned}
\mu(L)_{G(L/M)} &\cong \mu_{p^\infty}(L) \times (\overline{L}^*)_{G(L/M)} \\
&\cong \mu_{p^\infty}(L) \times \overline{M}^* \\
&\cong \mu(M).
\end{aligned}$$

Now let us prove that the right-hand vertical map is an isomorphism. Recall that we are in Case B. If $z \in {}_{p^\infty}K_2(M)$ we lift z to

$$(z_{d-a+1}, \ldots , z_d) \in \mathcal{W}_M \subset \oplus_{i=1}^a K_3^{ind}(M_0).$$

This a-tuple maps to

$$\sigma = z_{d-a+1} + \ldots + z_d \in \mathcal{U}_M = \bigcup_t Ker(N : K_3^{ind}(M(t)) \longrightarrow K_3^{ind}(M)).$$

Under the vertical map $(z_{d-a+1}, \ldots , z_d)$ maps to

$$(F_0x(z_{d-a+1}), \ldots , F_0x(z_d), xF_0x(z_{d-a+1}), \ldots , x^{s-1}F_0x(z_d)).$$

This d-tuple maps to its sum

$$(1 + x + \ldots + x^{s-1})F_0x(\sigma) \in \mathcal{U}_L = \bigcup_t Ker(N : K_3^{ind}(L(t)) \longrightarrow K_3^{ind}(L)).$$

Since $x^{s-1}F_0x(\sigma) = \sigma$ this element is

$$(x^{1-s} + x^{2-s} + \ldots + x^{-1} + 1)\sigma \in \mathcal{U}_L.$$

Note that $x = (F^{-a}, b)$ and $L(t) = M(ts)$ so that the image of $(x^{1-s} + x^{2-s} + \ldots + x^{-1} + 1)\sigma$ under $N_{L(t)/L} : K_3^{ind}(L(t)) \longrightarrow K_3^{ind}(L)$ is equal to $N_{L/M}(N_{L(t)/L}(\sigma)) = N_{M(ts)/M}(\sigma) = 0$.

Also, for large enough $e \in \mathbf{Z}$, $L(p^et) = M(p^ets)$ and there are isomorphisms of the form

$$H^1(G(L(p^et)/L); K_3^{ind}(L(p^et)) \cong {}_{p^\infty}K_2(L)$$

and

$$H^1(G(M(p^ets)/M); K_3^{ind}(M(p^ets)) \cong {}_{p^\infty}K_2(M).$$

The natural map, ${}_{p^\infty}K_2(M) \longrightarrow {}_{p^\infty}K_2(L)$, is multiplication by $s = [L : M]$. So also is the composition of this map followed by the norm. Hence the norm is an isomorphism between these groups. However, sending $\sigma \in Ker(N : K_3^{ind}(L(p^et)) \longrightarrow K_3^{ind}(L))$ to $(x^{1-s} + x^{2-s} + \ldots + x^{-1} + 1)\sigma$ induces the norm on cohomology

$$N : H^1(G(L(p^et)/L); K_3^{ind}(L(p^et)) \longrightarrow H^1(G(L(p^et)/M); K_3^{ind}(L(p^et)).$$

This shows that the right-hand vertical map is an isomorphism on p-primary torsion.

On prime-to-p torsion the right-hand map has the form

$$\overline{M}^* \cong \mathbf{Z}/(v^a - 1) \longrightarrow \overline{L}^* \cong \mathbf{Z}/(v^d - 1)$$

where $v = |\overline{K}|$. We may lift the generator, $1 \in \mathbf{Z}/(v^a - 1)$, to $1 \otimes (0,1) \in Ind_{G(M/W\cap M)}^{G(M/K)}((\mathbf{Q}/\mathbf{Z})(2)[1/p])\tilde{\oplus}\mathbf{Z}[1/p])$ which maps to $\sum_{i=1}^{s-1} b^j \otimes 1$ which, in turn, maps to $N_{\overline{L}/\overline{M}}(1)$, which is a generator of $(\overline{L}^*)^{G(L/M)} \cong \overline{M}^*$. This element corresponds to $1 \in (\overline{L}^*)_{G(L/M)}$ since the last map in the lower 2-extension is given by

$$\mathcal{W}_L^{G(L/M)} \stackrel{Norm^{-1}}{\longrightarrow} (\mathcal{W}_L)_{G(L/M)} \longrightarrow \mu(L)_{G(L/M)}.$$

Therefore we have proved the following result:

Theorem 4.6 *Let L/K be a Galois extension of p-adic local fields with group, $G(L/K)$. If $G(L/M) \lhd G(L/K)$ is a normal subgroup then the $G(L/M)$-invariant/ coinvariant sequence of Section 4.4 associated to the fundamental 2-extension of Theorem 3.14*

$$K_3^{ind}(L) \longrightarrow \tilde{\oplus}_{i=1}^d K_3^{ind}(L_0) \longrightarrow \tilde{\mathcal{W}}_L \longrightarrow TorsK_2(L) \cong \mu(L)$$

is quasi-isomorphic to the fundamental 2-extension associated to M/K

$$K_3^{ind}(M) \longrightarrow \tilde{\oplus}_{i=1}^a K_3^{ind}(M_0) \longrightarrow \tilde{\mathcal{W}}_M \longrightarrow TorsK_2(M) \cong \mu(M).$$

5 Construction of $\Omega_1(E/F, 2)$

5.1 *Five p-adic regulators*

Let L be a p-adic local field. A p-adic regulator is an homomorphism of the form

$$K_3^{ind}(L) \otimes \mathbf{Q}_p \longrightarrow L.$$

There are a number of such regulators to choose from and making comparison of the different types is very interesting and important. For the reader's convenience we shall pause to recall some of the available models.

(a) The cyclic homology regulator

Let L be a p-adic local field and let $\mathcal{GL}_n(L)$ denote the Lie algebra of $n \times n$ matrices with entries in L. From the localisation exact sequence we have an isomorphism of the form

$$K_3(\mathcal{O}_L) \cong K_3(L)$$

and a resulting Hurewicz map

$$H : K_3(L) \longrightarrow H_3(GL_\infty(\mathcal{O}_L); \mathbf{Z}).$$

From ([**40**], pp. 244-245; see also [**17**], Chapter V, Théorème 2.4.9) we have isomorphisms, for each n,

$$H^*(\mathcal{GL}_n(L); \mathbf{Q}_p) \cong (\lim_{\overleftarrow{t}} \lim_{\overrightarrow{s}} H^*(GL_n(\mathcal{O}_L/(\pi^s)); \mathbf{Z}/p^t)) \otimes \mathbf{Q}_p.$$

In any given dimension these finite-dimensional $\mathbf{Q}_p$-vector spaces stabilise with n and the above isomorphism is dual to

$$\mu : (\lim_{\overleftarrow{s}} H_*(GL_n(\mathcal{O}_L/(\pi^s)); \mathbf{Z}_p)) \otimes \mathbf{Q}_p \xrightarrow{\cong} H_*(\mathcal{GL}_n(L); \mathbf{Q}_p).$$

Write

$$H_m^{(p)}(GL_n(\mathcal{O}_L); \mathbf{Z}_p) = (\lim_{\overleftarrow{s}} H_m(GL_n(\mathcal{O}_L/(\pi^s)); \mathbf{Z}_p)).$$

In ([**21**], Theorem 6.2; see also [**22**]) we find homomorphisms of the form

$$Tr_*\lambda_* : H_*(\mathcal{GL}_\infty(L); \mathbf{Q}_p) \longrightarrow HC_{*-1}(L)$$

from Lie algebra homology to cyclic homology with a dimension shift of -1.

Finally, from the Hochschild/cyclic homology long exact sequence we see that $HC_2(L) \cong L$. Putting all these maps together yields the cyclic homology regulator

$$r = Tr_*\lambda_*\mu H : K_3(L) \longrightarrow L.$$

This regulator has the advantage of being given by explicit formulae on the chain-level. The disadvantage is the involvement of the elaborate and impenetrable theory of [**17**]. Here is an outline of the theory.

We may replace $GL_n(\mathcal{O}_L)$ by one of the groups

$$\Gamma_n(r) = ker(GL_n(\mathcal{O}_L) \longrightarrow GL_n(\mathcal{O}_L/(\pi^r)),$$

since

$$H_m^{(p)}(GL_n(\mathcal{O}_L); \mathbf{Z}_p) \otimes \mathbf{Q}_p \cong ((\lim_{\overleftarrow{s}} H_m(\Gamma_n(r)/\Gamma_n(r+s); \mathbf{Z}_p)) \otimes \mathbf{Q}_p)_{GL_n(\mathcal{O}_L/(\pi^r))}.$$

For $r \geq 1$, $\Gamma_n(r)$ is a p-divisible p-adic filtered group in the sense of ([**17**], p. 81). This means that there is a "valuation function"

$$\omega : \Gamma_n(r) \longrightarrow \mathbf{Q}_+ \cup \{\infty\}$$

given by $\omega(I + \pi_L^r X) = min_{i,j} v_L(X_{ij})$, where v_L is the valuation of L normalised so that $v_L(p) = 1$, $v_L(0) = \infty$ and that elements with large enough valuation have p-th roots in $\Gamma_n(r)$.

One uses ω to define a valuation (or, equivalently, a filtration, $\{A_\nu\}$)

$$w : \mathbf{Z}_p[\Gamma_n(r)] \longrightarrow \mathbf{Q}_+ \cup \{\infty\}$$

by setting ([**17**], p. 88) A_ν equal to the $\mathbf{Z}_p$-submodule generated by elements of the form ($\gamma_i \in \Gamma_n(r)$, $0 \leq h \in \mathbf{Z}$)

$$z = p^h(\gamma_1 - 1)(\gamma_2 - 1)\ldots(\gamma_n - 1)$$

with

$$h + \omega((\gamma_1) + \ldots + \omega((\gamma_n) \geq \nu.$$

Define $Al\Gamma_n(r)$ by

$$Al\Gamma_n(r) = \lim_{\overleftarrow{s}} \mathbf{Z}_p[\Gamma_n(r)/\Gamma_n(r+s)]$$

so that $Al\Gamma_n(r)$ is a filtered, p-adic valuation ring and filtered homological algebra applied to this yields

$$H_m^{(p)}(\Gamma_n(r); \mathbf{Z}_p) = Tor_{filt,m}^{Al\Gamma_n(r)}(\mathbf{Z}_p, \mathbf{Z}_p).$$

Let $\mathcal{L}Al\Gamma_n(r) \subset Al\Gamma_n(r)$ denote the set of all elements of the form $log(\gamma)$ with $\gamma \in \Gamma_n(r)$ where

$$\log(1+z) = \sum_{m=0}^{\infty} (-1)^{m+1} z^m/m$$

is the usual log-series, calculated in $Al\Gamma_n(r)$. Hence $\mathcal{L}Al\Gamma_n(r)$ consists of primitives in $Al\Gamma_n(r)$, with respect to the diagonal map induced $\Gamma_n(r) \longrightarrow \Gamma_n(r) \times \Gamma_n(r)$. Let $\mathcal{L}^*Al\Gamma_n(r)$ denote the primitives of valuation greater than or equal to $(p-1)^{-1}$. The exponential series is defined on $\mathcal{L}^*Al\Gamma_n(r)$ and is inverse to the logarithm, giving a bijection between $\mathcal{L}^*Al\Gamma_n(r)$ and $\Gamma_n(r)$ ([**17**], Chapter IV, Theorem 3.2.5, p. 134).

Since $\mathcal{L}^*Al\Gamma_n(r)$ is a Lie algebra, there is a canonical map

$$\rho : U\mathcal{L}^*Al\Gamma_n(r) \longrightarrow Al\Gamma_n(r)$$

which becomes an isomorphism after applying the p-saturation functor, Sat, of ([**17**], Chapter I, Section 2.2, p. 25)

$$\rho : SatU\mathcal{L}^*SatAl\Gamma_n(r) \xrightarrow{\cong} SatAl\Gamma_n(r)$$

([**17**], Chapter IV, Theorem 3.2.5, p. 134). Here U denotes the universal enveloping algebra. This is a p-adically topologised version of the familiar structure theorem for irreducible, cocommutative Hopf algebras over a field (c.f. [**39**], Theorem 13.0.1, p. 274)

The saturation, $SatA$, consists of those elements in $\mathbf{Q}_p \otimes A$ of valuation greater than or equal to zero. The filtered $\mathbf{Q}_p$-homology of $SatAl\Gamma_n(r)$ and $Al\Gamma_n(r)$ are equal ([**17**], Chapter V, Lemma 2.4.8, p. 177) and similarly for $U\mathcal{L}^*Al\Gamma_n(r)$. Since $\mathcal{L}^*Al\Gamma_n(r)$ is $\mathbf{Z}_p$-free the (filtered) Lie algebra homology of $\mathcal{L}^*Al\Gamma_n(r)$ with $\mathbf{Z}_p$ coefficients calculable via the Koszul resolution ([**4**], p. 270) is isomorphic to $H_*^{(p)}(\Gamma_n(r); \mathbf{Z}_p)$.

However, there is a natural algebra map from $\mathbf{Z}_p[\Gamma_n(r)]$ into the $n \times n$ matrix ring, $\mathcal{GL}_n(\mathcal{O}_L)$. This induces a map from the Lie algebra homology of $\mathcal{L}^*Al\Gamma_n(r)$ to $H_*(\mathcal{GL}_n(\mathcal{O}_L); \mathbf{Z}_p)$.

The explicit formulae of Tr_* and λ_* of [**21**] combined with the fact that $\mathcal{L}^*Al\Gamma_n(r)$ is generated by logarithms of elements of $\Gamma_n(r)$ may be used to establish the following result:

Proposition 5.2 *For all $r \geq 1$*

$$r(H_3(\Gamma_n(r); \mathbf{Z}_p)) \subset \pi_L^r \mathcal{O}_L \subset L.$$

(b) The dilogarithmic regulator

In [**2**] the dilogarithm is used to define a homomorphism of the form

$$K_3(\mathbf{C}) \longrightarrow \mathbf{C}$$

using the presentation for $K_3(\mathbf{C})$ given by the Bloch group, $B(\mathbf{C})$, ([**2**], Section 6.2, p. 62; [**38**], Section 1). In [**8**] p-adic polylogarithms are constructed and, in terms of $B(L)$, a homomorphism

$$K_3(\mathbf{C}_p) \longrightarrow \mathbf{C}_p$$

is constructed, where $\mathbf{C}_p$ is the completion of an algebraic closure of $\mathbf{Q}_p$. This homomorphism is Galois equivariant and therefore induces a regulator homomorphism

$$dilog : K_3(L) \longrightarrow L$$

which we shall call the *Coleman dilogarithm.*

(c) The étale regulator

There are a number of Chern classes defined on algebraic K-theory. Each of these may be used to construct regulator maps.

For example, in étale cohomology the Chern classes defined in ([**9**], [**35**], [**36**])

$$ch_{2r-1,p}^{\acute{e}t} : K_{2r-1}(L; \mathbf{Z}/p^s) \longrightarrow H^1_{\acute{e}t}(L; (\mathbf{Z}/p^s)(r))$$

induce

$$ch_{2r-1}^{\acute{e}t} : K_{2r-1}^{top}(L) \longrightarrow H^1_{\acute{e}t}(L; \mathbf{Z}_p(r))$$

where $K_{2r-1}^{top}(L) = \lim_{\overleftarrow{s}} K_{2r-1}(L; \mathbf{Z}/p^s)$. In dimension three

$$ch_3^{\acute{e}t} : K_3^{top}(L) \longrightarrow H^1_{\acute{e}t}(L; \mathbf{Z}_p(2))$$

is an isomorphism ([**18**], [**26**]).

(d) The syntomic regulator

Let p be an odd prime. Let E/F be a Galois extension of number fields and let $L = E_Q$ denote the completion of E at some prime, $Q \triangleleft \mathcal{O}_E$ over p.

Let $H^*(\mathcal{O}_L; S_\infty(m)_{\mathbf{Q}_p})$ denote the $\mathbf{Q}_p$-syntomic cohomology introduced in [**10**] (see also [**16**]). These groups are vector spaces over $\mathbf{Q}_p$. In [**12**] syntomic Chern classes are constructed of the form

$$ch_{2r-1}^{syn} : K_{2r-1}(L) \otimes \mathbf{Q}_p \longrightarrow H^1(\mathcal{O}_L; S_\infty(r)_{\mathbf{Q}_p}).$$

In [**13**] these Chern classes are refined to define "integral" Chern classes when $r < p$. Hence we have

$$ch_{2r-1}^{syn} : K_{2r-1}(L) \longrightarrow H^1(\mathcal{O}_L; S_\infty(r)).$$

There is a canonical isomorphism ([**10**], Chapter III, Sections 5-6) of the form

$$f_M : H^1(\mathcal{O}_L; S_\infty(r)_{\mathbf{Q}_p}) \stackrel{\cong}{\longrightarrow} H^1_{\acute{e}t}(L; \mathbf{Q}_p(r)).$$

such that ([**13**], Chapter II, Section 3.4; [**16**], Theorem 3.9)

$$f_M \cdot ch_{2r-1}^{syn} = ch_{2r-1}^{\acute{e}t} : K_{2r-1}(\mathcal{O}_L) \longrightarrow H^1_{\acute{e}t}(L; \mathbf{Q}_p(r)).$$

There is another canonical isomorphism

$$\alpha : H^1(\mathcal{O}_L; S_\infty(r)_{\mathbf{Q}_p}) \stackrel{\cong}{\longrightarrow} L.$$

In addition, if $L/\mathbf{Q}_p$ is unramified ([**16**], Lemma 1.1) then α induces an isomorphism of the form

$$\alpha : H^1(\mathcal{O}_L; S_\infty(r)) \xrightarrow{\cong} \mathcal{O}_L \subset L.$$

Since $ch_3^{\acute{e}t} : K_3^{top}(L) \longrightarrow H^1_{\acute{e}t}(L; (\mathbf{Z}_p)(2))$ is always an isomorphism, we see that the inverse of f_M induces an injection

$$f_M^{-1} : H^1_{\acute{e}t}(L; \mathbf{Z}_p(2))/Tors \longrightarrow \mathcal{O}_L$$

with finite cokernel. In ([**16**], Corollary 3.12) a formula is given for the order of the cokernel of f_M^{-1} which shows that f_M^{-1} is an isomorphism for almost all p, when $L/\mathbf{Q}_p$ is unramified. Therefore, for almost all p when $L/\mathbf{Q}_p$ is unramified, the isomorphism

$$\alpha \cdot ch_3^{syn} : K_3^{top}(L) \otimes \mathbf{Q}_p \longrightarrow L$$

maps $K_3^{top}(L)$ onto $\mathcal{O}_L$.

Results of [**13**] and [**34**] relate the syntomic regulator to the dilogarithmic regulator.

(e) The topological cyclic homology regulator

In [**3**] and [**14**] it is shown that the cyclotomic trace gives an isomorphism of the form

$$K_n^{top}(L) \xrightarrow{\cong} TC_n(\mathcal{O}_L)_p^{\wedge}$$

where TC_* denotes topological cyclic homology. The resulting regulators are related to the Coates-Wiles homomorphism [**1**].

We come now to the main result of this section.

Theorem 5.3 *Let E/F be a Galois extension of number fields. Then, in the formal expression of Section 1*

$$\Omega_1(E/F,2) = [X] + \sum_P Ind_{G(E_Q/F_P)}^{G(E/F)}(\Omega_1(E_Q/F_P, X_Q)) \in \mathcal{CL}(\mathbf{Z}[G(E/F)])$$

(the sum is over $P \triangleleft \mathcal{O}_F$) the terms, $\Omega_1(E_Q/F_P, X_Q)$, are zero for almost all p when $E_Q/\mathbf{Q}_p$ is unramified. In fact, the class-group element defined by this expression is independent of the choices of X and of all the choices involved in the construction of the local fundamental 2-extensions associated to the $\{E_Q/F_P\}$.

In addition, $\Omega_1(E/F,2)$ satisfies the following naturality properties with respect to an inclusion of a subgroup, $G(L/N) \subseteq G(L/K)$:

(i) *the induced map*

$$Res_{G(E/N)}^{G(E/F)} : \mathcal{CL}(\mathbf{Z}[G(E/F)]) \longrightarrow \mathcal{CL}(\mathbf{Z}[G(E/N)])$$

satisfies

$$Res_{G(E/N)}^{G(E/F)}(\Omega_1(E/F,2)) = \Omega_1(E/N,2)$$

and

(ii) *if $G(E/N) \triangleleft G(E/F)$ the induced map*

$$Inf_{E/N} : \mathcal{CL}(\mathbf{Z}[G(E/F)]) \longrightarrow \mathcal{CL}(\mathbf{Z}[G(N/F)])$$

satisfies

$$Inf_{E/N}(\Omega_1(E/F,2)) = \Omega_1(N/F,2).$$

Proof Firstly, let us examine the local invariants when $E_Q/\mathbf{Q}_p$ is unramified and

$$\alpha \cdot ch_3^{syn} : K_3^{top}(E_Q) \otimes \mathbf{Q}_p \longrightarrow E_Q$$

maps $K_3^{top}(E_Q)$ onto $\mathcal{O}_{E_Q}$. The discussion of Section 5.1(d) shows that this applies to all but finitely many primes, p. In this case, excluding $p = 2, 3$ if necessary, our local fundamental 2-extension becomes, up to quasi-isomorphism,

$$K_3(\overline{L}) \longrightarrow A' \longrightarrow B' \longrightarrow \mu(\overline{L}).$$

The right-hand module is cohomologically trivial and represents zero in the class-group ([**30**] p.352) and the left-hand module is equal to

$$K_3(\overline{L}) \cong (\mu(M) \otimes \mu(M))^{G(M/\overline{L})}$$

for some large finite field, M. Hence this module is also cohomologically trivial and represents zero in the class-group. Therefore the local Euler characteristic is zero if $E_Q/\mathbf{Q}_p$ is Galois and unramified. However, the general unramified case reduces quickly to the Galois case, by Theorem 4.6, so that only a finite number of the local Euler characteristics associated with E/F and X are possibly non-zero.

To show that $\Omega_1(E/F, 2)$ is well-defined one observes that the Euler characteristic in the class-group of the local fundamental 2-extension depends only on its quasi-isomorphism class and the argument of ([**30**], Section 7.2.7) shows that $\Omega_1(E/F, 2)$ is independent of the choice of the lattice, X.

For part (i), we have

$$X \subset \prod_Q \mathcal{O}_{E_Q}$$

which is a locally free $\mathbf{Z}[G(E/F)]$-module which may be used to define both $\Omega_1(E/F, 2)$ and $\Omega_1(E/N, 2)$.

For each prime, $P \triangleleft \mathcal{O}_F$ lying under $Q \triangleleft \mathcal{O}_E$, the double cosets

$$G(E/N)\backslash G(E/F)/G(E_Q/F_P)$$

parametrise the primes, $R \triangleleft \mathcal{O}_N$, divisible by Q and dividing P. Hence the Double Coset Formula, together with the naturality of the local fundamental 2-extension, implies

$$Res_{G(E/N)}^{G(E/F)}(Ind_{G(E_Q/F_P)}^{G(E/F)}(\Omega_1(E_Q/F_P, X_Q))$$

$$= \sum_{Q|R|P} Ind_{G(E_Q/N_R)}^{G(E/N)}(\Omega_1(E_Q/N_R, X_Q)).$$

Therefore the restriction map

$$Res_{G(E/N)}^{G(E/F)} : \mathcal{CL}(\mathbf{Z}[G(E/F)]) \longrightarrow \mathcal{CL}(\mathbf{Z}[G(E/N)])$$

satisfies

$$Res_{G(E/N)}^{G(E/F)}(\Omega_1(E/F, 2)) = \Omega_1(E/N, 2).$$

For part (ii), we first recall the behaviour of induced modules undertaking of fixed points. Suppose that $H \triangleleft G$, $J \subseteq G$ and let M be a $\mathbf{Z}[G]$-module. Then there is an isomorphism of $\mathbf{Z}[G]$-modules,

$$\mu : Ind_{HJ}^G(M^{J\cap H}) \xrightarrow{\cong} (Ind_J^G(M))^H$$

defined by $(g \in G, m \in M^{J \cap H})$

$$\mu(g \otimes_{HJ} m) = \sum_{h \in H/H \cap gJg^{-1}} hg \otimes_J m.$$

Now let us apply this fact to evaluate $Inf_{E/N}(\Omega_1(E/F,2))$ where $H = G(E/N) \lhd G(E/F) = G$. If $Q \lhd \mathcal{O}_E$ divides $P \lhd \mathcal{O}_F$ then $G(E/N) \cap G(E_Q/F_P) = G(N_R/F_P)$ where N_R is the completion of N in E_Q. By definition, we have

$$Inf_{E/N}(\Omega_1(E/F,2))$$

$$= [X^{G(E/N)}] + \textstyle\sum_P Ind_{G(E_Q/F_P)}^{G(E/F)}(\Omega_1(E_Q/F_P, X_Q))^{G(E/N)}$$

$$= [X^{G(E/N)}] + \sum_P Ind_{G(E/N)G(E_Q/F_P)}^{G(E/F)}(\Omega_1(E_Q/F_P, X_Q)^{G(E/N) \cap G(E_Q/F_P)})$$

$$= [X^{G(E/N)}]$$

$$+ \sum_{Q \mid R \mid P} Ind_{G(E_Q/F_P)/(G(E/N) \cap G(E_Q/F_P))}^{G(N/F)}(\Omega_1(E_Q/F_P, X_Q)^{G(E/N) \cap G(E_Q/F_P)})$$

$$= [X^{G(E/N)}] + \sum_{R \mid P} Ind_{G(N_R/F_P)}^{G(N/F)}(\Omega_1(N_R/F_P, X_R))$$

$$= \Omega_1(N/F, 2),$$

since $X_Q^{G(E/N) \cap G(E_Q/F_P)} = X_R$. This completes the proof of Theorem 5.3. □

References

[1] Bentzen, S. and Madsen, I. [1990], *Trace maps in algebraic K-theory and the Coates-Wiles homomorphism*, J. reine angew. Math. **411**, 171-195.

[2] Bloch, S., *Higher regulators, algebraic K-theory and zeta functions of elliptic curves*, Irvine University Lecture Notes.

[3] Böksted, M. and Madsen, I. [1994], *Algebraic K-theory of local fields: the unramified case*, Aarhus preprint #20.

[4] Cartan, H. and Eilenberg, S. [1956], *Homological algebra*, Princeton University Press.

[5] Chinburg, T. [1985], *Exact sequences and Galois module structure*, Annals of Mathematics **121**, 351-376.

[6] Chinburg, T., Kolster, M., Pappas, G. and Snaith, V. P. [1995], *Galois structure of K-groups of rings of integers*, C.R. Acad. Sci., Paris, t. 320, Série I, 1435-1440.

[7] Chinburg, T., Kolster, M., Pappas, G. and Snaith, V. P. [1997], *Quaternionic exercises in K-theory Galois module structure*, this volume.

[8] Coleman, R. F. [1982], *Dilogarithms, regulators and p-adic L-functions*, Inventiones Math. **69**, 171-208.

[9] Dwyer, W. G. and Friedlander, E. M. [1985], *Algebraic and étale K-theory*, Trans. A. M. Soc. **292 (1)**, 247-280.

[10] Fontaine, J-M. and Messing, W. [1987], *p-adic periods and p-adic étale cohomology*, in Current Trends in Arithmetical Algebraic Geometry (K. A. Ribet, ed.) Contemp. Math. A. M. Soc. **67**, 179-207.

[11] Fröhlich, A. [1983], *Galois Module Structure of Algebraic Integers*, Ergeb. Math. 3 Folge-Band 1, Springer-Verlag.

[12] Gros, M. [1990], *Régulateurs syntomiques et valeurs de fonctions L p-adiques I*, (with an appendix by M. Kurihara), Inventiones Math. **99**, 293-320.

[13] Gros, M. [1994], *Régulateurs syntomiques et valeurs de fonctions L p-adiques II*, Inventiones Math. **115**, 61-79.

[14] Hesselholt, L. and Madsen, I. [1995], *On the K-theory of finite algebras over Witt vectors of perfect fields*, Aarhus preprint #2.

[15] Kahn, B. [1993], *Descente Galoisienne et K_2 des corps de nombres*, K-theory **7 (1)**, 55-100.

[16] Kolster, M. and Nguyen Quang Do, Thong [1995], *Syntomic regulators and special values of p-adic L-functions*, McMaster University preprint.

[17] Lazard, M. [1965], *Groupes analytiques p-adiques*, Pub. Math. I.H.E.S. Paris **26**, 389-603.

[18] Levine, M. [1989], *The indecomposable K_3 of a field*, Ann. Sci. École Norm. Sup. **22**, 255-344.

[19] Lichtenbaum, S. [1987], *The construction of weight-two arithmetic cohomology*, Inventiones Math. **88**, 183-215.

[20] Lichtenbaum, S. [1990], *New results on weight-two motivic cohomology*, The Grothendieck Festschrift vol.III, (Birkhäuser), pp. 35-55.

[21] Loday, J-L. and Quillen, D. G. [1984], *Cyclic homology and the Lie algebra homology of matrices*, Comm. Math. Helv. **59**, 565-591.

[22] Loday, J-L. and Quillen, D. G. [1983], *Homologie cyclique et homologie de l'algèbre de Lie des matrices*, C. R. Acad. Sci. Paris t.**296**, 295-297.

[23] Merkurjev, A. S. [1983], *On the torsion of K_2 of local fields*, Annals of Maths. (2) **118**, 375-381.

[24] Merkurjev, A. S. [1986], *K_2 of fields and the Brauer group*, in Applications of algebraic K-theory to geometry and number theory (Part II), Contemp. Math. **55**, 529-546.

[25] Murkurjev, A. S. and Suslin, A. A. [1982], *K-cohomology of Severi-Brauer varieties and the norm residue homomorphism*, Izv. Nauk. SSSR **46**, 1011-1046 (Eng. trans. Math. USSR 21 (1983), 307-340).

[26] Murkurjev, A. S. and Suslin, A. A. [1990], *The K_3 group of a field*, Izv. Nauk. SSSR **54**, 339-356 (Eng. trans. Math. USSR 36 (1990), 541-565).

[27] Quillen, D. G. [1973], *Higher Algebraic K-theory I*, Battelle K-theory Conf. 1972, Lecture Notes in Math. **371**, 85-147.

[28] Serre, J-P. [1979], *Local Fields*, Grad. Texts in Math. **67**, Springer-Verlag.

[29] Snaith, V. P. [1989], *Topological Methods in Galois Representation Theory*; C. M. Soc. Monographs, Wiley.

[30] Snaith, V. P. [1994], *Explicit Brauer Induction (with applications to algebra and number theory)*; Cambridge Studies in Advanced Math. **40**, Cambridge University Press.

[31] Snaith, V. P. [1994], *Galois Module Structure*, The Fields Institute for Research in Mathematical Sciences Monograph Series, **2**, A. M. S., Providence, R.I.

[32] Snaith, V. P. [1996/7], *Local fundamental classes derived from higher K-groups in characteristic p*, McMaster University preprint #2.

[33] Snaith, V. P., *Local fundamental classes derived from higher-dimensional K-groups II*, this volume.

[34] Somekawa, M. [1995], *Log-syntomic regulators and p-adic polylogarithms*, Tokyo Inst. Tech., preprint.

[35] Soulé, C. [1979], *K-théorie des anneaux d'entiers de corps de nombres et cohomologie étale*, Inventiones Math. **55**, 251-295.

[36] Soulé, C. [1981], *On higher p-adic regulators*, Springer-Verlag Lecture Notes in Mathematics **854**, 372-401.

[37] Suslin, A. A. [1984], *On the K-theory of local fields*, J. P. A. Alg. (2-3) **34**, 301-318.

[38] Suslin, A. A. [1991], *K_3 of a field and the Bloch group*, Proc. Steklov Inst. Math. **4**, 217-239.

[39] Sweedler, M. E. [1969], *Hopf Algebras*, W. A. Benjamin.

[40] Wagoner, J. G. [1976], *Continuous cohomology and p-adic K-theory*, Springer-Verlag Lecture Notes in Mathematics **551**, 241-248.

Fields Institute Communications
Volume **16**, 1997

Local Fundamental Classes Derived from Higher-Dimensional K-Groups: II

Victor P. Snaith
Department of Mathematics
McMaster University
Hamilton, Ontario, Canada L8S 4K1
snaith@mcmaster.ca

Abstract. f L/K is a Galois extension of local fields then local class field theory constructs a 2-extension of Galois modules between L^* and $\mathbb{Z}$. Here we construct, modulo the Lichtenbaum-Quillen conjecture, correspondingly natural 2-extensions between $K_{1r+1}(L)$ and $K_{2r}(L)$ for all $r \geq 1$. In particular, for 2-adic local fields results of V. Voevodsky imply that there 2-extensions exist unconditionally.

1 Introduction

The purpose of this paper is to extend to dimensions higher than three the results concerning local fundamental classes coming from the higher dimensional K-groups of p-adic local fields. In brief, for each Galois extension of p-adic local fields, L/K, with Galois group $G(L/K)$ we construct a canonical element of

$$Ext^2_{\mathbf{Z}[G(L/K)]}(K_{2r-2}(L), K_{2r-1}(L))$$

which is natural with respect to passage to sub-groups and quotient groups and which induces an isomorphism in Tate cohomology

$$\hat{H}^i(G(L/K); K_{2r-2}(L)) \xrightarrow{\cong} \hat{H}^{i+2}(G(L/K); K_{2r-1}(L)).$$

The classical fundamental classes of local class field theory correspond to the case when $r = 1$ and the results of [**27**] deal with the case when $r = 2$.

This paper verifies that the results of [**27**] work in higher dimensions. As originally conceived, the paper would have been about higher étale K-theory groups but recent results of [**31**] have rather superannuated étale K-theory.

There should be a great deal more to be said about these fundamental classes. For example, they should exist (at least in dimension three) for any Galois extension of n-dimensional local fields with separable residue field extension. In other words, as explained to me by I. Fesenko, the results on K_2 of local fields which are used in [**27**] hold for higher dimensional local fields, using the Vostokov symbol to replace

1991 *Mathematics Subject Classification.* Primary 11R37; Secondary 19F05.
Research partially supported by an NSERC grant.

the norm residue symbol to analyse K_2 of such a field. When correctly extended these results should lead to fundamental classes involving the associated graded modules of a filtration on the K-groups in a similar, perhaps even identical, to the conjectured motivic $\Gamma(r)$-complexes of S. Lichtenbaum ([**14**], [**15**]).

The following hypothesis will be assumed throughout this paper.

Hypothesis: Let p be a prime and let F be a p-adic local field. Then the canonical map from K-theory to étale K-theory

$$K_i(F; \mathbf{Z}/p^m) \longrightarrow K_i^{ét}(F; \mathbf{Z}/p^m)$$

is an isomorphism for all $m \geq 1$ and all $i \geq 0$.

When dealing with p-adic fields we shall need this hypothesis for p, otherwise we merely obtain results for étale K-theory which, being only a p-adic gadget, is not sufficient for the arithmetical constructions we wish to make. However, for 2-adic fields we require *no hypothesis*, because of the following result of V Voevodsky.

Theorem 1.1 *Let F be any field of characteristic zero. Then the canonical map*

$$K_i(F; \mathbf{Z}/2^m) \longrightarrow K_i^{ét}(F; \mathbf{Z}/2^m)$$

is an isomorphism for all $m \geq 1$ and all $i \geq 0$.

The paper is arranged in the following manner. In Section 2, using [**31**], we collect all the relevant facts concerning the higher groups of local fields. In Section 3 we construct the canonical 2-extensions. In Section 4 we sketch the properties of the fundamental 2-extensions under changes of chosen parameters and under change of groups. In Section 5 we sketch how to use these fundamental classes to construct a higher dimensional K-theoretic Galois module structure invariant, lying in the class-group of the group ring of the Galois group, which generalises the second Chinburg invariant.

2 Higher K-Theory of Local Fields

2.1 Throughout this paper, if A is an abelian group, we shall denote by $\mathrm{Tors}(A)$ the torsion subgroup of A. Also we set ${}_nA = \{a \in A \mid na = 0\}$, ${}_{p^\infty}A = \cup_{m=1}^{\infty} {}_{p^m}A$, $A/n = A/nA = A \otimes \mathbf{Z}/n$ and $K_i(X)$ will denote the i-dimensional Quillen K-group of the ring or scheme, X.

Let L/K be a totally ramified Galois extension of p-adic local fields with Galois group, $G(L/K)$. Let K_0/K denote the maximal unramified extension and set $L_0 = K_0L$. Let $F \in G(K_0/K) \cong \hat{\mathbf{Z}} \cong G(L_0/L)$ denote the topological generator given by the Frobenius then we have an isomorphism of the form

$$G(L_0/K) \cong G(L/K) \times G(K_0/K) \cong G(L/K) \times \hat{\mathbf{Z}}.$$

For each integer, s, let $K(s) \subset K_0$ denote the fixed field of $s\hat{\mathbf{Z}}$ and set $L(s) = K(s)L$. Hence $G(L(s)/L) \cong \mathbf{Z}/s$.

For $r \geq 2$ the spectral sequence

$$E_2^{s,t} = H^s(G(L(s)/L); H^t(L(s); \mathbf{Z}_p(r))) \Longrightarrow H^{s+t}(L; \mathbf{Z}_p(r))$$

has only two non-zero rows, corresponding to $t = 0, 1$, so that we have a natural exact sequence of the form

$$0 \longrightarrow H^1(G(L(s)/L); H^1(L(s); \mathbf{Z}_p(r))) \longrightarrow H^2(L; \mathbf{Z}_p(r))$$

$$\longrightarrow H^2(L(s); \mathbf{Z}_p(r))^{G(L(s)/L)} \xrightarrow{d_2} H^2(G(L(s)/L); H^1(L(s); \mathbf{Z}_p(r))) \longrightarrow 0.$$

Let Ω_K denote the absolute Galois group of K.

Proposition 2.2 *In the notation of Section 2.1,*

(i) *The p-primary torsion subgroup, ${}_{p^\infty}K_{2r-1}(L)$, is finite for all $r \geq 1$. In fact,*

$${}_{p^\infty}K_{2r-1}(L) \cong ((\mathbf{Q}_p/\mathbf{Z}_p)(r))^{\Omega_L}.$$

(ii) *For $r \geq 2$, the p-primary torsion subgroup, ${}_{p^\infty}K_{2r-2}(L)$, is also finite.*

(iii) *There is a natural isomorphism (of finite groups) of the form*

$$H^1(G(L(s)/L); K_{2r-1}(L(s)) \otimes \mathbf{Z}_p) \cong H^1(G(L(s)/L); H^1(L(s); \mathbf{Z}_p(r))).$$

Proof Let π_L denote a uniformising element for L and let $\mathcal{O}_L$ denote the ring of integers in L. Then $K_*^{top}(L)$ is defined to be the inverse limit

$$K_*^{top}(L) = \lim_{\overleftarrow{n}} K_*(\mathcal{O}_L/(\pi_L^n)).$$

By ([**30**]; c.f. [**12**] Théorème 7.2) there is a natural map whose kernel and cokernel are uniquely divisible groups (possibly zero). Also

$$K_*^{top}(L) \cong K_*(\mathcal{O}_L/\pi_L) \oplus K_{*-1}(\mathcal{O}_L/\pi_L) \oplus X_*$$

where X_* is a finitely generated $\mathbf{Z}_p$-module in each dimension. By [**32**], $X_j \otimes \mathbf{Q}_p$ is zero when $j > 0$ is even and satisfies $dim_{\mathbf{Q}_p}(X_{2r-1} \otimes \mathbf{Q}_p) = [L : \mathbf{Q}_p]$ for $r \geq 1$. In addition, for $j > 0$, $K_j(\mathcal{O}_L/(\pi_L)$ is a finite group of order prime to p, by [**22**].

For part (i), consider the following commutative diagram of short exact sequences in which the right-hand vertical map is given by the inclusion.

$$\begin{array}{ccccc}
K_{2r}(L)/p^n & \longrightarrow & K_{2r}(L; \mathbf{Z}/p^n) & \longrightarrow & {}_{p^n}K_{2r-1}(L) \\
\Big\downarrow p & & \Big\downarrow & & \Big\downarrow \\
K_{2r}(L)/p^{n+1} & \longrightarrow & K_{2r}(L; \mathbf{Z}/p^{n+1}) & \longrightarrow & {}_{p^{n+1}}K_{2r-1}(L)
\end{array}$$

We need only consider the case when $r \geq 2$, since the result is well-known when $r = 1$. By [**31**] there exists a spectral sequence (the étale K-theory spectral Atiyah-Hirzebruch spectral sequence)

$$E_2^{s,-2t} = H^s(L; \mathbf{Z}/p^n(t)) \Longrightarrow K_{2t-s}(L; \mathbf{Z}/p^n).$$

Since the non-p-divisible part of $K_{2r}(L)$ consists of p-primary torsion

$$\lim_{\overrightarrow{n}} K_{2r}(L; \mathbf{Z}/p^n) \cong {}_{p^\infty}K_{2r-1}(L).$$

However, $E_2^{s,-2t}$ is non-zero only when $s = 0, 1, 2$ and

$$\lim_{\overrightarrow{n}} H^2(L; \mathbf{Z}/p^n(t)) = 0$$

when $t \neq 1$, by a result of Tate (c.f. [**13**], Section 1.1). Therefore the spectral sequence implies that

$${}_{p^\infty}K_{2r-1}(L) \cong lim_{\overrightarrow{n}} H^0(L; \mathbf{Z}/p^n(r)) \cong ((\mathbf{Q}_p/\mathbf{Z}_p)(r))^{\Omega_L},$$

as required.

For part (ii) we repeat the argument with $2r$ replaced by $2r+1$. By part (i) and the previous discussion

$$lim_{\overrightarrow{n}} K_{2r+1}(L)/p^n \cong \oplus_{i=1}^{[L:\mathbf{Q}_p]} \mathbf{Q}_p/\mathbf{Z}_p$$

and

$$lim_{\overrightarrow{n}} K_{2r+1}(L; \mathbf{Z}/p^n) \cong H^1(L; (\mathbf{Q}_p/\mathbf{Z}_p)(r+1)),$$

which is a $\mathbf{Z}_p$-module of co-rank equal to $[L : \mathbf{Q}_p]$. Hence the quotient group, isomorphic to ${}_{p^\infty}K_{2r}(L)$, is a finite p-group.

For part (iii) we observe that $K_{2r-1}(L(s)) \otimes \mathbf{Z}_p$ and $\lim_{\overleftarrow{n}} K_{2r-1}(L(s)) \otimes \mathbf{Z}/p^n$ differ by $\mathbf{Q}_p$-vector spaces and therefore have the same cohomology. On the other hand, by part (ii),

$$\lim_{\overleftarrow{n}} K_{2r-1}(L(s)) \otimes \mathbf{Z}/p^n \cong \lim_{\overleftarrow{n}} K_{2r-1}(L(s); \mathbf{Z}/p^n)$$

and the result follows from the spectral sequence. □

Proposition 2.3 *There is a natural isomorphism of the form*

$$H^2(L; \mathbf{Z}_p(r)) \cong {}_{p^\infty}K_{2r-2}(L)$$

for all $r \geq 2$.

Proof From Proposition 2.2 and the inverse limit of the exact sequences

$$0 \longrightarrow K_{2r}(L) \otimes \mathbf{Z}/p^n \longrightarrow K_{2r}(L; \mathbf{Z}/p^n) \longrightarrow {}_{p^n}K_{2r-1}(L) \longrightarrow 0$$

we see that

$${}_{p^\infty}K_{2r}(L) \cong \lim_{\overleftarrow{n}} K_{2r}(L; \mathbf{Z}/p^n)$$

which is isomorphic to $H^2(L; \mathbf{Z}_p(r+1))$, from the spectral sequence, since $H^0(L; \mathbf{Z}_p(r+1)) = 0$. □

Corollary 2.4 *There is a natural isomorphism of the form*

$$H^1(G(L(s)/L); K_{2r-1}(L(s))) \cong Ker(K_{2r-2}(L) \longrightarrow K_{2r-2}(L(s))^{G(L(s)/L)})$$

for all $r \geq 2$.

Proof By a transfer argument, the kernel of the natural map from $K_{2r-2}(L)$ to $K_{2r-2}(L(s))^{G(L(s)/L)}$ consists of s-torsion. By Section 2.1, Proposition 2.2(iii) and Proposition 2.3 we have only to verify the result at primes different from p.

If $\mathcal{O}_L/\pi_L \cong \mathbf{F}_q$, $\mathcal{O}_{L(s)}/\pi_{L(s)} \cong \mathbf{F}_{q^s}$ and if $\overline{\mathbf{F}}_q$ denotes the separable closure of $\mathbf{F}_q$ then, away from p,

$$\begin{aligned} &H^1(G(L(s)/L); K_{2r-1}(\mathcal{O}_{L(s)}/\pi_{L(s)})) \\ &\cong H^1(G(\mathbf{F}_{q^s}/\mathbf{F}_q); K_{2r-1}(\mathbf{F}_{q^s})) \\ &\cong H^1(G(\mathbf{F}_{q^s}/\mathbf{F}_q); H^0(\mathbf{F}_{q^s}; (\overline{\mathbf{F}}_q^*)(r))) \\ &\cong H^1(\mathbf{F}_q; (\overline{\mathbf{F}}_q^*)(r))) \\ &\cong 0. \end{aligned}$$

On the other hand, away from p,

$$Ker(K_{2r-2}(L) \longrightarrow K_{2r-2}(L(s))^{G(L(s)/L)})$$

$$\cong Ker(K_{2r-3}(\mathbf{F}_q) \longrightarrow K_{2r-3}(\mathbf{F}_{q^s})^{G(\mathbf{F}_{q^s}/\mathbf{F}_q)})$$

which is trivial, as required. □

Proposition 2.5 *For $r \geq 2$ there are natural isomorphisms*

$$_{p^\infty}K_{2r-2}(L) \cong H^2(L; \mathbf{Z}_p(r)) \cong Hom(H^0(L; (\mathbf{Q}_p/\mathbf{Z}_p)(1-r)), \mathbf{Q}/\mathbf{Z}).$$

For each extension, E/F, of p-adic local fields this isomorphism identifies the maps

$$i_{E/F} : {}_{p^\infty}K_{2r-2}(F) \longrightarrow {}_{p^\infty}K_{2r-2}(E)$$

and

$$N_{E/F} : {}_{p^\infty}K_{2r-2}(E) \longrightarrow {}_{p^\infty}K_{2r-2}(F)$$

with the Pontrjagin duals of the maps

$$N_{E/F} : H^0(E; (\mathbf{Q}_p/\mathbf{Z}_p)(1-r)) \longrightarrow H^0(F; (\mathbf{Q}_p/\mathbf{Z}_p)(1-r)),$$

and

$$i_{E/F} : H^0(F; (\mathbf{Q}_p/\mathbf{Z}_p)(1-r)) \longrightarrow H^0(E; (\mathbf{Q}_p/\mathbf{Z}_p)(1-r)),$$

respectively.

Proof The Poitou-Tate duality isomorphism is induced by the cup-product

$$H^2(L; (\mathbf{Z}/p^n)(r)) \otimes H^0(L; (\mathbf{Z}/p^n)(1-r)) \longrightarrow H^2(L; (\mathbf{Z}/p^n)(1)) \cong {}_{p^n}Br(L)$$

where $Br(L) \cong \mathbf{Q}/\mathbf{Z}$ is the Brauer group of L.

For $x \in H^j(F; (\mathbf{Z}/p^n)(r))$ and $y \in H^{2-j}(E; (\mathbf{Z}/p^n)(1-r))$ we have the projection formula

$$N_{E/F}(i_{E/F}(x) \cup y) = x \cup N_{E/F}(y)$$

so that the result follows from the fact that the norm

$$N_{E/F} : Br(E) \cong \mathbf{Q}/\mathbf{Z} \longrightarrow Br(F) \cong \mathbf{Q}/\mathbf{Z}$$

is the identity map. □

Corollary 2.6

(i) *For any extension of p-adic local fields, E/F,*

$$N_{E/F} : K_{2r-2}(E) \longrightarrow K_{2r-2}(F)$$

is surjective for all $r \geq 2$.

(ii) *Also*

$$\bigcup_s Ker(K_{2r-2}(L) \longrightarrow K_{2r-2}(L(s))) = {}_{p^\infty}K_{2r-2}(L).$$

Proof On $H^0(-; M)$ the natural map, $i_{E/F}$ is injective so that its Pontrjagin dual is surjective, which proves part (i).

For part (ii), the argument used in the proof of Corollary 2.4 shows that

$$\bigcup_s Ker(K_{2r-2}(L) \longrightarrow K_{2r-2}(L(s))) \subseteq {}_{p^\infty}K_{2r-2}(L).$$

The Frobenius, $F \in G(L(s)/L)$, may be lifted to $F \in \Omega_L$ which fixes the p-primary roots of unity. Since $\Omega_L =< \Omega_{L(s)}, F >$, we see that

$$H^0(L; (\mathbf{Q}_p/\mathbf{Z}_p)(1-r)) = (\mathbf{Q}_p/\mathbf{Z}_p)(1-r)^{\Omega_L}$$

$$= (\mathbf{Q}_p/\mathbf{Z}_p)(1-r)^{\Omega_{L(s)}} = H^0(L(s); (\mathbf{Q}_p/\mathbf{Z}_p)(1-r)).$$

Therefore the norm

$$N_{L(s)/L} : (\mathbf{Q}_p/\mathbf{Z}_p)(1-r)^{\Omega_{L(s)}} \longrightarrow (\mathbf{Q}_p/\mathbf{Z}_p)(1-r)^{\Omega_L}$$

will be zero when s is a large power of p. In this case,

$$_{p^\infty}K_{2r-2}(L) \longrightarrow {}_{p^\infty}K_{2r-2}(L(s))$$

will be zero, by Proposition 2.5. □

Definition 2.7 *For $r \geq 2$, let*

$$\kappa(s) = Ker(N_{L(s)/L} : K_{2r-1}(L(s)) \longrightarrow K_{2r-1}(L)).$$

The map, $i_{L(st)/L(s)} : K_{2r-1}(L(s)) \longrightarrow K_{2r-1}(L(st))$, is injective. This is seen on the torsion surgroup by means of Proposition 2.2(i) and the identification of the map on the prime-to-p torsion with the injective map on K_{2r-1} of the residue fields. For the torsion-free part it suffices to show that $K_{2r-1}^{top}(L(s)) \otimes \mathbf{Q}_p \longrightarrow K_{2r-1}^{top}(L(st)) \otimes \mathbf{Q}_p$ is injective and this follows from [**32**].

Therefore we may define

$$\mathcal{U} = \bigcup_s \kappa(s) \subseteq \bigcup_s K_{2r-1}(L(s)) = K_{2r-1}(L_0).$$

Similarly,

$$Ker(1-F : K_{2r-1}(L(s)) \longrightarrow K_{2r-1}(L(s)))$$

is isomorphic to $K_{2r-1}(L(s)^{G(L(s)/L)}) \cong K_{2r-1}(L)$. This follows by a transfer argument on the uniquely divisible part, from the isomorphism

$$H^0(G(L(s)/L); H^1(L(s); \mathbf{Z}_p(r))) \cong H^1(L; \mathbf{Z}_p(r))$$

on the p-adic part and from the exactness of the sequence

$$0 \longrightarrow K_{2r-1}(\mathcal{O}_L/\pi_L) \longrightarrow K_{2r-1}(\mathcal{O}_{L(s)}/\pi_{L(s)}) \stackrel{1-F}{\longrightarrow} K_{2r-1}(\mathcal{O}_{L(s)}/\pi_{L(s)})$$

on the prime-to-p torsion.

For $r \geq 2$, taking the direct limit of the sequences

$$0 \longrightarrow K_{2r-1}(L) \longrightarrow K_{2r-1}(L(s)) \stackrel{1-F}{\longrightarrow} \kappa(s)$$

$$\longrightarrow Ker(K_{2r-2}(L) \longrightarrow K_{2r-2}(L(s))^{G(L(s)/L)}) \longrightarrow 0$$

yields the following result, by Corollary 2.6.

Theorem 2.8 *In the notation of Section 2.1, there is a natural 2-extension of $\mathbf{Z}[G(L/K)]$-modules of the form*

$$K_{2r-1}(L) \longrightarrow K_{2r-1}(L(s)) \stackrel{1-F}{\longrightarrow} \mathcal{U} \longrightarrow {}_{p^\infty}K_{2r-2}(L).$$

Lemma 2.9 *In the notation of Section 2.1 and Proposition 2.5, for all* $r \geq 2$, *the norm*

$$N_{L(s)/L} : K_{2r-1}(L(s)) \longrightarrow K_{2r-1}(L)$$

is a surjective if $HCF(p, s) = 1$ *and otherwise the finite cokernel is a cyclic* $\mathbf{Z}_p$-*module.*

Proof By the argument of ([**27**], Lemma 2.4) we may reduce to showing that the norm

$$N_{L(s)/L} : K_{2r-1}(L(s))/p \longrightarrow K_{2r-1}(L)/p$$

has cokernel equal to zero or $\mathbf{Z}/p$ under the assumption that L contains a primitive p-th root of unity, ξ_p. Since $K_{2r-1}(L(s); \mathbf{Z}/p) \cong H^1(L(s); (\mathbf{Z}/p)(r))$, from the spectral sequence, we have a short exact sequence of the form

$$0 \longrightarrow K_{2r-1}(L(s))/p \longrightarrow K_{2r-1}(L(s); \mathbf{Z}/p) \longrightarrow_p K_{2r-2}(L(s)) \longrightarrow 0.$$

Since ${}_pK_{2r-2}(L) \cong \mathbf{Z}/p \cong {}_pK_{2r-2}(L(s))$, Proposition 2.5 implies that the norm, $N_{L(s)/L} : {}_pK_{2r-2}(L(s)) \longrightarrow {}_pK_{2r-2}(L)$ is an isomorphism. Therefore the cokernel of the norm, $K_{2r-1}(L(s))/p \longrightarrow K_{2r-1}(L)/p$, is equal to the cokernel of the norm

$$N_{L(s)/L} : H^1(L(s); (\mathbf{Z}/p)(r)) \longrightarrow H^1(L; (\mathbf{Z}/p)(r)).$$

However, since $\xi_p \in L^*$, this norm may be identified with the norm

$$N_{L(s)/L} : H^1(L(s); (\mathbf{Z}/p)(1)) \longrightarrow H^1(L; (\mathbf{Z}/p)(1))$$

whose cokernel was shown, in ([**27**], Lemma 2.4), to be equal to $\mathbf{Z}/s \otimes \mathbf{Z}/p$. □

Proposition 2.10 *Let* L/K *be a totally ramified Galois extension of* p-*adic local fields with Galois group,* $G(L/K)$. *In the notation of Section 2.1, set*

$$\mathcal{O}_{L_0} = \bigcup_n \mathcal{O}_{L(n)} \subset L_0$$

so that the natural map, $K_{2r-1}(\mathcal{O}_{L_0}) \longrightarrow K_{2r-1}(L_0)$, *is an isomorphism for all* $r \geq 2$. *Let* $\overline{L}_0$ *denote the residue field of* L_0. *Then the composition*

$$K_{2r-1}(L_0) \cong K_{2r-1}(\mathcal{O}_{L_0}) \longrightarrow K_{2r-1}(\overline{L}_0) \cong \mathbf{Q}/\mathbf{Z}(r)[1/p]$$

induces a Tate cohomology isomorphism

$$\hat{H}^i(G(L/K); K_{2r-1}(L_0)) \longrightarrow \hat{H}^i(G(L/K); \mathbf{Q}/\mathbf{Z}(2)[1/p])$$

for each i.

Since $G(L/K)$ *acts trivially on* $\mathbf{Q}/\mathbf{Z}(r)[1/p]$ *these groups are given by*

$$\hat{H}^i(G(L/K); K_{2r-1}(L_0)) \cong \begin{cases} \mathbf{Z}/[G(L/K) : G_1(L/K)] & \text{if } i \text{ is odd,} \\ 0 & \text{if } i \text{ is even.} \end{cases}$$

Proof This follows by an easy modification of the argument of ([**27**], Propositions 2.7 and 2.11).

Corollary 2.11 *Let* L/K *be a totally ramified Galois extension of* p-*adic local fields with Galois group,* $G(L/K)$, *as in Section 2.1. In Definition 2.7 the inclusion,* $\mathcal{U} \subset K_{2r-1}(L_0)$, *induces Tate cohomology isomorphisms*

$$\hat{H}^i(G(L/K); \mathcal{U}) \xrightarrow{\cong} \hat{H}^i(G(L/K); K_{2r-1}(L_0))$$

for all i.

In addition,

$$(1-F)_* : \hat{H}^i(G(L/K); K_{2r-1}(L_0)) \longrightarrow \hat{H}^i(G(L/K); \mathcal{U})$$

is the zero map.

Proof From Lemma 2.9, by the argument of ([**27**], Corollary 2.6), we have an exact sequence of $\mathbf{Z}[G(L/K)]$-modules of the form

$$0 \longrightarrow \mathcal{U} \longrightarrow K_{2r-1}(L_0) \longrightarrow K_{2r-1}(L) \otimes \mathbf{Q} \longrightarrow (\mathbf{Q}_p/\mathbf{Z}_p)^e \longrightarrow 0$$

where $e = 0$ or $e = 1$. From Proposition 2.10 and the cohomological argument of ([**27**], Theorem 2.10) one sees that $e = 0$ if p divides $[L : K]$. This establishes the cohomology isomorphism.

The fact that $(1 - F)_*$ is zero on cohomology follows from the fact that, if $t = [G(L/K) : G_1(L/K)]$, then t divides $q - 1$ where $\mathcal{O}_L/\pi_L = \mathbf{F}_q$ but $(1 - F)_*$ acts on

$$\hat{H}^{2s+1}(G(L/K); (\mathbf{Q}/\mathbf{Z})(r)[1/p]) \cong {}_t(\mathbf{Q}/\mathbf{Z})(r)[1/p]$$

by multiplication by $1 - q^r$, which is zero on t-torsion. □

3 Construction of the Canonical 2-Extensions

3.1 *Galois groups in the general case*

Let L/K be a Galois extension of p-adic local fields with Galois group, $G(L/K)$. We shall relax the total ramification condition of Section 2.1. Set $W = L \cap L_0$ so that W/K is the maximal unramified subextension of L/K. Hence L/W is totally ramified and the sequence

$$0 \longrightarrow K_{2r-1}(L) \longrightarrow K_{2r-1}(L_0) \stackrel{1-F}{\longrightarrow} \mathcal{U} \longrightarrow {}_{p^\infty}K_{2r-2}(L) \longrightarrow 0$$

of Theorem 2.8 is a 2-extension of $\mathbf{Z}[G(L/W)]$-modules.

We wish to construct a similar 2-extension of $\mathbf{Z}[G(L/K)]$-modules, following the method of ([**27**], Section 3) which was in turn an imitation of the fundamental class of local class field theory, as described in ([**25**], Section 7.1; [**26**], Section 1.2).

Notice that $G(L/W) = G_0(L/K)$ and $G_1(L/W) = G_1(L/K)$, by ([**23**] p.62), so that $t = [G(L/W) : G_1(L/W)] = [G_0(L/K) : G_1(L/K)]$. If $G(L/E) = G_1(L/K)$ then

$$G(L/K)/G(L/E) \cong G(E/K) = \{a, g \mid a^t = 1, g^d = a^c, gag^{-1} = a^v\}$$

where $< a >= G(E/W)$, $d = [W : K]$ and $v = |\mathcal{O}_K/\pi_K|$.

Now let us consider the Galois groups which will be involved in the construction. We have a map, given by the restriction of g to W and denoted by $(g|W)$, inducing a homomorphism

$$h_1 : G(K_0/K) \times G(L/K) \cong \hat{\mathbf{Z}} \times G(L/K) \longrightarrow G(W/K)$$

defined by $h_1(F^i, z) = (F^i|W) \cdot (z|W)^{-1}$. Here, as in Section 2.1, F denotes the Frobenius automorphism. To a pair, $(F^i, z) \in Ker(h_1)$, we may associate the Galois automorphism of $L_0 = LK_0$ which is equal to F^i on K_0 and to z on L. This induces an isomorphism

$$Ker(h_1) \stackrel{\cong}{\longrightarrow} G(L_0/K).$$

If $d = [W : K]$ define $F_0 \in G(L_0/K)$ by $(F_0|K_0) = F^d$ and $(F_0|L) = 1$, the identity map.

As explained in ([**25**], pp. 303-304; [**26**], pp. 9-10), there is an isomorphism of K-algebras

$$\lambda : L \otimes_K K_0 \longrightarrow \oplus_{i=1}^d L_0$$

given by the formula $\lambda(\alpha \otimes \beta) = (F^{d-1}(\beta)\alpha, F^{d-2}(\beta)\alpha, \ldots, \beta\alpha)$ and fitting into a commutative diagram of the form

$$\begin{array}{ccc} L \otimes_K K_{nr} & \xrightarrow{(1 \otimes F)} & L \otimes_K K_{nr} \\ \Big\downarrow \lambda & & \Big\downarrow \lambda \\ \oplus_{i=1}^d L_0 & \xrightarrow{\hat{F}} & \oplus_{i=1}^d L_0 \end{array}$$

where $\hat{F}(x_1, \ldots, x_d) = (F_0(x_d), x_1, \ldots, x_{d-1})$. Since K-theory is additive there is an induced isomorphism

$$\lambda : K_s(L \otimes_K K_0) \longrightarrow \oplus_{i=1}^d K_s(L_0)$$

for each $s \geq 0$. This isomorphism induces a $G(L/K)$-action on the right-hand group, induced by the Galois action on L. Since the Frobenius acts on the other factor, K_0, in the tensor product we may define a $\mathbf{Z}[G(L/K)]$-homomorphism

$$\hat{F} : \oplus_{i=1}^d K_s(L_0) \longrightarrow \oplus_{i=1}^d K_s(L_0)$$

by the same formula as before.

The explicit formula for the group action on these modules is given by the following result, proved in ([**27**], Lemma 3.2).

Lemma 3.2 *Suppose that $g \in G(L/K)$ satisfies $(g|W) = (F^j|W)$ for some $0 \leq j \leq d-1$. Then the action of g on $(u_1, u_2, \ldots, u_d) \in \oplus_{i=1}^d K_{2r-1}(L_0)$ is given by*

$$g(u_1, u_2, \ldots, u_d) = \hat{F}^{-j}((F^j, g)(u_1), (F^j, g)(u_2), \ldots, (F^j, g)(u_d)).$$

Definition 3.3 *Let $\mathcal{U} \subset K_{2r-1}(L_0)$ be as in Definition 2.7. Define a subgroup, $\mathcal{W} \subset \oplus_{i=1}^d K_{2r-1}(L_0)$, by*

$$\mathcal{W} = \{(u_1, u_2, \ldots, u_d) | \sum_{i=1}^d u_i \in \mathcal{U}\}.$$

Then $\mathcal{W} \subset \oplus_{i=1}^d K_{2r-1}(L_0)$ is a $\mathbf{Z}[G(L/K)]$-submodule and the image of $1 - \hat{F}$: $\oplus_{i=1}^d K_{2r-1}(L_0) \longrightarrow \oplus_{i=1}^d K_{2r-1}(L_0)$ lies in $\mathcal{W}$ (c.f. [**27**], Lemma 3.4).

Proposition 3.4 *In the notation of Section 3.1, for $r \geq 2$, there is a natural exact sequence of $\mathbf{Z}[G(L/K)]$-modules of the form*

$$0 \longrightarrow K_{2r-1}(L) \xrightarrow{\Delta} \oplus_{i=1}^d K_{2r-1}(L_0) \xrightarrow{1-\hat{F}} \mathcal{W} \xrightarrow{\pi} {}_{p^\infty}K_{2r-2}(L) \longrightarrow 0$$

where Δ denotes the diagonal map and π is the composition of the map of Theorem 2.8 with the addition of coordinates, $\mathcal{W} \longrightarrow \mathcal{U}$.

Proof This is proved by an easy modification of the proof of ([**27**], Proposition 3.5). □

Theorem 3.5 *With the notation of Section 3.1 the cohomology groups of the* $\mathbf{Z}[G(L/K)]$*-module,* $\oplus_{i=1}^{d} K_{2r-1}(L_0)$*, are given by*

$$H^i(G(L/K);\oplus_{i=1}^{d} K_{2r-1}(L_0)) \cong \begin{cases} \mathbf{Z}/t & \text{if } i \text{ is odd,} \\ 0 & \text{if } 0 < i \text{ is even,} \\ K_{2r-1}(W_0) & \text{if } i = 0. \end{cases}$$

Here $t = [G_0(L/K) : G_1(L/K)]$.

Proof The proof follows that of ([**27**], Theorem 3.6) with $q = |\mathcal{O}_K/\pi_K|^2 = v^2$ replaced by $q = v^r$ since F acts on $(\mathbf{Q}/\mathbf{Z})(r)[1/p]$ by multiplication by v^r. □

3.6 Suppose that L/K is a totally ramified Galois extension of p-adic local fields with group, $G(L/K)$ and set $t = [G(L/K) : G_1(L/K)]$. Choose a primitive t-th root of unity, ξ_t. If X is any $\mathbf{Z}[G(L/K)]$-module containing ${}_t(\mathbf{Q}/\mathbf{Z})(r)[1/p]$, depending on the choice of ξ_t, we shall define a new module $X\tilde{\oplus}\mathbf{Z}$, and an extension of the form

$$0 \longrightarrow X \longrightarrow X\tilde{\oplus}\mathbf{Z} \longrightarrow \mathbf{Z} \longrightarrow 0.$$

For example, this construction may be applied with X equal to any module in the chain

$$Tors(K_{2r-1}(L_0))[1/p] \cong (\mathbf{Q}/\mathbf{Z})(r)[1/p] \subset \mathcal{U} \subset K_{2r-1}(L_0).$$

Note that the action of $G(L/K)$ on $(\mathbf{Q}/\mathbf{Z})(r)[1/p]$ is trivial, since L/K is totally ramified, while the Frobenius, F of Section 2.1, acts like "multiplication" by v^r where $v = |\mathcal{O}_L/\pi_L|$.

As an abelian group $X\tilde{\oplus}\mathbf{Z}$ is just the direct sum, $X \oplus \mathbf{Z}$. As a $\mathbf{Z}[G(L/K)]$-module the embedding of X is given by sending $x \in X$ to $(x, 0)$. If $G_1(L/K) = G(L/E)$, as in Section 3.1, the action on $(0, 1) \in X\tilde{\oplus}\mathbf{Z}$ will factor through $G(L/K)/G(L/E) \cong G(E/K)$. Furthermore, if $G(E/K) \cong \mathbf{Z}/t$ with generator, g, then $g(0, 1) = (\xi_t, 1)$. This action is well-defined since $g^i(0, 1) = (i\xi_t, 1)$, in additive notation.

Similarly we may define a $\mathbf{Z}[G(L/K)]$-module, $X\tilde{\oplus}\mathbf{Z}[1/p]$, and an extension of the form

$$0 \longrightarrow X \longrightarrow X\tilde{\oplus}\mathbf{Z}[1/p] \longrightarrow \mathbf{Z}[1/p] \longrightarrow 0$$

by setting

$$X\tilde{\oplus}\mathbf{Z}[1/p] = \lim_{\rightarrow}(X\tilde{\oplus}\mathbf{Z} \stackrel{(1,v^r)}{\longrightarrow} X\tilde{\oplus}\mathbf{Z} \stackrel{(1,v^r)}{\longrightarrow} X\tilde{\oplus}\mathbf{Z} \stackrel{(1,v^r)}{\longrightarrow} \ldots),$$

the direct limit of iterations of the map sending (z, m) to (z, mv^r) where $v = |\mathcal{O}_L/\pi_L|$. This is a $\mathbf{Z}[G(L/K)]$-module, since $v^r \equiv 1$ (*modulo* t).

Proposition 3.7 *In the notation of Section 3.6, let* X *be any of the* $\mathbf{Z}[G(L/K)]$*-modules in the chain*

$$Tors(K_{2r-1}(L_0))[1/p] \cong (\mathbf{Q}/\mathbf{Z})(r)[1/p] \subset \mathcal{U} \subset K_{2r-1}(L_0).$$

Then $X\tilde{\oplus}\mathbf{Z}[1/p]$ *is a cohomologically trivial* $\mathbf{Z}[G(L/K)]$*-module.*

Proof The proof is an easy modification of that of ([**27**], Proposition 3.8). □

3.8 In the situation of Section 3.1 we shall define a $\mathbf{Z}[G(L/K)]$-module

$$\tilde{\oplus}_{i=1}^{d} K_{2r-1}(L_0).$$

As an abelian group this module is simply the direct sum

$$(\oplus_{i=1}^{d} K_{2r-1}(L_0)) \oplus (\oplus_{i=1}^{d} \mathbf{Z}[1/p]).$$

The summand, $\oplus_{i=1}^{d} K_{2r-1}(L_0))$, has the $\mathbf{Z}[G(L/K)]$-module structure described in Section 3.1 and Lemma 3.2. Thus $g \in G(L/K)$ lifts to $(F, F) \in G(L_0/K)$ and acts like

$$g(a_1, \ldots, a_d) = (F(a_2), F(a_3), \ldots, F(a_d), F_0^{-1}(F(a_1))).$$

Any element, $h \in G(L/W)$, acts component-by-component as $h(a_1, \ldots, a_d) = (h(a_1), \ldots, h(a_d))$.

Consider the $\mathbf{Z}[G(L/K)]$-submodule, $\oplus_{i=1}^{d}(\mathbf{Q}/\mathbf{Z})(r)[1/p] \subset \oplus_{i=1}^{d} K_{2r-1}(L_0)$. Since L/W is totally ramified, $G(L/W)$ acts trivially on this submodule. Hence $G(L/K)$ acts through the quotient, $G(W/K) =< g >$, with

$$g(a_1, \ldots, a_d) = (F(a_2), F(a_3), \ldots, F(a_d), F_0^{-1}F(a_1)) = (qa_2, qa_3, \ldots, qa_d, q^{1-d}a_1)$$

where $q = v^r$ and $v = |\mathcal{O}_K/\pi_K|$. An arbitrary element of the $\mathbf{Z}[G(L/K)]$-module

$$Inf_{G(W/K)}^{G(L/K)} Ind_{\{1\}}^{G(W/K)}((\mathbf{Q}/\mathbf{Z})(r)[1/p])$$

has the form $z = 1 \otimes b_d + g \otimes b_{d-1} + \ldots + g^{d-1} \otimes b_1$ and $g(z) = 1 \otimes b_1 + g \otimes b_d + \ldots + g^{d-1} \otimes b_2$. Hence, if we write $z = (b_1, \ldots, b_d)$, then $g(z) = (b_2, b_3, \ldots, b_d, b_1)$.

Define an isomorphism

$$\phi : \oplus_{i=1}^{d}(\mathbf{Q}/\mathbf{Z})(r)[1/p] \xrightarrow{\cong} Inf_{G(W/K)}^{G(L/K)} Ind_{\{1\}}^{G(W/K)}((\mathbf{Q}/\mathbf{Z})(r)[1/p])$$

by $\phi(a_1, \ldots, a_d) = (q^{1-d}a_1, q^{2-d}a_2, \ldots, q^{-1}a_{d-1}, a_d)$. This is an isomorphism of $\mathbf{Z}[G(L/K)]$-modules, since

$$\begin{aligned}\phi(g(a_1, \ldots, a_d)) &= \phi(qa_2, qa_3, \ldots, qa_d, q^{1-d}a_1)\\ &= (q^{2-d}a_2, q^{3-d}a_3, \ldots, a_d, q^{1-d}a_1)\end{aligned}$$

and

$$\begin{aligned}g(\phi(a_1, \ldots, a_d)) &= g(q^{1-d}a_1, q^{2-d}a_2, \ldots, q^{-1}a_{d-1}, a_d)\\ &= (q^{2-d}a_2, q^{3-d}a_3, \ldots, a_d, q^{1-d}a_1).\end{aligned}$$

We have an isomorphism of $\mathbf{Z}[G(L/K)]$-modules

$$Inf_{G(W/K)}^{G(L/K)} Ind_{\{1\}}^{G(W/K)}((\mathbf{Q}/\mathbf{Z})(r)[1/p]) \cong Ind_{G(L/W)}^{G(L/K)}((\mathbf{Q}/\mathbf{Z})(r)[1/p])$$

where $G(L/W)$ acts trivially on $(\mathbf{Q}/\mathbf{Z})(r)[1/p]$. Hence we may form the following push-out diagram, which defines the module $\tilde{\oplus}_{i=1}^{d} K_{2r-1}(L_0)$.

$$\begin{array}{ccc}
\oplus_{i=1}^{d}(\mathbf{Q}/\mathbf{Z})(r)[1/p] \cong Ind_{G(L/W)}^{G(L/K)}((\mathbf{Q}/\mathbf{Z})(r)[1/p]) & \longrightarrow & \oplus_{i=1}^{d} K_{2r-1}(L_0)\\
\downarrow & & \downarrow\\
Ind_{G(L/W)}^{G(L/K)}((\mathbf{Q}/\mathbf{Z})(r)[1/p]\tilde{\oplus}\mathbf{Z}[1/p]) & \longrightarrow & \tilde{\oplus}_{i=1}^{d} K_{2r-1}(L_0)
\end{array}$$

Proposition 3.9 *The* $\mathbf{Z}[G(L/K)]$*-module,* $\tilde{\oplus}_{i=1}^{d} K_{2r-1}(L_0)$*, of Section 3.8 is cohomologically trivial.*

Proof This proof is an easy modification of that of ([**27**] Proposition 3.10). □

3.10 Now consider the $\mathbf{Z}[G(L/K)]$-submodule, $\mathcal{W} \subset \oplus_{i=1}^{d} K_{2r-1}(L_0)$, of Definition 3.3. Since $(\mathbf{Q}/\mathbf{Z})(r)[1/p] \subset \mathcal{U}$ we have $\oplus_{i=1}^{d}(\mathbf{Q}/\mathbf{Z})(r)[1/p] \subset \mathcal{W}$. Also $G(L/W)$ acts component-by-component on $\mathcal{W}$ so that $\mathcal{W} \cong (\oplus_{i=1}^{d-1} K_{2r-1}(L_0)) \oplus \mathcal{U}$ as a $\mathbf{Z}[G(L/W)]$-module. Therefore, by Corollary 2.11, the inclusion induces a cohomology isomorphism of the following form

$$H^t(G(L/W); \mathcal{W}) \xrightarrow{\cong} H^t(G(L/W); \oplus_{i=1}^{d} K_{2r-1}(L_0)).$$

Comparison of cohomology spectral sequences implies that the natural map

$$\hat{H}^t(G(L/K); \mathcal{W}) \longrightarrow \hat{H}^t(G(L/K); \oplus_{i=1}^{d} K_{2r-1}(L_0))$$

is also an isomorphism.

Define $\tilde{\mathcal{W}}$ by the pushout diagram

$$\begin{array}{ccc} \oplus_{i=1}^{d}(\mathbf{Q}/\mathbf{Z})(r)[1/p] \cong Ind_{G(L/W)}^{G(L/K)}((\mathbf{Q}/\mathbf{Z})(r)[1/p]) & \longrightarrow & \mathcal{W} \\ \downarrow & & \downarrow \\ Ind_{G(L/W)}^{G(L/K)}((\mathbf{Q}/\mathbf{Z})(r)[1/p] \tilde{\oplus} \mathbf{Z}[1/p]) & \longrightarrow & \tilde{\mathcal{W}} \end{array}$$

so that we immediately obtain the following result from Proposition 3.9.

Proposition 3.11 *The* $\mathbf{Z}[G(L/K)]$*-module,* $\tilde{\mathcal{W}}$*, of Section 3.10 is cohomologically trivial.*

3.12 On $(\mathbf{Q}/\mathbf{Z})(r)[1/p]$ F_0 acts by multiplication by q^d where $q = v^r$, $v = |\mathcal{O}_K/\pi_K|$. Hence, if $a_i \in (\mathbf{Q}/\mathbf{Z})(r)[1/p]$, then

$$(1 - \hat{F})(a_1, \ldots, a_d) = (a_1 - q^d a_d, a_2 - a_1, \ldots, a_d - a_{d-1}).$$

Now consider the isomorphism of Section 3.8

$$\phi : \oplus_{i=1}^{d}(\mathbf{Q}/\mathbf{Z})(r)[1/p] \xrightarrow{\cong} Ind_{G(L/W)}^{G(L/K)}((\mathbf{Q}/\mathbf{Z})(r)[1/p])$$

where, on the right, $G(L/W)$ acts trivially on $(\mathbf{Q}/\mathbf{Z})(r)[1/p]$ and

$$\phi(a_1, \ldots, a_d) = (q^{1-d}a_1, q^{2-d}a_2, \ldots, q^{-1}a_{d-1}, a_d) = \sum_{i=0}^{d-1} g^i \otimes q^{-i} a_{d-i}.$$

Hence $\hat{F}$ translates to $\phi\hat{F}\phi^{-1}$, which is given by

$$\begin{aligned}
\phi(\hat{F}(\phi^{-1}(\sum_{i=0}^{d-1} g^i \otimes b_{d-i}))) &= \phi(\hat{F}(\phi^{-1}(b_1, \ldots, b_d))) \\
&= \phi(\hat{F}(q^{d-1}b_1, q^{d-2}b_2, \ldots, qb_{d-1}, b_d)) \\
&= \phi(q^d b_d, q^{d-1}b_1, q^{d-2}b_2, \ldots, qb_{d-1})) \\
&= (qb_d, qb_1, \ldots, qb_{d-1}) \\
&= q(1 \otimes b_{d-1} + g \otimes b_{d-2} + \ldots + g^{d-1} \otimes b_d) \\
&= qg^{d-1}(\sum_{i=0}^{d-1} g^i \otimes b_{d-i}) \\
&= qg^{-1}(\sum_{i=0}^{d-1} g^i \otimes b_{d-i}).
\end{aligned}$$

Define

$$\hat{F} : Ind_{G(L/W)}^{G(L/K)}((\mathbf{Q}/\mathbf{Z})(r)[1/p]\tilde{\oplus}\mathbf{Z}[1/p]) \longrightarrow Ind_{G(L/W)}^{G(L/K)}((\mathbf{Q}/\mathbf{Z})(r)[1/p]\tilde{\oplus}\mathbf{Z}[1/p])$$

by the formula ($b \in (\mathbf{Q}/\mathbf{Z})(r)[1/p]$, $m \in \mathbf{Z}[1/p]$)

$$\hat{F}(g^i \otimes (b, m)) = g^{i-1} \otimes (qb, v^{r-1}m).$$

If $u \in \mathbf{Z}$ satisfies $uv \equiv 1$ (modulo t) then $g^{-i}ag^i(0,1) = a^{u^i}(0,1) = (u^i\xi_t, 1) \in (\mathbf{Q}/\mathbf{Z})(r)[1/p]\tilde{\oplus}\mathbf{Z}[1/p])$, writing the first coordinate additively, as usual. Here a and g are as in Section 3.1. Then $\hat{F}$ is a $\mathbf{Z}[G(L/K)]$-module homomorphism since

$$\begin{aligned}
\hat{F}(a(g^i \otimes (0,1))) &= \hat{F}(g^i \otimes a^{u^i}(0,1)) \\
&= \hat{F}(g^i \otimes (u^i\xi_t, 1)) \\
&= g^{i-1} \otimes (qu^i\xi_t, v^{r-1})
\end{aligned}$$

while

$$\begin{aligned}
a(\hat{F}(g^i \otimes (0,1))) &= a(g^{i-1} \otimes (0, v^{r-1})) \\
&= g^{i-1} \otimes a^{u^i}(0, v^{r-1})) \\
&= g^{i-1} \otimes (v^{r-1}u^{i-1}\xi_t, v^{r-1})
\end{aligned}$$

and these are equal since $v^{r-1}u^{i-1} \equiv quu^{i-1} \equiv qu^i$ (modulo t). Also

$$\hat{F}(g(g^i \otimes (0,1))) = \hat{F}(g^{i+1} \otimes (0,1)) = g^i \otimes (0, v^{r-1})$$

while

$$g(\hat{F}(g^i \otimes (0,1))) = g(g^{i-1} \otimes (0, v^{r-1})) = g^i \otimes (0, v^{r-1}).$$

Theorem 3.13 *In the notation of Section 3.1 and Section 3.12, there is a 2-extension of* $\mathbf{Z}[G(L/K)]$*-modules of the form*

$$K_{2r-1}(L) \xrightarrow{\Delta} \tilde{\oplus}_{i=1}^{d} K_{2r-1}(L_0) \xrightarrow{1-\hat{F}} \tilde{\mathcal{W}} \xrightarrow{(\pi,\pi_2)} {}_{p^\infty}K_{2r-2}(L) \oplus K_{2r-3}(\mathcal{O}_L/\pi_L).$$

Note that the right-hand module is $Tors(K_{2r-2}(L))$.

Proof We have merely to evaluate the kernel and cokernel of the map

$$1 - \hat{F} : \tilde{\oplus}_{i=1}^{d} K_{2r-1}(L_0) \longrightarrow \tilde{\mathcal{W}}$$

where $\hat{F}$ was defined in Section 3.12. However, as abelian groups there are isomorphisms

$$\tilde{\oplus}_{i=1}^{d} K_{2r-1}(L_0) \cong (\oplus_{i=1}^{d} K_{2r-1}(L_0)) \oplus (\oplus_{i=1}^{d} \mathbf{Z}[1/p])$$

and

$$\tilde{\mathcal{W}} \cong \mathcal{W} \oplus (\oplus_{i=1}^{d} \mathbf{Z}[1/p]).$$

Furthermore, the resulting map on the quotients of these direct sums by the submodules given by the first summands has the form

$$\lambda : \oplus_{i=1}^{d} \mathbf{Z}[1/p] \longrightarrow \oplus_{i=1}^{d} \mathbf{Z}[1/p]$$

and is given by

$$\lambda(m_1, \ldots, m_d) = (m_1 - v^{r-1}m_d, \ldots, m_d - v^{r-1}m_{d-1}).$$

If $(m_1, \ldots, m_d) \in Ker(\lambda)$ then $m_1 = v^{r-1}m_d, m_2 = v^{r-1}m_1, \ldots$ and $m_d(1 - v^{(r-1)d}) = 0$ so that λ is injective. Also, we can solve the equation $\lambda(m_1, \ldots, m_d) = (x_1, \ldots, x_d)$ by choosing $x_d, x_1, \ldots, x_{d-2}$ successively but the equations are consistent if and only if

$$m_d(1 - v^{(r-1)d}) = v^{(r-1)(d-1)}x_1 + v^{(r-1)(d-2)}x_2 + \ldots + v^{r-1}x_{d-1} + x_d.$$

Therefore, if π_2 is defined to depend only on the coordinates in the second summand and to be given by

$$\pi_2(x_1, x_2, \ldots, x_d) = v^{(r-1)(d-1)}x_1 + v^{(r-1)(d-2)}x_2 + \ldots + v^{r-1}x_{d-1} + x_d$$

then $Ker(\pi_2) = Im(\lambda)$, which proves the exactness of the sequence once we recall from [**22**] that

$$K_{2r-3}(\mathcal{O}_L/\pi_L) \cong \mathbf{Z}/(q^d - 1) \cong \mathbf{Z}/(v^{(r-1)d} - 1).$$

Also, as Galois modules, we have

$$K_{2r-3}(\mathcal{O}_L/\pi_L) \cong (\mathbf{Q}/\mathbf{Z})(r-1)[1/p]^{\Omega_{\mathcal{O}_L/\pi_L}}.$$

Therefore it remains to verify that $G(L/K)$ acts on $\mathbf{Z}/(v^{(r-1)d}-1) \cong K_{2r-3}(\mathcal{O}_L/\pi_L)$ via the quotient to $G(W/K)$ whose generator acts by "multiplication" by v^{r-1}, where $v = |\mathcal{O}_K/\pi_K|$. Certainly, $G(L/E)$ acts trivially on $\tilde{\oplus}_{i=1}^{d} K_{2r-1}(L_0)$ and $\tilde{\mathcal{W}}$. Also the action by $a \in G(E/W)$ is trivial on the second coordinates in the direct sums. Finally the generator, $g \in G(W/K)$, acts via

$$\begin{aligned} g(x_1, \ldots, x_d) &= g(\sum_{i=0}^{d-1} g^i \otimes x_{d-i}) \\ &= \sum_{i=0}^{d-1} g^{i+1} \otimes x_{d-i} \\ &= (x_2, x_3, \ldots, x_d, x_1) \end{aligned}$$

and
$x_1 + v^{r-1}x_d + v^{2(r-1)}x_{d-1} + \ldots + v^{(r-1)(d-1)}x_2 \equiv v^{r-1}(v^{(r-1)(d-1)}x_1 + v^{(r-1)(d-2)}x_2 + \ldots + v^{r-1}x_{d-1} + x_d)$ (modulo $v^{d(r-1)} - 1$) so that g acts on $\mathbf{Z}/(v^d - 1)$ by multiplication by v^{r-1}, as required. □

4 Properties of the Canonical 2-Extension

4.1 We sketch the important properties on the canoncial 2-extensions of Theorem 3.13. The proofs are simple adaptations of those to be found in ([**27**], Section 4). Let us consider the effect of replacing ξ_t by ξ_t^b, with $HCF(b,t) = 1$, in the construction of the 2-extension of Theorem 3.13. The argument of ([**27**], Section 4.1) yields the following result.

Proposition 4.2 *In the notation of Section 4.1 there is a commutative diagram of $\mathbf{Z}[G(L/K)]$-isomorphisms of the form*

$$\begin{array}{ccccccc} K_{2r-1}(L) & \xrightarrow{\Delta} & \tilde{\oplus}_{i=1}^{d} K_{2r-1}(L_0) & \xrightarrow{1-\hat{F}} & \tilde{W} & \longrightarrow & Tors(K_{2r-2}(L)) \\ \downarrow \phi_b & & \downarrow \tilde{\Phi}_b & & \downarrow \tilde{\Phi}_b & & \downarrow 1 \\ K_{2r-1}(L) & \xrightarrow{\Delta} & \tilde{\oplus}_{i=1}^{d} K_{2r-1}(L_0) & \xrightarrow{1-\hat{F}} & \tilde{W} & \longrightarrow & Tors(K_{2r-2}(L)) \end{array}$$

in which the upper sequence is that of Theorem 3.13 defined using ξ_t while the lower sequence is that defined using ξ_t^b.

There is a similar quasi-isomorphism of 2-extensions of Theorem 3.13 corresponding to different choices of the generators, a and g, of $G(E/K)$.

4.3 Suppose that G is a finite group and that

$$A \longrightarrow B \longrightarrow C \longrightarrow D$$

is a 2-extension of $\mathbf{Z}[G]$-modules in which B and C are cohomologically trivial. Such a sequence defines an element of $Ext^2_{\mathbf{Z}[G]}(D, A)$ and we shall be interested in such extension classes up to quasi-isomorphism (i.e., up to the action of $Aut_{\mathbf{Z}[G]}(A) \times Aut_{\mathbf{Z}[G]}(D)$.

There are two natural operations associated to a subgroup, $J \subseteq G$. The first - passage to subgroups - is merely to consider the modules as $\mathbf{Z}[J]$-modules. The second - passage to quotient groups - is more complicated and applies to the case of a normal subgroup, $J \triangleleft G$. In this case, let A^J, A_J denote the J-invariants and J-coinvariants of A, respectively ([**26**], p. 3). We have a commutative diagram of $\mathbf{Z}[G/J]$-modules in which the rows and columns are exact and N_J denotes the norm, $N(x) = \sum_{g \in J} gx$,

$$\begin{array}{ccccccc} & & B_J & \longrightarrow & C_J & \longrightarrow & D_J \longrightarrow 0 \\ & & \cong \Big\downarrow N_J & & \cong \Big\downarrow N_J & & \\ 0 \longrightarrow & A^J \longrightarrow & B^J & \longrightarrow & C^J & & \end{array}$$

resulting in the associated J-invariant/coinvariant 2-extension

$$A^J \longrightarrow B^J \longrightarrow C^J \longrightarrow D_J$$

in which the $\mathbf{Z}[G/J]$-modules, B^J and C^J, are cohomologically trivial.

Following the argument of ([**27**], Theorem 4.6) establishes the passage-to-quotients part of the following result and the passage-to-subgroups is relatively straightforward.

Theorem 4.4 *Let L/K be a Galois extension of p-adic local fields with group, $G(L/K)$.*

(i) *If $G(L/M) \triangleleft G(L/K)$ is a normal subgroup then the $G(L/M)$-invariant/ coinvariant sequence of Section 4.3 associated to the fundamental 2-extension of Theorem 3.13.*

$$K_{2r-1}(L) \stackrel{\Delta}{\longrightarrow} \tilde{\oplus}_{i=1}^{d} K_{2r-1}(L_0) \stackrel{1-\hat{F}}{\longrightarrow} \tilde{\mathcal{W}}_L \stackrel{(\pi,\pi_2)}{\longrightarrow} Tors(K_{2r-2}(L))$$

is quasi-isomorphic to the fundamental 2-extension associated to M/K

$$K_{2r-1}(M) \stackrel{\Delta}{\longrightarrow} \tilde{\oplus}_{i=1}^{d} K_{2r-1}(M_0) \stackrel{1-\hat{F}}{\longrightarrow} \tilde{\mathcal{W}}_M \stackrel{(\pi,\pi_2)}{\longrightarrow} Tors(K_{2r-2}(M)).$$

(ii) *If $G(L/E) \subseteq G(L/K)$ the restriction to $G(L/E)$ of the canonical 2-extension of Theorem 3.13 associated to L/K is quasi-isomorphic to the canonical 2-extension associated to L/E.*

5 The Galois Module Structure Invariants $\Omega_r(L/K, 2)$

5.1 Let L/K be a Galois extension of number fields. The second Chinburg invariant ($\Omega_0(L/K, 2)$ in our notation) is constructed from a prjective $\mathbf{Z}[G(L/K)]$-module, X, and the classical fundamental classes of local class field theory. One may use the fundamental classes of Theorem 3.13 in place of the classical one to construct Galois module structure invariants

$$\Omega_{r-1}(L/K, 2) \in \mathcal{CL}(\mathbf{Z}[G(L/K)])$$

lying in the class-group of $\mathbf{Z}[G(L/K)]$. These invariants measure the Galois module structure in dimensions $2r-1$ and $2r-2$ and are expected to be connected with the values of the L-functions at $s = 1 - r$ (or, equivalently, $s = r$). This expectation is based upon the clues such appearance of these special L-function values in the examples of ([**25**], Theorem 7.4.60; [**26**], Section 7.2; [**4**]; [**5**]).

In the construction of the original Chinburg invariant in ([**25**], Section 7.2) – denoted there by $\Omega(L/K, 2)$ rather than $\Omega_0(L/K, 2)$ – one chooses a locally free, hence projective, module X in the following manner.

For each finite prime, $P \triangleleft \mathcal{O}_K$, choose a prime lying over it, $Q \triangleleft \mathcal{O}_L$. Therefore the decomposition group of Q is isomorphic to $G(L_Q/K_P)$. We shall say that P is *tame* if L_Q/K_P is tamely ramified (i.e., $G_1(L_Q/K_P) = \{1\}$) and that P is *wild* otherwise. By a theorem of E. Noether, if P is tame then $\mathcal{O}_{L_Q}$ is a free $\mathcal{O}_{K_P}[G(L_Q/K_P)]$-module of rank one. Therefore we may choose an adèle

$$(a_P) = a \in \prod_P \mathcal{O}_{L_Q}$$

(the product being taken over finite primes of K with Q being the chosen prime over P) such that

(i) $a_P \in \mathcal{O}_{L_Q}$, $K_P[G(L_Q/K_P)]a_P = L_Q$ for each P, and
(ii) $\mathcal{O}_{K_P}[G(L_Q/K_P)]a_P = \mathcal{O}_{L_Q}$ for each tame P.

There are isomorphisms of the form

$$L \otimes_K K_P \cong \prod_{R|P} L_R \cong Ind_{G(L_Q/K_P)}^{G(L/K)}(L_Q)$$

and

$$\mathcal{O}_L \otimes_{\mathcal{O}_K} \mathcal{O}_{K_P} \cong \prod_{R|P} \mathcal{O}_{L_R} \cong Ind_{G(L_Q/K_P)}^{G(L/K)}(\mathcal{O}_{L_Q})$$

so that

(i) $K_P[G(L/K)]a_P \cong \prod_{R|P} L_R$ for each P, and
(ii) $\mathcal{O}_{K_P}[G(L/K)]a_P \cong \prod_{R|P} \mathcal{O}_{L_R}$ for each tame P.

We shall abbreviate $\mathcal{O}_{K_P}[G(L_Q/K_P)]a_P$ to X_Q and set $X = \mathcal{O}_K[G(L/K)]a$. This is to be interpreted as meaning that X is the intersection of L with the product of its P-completions, X_P, where both are considered as subgroups of the adèles. Hence X is a locally free $\mathcal{O}_K[G(L/K)]$-module whose P-completion is

$$X_P = \mathcal{O}_{K_P}[G(L/K)]a_P = Ind_{G(L_Q/K_P)}^{G(L/K)}(X_Q).$$

In addition, we shall assume henceforth (replacing X by mX for a suitable integer, $m \in \mathbf{Z}$, if necessary) that the Q-adic exponential defines an isomorphism

$$exp : X_Q \xrightarrow{\cong} 1 + X_Q \subset \mathcal{O}_{L_Q}^*$$

for all wild L_Q/K_P. This assumption is only needed in the construction of the invariant $\Omega_0(L/K, 2)$. Since X is locally free, it is cohomologically trivial and so also is X_Q for each Q. Hence X defines a class, $[X] \in \mathcal{CL}(\mathbf{Z}[G(L/K)])$.

From Theorem 3.13 we have a canonical class in

$$Ext^2_{\mathbf{Z}[G(L_Q/K_P)]}(Tors(K_{2r-2}(L_Q)), K_{2r-1}(L_Q)).$$

From the isomorphism $K_{2r-1}(L) \cong K_{2r-1}(\mathcal{O}_{L_Q})$ we have a canonical map

$$\phi_{2r-1} : K_{2r-1}(L) \longrightarrow \lim_{\overleftarrow{s}} K_{2r-1}(\mathcal{O}_{L_Q}/(\pi_{L_Q}^s)) = K_{2r-1}^{top}(L_Q).$$

The target group, $K_{2r-1}^{top}(L_Q)$, is a finitely generated $\mathbf{Z}_p[G(L_Q/K_P)]$-module and ϕ_{2r-1} induces an isomorphism in p-adic cohomology. In addition, we have

$$\rho_{2r-1} : K_{2r-1}(L_Q) \cong K_{2r-1}(\mathcal{O}_{L_Q}) \longrightarrow K_{2r-1}(\mathcal{O}_{L_Q}/\pi_{L_Q})$$

induced by reduction modulo π_{L_Q}, where $\mathcal{O}_{L_Q}/(\pi_{L_Q})$ is the residue field of L_Q.

The map

$$(\phi_{2r-1}, \rho_{2r-1}) : K_{2r-1}(L_Q) \longrightarrow K^{top}_{2r-1}(L_Q) \oplus K_{2r-1}(\mathcal{O}_{L_Q}/(\pi_{L_Q}))$$

induces isomorphisms in Tate cohomology in all dimensions. Hence, pushing out via $(\phi_{2r-1}, \rho_{2r-1})$, we obtain a canonical 2-extension of the form

$$K^{top}_{2r-1}(L_Q) \oplus K_3(\mathcal{O}_{L_Q}/(\pi_{L_Q})) \longrightarrow A \longrightarrow B \longrightarrow Tors(K_{2r-2}(L_Q))$$

in which A and B may be chosen to be cohomologically trivial with B finitely generated over $\mathbf{Z}[G(L_Q/K_P)]$ and the p-adic part of A is finitely generated over $\mathbf{Z}_p[G(L_Q/K_P)]$.

Next we choose our favourite regulator. A regulator is a Galois-equivariant isomorphism of the form

$$K^{top}_{2r-1}(L_Q) \otimes \mathbf{Q}_p \cong L_Q$$

where p is the residue characteristic. There are many to choose from and several are described in ([**27**], Section 5), which may be derived (sometimes a little work is needed) from ([**1**], [**2**], [**3**], [**6**], [**7**], [**8**], [**9**], [**10**], [**11**], [**13**], [**16**], [**17**], [**28**], [**29**], [**32**]). For example, we shall settle for the regulator given by the syntomic Chern classes. More precisely, we have canonical isomorphisms, given by étale and syntomic Chern classes, of the form

$$ch^{\acute{e}t}_{2r-1} : K^{top}_{2r-1}(L_Q) \xrightarrow{\cong} H^1_{\acute{e}t}(L_Q; \mathbf{Z}_p(r))$$

and

$$K^{top}_{2r-1}(L_Q) \otimes \mathbf{Q}_p \cong H^1_{\acute{e}t}(L_Q; \mathbf{Q}_p(r)) \xleftarrow{f_M} H^1(\mathcal{O}_{L_Q}; S_\infty(r)_{\mathbf{Q}_p}) \xrightarrow{\alpha} L_Q$$

such that, if $L_Q/\mathbf{Q}_p$ is unramified, $K^{top}_{2r-1}(L_Q)/Tors$ is identified with the ring of integers, $\mathcal{O}_{L_Q} \subset L_Q$ for almost all p.

Let X denote the finitely generated, locally free $\mathbf{Z}_p[G(L/K)]$-module introduced previously, as used in the construction of the second Chinburg invariant, $\Omega_0(L/K, 2)$, ([**25**], Section 7.2). By construction, $X_Q \subseteq \mathcal{O}_{L_Q}$ with equality if L/K is tamely ramified at $Q \triangleleft \mathcal{O}_L$. We may assume that X_Q has been lifted to a p-adic lattice

$$X_Q \subseteq K^{top}_{2r-1}(L_Q)$$

mapping isomorphically to $X_Q \subseteq \mathcal{O}_{L_Q}$ under the syntomic regulator map. Dividing out by X_Q yields a 2-extension of $\mathbf{Z}[G(L_Q/K_P)]$-modules of the form

$$(K^{top}_{2r-1}(L_Q)/X_Q) \oplus K_{2r-1}(\mathcal{O}_L/Q) \longrightarrow A/X_Q \longrightarrow B \longrightarrow Tors(K_{2r-2}(L_Q))$$

in which the two middle modules are finitely generated and cohomologically trivial $\mathbf{Z}[G(L_Q/K_P)]$-modules. Hence we may define an Euler characteristic

$$\Omega_{r-1}(L_Q/K_P, X_Q) = [A/X_Q] - [B] \in \mathcal{CL}(\mathbf{Z}[G(L_Q/K_P)]).$$

We define

$$\Omega_{r-1}(L/K, 2) = [X] + \sum_P Ind^{G(L/K)}_{G(L_Q/K_P)}(\Omega_{r-1}(L_Q/K_P, X_Q)) \in \mathcal{CL}(\mathbf{Z}[G(L/K)]).$$

In the sum Q is a chosen prime over $P \triangleleft \mathcal{O}_K$. As explained in ([**27**], Section 5), only finitely many of the local invariants in the sum are non-zero (at least, this is true for the syntomic regulator) and $\Omega_{r-1}(L/K, 2)$ is independent of all choices involved in the construction.

The properties of these invariants are summarised in the following result, whose proof is analogous to that of ([**27**], Theorem 5.3), which is the special case when $r = 2$.

Theorem 5.2 *Let L/K be a Galois extension of number fields. Then, in the formal expression of Section 5.1*

$$\Omega_{r-1}(L/K,2) = [X] + \sum_{P} Ind^{G(L/K)}_{G(L_Q/K_P)}(\Omega_{r-1}(L_Q/K_P, X_Q)) \in \mathcal{CL}(\mathbf{Z}[G(L/K)])$$

(the sum is over $P \lhd \mathcal{O}_K$) the terms, $\Omega_{r-1}(L_Q/K_P, X_Q)$, are zero for almost all p when $L_Q/\mathbf{Q}_p$ is unramified. In fact, the class-group element defined by this expression is independent of the choices of X and of all the choices involved in the construction of the local fundamental 2-extensions associated to the $\{L_Q/K_P\}$.

In addition, $\Omega_{r-1}(L/K,2)$ satisfies the following naturality properties with respect to an inclusion of a subgroup, $G(L/N) \subseteq G(L/K)$:

(i) the induced map

$$Res^{G(L/K)}_{G(L/N)} : \mathcal{CL}(\mathbf{Z}[G(L/K)]) \longrightarrow \mathcal{CL}(\mathbf{Z}[G(L/N)])$$

satisfies

$$Res^{G(L/K)}_{G(L/N)}(\Omega_{r-1}(L/K,2)) = \Omega_{r-1}(L/N,2)$$

and

(ii) if $G(L/N) \lhd G(L/K)$ the induced map

$$Inf_{L/N} : \mathcal{CL}(\mathbf{Z}[G(L/K)]) \longrightarrow \mathcal{CL}(\mathbf{Z}[G(N/K)])$$

satisfies

$$Inf_{L/N}(\Omega_{r-1}(L/K,2)) = \Omega_{r-1}(N/K,2).$$

References

[1] Bentzen, S. and Madsen, I. [1990], *Trace maps in algebraic K-theory and the Coates-Wiles homomorphism*, J. reine angew. Math. **411**, 171-195.

[2] Bloch, S., *Higher regulators, algebraic K-theory and zeta functions of elliptic curves*, Irvine University Lecture Notes.

[3] Böksted, M. and Madsen, I. [1994], *Algebraic K-theory of local fields: the unramified case*, Aarhus preprint #20.

[4] Chinburg, T., Kolster, M., Pappas, G. and Snaith, V. P. [1995], *Galois structure of K-groups of rings of integers*, C. R. Acad. Sci., Paris, t. 320, Série I, 1435-1440.

[5] Chinburg, T., Kolster, M., Pappas, G. and Snaith, V. P. [1995], *Quaternionic exercises in K-theory Galois module structure*, this volume.

[6] Coleman, R. G. [1982], *Dilogarithms, regulators and p-adic L-functions*, Inventiones Math. **69**, 171-208.

[7] Dwyer, W. G. and Friedlander, E. M. [1985], *Algebraic and étale K-theory*, Trans. A. M. Soc. **292** (1), 247-280.

[8] Fontaine, J-M. and Messing, W. [1987], *p-adic periods and p-adic étale cohomology*, in Current Trends in Arithmetical Algebraic Geometry (K. A. Ribet, ed.), Contemp. Math. A. M. Soc. **67**, 179-207.

[9] Gros, M. [1990], *Régulateurs syntomiques et valeurs de fonctions L p-adiques I* (with an appendix by M Kurihara), Inventiones Math. **99**, 293-320.

[10] Gros, M. [1994], *Régulateurs syntomiques et valeurs de fonctions L p-adiques II*, Inventiones Math. **115**, 61-79.

[11] Hesselholt, L. and Madsen, I. [1995], *On the K-theory of finite algebras over Witt vectors of perfect fields*, Aarhus preprint #2.

[12] Kahn, B. [1993], *Descente Galoisienne et K_2 des corps de nombres*, K-theory **7** (1), 55-100.

[13] Kolster, M. and Nguyen Quang Do, Thong [1995], *Syntomic regulators and special values of p-adic L-functions*, McMaster University preprint.

[14] Lichtenbaum, S. [1987], *The construction of weight-two arithmetic cohomology*, Inventiones Math. **88**, 183-215.

[15] Lichtenbaum, S. [1990], *New results on weight-two motivic cohomology*, The Grothendieck Festschrift vol.III (Birkhäuser), pp. 35-55.

[16] Loday, J-L. and Quillen, D. G. [1984], *Cyclic homology and the Lie algebra homology of matrices*, Comm. Math. Helv. **59**, 565-591.

[17] Loday, J-L. and Quillen, D. G. [1983], *Homologie cyclique et homologie de l'algèbre de Lie des matrices*, C. R. Acad. Sci. Paris t.**296**, 295-297.

[18] Merkurjev, A. S. [1983], *On the torsion of K_2 of local fields*, Annals of Maths. (2) **118**, 375-381.

[19] Merkurjev, A. S. [1986], *K_2 of fields and the Brauer group*, in Applications of algebraic K-theory to geometry and number theory (Part II), Contemp. Math. **55**, 529-546.

[20] Murkurjev, A. S. and Suslin, A. A. [1982], *K-cohomology of Severi-Brauer varieties and the norm residue homomorphism*, Izv. Nauk. SSSR **46**, 1011-1046 (Eng. trans. Math. USSR 21 (1983) 307-340).

[21] Murkurjev, A. S. and Suslin A. A. [1990], *The K_3 group of a field*, Izv. Nauk. SSSR **54**, 339-356 (Eng. trans. Math. USSR 36 (1990), 541-565).

[22] Quillen, D. G. [1972], *On the cohomology and K-theory of the general linear groups over a finite field*, Annals of Math. **96**, 552-586.

[23] Serre, J-P. [1979], *Local Fields* , Grad. Texts in Math. **67**, Springer-Verlag.

[24] Snaith, V. P. [1989], *Topological Methods in Galois Representation Theory*, C. M. Soc. Monographs, Wiley.

[25] Snaith, V. P. [1994], *Explicit Brauer Induction (with applications to algebra and number theory)*, Cambridge Studies in Advanced Math. **40**, Cambridge University Press.

[26] Snaith, V. P. [1994], *Galois Module Structure*, The Fields Institute for Research in Mathematical Sciences Monograph Series, **2**, A. M. S., Providence, R.I.

[27] Snaith, V. P. [1997], *Local fundamental classes derived from higher-dimensional K-groups*, this volume.

[28] Somekawa, M. [1995] *Log-syntomic regulators and p-adic polylogarithms*, Tokyo Inst. Tech., preprint.

[29] Soulé, C. [1981], *On higher p-adic regulators*, Springer-Verlag Lecture Notes in Mathematics **854**, 372-401.

[30] Suslin, A. A. [1984], *On the K-theory of local fields*, J. P. A. Alg. (2-3) **34**, 301-318.

[31] Voevodsky, V. [1996], *Lecture at the Great Lakes K-theory Conference*, The Fields Institute for Research in Mathematical Sciences, Toronto, March 1996.

[32] Wagoner, J. B. [1976], *Continuous cohomology and p-adic K-theory*, Springer-Verlag Lecture Notes in Mathematics **551**, 241-248.

Fields Institute Communications
Volume **16**, 1997

Weak Approximation, R-Equivalence and Whitehead Groups

Nguyễn Quốc Thắng
Department of Mathematics and Statistics
McMaster University
Hamilton, Ontario, Canada L8S 4K1
Current Address: Department of Mathematics
Israel Institute of Technology
Technion, Haifa 3200, Israel
nguyen@techunix.technion.ac.il

Abstract. We express the obstruction to weak approximation in a connected reductive group defined over a global field k via its R-equivalence group over k and its completions, extending a similar result of Colliot-Thélène and Sansuc for tori over number fields. A similar result for Whitehead groups is given.

Introduction

Let G be a connected linear algebraic group defined over a number field k. Denote by

$$\mathrm{A}(G) = \prod_v G(k_v)/Cl(G(k))$$

the obstruction to weak approximation in G,

$$\text{III}(G) = \mathrm{Ker}\ (\mathrm{H}^1(k, G) \to \prod_v \mathrm{H}^1(k_v, G))$$

the Tate - Shafarevich group of G where $\mathrm{H}^1(.,.)$ denotes Galois cohomology and k_v denotes the completion of the field k at a valuation v. We denote also by $G(k)/R$ the group of R-equivalences of $G(k)$, $RG(k)$ the subgroup of $G(k)$ consisting of all elements of $G(k)$ which are R-equivalent to 1 in $G(k)$ (see [3], [7]). One defines similar groups in the case of local fields k_v which are completions of k.

If $G = T$ is a k-torus, then Voskresenskiĭ [14] showed that there is an exact sequence connecting arithmetic, geometric and cohomological invariants of T

$$1 \to \mathrm{A}(T) \to \mathrm{H}^1(k, Pic\bar{X})^* \to \text{III}(T) \to 1, \qquad (1)$$

where X is a smooth compactification of T over k, $\bar{X} = X \times_k \bar{k}$ and $()^*$ means taking the character group with values in $\mathbf{Q}/\mathbf{Z}$. Later on, Sansuc [11] showed that this sequence also holds for arbitrary connected k-group G over a number field k without factors of type E_8, thus also for any connected k-group by combining this with the result on the Hasse principle for E_8 by Chernousov. In their fundamental paper ([3], Proposition 19) Colliot-Thélène and Sansuc have established, among other

1991 *Mathematics Subject Classification.* Primary 20G10; Secondary 11E72, 18G50.

things, the following exact sequence which connects various important arithmetic and geometric invariants of algebraic tori defined over a number field k. Namely we have

$$1 \to \text{III}(S) \to T(k)/R \xrightarrow{\beta} \prod_v T(k_v)/R \to \text{A}(T) \to 1, \qquad (2)$$

an exact sequence where S is the Neron - Severi torus of T which comes from a flasque resolution of T.

This exact sequence gives us the information about the kernel of β : it is the group $\text{III}(S)$, and the cokernel of β : it is the group $\text{A}(T)$. It is a natural question to ask if one can extend (2) to the case of connected reductive groups over global fields.

We show in this note that the right handside of (2) can be extended to the case of connected reductive k-groups G over global fields. For a finite set S of valuations of k we denote $\text{A}(S,G) = \prod_{v\in S} G(k_v)/Cl(G(k))$ the obstruction to weak approximation in S. Then we have

Theorem *For any connected group G defined over a global field k, assumed reductive if char.$k > 0$, and for any finite set of valuations S of k, we have the following isomorphisms of finite abelian groups*

$$\text{A}(S,G) \simeq \text{Coker } (G(k)/R \to \prod_{v\in S} G(k_v)/R),$$

$$\text{A}(G) \simeq \text{Coker } (G(k)/R \xrightarrow{\alpha} \prod_v G(k_v)/R).$$

Now let G be a semisimple k-group with isotropic simple components. We denote by $G(k)^+$ the normal subgroup of $G(k)$ generated by k-points of unipotent radicals of parabolic k-subgroups of G. As in the case of almost simple groups, we call (after Tits [13]) the factor group $G(k)/G(k)^+$ *the Whitehead group of G over k* and denote it by $W(G,k)$. Notice that in this case, G is almost simple and simply connected group scheme of Chevalley type over a ring R, the notion of Whitehead group of G reduces to the values of K_1-functor of G over the ring R. We have

Theorem *Let G be a semisimple group defined over a global field k with isotropic simple components and let S be any finite set of valuations of k. Then we have the following isomorphisms of finite abelian groups*

$$\text{A}(S,G) \simeq \text{Coker } (W(G,k) \to \prod_{v\in S} W(G,k_v)),$$

$$\text{A}(G) \simeq proj.lim_S(\text{Coker } (W(G,k) \xrightarrow{\alpha'} \prod_{v\in S} W(G,k_v)))$$

$$\simeq \text{Coker } (W(G,k) \to \prod_v W(G,k_v)).$$

We give some preliminary results in Section 1 and we prove the main result in Section 2. In Section 3 we prove a similar result for the Whitehead group of G. The discussion of some possible extensions of the results of this paper to an arbitrary field and also the question of describing the kernel of the map α and α' above have independent interest and will be considered in another place. (In the preprint [T] we gave a possible cohomological interpretation of the group $\cap_v RG_v$, which is the kernel of the map $G(k)/R \to \prod_v G(k_v)/R$.)

1 Some Auxiliary Results

We need the following auxiliary results in the arithmetic of algebraic groups.

Theorem 1.1 *Any semisimple simply connected group defined over a global field k satisfies weak approximation.*

This result is due to Kneser and Harder. We refer to [11] for historical remarks and references.

The following lemma is due to Ono [8] while he introduced "cross-diagram" which allows one to reduce many problems in reductive groups to the case of a reductive group with simply connected semisimple parts. (Later on, Langlands [6] also came to the same construction and in the literature, some authors call such central extension of G a z-extension of G.)

Lemma 1.2 *[8] Let G be a connected reductive group defined over a field k. Then there exists a connected reductive group H with simply connected semisimple parts, an induced torus Z and a central extension*

$$1 \to Z \to H \to G \to 1,$$

all defined over k.

Proposition 1.3 *Let G be a unirational group over a field k and S a finite set of discrete valuations of k. Then via the diagonal embedding, the closure $Cl(G(k))$ of $G(k)$ in $\prod_{v \in S} G(k_v)$ in the product topology contains an open subgroup of the latter.*

This result is implicitly due to Kneser, who used it in the study of strong approximation. The proof follows from the Implicit Function Theorem.

The following result perhaps is well-known to experts but it is given here since I could not find an appropriate reference.

Proposition 1.4 *Let G be a connected reductive group defined over a field k. If the simply connected covering $\tilde{G}$ of the semisimple part G' of G has weak approximation with respect to a finite set S of discrete valuations of k then the closure $Cl(G(k))$ of $G(k)$ in $G_S := \prod_{v \in S} G(k_v)$ is an open normal subgroup containing the commutator of G_S.*

Proof The fact that $Cl(G(k))$ is an open subgroup follows immediately from a similar result (due to Harder [4], Hauptsatz) for semisimple groups. Next for simplicity we assume that G is semisimple and that the canonical central isogeny $\pi : \tilde{G} \to G$ is separable. We denote by the same π the induced map $\tilde{G}_S := \prod_{v \in S} \tilde{G}(k_v) \to G_S := \prod_{v \in S} G(k_v)$. Then one sees that

$$[G_S, G_S] \subset \pi(\tilde{G}_S) = \pi(Cl(\tilde{G}(k))) \subset Cl(G(k)).$$

The proposition is proved. □

We recall [3] that a torus T defined over a field k and splitting over a finite Galois extension L/k is called *flasque* if $\mathrm{H}^{-1}(H, X^*(T)) = 0$ for any subgroup $H \subset Gal(L/k)$. Here $X^*(T)$ stands for the character group of T.

Theorem 1.5 *([3], Section 5) Let T be a torus defined over a field k. Then there exists a flasque resolution of T, i.e., an exact sequence of k-tori*

$$1 \to S \to N \to T \to 1,$$

where N is an induced torus and S is a flasque torus. Moreover $T(k)/R \simeq \mathrm{H}^1(k, S)$.

We will also need the following classical results about Galois cohomology of simply connected groups over global function fields and their completions.

Theorem 1.6 *[5] For any simply connected semisimple group G defined over a global function field k,* $\mathrm{H}^1(k, G) = 0$.

Theorem 1.7 *[1] For any simply connected semisimple group G defined over a complete non-archimedean field k with residue field of cohomological dimension* ≤ 1, $\mathrm{H}^1(k, G) = 0$.

2 Main Theorem and the Proof

First we need the following,

Lemma 2.1 *Let G be connected reductive, and H a z-extension of G, all defined over a field k. Then there are canonical isomorphisms of groups and bijection of factor sets, respectively*

$$G(k)/R \simeq H(k)/R, \ \mathrm{A}(S, G) \simeq \mathrm{A}(S, H),$$

where $\mathrm{A}(G, S)$ denotes the factor $\prod_{v \in S} G(k_v)/Cl(G(k))$ with S a finite set of discrete valuations of k and $Cl(G(k))$ the closure of $G(k)$ via diagonal embedding in the product topology of $\prod_{v \in S} G(k_v)$.

Proof Let $1 \to Z \to H \xrightarrow{\pi} G \to 1$ be the corresponding z-extension. Since Z is an induced torus, it has trivial cohomology so π is surjective while restricted to $H(K)$ for any field extension K of k. It is clear that $\pi(RH(k)) \subset RG(k)$, hence π defines a surjective homomorphism of groups

$$\pi' : H(k)/R \to G(k)/R.$$

We show that π' is injective. For this it suffices to show that if $x \in H(k)$ such that $\pi(x)$ is R-equivalent to 1 in $G(k)$ then x is R-equivalent to 1 in $H(k)$. Indeed, let $\phi : \mathbf{P}^1 \to G$ be a k-rational map with $\phi(0) = 1$, $\phi(1) = \pi(x)$. We know that there is a k-section $\zeta : G \to H$, so $\phi' = \zeta \circ \phi : \mathbf{P}^1 \to H$ is a k-rational map such that $\phi'(0) = 1, \pi(\phi'(1)) = \phi(1) = \pi(x)$, thus $x \in \phi'(1)Z(k)$. Since Z is k-rational, we are done.

To prove the second bijection, it suffices to prove the following equality

$$\pi^{-1}(Cl(G(k))) = Cl(H(k)). \tag{3}$$

One inclusion is obvious. We provide two arguments to prove the other inclusion.

First, it is not hard to see that the induced map $\pi : H_S \to G_S$ is an *open* map. Since $Cl(H(k))$ is open in H_S, $\pi(Cl(H(k)))$ is an open subgroup of G_S. It is clear that

$$Cl(G(k)) = G(k)Cl(\pi(H(k))) \subset \pi(H(k))Cl(\pi(H(k))) \subset \pi(Cl(H(k)))$$

and we are done.

Second, let $h \in \pi^{-1}(Cl(G(k)))$. Then $\pi(h) = lim_n g_n$, with $g_n \in G(k)$. There are $h_n \in H(k)$ such that $g_n = \pi(h_n)$, so $lim_n \pi(h_n^{-1}h) = 1$. Let U_n be a system of open neighbourhoods of 1 in $G_S = \prod_{v \in S} G(k_v)$ such that $U_n \subset U_{n-1}$, $\cap_n U_n = \{1\}$ and $\pi(h_n^{-1}h) \in U_n$. Thus for all n we have

$$\pi^{-1}(1) = \pi^{-1}(\cap_n U_n) = \cap_n \pi^{-1}(U_n) = \prod_{v \in S} Z(k_v),$$

and

$$h_n^{-1}h \in \pi^{-1}(U_n). \tag{4}$$

It is well-known by a result of Grothendieck - Rosenlicht that connected reductive groups are unirational over the field of definition. Thus by Proposition 1.4, $Cl(H(k))$ is an open subgroup of H_S, hence $Cl(H(k))\prod_{v\in S} Z(k_v)$ is an open neighbourhood of $\prod_{v\in S} Z(k_v)$ in $\prod_{v\in S} H(k_v)$. Therefore for n large we have

$$\pi^{-1}(U_n) \subset Cl(H(k)) \prod_{v\in S} Z(k_v) \subset Cl(H(k)),$$

since Z has weak approximation property. Thus for n large we derive from (4)

$$h \in h_n\pi^{-1}(U_n) \subset h_n Cl(H(k)) = Cl(H(k))$$

and the lemma follows. □

Lemma 2.2 *Let G be a connected reductive group defined over a field k, H a z-extension of G. For a finite set S of valuations of k, let $RG_S = \prod_{v\in S} RG(k_v)$. Then $RG_S \subset Cl(G(k))$ if and only if $RH_S \subset Cl(H(k))$.*

Proof We have $\pi(RH_S) \subset RG_S$ where we denote by the same symbol π the homomorphism $H_S \to G_S$ induced from $\pi : H \to G$. Hence

$$RH_S \subset \pi^{-1}(RG_S) \subset \pi^{-1}(Cl(G(k))) = Cl(H(k))$$

by Lemma 2.1.

Conversely, by Lemma 2.1 we have $\pi(RH_S) = RG_S$, thus

$$RG_S \subset \pi(Cl(H(k))) = Cl(G(k)).$$

The lemma is proved. □

The following was first mentioned in [15]. In this case the finite set of valuations consists of only one element, but the proof for the general case of any finite set of valuations is as in ([3], Theorem 1.5).

Lemma 2.3 *Let T be a torus defined over a field k, S a finite set of valuations of k. Then $RT_S \subset Cl(T(k))$.*

Proof Consider the following flasque resolution of T

$$1 \to S \to N \xrightarrow{p} T \to 1.$$

Since this is also a flasque resolution of T over k_v, from the proof of Theorem 1.5 (in [3]) it follows that

$$p(N(k)) = RT(k),\ p(N(k_v)) = RT(k_v)$$

for any $v \in S$ and since N has weak approximation over k, it follows that

$$\prod_{v\in S} RT(k_v) = p\left(\prod_{v\in S} N(k_v)\right) = p(Cl(N(k))) \subset Cl(T(k)).$$

The lemma is proved. □

For arbitrary k-group it is not known if Lemma 2.3 is true in general, but in the important case of connected reductive groups over global fields we have the following result, which answers a (implicit) question of Voskresenskiĭ [15] in this case.

Theorem 2.4 *Let k be a global field and G a connected reductive group over k. For any finite set S of valuations of k we have*

$$RG_S \subset Cl(G(k)).$$

Proof By 2.3 we may assume that G is not a torus. By 1.2 and 2.2 we may assume that the semisimple part G' of G is simply connected. By 1.1 we may assume that G is not semisimple. It is clear that if the theorem holds for S and S_0 is any nonempty subset of S then $RG_{S_0} \subset Cl(G(k))$, where Cl is taken in $\prod_{v\in S_0} G(k_v)$. Therefore, by enlarging S we may also assume that S contains all archimedean valuations of k. Let $G = G'T$, where T is a nontrivial central torus. We have the following exact sequence of k-groups

$$1 \to G' \to G \xrightarrow{p} T' \to 1,$$

where T' is a factor of T. Consider the following commutative diagram, where the rows are exact

$$\begin{array}{ccccc} G(k) & \longrightarrow & T'(k) & \xrightarrow{\delta} & \mathrm{H}^1(k, G') \\ \downarrow & & \downarrow & & \downarrow \gamma \\ G_S & \longrightarrow & T'_S & \xrightarrow{\delta_S} & \prod_{v\in S} \mathrm{H}^1(k_v, G') \end{array}$$

where $G_S = \prod_{v\in S} G(k_v)$, $T'_S = \prod_{v\in S} T'(k_v)$. Since G' is simply connected, it is well-known (the Hasse principle for simply connected groups over global fields, see [5], [9]) that γ is bijective. Let $g \in RG_S$. Then $p(g) \in RT_S \subset Cl(T'(k))$ by 2.3. Therefore there is a sequence (t_n), $t_n \in T'(k)$ such that

$$p(g) = lim_n t_n,$$

hence

$$\delta_S(p(g)) = \delta_S(lim_n t_n) = 1.$$

It is known that the product $\prod_{v\in S} \mathrm{H}^1(k_v, G')$ is finite, for large n we have

$$\delta(t_n) = 1,$$

i.e., for large n, there are $g_n \in G(k)$ such that $t_n = p(g_n)$. Therefore

$$lim_n p(g_n^{-1} g) = 1.$$

By the same argument we used in the proof of 2.1, and by using weak approximation property in simply connected groups (see 1.1) we see that for large n

$$g_n^{-1} g \in G'_S Cl(G(k)) \subset Cl(G(k)),$$

i.e., $g \in Cl(G(k))$. □

To deduce the relation between weak approximation and R-equivalence we need the following well-known

Lemma 2.5 *With the same notation as in 2.1, RG_S is an open subgroup of $\prod_{v\in S} G(k_v)$.*

Proof It follows from the fact that G is unirational and that affine spaces have only trivial R-equivalent classes. □

Proposition 2.6 *[15] Let G be a connected reductive group over a local field k. Then $G(k)/R$ is a finite abelian group.*

Now we have

Theorem 2.7 *Let k be a global field and G a connected k-group, which is reductive if char.$k > 0$. For any finite set S of valuations of k we have an isomorphism of finite abelian groups*

$$\mathrm{A}(S,G) \simeq \mathrm{Coker}\ (G(k)/R \to \prod_{v\in S} G(k_v)/R).$$

In particular, the obstruction $\mathrm{A}(S,G)$ has a natural group structure inherited from that of Coker $(G(k)/R \to \prod_{v\in S} G(k_v)/R)$.

Proof By 1.1 and 1.4, $\mathrm{A}(S,G)$ is an abelian group. By 2.4 we have

$$RG_S \subset Cl(G(k)),$$

and by 2.5

$$Cl(G(k)) = RG_S G(k).$$

By 2.7 the groups $G(k_v)/R$ are all finite abelian. Thus

$$\mathrm{Coker}\ \left(G(k)/R \to \prod_{v\in S} G(k_v)/R\right) = \prod_{v\in S} G(k_v)/Cl(G(k)) = \mathrm{A}(S,G).$$

The theorem is proved. □

It is well-known (see [3], p. 205, [11], Theorem 3.3) that for almost all v, $G(k_v)/R = 1$ and there is a finite set S_0 of valuations of k such that G has weak approximation over k with respect to any finite set of valuations of k outside S_0. (In [11] this assertion was proved for number fields, but the proof is valid also for global function fields k, by taking into account Harder's result about the triviality of $\mathrm{H}^1(k,G)$ for any simply connected group G, see 1.6). Thus we have

Theorem 2.8 *Let the notation be as in 2.7 and the introduction. Then we have the following isomorphisms of finite abelian groups*

$$\mathrm{A}(G) \simeq \mathrm{Coker}\ \left(G(k)/R \to \prod_{v} G(k_v)/R\right). \tag{5}$$

3 An Analogue Result for Whitehead Groups

We keep the notation used in the introduction and we prove in this section the following

Theorem 3.1 *Let k be a global field and G a semisimple k-group with isotropic simple factors. Let $W(G,L)$ be the Whitehead group of G over a field extension L of k. For any finite set S of valuations of k we have the following isomorphisms of finite abelian groups*

$$\mathrm{A}(S,G) \simeq \mathrm{Coker}\ \left(W(G,k) \to \prod_{v\in S} W(G,k_v)\right),$$

$$\mathrm{A}(G) \simeq proj.lim_S \left(\mathrm{Coker}\ \left(W(G,k) \to \prod_{v\in S} W(G,k_v)\right)\right)$$

$$\simeq \mathrm{Coker}\ \left(W(G,k) \to \prod_{v} \mathrm{Coker}\ W(G,k_v)\right).$$

Proof We present two proofs of the first two isomorphisms in the theorem.

1. Let $\tilde{G}$ be the simply connected covering of G and $\pi : \tilde{G} \to G$ the canonical central isogeny. We have to distinguish two cases: π is separable or inseparable isogeny. It follows from a well-known theorem of Riehm [R] that if π is separable, then over local fields, $G(k_v)^+$ is a normal thus also open subgroup of $G(k_v)$ for all v and from the well-known fact that $\tilde{G}(k_v) = \tilde{G}(k_v)^+$ (the Kneser - Tits conjecture for $\tilde{G}$, see e.g. [9]) and the finiteness of $\mathrm{H}^1(k_v, F)$, where $F = \mathrm{Ker}\ \pi$, we see that $W(G, k_v)$ is a finite abelian group for any v.

Since $G(k_v)^+ \subset RG_v$ and (by Theorem 2.4) $RG_v \subset Cl(G(k))$, hence $\prod_{v \in S} G(k_v)^+ \subset Cl(G(k))$. The rest of the proof follows as in the proof of Theorem 2.7. The second isomorphism follows from the first.

If π is inseparable isogeny, then $char.k > 0$ and instead of usual Galois cohomology, we may consider the *inseparable* or *flat* Galois cohomology $\mathrm{H}^1_{fl}(k, F)$, $\mathrm{H}^1_{fl}(k_v, F)$ of $F = \mathrm{Ker}\ \pi$ (see [2], [5]) and we have the following exact sequence [13]

$$W(\tilde{G}, k) \to W(G, k) \to \mathrm{H}^1_{fl}(k, F) \to \mathrm{H}^1(k, \tilde{G}) = 0,$$

$$W(\tilde{G}, k_v) \to W(G, k_v) \to \mathrm{H}^1_{fl}(k_v, F) \to \mathrm{H}^1(k_v, \tilde{G}) = 0.$$

The above triviality of cohomology of $\tilde{G}$ is the theorem of Harder (1.6) (for k), and a theorem of Bruhat - Tits (1.7) (for k_v).

Since π is bijective on k- and k_v-points, it follows that G has weak approximation property with respect to any S. Since over local fields and global function fields K, the Kneser - Tits conjecture holds for simply connected semisimple groups, i.e., $\tilde{G}(K) = \tilde{G}(K)^+$ (see [9]) we deduce from exact sequences above that

$$W(G, K) \simeq \mathrm{H}^1(K, F).$$

Thus one sees that the natural map

$$W(G, k) \to \prod_{v \in S} W(G, k_v)$$

is surjective. The first and therefore also the second isomorphism in the theorem follow.

2. We need the following simple lemma.

Lemma *Assume that in the following commutative diagram of topological groups*

$$\begin{array}{ccccccc}
A & \longrightarrow & B & \longrightarrow & C & \longrightarrow & 1 \\
\downarrow{\scriptstyle \alpha} & & \downarrow{\scriptstyle \beta} & & \downarrow{\scriptstyle \gamma} & & \\
A' & \longrightarrow & B' & \longrightarrow & C' & \longrightarrow & 1
\end{array}$$

all rows are exact and C and C' are endowed with discrete topology. If the image of α is dense in A' then we have the following bijection

$$B'/Cl(\mathrm{Im}\ (\beta)) \simeq C'/\mathrm{Im}\ (\gamma).$$

The proof is simple so we omit it. By considering the following commutative diagram

$$\begin{array}{ccc} G(k)/G(k)^+ & \longrightarrow & G(k)/\pi(\tilde{G}(k)) \\ \downarrow & & \downarrow \\ G(k_v)/G(k_v)^+ & = & G(k_v)/\pi(\tilde{G}(k_v)) \end{array}$$

we conclude that

$$\text{Coker}\left(W(G,k) \to \prod_{v\in T} W(G,k_v)\right)$$
$$= \text{Coker}\left(G(k)/\pi(\tilde{G}(k)) \to \prod_{v\in T} G(k_v)/\pi(\tilde{G}(k_v))\right),$$

where T is any set of valuations of k. By applying the above lemma to

$$A = \tilde{G}(k), B = G(k), C = G(k)/\pi(\tilde{G}(k)),$$

$$A' = \prod_{v\in T} \tilde{G}(k_v), B' = \prod_{v\in T} G(k_v), C' = \prod_{v\in T} G(k_v)/\pi(\tilde{G}(k_v)),$$

the theorem follows. □

Acknowledgements

I would like to thank J. -L. Colliot-Thélène for useful remarks and criticism. Thanks are due to M. Kolster, C. Riehm, V. Snaith for encouragement and support and the support by Department of Mathematics and Statistics, McMaster University is greatly acknowledged.

References

[1] Bruhat, F. et Tits, J. [1987], *Groupes algébriques sur un corps local : Chapitre III. Complément et applications à cohomologie galoisienne*, J. Fac. Sci. Univ. Tokyo, Ser. IA, v. 34, 671 - 698.

[2] Cartier, P. [1966], *Inseparable Galois cohomology*, Proc. Sym. Pure Math. A.M.S., v. 9, 183 - 186.

[3] Colliot-Thélène, J.-L. et Sansuc, J.-J. [1976], *La R-équivalence sur les tores*, Ann. Sc. E. N. S. t. **10**, 175 - 230.

[4] Harder, G. [1968], *Eine Bemerkung zum schwachen Approximationssatz*, Arch. der Math. **19**, 465 - 471.

[5] Harder, G. [1975], *Über die Galoiskohomologie halbeinfacher Matrizengruppen III*, J. reine. angew. Math. **274/275**, 125 - 138.

[6] Langlands, R. [1979], *Stable conjugacy : definition and lemmas*, Can. J. Math. **31**, 300 - 325.

[7] Manin, Yu. I. [1986], *Cubic forms : Algebra, Geometry, Arithmetic*, North Holland, 2nd ed.

[8] Ono, T. [1965], *On the relative theory of Tamagawa numbers*, Ann. Math. **82**, 88-111.

[9] Platonov, V. and Rapinchuk, A. [1994], *Algebraic Groups and Number Theory*, Academic Press.

[10] Riehm, C. [1970], *The congruence subgroup problem over local fields*, Amer. J. Math. **92**, 771 - 778.

[11] Sansuc, J.-J. [1981], *Groupes de Brauer et arithmétique des groupes algébriques linéaires sur un corps de nombres*, J. reine. angew. Math. Bd. **327**, 12 - 80.

[12] Thang, Nguyen Q. [1996], *Weak approximation and Manin group of R-equivalences*, McMaster University, preprint.

[13] Tits, J. [1976/77], *Groupe de Whitehead des groupes algébriques simples sur un corps*, (d'après V. Platonov et al.), Sémin. Bourbaki 29-année, No. 505.
[14] Voskresenskiĭ, V. E. [1976], *Algebraic Tori*, Moscow, in russian.
[15] Voskresenskiĭ, V. E. [1981], *Problems of R-equivalence in semisimple groups*, J. Soviet Math. **17**, 1975 - 1987.

Appendix I

The Second Great Lakes Conference on Algebraic K-Theory

Program List of Speakers and Topics

T. Chinburg	Galois Structure of deRham Cohomology
J.-L. Colliot-Thélène	Brauer Groups and Zero-Cycles on Varities over Local Fields
E. Friedlander	Cohomology of Finite Group Schemes over a Field
S. Lichtenbaum	Motives of Curves
R. McCarthy	A Chain Complex for the Spectrum Homology of the Algebraic K-Theory of an Exact Category
V. Voevodsky	Cohomological Operations in Motivic Cohomology
C. Weibel	The Work of Bob Thomasan (1952 - 1995)

Appendix II

List of Participants

Andersen, Kasper K. S., The Fields Institute, Toronto, Ontario, Canada
Anton, Marian F., Notre Dame University, Indiana, USA
Arlettaz, Dominique, University of Lawsanne, Lausanne, Switzerland
Bell, Robert, Cooper Union, New York, NY USA
Chinburg, Ted, University of Pennsylvania, Philadelphia, PA USA
Cohen, Fred, The University of Rochester, Rochester, NY USA
Colliot-Thèléne, Jean-Louis, Universitè de Paris-Sud, Paris, France
Dean, Andrew, The Fields Institute, Toronto, Ontario, Canada
Dumey, Luc, The Fields Institute, Toronto, Ontario, Canada
Elliott, George A., The Fields Institute, Toronto, Ontario, Canada
Falk, Geoffrey T., The Fields Institute, Toronto, Ontario, Canada
Friedlander, Eric M., Northwestern University, Evanston, IL USA
Gao, Wenfeng, University of Western Ontario, London, Ontario, Canada
Geller, Sue C., Texas A & M University, College Station, TX USA
Grayson, Daniel R., University of Illinois at Urbana-Champaign, Urbana, IL USA
Hesselholt, Lars, Massachusetts Institute of Technology, Cambridge, MA USA
Holdener, J., The Fields Institute, Toronto, Ontario, Canada
Hoobler, Raymond T., City College of New York, New York, NY USA
Hooper, Jeff, Cambridge University, Cambridge, England
Jardine, Rick, University of Western Ontario, London, Ontario, Canada
Jiang, Xinhui, The Fields Institute, Toronto, Ontario, Canada
Kallel, Sadok, The Fields Institute, Toronto, Ontario, Canada
Karkhanis, Simba, University of Western Ontario, London, Ontario, Canada
Khalkhali, Masoud, University of Western Ontario, London, Ontario, Canada
Kochman, Stan, York University, Toronto, Ontario, Canada
Kolster, Manfred, McMaster University, Hamilton, Ontario, Canada
Kuku, Aderemi, International Centre for Theoretical Physics, Trieste, Italy
Lambrechts, Pascal, The Fields Institute, Toronto, Ontario, Canada
Landsburg, Steven E., University of Rochester, Rochester, NY USA
Lichtenbaum, Stephen, Brown University, Providence, RI USA
Mahowald, Mark, Northwestern University, Evanston, IL USA
Mandell, Michael A., University of Chicago, Chicago, IL USA
McQuillan, Dan, University of Western Ontario, London, Ontario, Canada
Michler, Ruth, University of North Texas, Denton, TX USA
Mináс, Jan, University of Western Ontario, London, Ontario, Canada

Patashnick, Owen, University of Chicago, Chicago, IL USA
Richard, Kane, The Fields Institute, Toronto, Ontario, Canada
Roberts, Leslie, Queen's University, Kingston, Ontario, Canada
Rodriguez, Jose L., The Fields Institute, Toronto, Ontario, Canada
Santos, Marta, The Fields Institute, Toronto, Ontario, Canada
Schere, Jerome, The Fields Institute, Toronto, Ontario, Canada
Schuster, Bjorn, The Fields Institute, Toronto, Ontario, Canada
Snaith, Victor, McMaster University, Hamilton, Ontario, Canada
Thắng, Nguyên Quôc, Israel Institute of Technology, Technion, Haifa, Israel
Vandembroucq, Lucile, The Fields Institute, Toronto, Ontario, Canada
Voevodsky, Vladimir, Harvard University, Cambridge, MA USA
Walker, Mark, University of Ilinois, Urbana, IL USA
Weibel, Charles, Rutgers University, New Brunswick NJ USA
Xicotencatl, Miguel A., The Fields Institute, Toronto, Ontario, Canada
Yui, Noriko, Queen's University, Kingston, Ontario, Canada
Zelov, Vitaly, The Fields Institute, Toronto, Ontario, Canada